Historical & Cultural Astronomy

Historical & Cultural Astronomy

More information about this series at http://www.springer.com/series/15156

Marion Dolan

Astronomical Knowledge Transmission Through Illustrated *Aratea* Manuscripts

Springer

Marion Dolan
Deerfield Beach, FL
USA

ISSN 2509-310X ISSN 2509-3118 (electronic)
Historical & Cultural Astronomy
ISBN 978-3-319-56783-9 ISBN 978-3-319-56784-6 (eBook)
DOI 10.1007/978-3-319-56784-6

Library of Congress Control Number: 2017939309

Printed on acid-free paper

This Springer imprint is published by Springer Nature
The registered company is Springer International Publishing AG
The registered company address is: Gewerbestrasse 11, 6330 Cham, Switzerland

Pisces, the Fishes f. 3v Leiden Universiteitsbibliotheek MS Voss. lat. Q.79, Courtesy of Leiden University Library

Still farther ahead and more in the approaches to the south are the Fishes;but one is always more prominent than the other and hears more the fresh onset of the north wind. From both of them stretch, as it were, chains from the tail-parts, coming together in an unbroken line on both sides.One beautiful bright star occupies this position, and they call it the celestial Knot.Let Andromeda's left shoulder be your guide to the more northerly Fish, for it is very close to it. *Phaenomena*, Kidd (239–247)

Orion, the Hunter f. 58v Leiden Universiteitsbibliotheek MS Voss. lat. Q.79, Courtesy of Leiden University Library

Aslant from the cut-off figure of the Bull lies Orion himself. Anyone whose glance misses him when he is positioned high up on a clear night may be sure he can never sight anything better to identify when he gazes up at the sky. *Phaenomena*, Kidd (322–325)

Preface

This book developed from a strong attraction to the incredible illuminated manuscripts produced during the Middle Ages and the early Renaissance. With a background in the history and philosophy of science, and the history of astronomy, in particular, astronomical manuscripts inspired a more concentrated interest and desire to understand the long transmittal of astronomical knowledge. When beginning this study, 10–12 years ago as a doctoral dissertation, the amount of scholarly information published on medieval astronomical manuscripts was rather scant and most focused on specific details of particular manuscripts. Upon deciding to revise and update the initial research, I was blissfully unaware of the enormous amount of academic interest and current research readily available on astronomical manuscripts and associated art. Not only have there been dozens and dozens of scholarly articles written and scientific books produced on all aspects of medieval astronomy, many of these illuminated manuscripts are now freely available for viewing and research on line.

The author's interest, more precisely fascination, with astronomy and its historical transmission through the millennia began quite unexpectedly. Having lived among the towering smokestacks of Pittsburgh, with uninterrupted steel mills lining the rivers and emitting thick ashen clouds, with coke ovens smoldering, and industrial trains piled high with coal and steel lumbering in every direction, the night skies glowed gray suspending the aftermath of the day's heavy pollution. An evening's entertainment was often spent watching steel carts pour fiery residue down mountainous dumps of slag which spewed and sparkled brighter than fireworks; barely a star could be seen. As an adult, a trip to Jamaica transformed that usual expectation of a cloudy, steel-gray sky that hardly differed, day from night. After a pleasant island dinner with friends and steel drums setting the mood, we stepped outside to stroll along the Caribbean beach. Overhead the crystal-clear moonless sky danced with millions of glittering, breathtaking lights. Having never before experienced such a smoke-free starry night, the nocturnal vision was a shocking and exhilarating sight that demanded gazing late into the night. That sudden revelation of a previously unknown world ignited a lifelong study of celestial phenomena.

The resultant monograph recounts the process of knowledge transmission through time, through cultures, and through exceptional individuals, not focusing on the physical data or textual material but on the actual means of this transference. Many medieval scholars spent their lives gathering and protecting earlier knowledge; some, like Bede in the seventh century, observed and experimented, then recorded their ideas; some merely copied the ancient texts in danger of being lost. As an art historian, the role that images and diagrams, preserved in medieval manuscripts, played in that long period of transmittal from ancient Greece to the Renaissance became a vital concern along with the determination of exactly how the illustrations of constellations functioned in transmitting that knowledge. There existed times of stagnation and times of remarkable progress in the advancement of astronomical knowledge. One of the most important mechanisms in the transmission of astronomy can be credited to the poem of Aratus of Soli and the *Aratea* tradition with its texts and illustrations which became the main vehicle for conducting this search.

The history of astronomy does not begin in Greece but much, much earlier; the formal story could begin in Mesopotamia, but that part of the process has not yet been told. Hundreds of thousands of cuneiform tablets survive but remain to be read, analyzed, and published. The informal story of the role of cultural astronomy from prehistoric ages, and how detailed celestial knowledge and understanding accumulated bit by bit, can never be told in any detail. No wonder astronomy, that fascination with the motions of the Sun, Moon, planets, and stars, became the first science and, as demonstrated by the long investigation of worldwide astronomical myths by historians De Santillana and Von Dechend in *Hamlet's Mill*, the cosmos was the most dominate aspect of almost every early culture's myths, religious rites, most secret knowledge, and kingship rituals.[1]

Many scientific books have been written tracing the transmission of astronomy but not in light of the illustrated manuscripts, the material means by which this knowledge was held, protected, and passed on, which of course, after oral transmission, was on papyrus and parchment. This study investigates that process, pulling together and organizing a synopsis of the research currently available, concentrating on the role of astronomical art in the transmission of that knowledge.

The story of the over two-thousand-year transmission of the *Aratea* manuscripts bridges many disciplines and includes much more than the history of astronomy. The survey encompasses aspects of classical poetry, manuscript production, historical and religious events, artistic advancements, and cultural clashes when West finally meets East. The highly advanced Arabic manuscripts introduced both new ideas and ancient accomplishments drawn from the writings of Greece, Mesopotamia, India, and Persia and turned the state of astronomical knowledge in the Latin West upside down. Another crucial factor in this long story of astronomy

[1] *Hamlet's Mill: An Essay on Myth and the Frame of Time* (1969) was not well-received initially but has garnered support through time and with new anthropologic discoveries and research in cultural astronomy or archaeoastronomy.

is apparent in the influence of astrology, as this practice pushed in and out of the scientific traditions, especially influential in the history of medicine. Researching these closely connected fields has shown that a far greater wealth and range of material relevant to the *Aratea* investigation exists than imagined at the outset. It would be impossible to give full justice to every aspect, follow every thread, or discuss all relevant material effectively in a volume of this length. What follows is by no means a comprehensive study of knowledge transmission, but concentrates on the most significant manuscript traditions. This book grew out of that varied research and provides a basic overview of these interrelated fields; the more complex, in-depth details of each discipline are left for specialists more qualified; their knowledge exceeds anything written here. The writing is intended for students of the past, for an educated, nonspecialist reading audience interested in astronomy and the process of its transmittal.

In following the tortuous path of scientific history, this survey provides a brief description of the various medieval authors who played an influential role in the transmission of astronomical knowledge and mythology addressing the essential nature of each one's contribution. In more detail, it traces the history of the *Phaenomena* of Aratus from its composition in ancient Greece, to its reception and Latin translations in imperial Rome, and then its trajectory through the thousand years of the Middle Ages. The textual format and constellation illustrations of the *Aratea* became significantly modified and interest very nearly died out through the passage of time and through societal advances and declines. The medieval transmission culminated in an explosion of illustrated astronomical manuscripts, including the original poem of Aratus and the *Aratea,* during the fifteenth-century Italian Renaissance and has never ended even through the age of printing and still instigates investigation in the present day.

"The long record of achievement in astronomy has very few intellectual parallels in the whole of human history" (North 2008: 3). This monograph presents a general explanation of how we attained possession of these texts and illustrations today, considering the importance they held to those who cherished and retained that knowledge, almost as sacred as religious texts, and why these manuscripts still speak to us.

Chapter 1 discusses the history and text structure of the ancient poem, the *Phaenomena*, and explains the text of its Latin translators, Cicero, Germanicus, and Avienus and their reception in ancient Greece and Rome. This section also speaks to the utilization of celestial symbolism among the nobility to enhance political and personal power beginning with Caesar Augustus in imperial Rome. Chapter 2 examines the many early sources and motifs found in astronomical art that were available as models for manuscript illustrations and comparisons with the few extant classical illustrated manuscripts. Chapter 3 lists the additional early sources of astronomical information in the Middle Ages and follows the thread of transmission through various authors, their content, survival numbers, and the illuminations of their manuscripts from classical antiquity through the early Renaissance. Chapter 4 investigates the translations of Greek texts in Arabic educational centers in Persia and Baghdad in the ninth century. The enormous Islamic contributions to the development of themes, information, experimentation, and instruments pertaining to astronomical and astrological knowledge are discussed. The writings of some vital individual Islamic astronomers and astrologers are included along with the advancements brought to astronomical art by their manuscript traditions.

Continuing on, Chap. 5 surveys the translations of those Arabic astronomical writings into Latin by medieval scholars, mostly monks with the aid of Islamic and Jewish translation teams that began in the twelfth century, mainly in Spain and Sicily. How those translations introduced the advanced knowledge of Islamic astronomers to Western educational centers is also discussed. In addition, a few examples of astrological manuscripts commissioned in royal courts of Europe are included. Chapter 6 supplies essential historical information and conditions that influenced the production of astronomical manuscripts. In particular, this section concentrates on the *Aratea* manuscripts produced by the Carolingian kings and their appropriation of astronomical symbolism to re-enforce their power and legitimacy, and presents other examples of royal astronomical art. Chapter 7 concentrates on the history and illustrations of a few exemplary *Aratea* manuscripts from the ninth, twelfth, and fifteenth centuries and discusses their contents and influences. Chapter 8 analyzes and draws conclusions on the various functions of the *Aratea* manuscripts, including their role in conveying scientific knowledge, classical poetry, ancient myths, and the iconography of its constellation illustrations. Appendix A lists the surviving *Aratea* manuscripts by country and by century; Appendix B provides elementary descriptions of the manuscripts; Appendix C includes pertinent verses of the *Phaenomena,* describing each constellation and discussing the mythology of the 46 or so constellations that Aratus identified and described so eloquently.

In this work, I have used a simplified system to transcribe Arabic names and words, giving common English equivalents where possible and avoiding special characters or placing blank spaces instead of vowels, i.e., al-Sufi rather than _l-S_f_. I hope that specialists will pardon my sometimes sweeping generalizations for the sake of making this important subject matter more accessible to a wide audience.

In bringing this large selection of intriguing material together, much has been gained from the work of scholars present and past whose publications I have pondered and quarried. I owe a large debt of gratitude to my mentors, Profs. Alison Stones for her illustrated manuscript expertise, Mark Possanza for his expert knowledge of classical poetry and the *Aratea*, and Peter Machamer for his vast proficiency in the history and philosophy of science, and to colleagues at the University of Pittsburgh: Chair of the History of Art department Barbara McCloskey, visual resource expert Veronica Gazdig, and library specialist Marcia Rostek. I am also indebted to my four beautiful daughters who encouraged my work and to the historian of astronomy, Clifford Cunningham Ph.D., who inspired me and offered much assistance in bringing this project to completion. I alone am responsible for its errors.

The author would like to express her sincere thanks to the following institutions who kindly granted permission to reproduce their illustrations in this book: University of Leiden; University of Pittsburgh Medart website; Ghent University Library, Morgan Library Museum, and Biblioteca Apostolica Vaticana.

Deerfield Beach, USA Marion Dolan

Reference

North, J. (2008) *Cosmos: an Illustrated History of Astronomy and Cosmology*, Chicago: University of Chicago Press.

Contents

List of Figures

Chapter 1
Background and Objectives

Abstract This initial section introduces the Greek poem of Aratus, *the Phaenomena*, and discusses its reception and commentators in ancient Greece; it also explains the background and structure of the original poem. After the poem was translated into Latin, the *Phaenomena* became equally popular in imperial Rome and attracted many commentators there. The Latin translations of Aratus' poem, called the *Aratea*, composed by three authors: Cicero, Germanicus, and Avienus are also described, analyzed, and compared. Astrology had been formulated and systematized by Hellenistic astronomers in the last few centuries BCE, mainly in Alexander, and spread quite quickly, attaining popular acceptance in the main cities of the Roman Empire. Celestial symbolism and artworks became a significant resource in promoting royal propaganda; astronomy, astrology, and imperial apotheosis flourished as popular themes for artistic iconography. Clarifying the misinterpretations of the *Phaenomena* and the *Aratea* is most important as many false claims have been attached to these poems. The various scholia attached to the textual material of the *Aratea* are cited. The last section of this chapter discusses the *Aratus Latinus*, a later and poorly produced translation of the Greek poem that circulated in illustrated constellation cycles, often wrongly labeled as Aratus or *Aratea* text.

How fortunate that we in the twenty-first century retain possession of the earliest detailed description of the positions, myths, and diurnal motions of the numerous constellations that fill the night sky, when so much of ancient literature has been lost, or survives only in mere fragments. This monograph surveys and explores the transmission of the most ancient astronomical poem that survives intact from its beginnings in Hellenistic Greece and passes through the monasteries, cathedral schools, and royal courts of medieval Europe in various forms and translations, only to acquire a new life in the Italian Renaissance. For comparison and background, this work also presents other astronomical writings and the astronomers or scholars who were instrumental in transmitting, preserving, and advancing our knowledge of the heavens, its constellations and myths in the Middle Ages. Most importantly, it

M. Dolan, *Astronomical Knowledge Transmission Through Illustrated Aratea Manuscripts*, Historical & Cultural Astronomy,
DOI 10.1007/978-3-319-56784-6_1

gathers information on the various manuscript traditions; how they were illustrated and how many survive.

The heart of this study, an astronomical poem named the *Phaenomena* composed by Greek writer Aratus of Soli (c. 310–240/239 BCE), survives in its entirety. The clever and enlightening verses of Aratus' astronomical poem that describe 46 constellations, basic celestial features, and *Weather Signs* were greatly admired in the Hellenistic world. The poem was produced amidst the Greek literary culture in Athens at the court of the Macedonian palace rather than in Hellenistic Egypt, where astronomical writings and research prevailed and astrological authors were establishing their predictive craft. When the primary center of literature and poetry shifted from Greece to Rome, the popularity of this Greek poem continued unabated. The Greek ode to the heavens was translated into Latin by many admirers, so that its contents could be enjoyed by educated readers in the Roman World. The Latin translations of the *Phaenomena*, commonly called the *Aratea*, never lost their status in the Latin West. Translated into French and English, as well as many other languages, the *Phaenomena* and the *Aratea* continue to be studied and valued to the present day, maintaining a life span of 2300 years, second only to Homer's *Iliad* and *Odyssey*.

The Latin poems, created by admired Arateans, were copied by the Carolingian 'court schools' and monastic scriptoria in the ninth century; the narrative descriptions of the constellations were enhanced with exceptional illustrations by Carolingian artists noted for their sophistication and the quality of their paintings. In fact, Kurt Weitzmann named the *Aratea* "the most frequently illustrated text from antiquity" (Weitzman 1959: 25). When undertaking this investigation of *Aratea* manuscripts produced during the Middle Ages, I expected to examine only a dozen or so surviving codices; but surprisingly a total of at least 60 *Aratea* manuscripts are still extant in European and American libraries.

Due to the diligence and devotion of the numerous scholars active during Charlemagne's reign (768–814) and to the Holy Roman Emperor's supreme efforts and achievement in the restitution of literacy, culture, and art, the earliest surviving manuscripts of the *Aratea*, date to the ninth centuries. The extant writings of most other ancient Greek and Latin literary and scientific authors date to this prolific period as well; the earliest volumes in each discipline were produced in the eighth and ninth centuries. In his various European courts, Charlemagne assembled the most renowned scholars who sought to recover and restore the poetry, literature, philosophical and scientific writings passed down from the classical world and to revive classical architecture and decorative articulation, inspiring the so-called Carolingian Renaissance. Monastic scriptoria scattered throughout the Carolingian Empire were assigned the daunting task of reproducing essential religious and secular manuscripts that had been sitting neglected and deteriorating in the few early medieval libraries. Of the numerous manuscripts extant from the ninth century, thirteen illustrated *Aratea* manuscripts survive, many incomplete. In turn those Carolingian manuscripts were replicated over and over throughout the medieval and early Renaissance eras but actually none were exactly alike.

The poetic verses of the *Aratea*, along with the imaginative body of illustrations that brought the constellation myths to life, were copied from these exemplars, emulating but seldom equaling the superb Carolingian originals. Fully illustrated manuscripts of the admired astronomical poem survive from each century ranging from the ninth through the fifteenth.[1] The traditional constellation images that we easily recognize today were encapsulated by Aratean text; these images impacted astronomical art in frescoes, mosaics, sculptures, and especially, illuminated manuscripts. The illustrations of the surviving *Aratea* manuscripts comprise the main focus of this study along with their influence, distribution patterns, and companion texts. This monograph will not discuss textual details or intrinsic literary qualities of the Aratean translations, nor analyze the intricacies of Augustan poetry as these particular aspects have been thoroughly scrutinized by more qualified textual scholars. This examination of surviving *Aratea* manuscripts and their function in the transmission of astronomical knowledge was undertaken to assist in bringing some clarification to a very complicated sphere of knowledge.[2] Thanks to our electronic age, many detailed studies of specific areas in the history of astronomy can be discovered, read and enjoyed online. This overview and summarization is directed for interested readers concerned with the history of astronomy (and astrology), astronomical art, mythology, and illuminated manuscripts.

The Transmission of *Aratea* Manuscripts

Throughout the 1500-year-span of *Aratea* manuscript production, the Latin texts of the three principal translations of the poem were transmitted nearly complete but over the centuries have accumulated a great number of changes, losses, and additions. The cycle of illustrations depicting the constellations and complementing the verses also demonstrates substantial variations; these pictorial changes developed and evolved throughout the lengthy medieval period. From its Carolingian beginnings, through the ever-evolving artistic styles of the Anglo-Saxon, Romanesque, Gothic eras and the early Italian Renaissance, astronomical art reflects societal preferences and distinctive identifiable traits. Classical elements of Roman painting, retained from illustrations in Late Antique scrolls, can be detected in ninth-century manuscripts and in later ones as well. Fortunately, the iconographic characteristics and graphic aspects of Greco–Roman mythological depictions in frescoes, sculptures, and relief carvings were borrowed by observant medieval illustrators. Many

[1]A listing of the surviving illustrated manuscripts of the *Aratea* can be found in Appendix A, arranged by century and by country in which the manuscript is currently held.

[2]After reading many conflicting 'facts' and opinions concerning this astronomical poem, this investigation was began in order to clarify the function of the *Aratea* tradition in my own mind. Became intrigued by the broad influence this small poem had in the fields of medieval astronomy, literacy, mythology, and art, the survey of surviving *Aratea* manuscripts became my PhD dissertation.

Carolingian astronomical manuscripts circulated through monastic scriptoria and were used as exemplars during the tenth and eleventh centuries. Later illustrations preserved those classical features; Hercules, the Kneeler, still brandished his deadly club, threatening the dragon that guarded the Golden Apples of the Hesperides; Orion, the handsome hunter, displayed his triple-starred belt, while his faithful dog Canis stood at his feet; the brightest star in the sky Sirius marked his head.

During the twelfth and thirteenth centuries, the translation of great numbers of astronomical and astrological texts from the Arab World stimulated further fascination with the heavens in the Latin West. The discovery of astronomical and scientific treatises by ancient Greek authors, previously considered lost, brought scores of Latin translators to the numerous libraries of the former Islamic cities of Toledo, Cordoba, and Palermo. The long-sought astronomical writings of Aristotle and Ptolemy were among the first Arabic texts sought by Latin scholars and anxiously translated. During this translation process, many illustrations of the planetary personifications and constellation figures, pictured in later Latin *Aratea* manuscripts, take on distinctive Islamic characteristics, introducing more exotic features and costumes to the mythological characters. The male heroes wear garments more appropriate for an Islamic warrior rather than a medieval or classical hero. A significant change can be seen in the depiction of the individual stars that form the constellation illustrations in some manuscript traditions. Instead of a random sprinkling of stars within or around the constellation image seen previously, the individual stars copied from Arabic illustrations are now positioned with more precise accuracy including their relative placement, brightness and even in some instances their magnitude.

Rather than declining in popularity at the waning of the Middle Ages, the *Aratea* manuscripts enjoyed an enormous boost in production during the Italian Renaissance, as the study of astronomy and astrology blossomed and fascination with the gods and goddesses of the classical past flourished.[3] Illustrations of constellations found in fifteenth-century manuscripts display more colorful, costly, and detailed picture cycles appropriate to the extravagant lifestyles and luxurious fashions of affluent European courts. The popularity of the illuminated astronomical poem of Aratus continued even into the era of printed books that took over manuscript production in the mid-fifteenth century. This continuation can be attested by the numerous incunabula and early-printed copies of both the *Phaenomena* of Aratus and its Latin translations. These printed versions of the *Aratea* are often illustrated with woodblock prints copied from painted images in early illuminated manuscripts. Incunabula editions of the *Phaenomena* and the Latin *Aratea* can be found in most major libraries today.

When considering this extensive *Aratea* manuscript tradition, the presentation of the texts and accompanying pictorial cycles have no standard format, other than the

[3]See Alexander (1994). *The Painted Page: Italian Renaissance Book Illumination*, Godwin (2002). *The Pagan Dream of the Renaissance and* Seznec (1953). *The Survival of the Pagan Gods*. Although some of Selznec's theories are now being questioned and disregarded.

general order of the constellations, which in most cases follows the original arrangement as designed by Aratus and based on a model by Eudoxus. Latin *Aratea* manuscripts are usually amplified with added text, commentaries, and *scholia* (medieval Latin commentaries). Typically, the constellation pictures precede the descriptions and appear below or intermingled with the corresponding verses of the poem, rather than framed or set apart.

Most often, medieval Aratean texts are not 'pure' translations of the *Phenomena*, but include verses or treatises of other astronomical writers, such as Hyginus (64 BCE-CE17) or Pliny the Elder (CE 23/4-79). In these astronomical compilations, a huge variance can be detected in the arrangement and calligraphy of the texts, and also in the quality of the illustrations and artistic ability of the artists. Some *Aratea* manuscripts are luxurious copies of the highest caliber, produced for the prestigious library of a cardinal or royal court; others are modest, well-worn texts with poorly executed sketches for scholars, monks, or students; between these two extremes, many variations can be found.

In order to fully appreciate the *Aratea* tradition, it is necessary to recognize the three different Latin translations and aspects of Aratus' original Greek poem as well. In addition to the surviving manuscripts of the *Aratea* poems, this study surveys and compares other widely read illustrated manuscript traditions produced during the medieval period that conveyed astronomical information, since these texts were often intermingled even on the same folio. The comparison of the *Aratea* texts and illustrations with other scientific traditions allows a better understanding of the role of *Aratea* manuscripts in the transmittal of astronomical knowledge. Compiling, cataloging, and examining the extant astronomical manuscripts offers an overview of this extensive poetic and pictorial theme. By inference, this compilation may also provide some idea about how much does not survive.

As the number of extant *Aratea* manuscripts proved to be surprisingly large, this study devotes individual attention to only eight selected manuscripts: two from the ninth century, two from the twelfth, and four from the fifteenth. Each of the selected manuscripts was carefully chosen for its particular qualities and its exemplary time period and will be presented in relation to its 'scientific', poetic and cultural aspects. This investigation also establishes criteria for identifying the authors of other astronomical texts and discusses the various types of astronomical pictorial cycles.

Until recently art historians have had few resources or informational bases to consult when researching the topic of astronomical illustrations in general or of the complicated *Aratea* manuscripts in particular. This survey of surviving *Aratea* manuscripts, along with the many recent studies, will assist in bringing a broader understanding of the functions of this essential body of information in the history of astronomy, written originally in Greek, translated into Latin during the supremacy of the Roman Empire, and conveyed through the Medieval and Renaissance Eras through its text and images.

Significance of the Astronomical Poem

The poetry of the *Phaenomena* has survived periods of great interest by text scholars along with periods of neglect; but its bibliography is quite extensive. A thorough textual analysis of the original Greek poem was carried out in the nineteenth century; during the 1990s annotated editions and translations of the text were published by both Douglas Kidd and Jean Martin.[4] The text of the original Greek poem has been thoroughly studied and commented upon by both literary specialists and historians of astronomy. Remarkably, after over two millennia scholars are still mining material from the rich depths of the *Phaenomena*. A new English translation of Aratus' poem appeared in 2015 by R. Hard, *Eratosthenes and Hyginus, Constellation Myths with Aratus' Phaenomena* which provides lengthy commentaries with insights into the complicated and conflicting constellation myths partially created and mapped in ancient Greece. *Aratus and the Astronomical Tradition* by Emma Gee in 2013 relates the historical importance of the poem and explains how Aratus was relevant to his time and through history and why that still matters to scholars today. In 2010 Aaron Poochigian, published *Aratus Phaenomena, translated with an introduction and notes,* a new translation of the ancient poem, which Robin Hard describes as "enjoyable to read, but is often very free" (Hard 2015: xli). An earlier translation from Greek was published in 1976 by S. Lombardo, *Aratus' Phaenomena: an introduction and translation,* as a doctoral dissertation. In addition to the English versions, the poem has been translated into many other languages. The journal articles published on the original astronomical poem are too numerous to mention in their entirety, but pertinent essays will be cited as introduced and their thoughts are repeated or questioned.

The Latin translations of the *Phaenomena* that survive from three ancient Roman authors have also been thoroughly studied by classical text scholars in respect to their place in both Hellenistic literary culture and in Augustan poetry. The bibliography for each chapter cites the large number of books and articles written by classical and textual scholars, but specialists in these areas have little interest in the illustrations and rarely even mention if pictures have been included.

The illustrated manuscripts of the *Aratea*, even though they constitute the oldest surviving corpus of astronomical images, have had less attention as a whole tradition until recently. The monumental compilation of astronomical and mythological illustrated manuscripts by Saxl, Meier, and McGurk includes mention of some *Aratea* manuscripts and provides a few sample miniatures of the constellations but is outdated.[5] Their collaborative work discussed astronomical picture

[4]Martin (1998) *Aratos: Phénomènes*, Texte, établi, traduit et commenté. 2 vol., and Douglas Kidd (1997) *Aratus Phaenomena.* See also: Erren (1971) Aratus, *Phainomena. Sternbilder und Wetterzeichen*, ed. and tr. (German), Munich.

[55]For full citation of Saxl and Meier's study see in this Chapter bibliography. See also Lippincott (2016) *The Saxl Project* website for a continuation of Saxl's original research on astronomical art illuminations.

cycles in general, and includes some basic information and analysis on a few of the surviving manuscripts. In 2012 a much-needed expansive survey of astronomical art was published in German in two volumes by Blume, M. Haffner, and W. Metzger, *Sternbilder des Mittelalters: Der gemalte Himmel zwischen Wissenschaft und Phantasie,* Part One. The 68 illustrated manuscripts categorized in this meticulous survey are analyzed, advancing the knowledge and understanding of astronomical and scientific principles from the ninth through the twelfth centuries. This first publication includes an impressive number of illustrations of full-sky celestial maps, individual constellations and the Sun, Moon and five planets found in astronomical manuscripts produced in European scriptoria. A second set of volumes, Part Two, also in German, surveying the later Middle Ages and early Renaissance by the same three specialists was published in 2016. These three volumes include scholarly articles and detailed information on 143 illuminated manuscripts produced from 1200 to 1500. *Sternbilder des Mittelalters* provides new and invaluable research assistance for scholars investigating not only the transmission of medieval astronomy and astrology, but also mythology, classicism, history, historiography, education, science, and medicine.

Research on the topic of astronomical art has flourished in recent years; an extensive study in astronomical art was published in 2013 by Elly Dekker, *Illustrating the Phaenomena: Celestial Cartography in Antiquity and the Middle Age,* in which she catalogs and describes every surviving celestial globe and many full-sky maps dotted with illustrations of the constellations that are found in medieval manuscripts. In her comprehensive account she demonstrates how the design and construction of the celestial globes and the pictorial images of the heavens both change and remain the same through hundreds of years. These various studies demonstrate how the miniatures illustrate the lengthy textual traditions which are intertwined with the history of science.

A small number of the illustrated astronomical manuscripts of the *Aratea* have been given detailed attention by art historians and most are described concisely in their individual library or museum catalogs. Some of the most outstanding illuminations from *Aratea* manuscripts have appeared in journal articles, exhibition catalogs, and in art historical survey books. Various journal articles have been published discussing features of the well-known *Aratea* manuscripts; the most publicized and highly regarded Latin version of the *Phaenomena* is the Leiden *Aratea*. An important journal article by Dekker (2010), "The Provenance of the Stars in the Leiden *Aratea* Picture Book" examined the placement and number of individual stars in each of the 46 constellations illustrated in the Carolingian manuscripts; this will be discussed in Chap. 7. The Warburg and Courtauld Institute, University of London has instituted a new digital resource to provide complete access to over 250 medieval and Renaissance illustrated astronomical and astrological manuscripts. Under the direction of Kristen Lippincott, the Saxl Project, named for the pioneering work on astronomical manuscripts carried out by art historian Fritz Saxl (1890–1948) is meant to continue and update his multi-volume work that was expanded by Hans Meier and Patrick McGurk.

The history and historiography of the texts is complicated as it includes the original Greek and three different Latin translations of the poem and an accumulation of various *scholia*.[6] For many years the study of the artwork, maps, and diagrams in astronomical manuscripts seemed to have fallen between the seams of art and science, but current interest has attracted researches that are being fully investigated in each particular area. Many uninvestigated areas in astronomical art remain ripe for future research. Although the textual component of the poem comprises the major element of most studies, the art historical aspects of this significant literary and educative value of the illustrations are the main focus of this monograph.

Essential questions need to be discussed beyond the quantity, quality, and dates of the surviving illustrated manuscripts. What astronomical models were available for medieval illuminators to use as exemplars? Are there patterns to be discerned in regard to the illustrations of each author? Do particular styles of paintings accompany each individual author of astronomical material? Are the illustrators reading the poem and creating images in accordance with their reading, or primarily copying earlier exemplars? Were the *Aratea* manuscripts, as often stated, actually the main source of astronomical information in the Middle Ages? Who were the readers of *Aratea* manuscripts? Why did this astronomical poem remain so popular? What were the functions of the *Aratea*? This study examines, defines, and interprets the extant *Aratea* manuscripts to help in clarifying many of these complex issues, both within the individual examples and as a 2300-year-old illustrated poetic tradition.

Significance of the Research

Although the text of the original Greek didactic poem composed by Aratus in 275 BCE survives intact, his initial version does not constitute the main focus of this investigation. For the most part, the surviving copies of Aratus' poem from the Hellenistic and Byzantine worlds are not illustrated and did not circulate in the Latin West; therefore the few surviving Greek manuscripts of the *Phenomena* that survive (including one at the Vatican Library) are not included in this study.

Aratus was originally appointed to write a poem about the stars in the heavens, but what he composed was more than just a guide for learning the names and positions of the constellations; he combined the stars with philosophical ideas of divine and human order. Important elements of astronomical data and measurement; the rising and setting of the constellations, the four celestial circles: the

[6]D.B. Gain (1976), in *The Aratus ascribed to Germanicus Caesar* gives the stemma of Germanicus manuscripts and Soubiran (1972) in *Cicèron: Aratea Fragments Poètiques*, lists surviving manuscripts of Cicero, but neither are interested in the illuminations. Kidd, Martin and Mair, Aratus, *Phaenomena* translations each present an account of the Greek manuscripts and the early-printed copies.

Equator, the Tropics of Cancer and Capricorn, and the Ecliptic, are also defined in the *Phaenomena*. Learning these various details would aid a layperson's understanding of the structure and diurnal motions of the cosmos. Aratus says he gathered this essential knowledge established by the benevolent god Zeus to assist those on earth and to orient their lives when on land or on sea: for travel, for tracking time, to know when to pray, when to celebrate, and to assist sky watchers in committing to memory the innumerable celestial star patterns.

Necessarily this study examines Aratus' poem for contextual material, as he was the original composer of the*Phaenomena*, but not the primary author of concern. The medieval manuscript tradition under investigation focuses on the Latin translations of only two of the three famous Roman authors. The Latin translations of Aratus' poem circulated in the medieval world for more than 800 years, as it was the ideal vehicle for the assimilation of astronomical information into abstract cosmology, encouraging the direct connection with cosmological data passed down from the Golden Age of Greece and Rome. The picture cycle of the heavenly constellations that accompanies the Latin astronomical poem developed during the late eighth century. All surviving *Aratea* manuscripts include most constellation illustrations—that cast of mythological characters that faithfully encircle the dark sky, never failing to reappear for their cyclical appearances (Fig. 1.1).

Most *Aratea* manuscripts were decorated not only with remarkable constellation pictures, but also with depictions of the Sun and Moon and planetary deities, celestial maps, and the full cycle of constellations; all aspects that broadened their

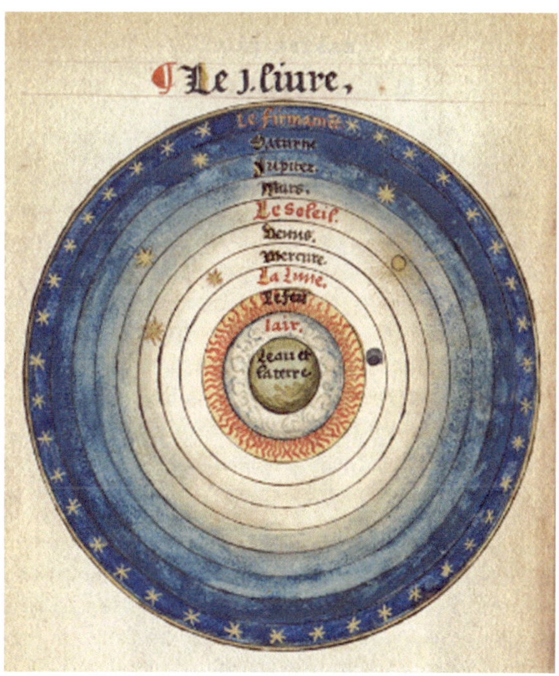

Fig. 1.1 Medieval Celestial Sphere, Zones of the Seven Planets and Fixed Stars, Public Domain

popularity. Typically a constellation-filled map of the Northern and Southern Celestial Hemisphere, divided into five zones, appeared on the initial folios of an *Aratea* manuscript, followed by the constellation descriptions and illustrations. Although the constellations represent personifications and honored beings or creatures from Greek mythology, Aratus includes few detailed myths. In their translations, the Roman authors of the *Aratea* amplified the number of astral myths and imaginative lore that appeared in Aratus' original poem.

Evidence survives demonstrating the popularity of the *Phaenomena* and the *Aratea* among contemporaries; the poem provided entertaining verses for educated Hellenistic and Roman elite to read or more correctly to hear, as poetry was meant to be intoned. But historians can merely speculate on other functions of the poem during classical times. Possibly the students' educational program, based on the seven liberal arts, required the study of this poem to fulfill their knowledge of one of the Quadrivium, astronomy. Perhaps itinerant merchants and peripatetic sailors slowly memorized the poem, section by section, season by season, impressing the vital star formations on their minds. Similar functions are conceivable during the long medieval period along with devout monks studying and teaching astronomy in book-lined libraries. These usages cannot be answered decisively, but investigations of various issues can specify the quantity, quality, content, and condition of the extant manuscripts. This particular research aspires to cast a brighter light upon its celestial topics.

Biography of Aratus of Soli

> Aratus shall always endue with the sun and the moon
> Cum sole et luna semper Aratus erit (Ovid Amores 1.15.16)

Aratus (Gr. Aratos, c.310–240 BCE), one of the greatest Hellenistic poets, was born in Soli, a city in Cilicia, now Southern Turkey. Our main sources for reliable information on his life have been preserved in manuscripts of the *Phaenomena* and the *Aratea* (Kidd 1987:3)[7] which report that he studied under both Menecrates in Ephesus and Philitas in Cos. As a disciple of the Peripatetic philosopher Praxiphanes, Aratus became acquainted with Zeno, the Stoic philosopher, with whom he studied in Athens in the 280s or 270s; he also knew Callimachus (310/315–240 BCE) of Cyrene, one of the most influential poets of the age. Aratus' literary output was greatly admired by his contemporaries and other Greek authors and which included hymns, epigrams, as well as didactic poems, but the astronomical poem *Phaenomena* became the most celebrated

[7]For the text and full description of the four early biographical sources see J. Martin HT 137–95. Martin attributes the *vita* to Theon of Alexandria, a grammarian of the first-century BCE. A biography also appears in A. W. & G. R. Mair, 185–189.

of his many works, the only work to survive intact. The ruler of Macedonia, King Antigonus II Gonatas (Knock-Kneed) summoned Aratus to his royal court and commissioned him to compose a poem formalizing the positions and motions of the heavenly constellations. Aratus' composition of the *Phaenomena* can be dated between 276 and 274 BCE, because the Macedonian literary circle, to which Aratus belonged, was broken up after that time. The dissolution of that Macedonian circle, caused by the invasion of Pyrrhus, allowed Aratus to leave for the Court of Antiochus I Soter in Syria where he completed an edition of Homer's *Odyssey* (Mair 1969: 189). He eventually returned to Pella, the capital of the Macedonian Kingdom where he died in 240 BCE (Figs. 1.2 and 1.3).

For assistance and accuracy in his descriptions of the constellations and celestial coordinates, Aratus consulted an earlier astronomical compendium, also called the *Phaenomena*, composed by Greek astronomer Eudoxus of Cnidos (ca.400-347 BCE), a renowned Greek astronomer, mathematician, geographer and philosopher. Eudoxus, a student of Plato and a contemporary of Aristotle, also established an academy of learning in Athens (Dicks 1985: 151). Although none of the writings of Eudoxus survive, contemporaries have commented on his learned treatises relating to astronomy, music and medicine; historians assert that formal Greek mathematical astronomy began with Eudoxus (Tester 1987: 11). A few portions of Eudoxus' *Phaenomena* have been preserved within Hipparchus' commentary on Aratus'

Fig. 1.2 Aratus of Soli marble bust, Public Domain

Fig. 1.3 Eudoxus of Cnidos Ancient Greek papyrus fragment, Public Domain. It has been suggested that crescent with an arrow in its midst refers to the necessity of the sun, moon and earth to be in a straight line for an eclipse to occur.

poem.[8] Cicerorecorded the tradition that Eudoxus had created a celestial globe covered with accurate images of the constellations (Dicks 1985: 153); both his treatises and globe are known only from early references by other Greek authors (Kidd 1987: 17).[9] Unfortunately neither his astronomical descriptions nor celestial images have endured. Historians consider Eudoxus to be the first astronomer to attempt both a mathematical description of the celestial sphere and a formal written arrangement of the constellations (Kidd 1987: 14).[10] In his *Phaenomena*, Aratus avoids the mathematical and objective perspective of Eudoxus leans to that of the subjective observer. Upon investigation of Aratus' poem explaining the constellations and heavenly motions, experts assume that for guidance he made use of the pictures seen on Eudoxus' celestial globe as well as his written account (Martin 1989: XCVI).

[8]See *Hipparchi in Arati et Eudoxi Phaenomena commentariorum libri tres*, ed. And tr. (German) K. Manitius, Leipzig: 1894.

[9]Kidd suggests that the globe may actually have been an armillary sphere as there is evidence that these existed in the fourth-century BCE; as no evidence for a true celestial globe has been discovered from that early date. Either type of globe could depict illustrations of the constellations. Check Dekker??

[10]Eudoxus studied with the Pythagoreans and spent 16 months with Egyptian priests studying at Heliopolis where he also made astral observations. Eudoxus also spent time in Athens and had the opportunity to discuss his findings and writings with both Plato and Aristotle.

Utilizing the scientific writings and heavenly images of Eudoxus as guides for the technical material, and considering the writings of the two great epic poets Homer and Hesiod as models for his poetry, Aratus designed his masterwork, the *Phaenomena*. Even though Hesiod's *Works and Days* (early seventh century) was written to be performed orally, Aratus' didactic epic had a literary basis, that earlier Greek poem is considered Aratus' most important model. The narrator of a didactic poem speaks to the listener or reader in the second-person singular: in this poem there is no specific person or group that he is addressing but a more universal audience. When displaying his erudition and familiarity with Greek poetry, Aratus makes frequent references to Homer's *Iliad* and *Odyssey*. Although not a practicing astronomer himself, Aratus proved that his lack of technical or mathematical expertise caused no difficulties when writing on the constellations and orderly design of the heavens with intelligence and eloquence. Aratus claims to have no special knowledge, but his poem provides a summary of the entire celestial phenomena as it was understood by Greek astronomers in the third-century BCE.

Because Aratus lacked the expertise of a practiced stargazer, the didactic poet did get his facts wrong in several sections; as a result erudite astronomers, such as Hipparchus (fl.150–125 BCE), were anxious to display their own expertise by pointing out his scientific errors (North 1994: 98).[11] Even before Hipparchus wrote his critique, Attalus of Rhodes had also composed a commentary on Aratus' astronomical errors. In spite of his few technical inaccuracies, Aratus' *Phaenomena* brought him immediate success and enormous esteem among Hellenistic poets, inspiring numerous commentaries, criticisms, and exegesis. The words of Aratus were even mentioned by St Luke in the New Testament, Acts of the Apostles, when Luke quotes a sermon in Athens by St. Paul (Farber 1993: 3)[12] and again by St. Jerome.[13] For his great poetic accomplishment, Aratus was recognized with a monument in his birthplace of Soli and his aged image appeared on local coins (Mair 1969: 189). Not overlooked by modern astronomers, Aratus was memorialized in the present time when a crater on the Moon was named in his honor.

---

[11]The sole surviving work of Hipparchus is *A Commentary on the Phaenomena of Aratus and Eudoxus*, in which he computes many simultaneous risings and settings of stars. It is thought that Hipparchus worked mainly on the island of Rhodes. Hipparchus was the "first Greek astronomer who is known to have systematically applied arithmetical methods to geometrical astronomical models".

[12]St Luke writes of St. Paul's visit to Athens in CE 51, where Paul wishes to convert, yet not to ridicule his educated audience. Paul, speaking of God, quotes from the poem of Aratus, who spoke of Zeus. Aratus' words from the fifth line of the proem of the *Phaenomena* are recorded by Luke and attributed to Paul in the second half of Acts, 17.28: *For we are indeed his offspring…*

[13]St. Jerome also comments on St Paul's reference to the poem saying "*quod hemistichium in Phaenomenis Arati legitur, quem Cicero in latinum transtulit, et Germanicus Caesar, et nuper Avianus et multi quos enumerare perlongum est*", *Comment. in Epistolam ad Titum* quoted by Vicor Buescu in *Cicéron: Les Aratea* (1941) p. 20 as mentioned by Dodwell, p. 61.

The *Phaenomena*: Text and Structure of the Poem

The *Phaenomena* of Aratus lives on not only as a brilliant work of poetry; his hexameter poem survives as the oldest methodical record of the heavenly constellations, including a precise verbal description of their relative positions. Aratus modeled his didactic epic on the folk wisdom of Hesiod's *Works and Days*, and specifies influence from the current ideas and writings of Stoic philosophers as well. The *Phaenomena* has been organized by text scholars into five sections: the Proem (1–18); the Constellations and Planets (19–461); the Measurement of Time (462–757); the *Weather Signs* (758–1141); and the Conclusion (1142–1154) Kidd (Kidd, 1987:5–7).[14]

The Proem opens by honoring the mighty god Zeus, recognized as the designer of the celestial phenomena that he in his paternal kindness organized into nocturnal sign for the benefit of all humanity. Following the invocation, Aratus launches into a systematic presentation of each stellar constellation establishing their name and relative placement; following the description of the 46 constellations, the five planets are acknowledged only briefly. As the poem advances, Aratus balances the entire cosmic system by applying reference lines onto the celestial sphere: the celestial equator, the ecliptic, and the two tropics, and then relates them to the Milky Way. Once the basic structure of the universe has been clarified, Aratus introduces time by thrusting the celestial system into motion. He describes each constellation figure as it rises in the east and sets dipping below the western horizon; after which he adds the motions of the sun, rising in turn with each sign of the zodiac and then the quickly changing phases of the moon. Finally, he descends from the delineation of the celestial patterns to attend to terrestrial phenomena, the *Weather Signs*; here he links the wind, rain, clouds, and storms with the activities and behavior of earthly animals. In conclusion Aratus advises his readers to keep a watchful eye on these providential sign, so that they will never draw a baseless conclusion.

Through the generosity of a watchful god, the various activities throughout the natural world become available for humanity's advantage by observing providential signs. In his didactic verses, Aratus dramatizes a poetic vision of a starry sky, not as meaningless points of light, but as a dynamic arrangement, that demands our notice, involvement and interpretation. "*Inevitably one discovers upon careful reading of the Phaenomena that beneath the routine of astronomical description and the veneer of plainness and simplicity the Phaenomena is a complex literary cosmos that mirrors its lofty subject, Zeus' cosmos*" (Possanza 2004: 21). Aratus invites the reader to interact with the nightly narratives; he desires that the viewers 'take pains' to learn the patterns of the stars and attune themselves with the semiotic patterns of celestial and earthly phenomena.[15]

[14]This division refers to the original poem in Greek, not the Latin versions, but the Arateans follow the same basic outline in their poems that Aratus designed initially. See Possanza (2004), 8–9 for a comparison of Aratus and Germanicus verses.

[15]All *Phaenomena* excerpts are quoted from Kidd.

and the stars which, when sighted in Poseidon's realm or
in that of Zeus himself, give clearly defined signs to men.
Therefore take pains to learn them [758]

The Proem

The Proem is intended as a hymn to Zeus the Father-God, who designed the constellations to define the year's rhythmic cycle. The stars are not positioned merely as guides for determining time and the seasons, but also as signs of the existence and power of almighty Zeus. But the Zeus of Aratus rules not as the vindictive, impersonal god of Hesiod; rather he represents the origin of all forms of life, the benevolent vital force of the Stoics, permeating the complete cosmos (Hornblower 1998: 691–2).[16] Like a loving father, Zeus supports humans in their struggle for existence on earth, especially by providing guidance through signs in the heavens.

Let us begin with Zeus, whom we men never leave unspoken.
Filled with Zeus are all highways and all meeting places of people
Filled are the seas and harbors;
In all circumstances we are all dependent on Zeus.

For we are also his children, and he benignly gives helpful signs to men,
and rouses people to work, reminding them of their livelihood,
tells when the soil is best for oxen and mattocks,
and tells when the seasons are right both for planting trees
and for sowing every kind of seed.

For it was Zeus himself who fixed the signs in the sky,
making them into distinct constellations,
and organized stars for the year
to give the most clearly defined sign of the seasonal round to men,
so that everything may grow without fail.

That is why men always pay homage to him first and last.
Hail, Father, great wonder, great boon to men, yourself and the earlier race!
And hail, Muses, all most gracious!
In answer to my prayer to tell of the stars in so far as I may,
Guide all my singing (Kidd 7: 1–18).

Aratus initiates the Proem utilizing a reverential tone that invokes and honors Father Zeus, establishing the theme for the verses that follow. He concludes with a humble prayer invoking the Muses, inviting their guidance and inspiration in his 'singing' this tale of the stars.

---

[16]The Stoics were a philosophical movement based on the idea of a harmonious, well-organized universe designed by a providential deity, making itself known to humanity through the sign and symbols of nature.

The Constellations

The vast expanse of stars and their random dispersal throughout the night sky preclude any means of organizing and discussing them without arranging the heavens in some comprehensible manner. Ancient cultures, unaware of the coordinates, right ascension and declination utilized in modern astronomical considerations, simply arranged the brightest stars into simple patterns; then they assigned easily remembered pictures and mythological stories to each recognizable star assemblage. The constellation patterns and myths did not originate with the astronomer Eudoxus but reach far back into earlier ages; current research suggests some familiar patterns may date back as far as the Paleolithic painted caves (Fig. 1.4).[17]

The constellations described by Aratus had already been formalized and personified by the ancient Sumerians, Babylonians and Greeks and even earlier cultures (North 2008: 46).[18] The main body of the poem, the Constellations (19–453), supplies factual information elaborating on 46 individual constellations by describing their positions and appearances, so that they could be recalled without difficulty. The poem continues in an orderly fashion with a brief account of each star pattern, occasionally including the mythology surrounding the celestial figures. Aratus' descriptive verses and a summary of pertinent information explaining the significance of each constellation with sample illustrations can be found in Appendix C.

When creating his poem, Aratus did not simply turn a list of constellations into verse, but applied his particular talent and ingenuity. Through his poetic descriptions, Aratus created word pictures, without actually discussing the number, position, or characteristics of individual stars within the configurations.[19] By directing the reader toward a specific place in the sky, adjacent to another constellation, and then describing this group of stars, Aratus enables the reader to compose a picture inspired by the magic of his words. The mental images created by the words did not

---

[17]According to German researcher Dr. Michael Rappenglueck, of the U. of Munich, some cave art shows that our ancestors were more knowledgeable than many believe. He claims that paintings in both Lascaux cave and Cueva di El Castillo cave, Spain depict ancient star maps. "The Pleiades in the "alle des Taureaux" Grottedes Lascaux" (1997). Other astronomers have also put forth similar theories, i.e., Frank Edge and Wheston Price.

[18]"Even if their meaning has shifted somewhat, the Sumerian names for the constellations of the Bull, the Lion and the Scorpion have all passed down to our own day" (North).

[19]Other writers enumerate the relative positions and the numbers of stars on head, feet, or tail of a figure. With the introduction of the Arabic texts and illustrations during the twelfth century, the placement of the stars within the constellation becomes more precise.

Fig. 1.4 Bull with Group of Seven Dots, Hall of Bulls, Lascaux, France, Public Domain

appear as painted pictures in the Greek manuscript tradition; except for a few minor exceptions, as no illustrations survive in Greek copies of the *Phaenomena*. The two exceptions, both from the fifteenth century, are listed in Appendix A, discussed in Appendix B.[20]

Through time, these mythological and symbolic images created from star patterns, became icons packed with layers of meaning, as astronomical knowledge continued to be assembled by astronomers and astrologers. The most important of these star formations were the twelve signs of the zodiac, the constellations that circle the heavens on a specific path called the ecliptic, through which the Sun, Moon, and the five known planets traverse on their yearly sojourns. Aratus did not separate the signs of the zodiac as a distinctive group, but included each sign as he approached its particular area of the sky. The zodiacal constellations are of special consideration in this study as they, along with the Labors of the Month, were utilized in artworks as the prime iconographic symbols for the months and seasonal occupations of the year.

With the spread of Christianity throughout the medieval period, the zodiacal signs assumed additional symbolism denoting the omnipotence of God as the creator and ruler of all things both in Heaven and on Earth. Subsequently the signs

[20]The two illustrated copies in Greek are held at Biblioteca Apostolica Vaticana Cod. Vat. Gr.1087 (1320–1330) and El Escorial, Scorialensis Σ III 390 (1490).

Fig. 1.5 West Façade, north portal zodiacal archivolts, Chartres Cathedral, Twelfth Century, Courtesy University of Pittsburgh Medart Website

of the zodiac also came to personify distinctive personality types as characterized by astrology, but these relationships did not become widespread until the later Middle Ages after the absorption in the West of the astrology defined by Arabic sources. Thus bundled with layers of meaning, illustrations of the signs of the zodiac appeared as symbols not only in astronomical manuscripts, but also in calendars, medical texts and Books of Hours, and in numerous other art forms: mosaics, frescoes, tapestries, stained glass windows, metal-works, and sculptures (Figs. 1.5 and 1.6).[21]

The Planets

In contrast to their prime importance in astronomical and astrological texts of the later Middle Ages and early Renaissance, Aratus' *Phaenomena* conveys scant attention to the erratic movements of the five known planets, only seven lines (454–461) are devoted to these wandering stars. Aratus himself tells us that he does not

[21]The specific structure of the constellations of Aratus' poem and a discussion of their significance is found in Appendix C.

Fig. 1.6 Virgo and Scorpio, Zodiacal Window, Chartres Cathedral, Early Thirteenth Century, Courtesy of University of Pittsburgh Medart Website

feel competent to discuss their ever-changing positions; he declines to even name them individually. The 'signs' in the heavens are reliable because they repeat in regular and eternal recurrence. "Confidence comes from the repeated pattern *(kósmos)* of successful 'sign-reading'; in this, the argument runs, the planets fail us" (Hunter 1995: 6). The planets were not ignored due to a lack of interest, but because they were irrelevant for his purposes. Because of their unpredictability, lengthy rotation periods and retrograde motion, the planets provided no assistance for telling the time of year or the season (Kidd 1987: 343).[22] The problem of understanding the unpredictable motions of the planets was not a serious threat to an early astronomer, but within the theme of his poem which emphasized the divine order of the cosmos designed by Zeus, Aratus places particular importance on their 'quirkiness' (Hunter 1995: 6).

> But there are five other stars among them, but quite unlike them,
> that circulate all the way through the twelve figures of the Zodiac.
> You cannot in this case identify where these lie by looking at other stars,
> for they change their positions. The years of their orbits are long,
> and at long intervals are their configurations when they come from afar into conjunction.
> I am not at all confident in dealing with them: I hope I may be adequate
> in expounding the circle of the fixed stars and their guide-constellations in the sky (455–62).

---

[22]From the time of Eudoxus, the interpretation and explanation of the erratic movements of the planets was a central concern for Greek and Hellenistic astronomers.

The other two ‘planets’, the Sun and Moon, were of much greater interest to Aratus, since their positions and appearances are instrumental for calendric purposes and for weather prediction. Consequently he discusses the Sun and Moon in the final section of the poem, the *Weather Signs*.

The Measurement of Time

After the standard 46 constellations of the celestial sphere had been identified and described, Aratus then proceeded to explain how to use them to measure the passage of time. The third section of the poem (lines 462–757) contains the more technical information on how the universe functions (Kidd 1997: 347).[23] The poem discusses the celestial circles, and also the simultaneous risings and settings of the constellations. The identification of the signs of the zodiac helped to determine the time of night since six signs rise in the East and set in the West each night. Aratus explains the star formations or asterisms that accompany the signs of the zodiac, those that rise at the same time, called the *paranatellonta* (lines 559–732) and he tells how these astral events indicate the periods of time.[24]

Aratus describes the waxing and waning of the moon as well, and advises his readers to be observant of the moon’s movements; as it spins its course quickly, the moon passes into a new zodiacal sign every 2 hours. Vigilant attention to the moon’s changing appearance can provide helpful clues for predicting the weather. His account of the methods to estimate the passage of time by observing the positions of the sun, moon and constellations concludes the section just before the *Weather Signs*. The Measurement of Time would have been particularly helpful for sailors or sea voyagers who had no landmarks to orient their earthly position. Aratus tells his readers that although we are blessed with numerous signs, we “do not yet have knowledge of everything from Zeus, but much still is hidden” (769–70).

The Weather Signs

The fourth part of Aratus’ *Phaenomena,* known as the *Weather Signs* (lines 758–1141), was sometimes referred to as the *Diosemeia,* or Signs from Zeus. This section of the *Phaenomena* was based on a lost treatise of the fourth century, perhaps by Theophrastus, known mainly from a surviving prose version that dated

[23]Aratus achieves this through an established system of four circles that mathematicians have drawn on the celestial sphere to give points of reference for determining the apparent position of the sun in reference to the fixed stars at any time of the year.

[24]Illustrations of the *paranatellonta* do not appear in astronomical manuscripts until after the translations of the Arabic texts, mainly in the twelfth century. The *paranatellonta* will be discussed with the Fendulus translations of Abu Ma’shar in Chap. 4.

after the time of Aratus (Hunter 1995: 1). Here Aratus switched his focus from the sign appearing in the heavens to the sign that Zeus provides on earth. These verses explain the phenomena of nature and tell how storms, lightening, winds or even good weather can be predicted by observing the behavior of birds and animals in addition to the rhythmic events in the heavens. The animals interact with the environment and their actions reflect the processes of nature, but do not necessarily cause them. Their activities, along with other phenomena such as clouds or storms, act as indicators or warnings to assist humanity; even tiny mice can be prophets. The *Weather Signs* provide particular details on what type of weather would follow the typical or uncommon actions of birds and beasts. This attunement with nature makes it less risky for farmers and sailors to forecast favorable conditions. For example, the appearance of Arcturus is useful in predicting rough weather at sea (Kidd 1987: 431). The *Weather Signs* deliver entertaining, informative and historical data since these sign relate to current folklore and weather lore. Since the *Weather Signs* focus on earthly phenomena, this section rarely appears in the *Aratea* manuscripts, and was never illustrated; therefore this last portion of the *Phaenomena* will not be included in this investigation.

The Conclusion

Aratus sums up the principal lessons of his great work with a comforting conclusion in the final section of the poem (lines 1142–1154). He directs his readers not to make light of the God-given sign, and to watch carefully for confirmation. If two indicators point to the same situation, have hope, if three sign are seen, have confidence, for these sign in the heavens and on the Earth have been provided through the benevolence of Zeus. "There are so many signs in nature that it would be impossible to tell them all and in any case Aratus wants us to observe for ourselves and to find our own 'signs'" (Hunter 1995: 10). Aratus' final lines give hope and reassurance to those who would follow his guidance: There are so many signs in nature that it would be impossible to tell them all, and in any case Aratus wants us to observe for ourselves and to find our own 'signs'.

> If you have watched for these sign, all together for the year,
> you will never make an uninformed judgement on the evidence of the sky (1154).

Reception and Influence of the *Phaenomena* in the Hellenistic World

Aratus' *Phaenomena* enjoyed immediate fame among Greek readers not only for its concise and helpful astronomical information, but also for its literary style, honored by a society that valued its rich poetic traditions (Lombardo

1976: 1).[25] The poem was carefully composed and successful in its goal to educate and inspire appreciation for the beneficial signs placed in the heavens and on the earth by a kind and sympathetic god. Theories of the universe were much debated in the ancient world; there was no universal agreement among philosophers. Various ideas circulated and contemplation of the celestial design and order was thought to lead to knowledge of the divine (Gee 2013: 13). Memorizing the verses of the didactic poem became a convenient way to become familiar with the constellations and the harmonious predictability of their nightly or seasonal appearances. Aratus' genius for creating verse enabled him to turn a dry prose work of technical astronomy into a sensitive and easily learned poem; his ability to create puns and literary witticisms provided entertainment for his more erudite readers. The *Phaenomena* fulfilled the desire for technical knowledge in an acceptable literary form, and also presented a system of physics that explained the "unified and divinely governed cosmos" (Gee 2000: 90).[26]

> The totality of Greek life cannot be understood unless the oral side of Greek society is appreciated as well as the written and, if possible, the relationship between the two. Of course literacy in ancient Greek society was not extensive; only a limited small audience could understand the literary qualities of the poem, and an even smaller number could appreciate Aratus' subtle references to Homer and Hesiod. But since poetry is created to be heard as well as to be read, all listeners - young or old - could enjoy listening to recitations of the stories and myths, and could commit key sections to memory (Thomas 1991: 105).

Knowledge of the heavens was essential to those living in earlier times, a necessity which modern society can never truly understand. Familiarity with the rhythms and cycles of the fixed stars and shifting planets was crucial for functioning in a world without lights, clocks, or an accurate calendar. Awareness of the celestial movements was indispensable for travel on land and sea, to determine the time and direction, during both day and night; Aratus provided a basic list of quantifiable data describing the cosmos. Gee names Aratus as "the first author to make the union between astronomy (data) and cosmology (imagination) in a form accessible to many readers" (Gee 2013: 17). Therefore the *Phaenomena* was effective not only for literary reasons, but for providing fundamental and essential scientific knowledge in an enjoyable and entertaining format.

The mythological stories of the celestial inhabitants were well known in the Hellenistic and Roman worlds, but visualizing the characters nightly acting out their roles in a precise, predictable routine enabled the constellations to be more easily imprinted on the viewer's mind. By dividing the vast sky into compact sections, the poem organizes the massive expanse of individual stars that would otherwise be impossible to identify or retain mentally. The intimate association of pictures and words created by the poem helped the reader to determine the time of night, season

---

[25]Lombardo calls the *Phaenomena* "one of the most successful poems in the classical world."

[26]Gee explains how Aratus incorporates the Stoic philosophy into the poem. "Aratus as a Stoic poet, who works with the elements of language, "builds the cosmos".

of the year, and the cardinal directions. The celebrated opus of Aratus inspired other writers to imitate his success and to comment on his creation.

The fame of the *Phaenomena* spread rapidly from Greece, extending to Alexandria and the entire Hellenistic world; commentaries on the *Phaenomena* followed just as quickly. Aratus attracted two types of early commentaries: the grammarians, who were concerned with the literary aspects, and astronomers who were involved with the scientific aspects, and a few who addressed both issues (Kidd 1987: 18) Twenty-seven separate commentaries were written by ancient scholars on the work. In the middle of the second-century BCE, Attalus of Rhodes produced an edition of the poem with an important commentary (Mair 1969: 195).[27] About the same time his contemporary, Hipparchus (c.190–129 BCE), the most influential astronomer before Ptolemy, (North 2008:52) wrote his *Commentary on Eudoxus and Aratos,* which is still extant (Mair 1969: 195).[28]The *Commentary* of Hipparchus was addressed to a fellow astronomer; in it he criticized and corrected errors found in the astronomical positions set forth by Eudoxus and Aratus. (Kidd 1987: 18–19)[29] Hipparchus was troubled by the descriptions of the rising and setting positions of the constellations given by Aratus.[30] A large group of exegetical *scholia* devoted to the *Phaenomena* also survives from the Hellenistic era. (Mair 1969: 191)[31] These interpretive writings were designed to be read along with the poem, what Gee calls the 'paratextual' tradition (Gee 2013: 7). Aratus' poem is one of the most comprehensively annotated works of the Classical world. The survival of much of this related textual material from early writers and commentators provides a fuller picture of the impact and influence of the original writing. The inclusion of literary devices such as acrostics, by Aratus show that even though most poetry in ancient Greece was oral, the *Phaenomena* was enjoyed for both its oral and textual proficiency. The dialogues of Plato emphasized the oral component to the process of communicating of ideas and gaining knowledge through orality, demonstrating that orality and literacy were both crucial in Greek culture. "Thus Greek history and culture can be seen partly in terms of the progressive weakening of the oral mentality, increasing individualism, rationality, and

[27]According to Hipparchus, Attalus wrote the most careful commentary.

[28]Hipparchus is most noted for the discovery and quantification of the Precession of the Equinoxes. He wrote three books of commentary on the *Phaenomena* and determined that it was in concordance with Eudoxus on all points.

[29]The detailed astronomical commentary by Hipparchus is intact; he chiefly addresses the astronomy in the poem; but he is also competent as a textual critic. He makes necessary corrections to improve the astronomical orientations, since the poem had become a primary source for study of the heavens. For a full discussion of Hellenistic commentaries and scholia see Kidd 44–48.

[30]It has recently been established that the star charts used by Aratus actually date to around 2800 BCE. (This discovery will be discussed in Appendix C) Hipparchus' investigation into the astronomical inaccuracies of Aratus, may have led to his most famous contribution to science, the 'discovery' of the Precession of the Equinoxes. This information is discussed in more detail in Appendix C.

[31]The Scholia are based mainly on a commentary of mathematician Theon of Alexandria (*c.* 335–405 CE).

the discovery of self, as literacy undermines it" (Thomas 1992: 26). These interactive discussions of the astronomy of Aratus in publications, as well as oral interpretations that have left no trace, offer an added benefit that can be useful in determining how the Hellenistic commentaries influenced and improved the translations of the three most significant Latin Arateans.

The Latin Arateans

The numerous Latin translators of the *Phaenomena* of Aratus provided a bridge between the poetic traditions of Greece and Rome. The controversies over orality and literacy were not a consideration for Roman society; during the late Republic and Empire, the culture was much more involved with books, literature, and legal documents than classical Greece. Latin literature became heir to vast holdings of Hellenistic scholarship from all parts of the widespread Greek-speaking world. Writing had become much more assimilated to Roman daily life than it had in Greece. (Thomas 1992: 158) Public monuments were inscribed; news, maps, information, and calendric notations were displayed in the forum for a relatively widespread reading public.

After achieving great success among the poets and literati in the Greek world, Aratus became even more popular in Roman circles in the first-century BCE, when his *Phaenomena* was translated into Latin by Cicero, Ovid and Varro Atacinus, and his work was imitated by Virgil and other Roman poets. (Kidd 1987: 46) In the early years of the first-century CE, Germanicus Caesar added his imperial influence to the poem's reputation and dissemination with his rendition. Another translation was made in the third century by Emperor Gordian I (r. CE 238), who only reigned a few months with his son, Gordian II, but not a single verse of his version survives. In the fourth century, a rendition composed by Roman poet Rufus Festus Avienus demonstrates that Aratus' popularity had not faded hundreds of years later.

Each of the three principal translators of the *Phaenomena* contributed his own expertise and individual style while creating his singular conception of the astronomical poem. Each author brings to the work his own distinct qualities, his particular poetic talent, and distinct interpretation of the source material passed down from Aratus. Although each Latin translation was written independently and during different historic eras, each author was influenced by the poem's reputation, as well as the large compendium of criticisms, commentaries, and scholia that was received from readers in both the Hellenistic and Roman worlds. Astronomical amendments, scientific updates, and mythological interpretations were inserted as the Latin poets felt necessary. Translating the *Phaenomena* would have presented a challenging ordeal for a Latin poet for many reasons: because of its 'canonical status' and 'refined style'; because it was a technical guide to the stars; and most significantly because the poem was, and still remains, a 'sophisticated literary work' conveying religious, scientific, and philosophical themes (Possanza 2004: 98). In spite of the challenges, the *Phaenomena* became the only writing from Greek literature to inspire Latin translations spanning three centuries. The poem of each of the three principal Latin Arateans will be examined in more detail.

Marcus Tullius Cicero and the *Aratea*

Cicero (106–43 BCE) was the earliest and most renowned Roman to Latinize the *Phaenomena* translating most of the poem from Greek into Latin while still a young man, ca. 87 BCE (Buescu 1941: 29), 8 years before his first legal case. He may have translated Aratus' *Prognostica* or *Weather Signs* a few years later. Perhaps he felt it necessary to translate Aratus' *Phaenomena* because at that time no functional astronomical manual existed in Latin. Cicero became a distinguished statesman, persuasive orator, skillful lawyer, articulate writer, and a dominant member of Roman society, a controversial voice in the politics of his time. Many of Cicero's speeches and private letters are extant, but few of his poetic efforts survive. Of his known poetry, his Latin version of the *Phaenomena* was the composition most respected by his contemporaries. Cicero had a limited proficiency in poetry, and in spite of his reputation as an excellent speech writer, his colleagues did not especially admire his expertise in the art of poetry. But he did create his version of the *Phaenomena* when still an adolescent. Cicero maintained his interest in the astronomical poem for he mentions it again when a mature philosopher, in the second book of his important religious and philosophical dialogue, *De natura deorum*, written in 45 BCE (Gee 2001: 520) (Fig. 1.7).

Cicero's translation of the poem exists in fragments totaling only 469 lines out of the original 1154 of Aratus (including the *Weather Signs*). Almost nothing of Cicero's Proem is known; therefore it cannot be compared with either the original or the other translations. Before line 230, survival of his version is very limited with a mere 63 lines extant. After that a large section from line 230 to 400 exists complete. The poem then breaks off, and various portions of the remaining verses survive; of the later verses, between lines 479 and 700, we have barely one line. There is little extant of the last section of the poem, the *Weather Signs* and Prognostications, just a meager 27 lines.[32] It is probable that Cicero's *Aratea* existed in its entirety until the sixth century (Gee 2007: 585).[33]

> Between the sixth and ninth centuries, the initial and final folia of the archetype, including the Prognostica became detached, leaving by c. 800 the work we know. H of Cicero (Harley 647) probably represents faithfully an intermediate copy made between the sixth and eighth centuries, by the latter date the text was probably already fragmentary (Gee 2007: 585).

---

[32]Perhaps the *Prognostications* were removed during the Middle Ages by zealous monks, as predictions and omens were considered superstition and condemned by the Church.

[33]Gee suggests that the fragment of astronomical poetry credited to Marcus' brother Quintus may have been a listing of the sign of the zodiac from Marcus' *Aratea*, namely from the *Weather Signs* which in his version are mostly missing.

Fig. 1.7 Cicero Bust, Ancient Rome, marble, Public Domain

Poetic fragments referring to astronomy were found among the *Eclogues* of Ausonius (CE c.490) that were attributed by its superscription to the brother of the orator, Quintus Tullius Cicero (Gee 2007: 565). Gee tentatively suggests that these fragments of astronomical poetry were attributed to Quintus Tullius Cicero erroneously and may have been written instead by the famous orator and may be some of the lost lines of his *Aratea*. She cautiously presents her reasons for proposing this unexpected change in attribution. The poetic fragments contain one of the oldest listings of the zodiacal sign in Latin after that of Marcus Cicero. If the verses belong to the *Aratea*, they would be a second listing of the zodiac which could have been a part of the *Prognostica*. This suggestion does not conflict since the list of Quintus admires the stars as weather signs (Gee 2007: 579).

Cicero's translation remains close to the original Greek; but he adds brief descriptions, especially of the animal formations depicted in the sky, such as Aries, the Ram and Cygnus, the Swan. Like the original, his version has a lengthy section that describes the constellations, the circles that define the path of the sun, and the rising and setting of the constellations. Cicero includes descriptions of the mythological stories associated with the various constellations, even in some areas expanding the lore of the original poem[34] (Gee 2000: 87). To improve on the rather technical and mechanical format of Aratus, Cicero's translation of the *Phaenomena* appears more dramatic than the slightly bland original, displaying 'stylistic dissonance' and a 'fondness for the epico-tragic diction of early Latin poetry' (Possanza 2004: 10). Taking into account that Cicero composed his poem before the development of the radical changes brought to the art by sophisticated Augustan poetics, his poetry seems wordier with more details and less compact in form. He did not enjoy the benefits that Germanicus had available from reading Augustan poetic compositions, nor the opportunity to study and draw inspiration from the highly sophisticated efforts of his contemporaries.

Cicero wished to enhance the more didactic aspect of the poem; he made frequent reference to the Latin equivalents of the Greek names of the constellations. When describing a constellation, discussing its brightness, its position, and its motion through the sky, he repeatedly used 'formulaic phrases'. As stated by Possanza in his discussion of Cicero's description of the zodiac,

> …in order to construct a numerical correspondence between the description itself and what is being described: there is one verse for each of the twelve zodiacal constellations, in seven of those verses the name of the constellation occupies the last position in the line, a structural device which may have aided in the memorization of the sequence (Possanza 2004: 177).[35]

Cicero's purpose is more directed and instructive, sometimes following Aratus line for line, but the intention of this poet-translator seems directed toward a less spiritual or philosophical end than the original. "Cicero's poem is an enjoyable read, in its own right, with its vividly cartographic, and relational, but at the same time intimate, descriptions of the stars, and its games with literary models, not just Aratus, but the Aratean scholiastic and mythographical tradition…" (Gee 2001: 523). By expanding and reconceiving the poem of Aratus, Cicero was not just producing a crib for understanding the original but creating an independent piece of literature in the Latin language (Poochigian 2010: xxiv).

Overall Cicero's version of the *Phaenomena* has received less attention from literary scholars than Germanicus', and is less admired as a work of epic poetry. The number of extant manuscripts of Cicero's *Aratea* is greatly exceeded by those of the more esteemed and more studied author Germanicus Caesar. But in spite of its poetic limitations and fragmented condition, this survey of surviving Aratean manuscripts will demonstrate that Cicero's translation remained quite popular in monastic and royal libraries throughout the Middle Ages. Manuscripts of Cicero's *Aratea* were widely copied and attractively illustrated by medieval scribes and artists.

---

[34]A full listing and discussion of the extant Ciceronian Aratean manuscripts can be found in Appendices A and B.

[35]Possanza also discusses a poem on the *Signs of the Zodiac* by Marcus Tullius Cicero's brother Quintus.

Germanicus Julius Caesar and the *Aratea*

Considering the three surviving Latin versions of Aratus' *Phaenomena,* the translation by Germanicus attracted the most attention and has inspired an enormous number of literary and historical articles. Germanicus Julius Caesar (16 or 15 BCE- CE 19) was the son of Nero Claudius Drusus and Antonia, the younger daughter of Marc Antony and Octavia. After the death of his father in CE 4, Germanicus was adopted that same year by his uncle Tiberius (43 BCE–CE 37), who himself had been adopted by his stepfather Emperor Augustus.[36] Therefore, Germanicus was a member of the semi-divine Julio-Claudian dynasty in direct line for succession as third Imperator of the Roman Empire. Admired as an excellent soldier and brilliant military leader, the young, charismatic prince was loved by both the Roman legions and the Roman people as well. In contrast to the disliked, reclusive Emperor Tiberius, who came to power in CE 14, Germanicus was well-respected, of virtuous character, emotional and affable; but because of his untimely death, Germanicus never reached the imperial throne. (Gleadow 1968: 65).[37] On 10 October, CE 19 while in the East on military campaign, Germanicus was stricken with a mysterious illness and died prematurely near Antioch. Although highly suspected by most and considered so by his wife Agrippina the Elder, whether he succumbed as a victim of poison will never be known for certain. Only 33 at the time of his sudden death, he had already fathered nine children by Agrippina.[38] Germanicus was honored with coins struck in his memory by both his son Emperor Caligula and his brother Emperor Claudius (Figs. 1.8 and 1.9).

Some controversy and confusion exists as to whether the extant Latin translation of the *Phaenomena* attributed to Germanicus has been proven positively to be his creation; as a few scholars have offered a suggestion that the author may have actually been Emperor Tiberius; a case could be argued for each (Gain 1997: 17).[39]

---

[36]Augustus's two grandsons, who were in line for succession to the monarchy, both died young, requiring him to adopt Tiberius to assure the Julian family would remain in power. Augustus adopted Tiberius in CE 4 and demanded that Tiberius adopt Germanicus at the same time.

[37]Gleadow quotes Suetonius, who wrote of Tiberius; "He was rather neglectful of the gods and religions, being addicted to astrology, and fully persuaded that all things are brought about by fate. Yet he was immoderately afraid of thunder, and if the sky became threatening, he never failed to put on a laurel wreath, because of the common saying that that kind of leaf is never touched by lightening." Suetonius Tib. Ixix, Cp. F. H. Cramer, "Astrology in Roman Law and Politics" (*Amer. Philos. Soc.*, 1954).

[38]Agrippina was the daughter of Marcus Agrippa and Julia, the daughter of Augustus. Nearly all of Rome, and especially Agrippina blamed Tiberius for the death of her beloved husband. She was eventually banished from Rome and is said to have starved to death in exile. Their son Gaius, nicknamed Caligula, was proclaimed emperor on 16 March 37; Germanicus' brother Claudius became emperor in CE 41, after the assassination of Caligula. Emperor Nero was Germanicus' grandson.

[39]Gain gives a thorough account of the arguments for both Tiberius and for Germanicus as possible authors using contemporary attributions which are vague and could apply to either. His final opinion is that the available evidence does not allow one to know conclusively which of the two the actual author was.

Fig. 1.8 Germanicus Bust, Ancient Rome, marble, Public Domain

Fig. 1.9 Germanicus Ancient Roman Coin, struck in 37 CE under Caligula, Public Domain

Contemporary sources claim that Tiberius became beguiled by the developing 'science' and predictability of Hellenistic astrology was the first Roman emperor to retain an official court astrologer (Tester 1987: 41). Tiberius employed Thrasyllus, a respected scholar from Rhodes (fl. 1st century CE) as his authorized astrologer. Tiberius began to consult with Thrasyllus for advice even before he became emperor, during his self-imposed exile on the island of Rhodes. After Tiberius initiated the position, most Roman emperors also retained an official astrologer.

Like his adopted son Germanicus, Tiberius also had a fervent interest in astronomy and mythology,[40] known to be fluent in Greek and devoted to literary studies (Zanker 1988: 227).[41] Most experts say that not enough evidence exists to make a positive identification of the author, but most accept the indications that Germanicus seems the more likely candidate. (Gain 1997: 18) All secondary literature on this version of the *Aratea* is credited to Germanicus. For this study, I will assume that the author is Germanicus, son of Drusus, adopted son of Tiberius, grandson of Marc Antony (Possanza 2004: 15).[42]

Well educated in the classics, Germanicus belonged to the Roman literary culture and studied both astronomy and astrology. He was eloquent and studious, writing comedies in Greek (all lost) as well as Latin epigrams. Evidence shows that, like Cicero, Germanicus translated the *Phaenomena* as a young man, between CE 4 and 7, "as it was a good way to cut one's poetic teeth" (Gee 2000: 69).[43] During this time he would have been between 18 and 21 years old, at the close of his years as a student, but before assuming the obligations of a political and military career mandatory for a prominent member of the ruling family.

Germanicus' Latin translation derives from Aratus, but differs dramatically from an exact copy. His version, composed almost 300 years after the original and eighty-some years after Cicero's translation, was centered on a different vision of the heavens and was directed to a more diverse audience, residing in a vastly different society. Germanicus transformed the Hellenistic text of Aratus into a Latin poem, the epitome of well-educated Augustan literary society. He exchanged the Greek myths, gods, and customs that Aratus had gathered in the *Phaenomena* to those preferred in Roman culture instead. Like Aratus, he delved into the famous Greek epics of Homer and Hesiod for inspiration, but Germanicus had the

---

[40]Gain attributes this to Suetonius 69.1.

[41]Zanker states that while in voluntary exile on Rhodes, Tiberius lived as a Greek and gathered around him a circle of poets and philosophers. Later Tiberius moved to his retirement home on the Isle of Capri with only a few Greek intellectuals for companionship.

[42]In this decision, I follow the respected opinion of Professor Possanza whose authoritative study, *Translating the Heavens: Aratus, Germanicus, and the Poetics of Latin Translation*, postulates this acceptance of Germanicus as the author rather than Tiberius. See his Appendix A, Authorship and Date, 219–43.

[43]Gee shows evidence that Ovid also translated the *Phaenomena* at an early age. This would have been an excellent exercise as students could learn to translate Greek into Latin, learn versification; mythology and astronomy at the same time, only two brief fragments of Ovid's version remain. Germanicus' poem was influential in Ovid's other writings, as Gee links his *Fasti* with Aratus' astronomy.

additional advantage of drawing from the Hellenistic poets, Callimachus, and Nicander, and also the outstanding Latin poets, Virgil and Ovid (Possanza 2004: 2–5).[44] Germanicus based his translation on the Greek poem, but reinterprets its original intentions and then supplements the verses with allusions to current works by contemporary Roman writers. He responds to Cicero's translation, but avoids the boring, formulaic, and repetitious style of his *Aratea*. Through his creative writing, Germanicus makes subtle but perceptible changes in order to create a newly focused poem, quite similar in its subject matter, the 46 heavenly constellations, but completely different in his objective from the original.

His Latin Proem imitates the 'solemn and elevated tone' of Aratus' hymn to Zeus; his verse construction, supplication and content retain that similarity. But Germanicus does not wish to promote Stoicism nor to honor the god Zeus who, as Aratus affirms, placed the stars in particular patterns as sign and guides to assist those living on Earth. His stated purpose for the stars' arrangement was instead to honor Emperor Augustus for establishing his reign of peace throughout the Roman Empire (Possanza 2004: 107–8)

The astronomical section of Germanicus' hexameter poem survives complete except for several small lacunae listed by the editors; but his translation of the *Weather Signs* is not extant (Possanza 2004: 8–10) When comparing Germanicus' structure of the *Aratea* with Aratus' poem, the similarities seem very obvious. Looking to the original line by line, his *Aratea* survives intact to line 725, which seems comparable to Aratus' line 732, but with "many omissions, expansions, contractions or additions" (Gain 1997: 13).[45] Basically the structure of the two poems is quite similar, but there are pronounced differences in their content. In addition to the greater amount of mythical material that Germanicus includes, he also expands the quantity of valuable information describing the five planets, which for the most part Aratus chose to ignore, calling their motions too complex. The inclusion of planetary material in the *Aratea* becomes vital for the pictorial program, as the more inclusive text invites illustrations of the heroic deity represented by each planet. Germanicus' lengthy descriptions of the celestial constellations are followed by the *Prognostications*, but incomplete fragments survive from this last part of the poem.[46]

The concluding section of both Aratus' and Germanicus' poems embraces the same theme, the influence of the phenomena of the heavens on atmospheric conditions. But again, the intention of Germanicus differs greatly from that of Aratus; the stars and planets of Germanicus are not just predictors of the weather; but instead are the active causal agent. This idea differs significantly from Aratus' *Weather Signs* which point to interaction with earthly animals in a semiotic manner

---

[44]Possanza presents specific examples of how Germanicus cleverly incorporates the vocabulary and phrases of the other poets, bringing attention to both their work and his erudition.

[45]Germanicus' proem is sixteen lines compared to the eighteen lines of Aratus' proem. Germanicus describes the constellations on lines 24–433 compared to Aratus, lines 26–450; his Risings and Settings appear on lines 573–725 compared to Aratus' 559–732.

[46]This final section is not based on Aratus' *Weather Signs,* but on astro-meteorology; since it is fragmentary, its source and totality cannot be known.

to help humanity. Over all a grand total of 948 lines survive in various Germanicus manuscripts—almost double Cicero's 469 lines, but less than Avienus' 1878 lines, by far the lengthiest *Aratea* translation.

For the most part Germanicus mirrors Aratus' poetic construction, format and content; but he eliminates some segments, elaborates others, and often corrects observational errors that had been pointed out by commentators, such as Hippocrates. Germanicus often prefers to paraphrase in his text, rarely conforming to Aratus' poem word for word (Gain 1997: 13).[47] His translation differs most significantly in his treatment of the constellations, the principal subject matter. Germanicus concentrates his attention on the heavenly figures formed by the stars, showing greater interest in the myths; whereas Aratus focused more attention on the relationship of the star patterns to one another. The few missing pieces of Germanicus' poem were attached sometime before the fourth-century CE, when the first of a series of *scholia* was added providing supplemental myths and descriptions.

While Aratus included 44 constellations in his *Phaenomena*, his discourse on the associated fables is very limited, for he provides an explanatory catasterism myth for only 14. Since Aratus based his poem on the writings of an astronomer not an essayist, his attention was more concerned with the relative placements of the constellations than with the myths and legends attached to them. In contrast, Germanicus recounted myriad sources of the archetypal stories about the humans, animals and objects that had been memorialized in the sky. Germanicus provides brief allusions to the myths of the zodiacal constellations at lines 531–564, including the lengthy tale of Virgo/Dike. "Germanicus doubles the number of star myths, and develops the erotic element in those myths, even when it creates serious embarrassment for Aratus's providential Zeus" (Gee 2000: 208). According to Possanza (Fig. 1.10),

> …the *Aratea* has more in common with the aetiological, erotic treatment of the stars found in elegiac poetry of Callimachus's Lock of Berenice, together with Catullus's translation, and in Ovid's *Fasti* than it does with Aratus' religious and philosophical meditation on the constellations as proof of the existence of the divine order of the universe (Possanza 2004: 169).

Germanicus displays his familiarity with the popular Latin authors for he was writing at the close of the Augustan era, when the earlier classical poets, especially Vergil, were already being enshrined. Animated by Germanicus' poetic ability, these ancient myths were reborn; his vibrant and colorful descriptions inspired the development of a pictorial cycle to embellish the stories with mythological and astronomical art. At what moment Germanicus' poem first acquired its numerous illustrations is impossible to determine, but the images found in the earliest manuscripts that survive from the ninth century retain classical elements and specific iconography that can be traced to Hellenistic and Roman art. The pictures of the constellation figures would not be out of place on a frescoed wall or mosaic floor from Pompeii. Therefore, it is safe to suggest that the Late Antique exemplars copied by artists in Carolingian scriptoria

[47]The text of Germanicus is the focus of a study by Gain in which he traces the source manuscripts and the relationship of their copies. He points out the errors and inconsistencies of the various editions. Gain also includes a full English translation of Germanicus' poem.

Fig. 1.10 Aries, f. 34, Leiden *Aratea*, c. 820, University of Leiden Library, MS Voss. Lat. Q. 79, Courtesy of University of Leiden Library. Aries is encircled by the zodiacal ring, as it is the first sign of the zodiac

almost certainly were illustrated as well. The illuminated manuscripts of Germanicus' *Aratea* survive in a greater number of copies, demonstrating that his version was far more popular than either Cicero or Avienus.[48]

Rufus Festus Avienus and the *Aratea*

The third surviving Latin translation of Aratus' *Phaenomena* was composed by Avienus, (fourth-century CE) a Roman aristocrat and successful poet, who lived in the second half of the fourth century. Few facts are known for certain concerning

[48]A listing of the surviving illustrated *Aratea* of Cicero and Germanicus is included in Appendix A.

his life, other than he was a native of Volsinii in Etruria and became a very well-educated author. He was appointed proconsul of Africa and was living there in CE 366, (if that is our Avienus). Unlike Cicero, Ovid, Germanicus and Emperor Gordian I, Avienus created his version of the *Phaenomena* later in life (Poochigian xxiii), currently dated between 310 and 386 CE. Though living in the Early Christian era, Avienus staunchly supported the traditional pagan religious customs; no hint of Christian doctrines appears within his poetic work. Avienus not only translated but greatly enhanced Aratus' astronomical poem; his version contains a substantial 1878 lines, 728 more than the original. The lengthy increase cannot be attributed to the introduction of new themes, but demonstrates the expanse of Avienus' dramatic poetic language. He sometimes paraphrases making his poem more verbose; but generally augmenting Aratus' adaptation of the myths and including more current scientific theories. Moreover certain evidence indicates that Avienus may have used Germanicus' *Aratea* poem as his model rather than Aratus' original Greek *Phaenomena*.[49] Avienus takes fewer liberties with his source-text than Germanicus did, perhaps because he had Cicero's version of the poem as a model while Avienus did not (Gee 2007: 257).

When examining *Aratea* survival patterns from the Middle Ages, Avienus' version of the astronomical poem was not as successful; many more manuscripts of Cicero and Germanicus are extant; several obvious reasons could contribute to this limited interest in his version. First and foremost Avienus' poem survives amid most illustrious company; his predecessors were renowned names in Roman society; to read or own a copy of either of their poems would be more impressive and desirable for an educated or royal patron. In addition Avienus' extensive elaborations of the constellation myths included many unwarranted details. For example, the catasterism myth that developed around the constellation of Virgo/Dike consists of a lengthy forty lines in Germanicus; but the retelling by Avienus expanded this account even further to eighty lines, twice as long (Possanza 2004: 128–145).[50] Aratus himself provides his longest narrative when relating this particular myth as he dedicates 41 lines to Virgo.[51]

As noted, Aratus and the Aratean authors have attracted a significant number of scholarly discourses devoted to uncovering the wealth of material imbedded within their poems, especially their divergent interpretations of the Virgo/Dike myth. In an

---

[49]One bit of evidence supporting this theory concerns the naming of the Pleiades; all sources list the names of the stars in different orders. But Avienus lists the names very near to the order of Germanicus, suggesting possible imitation.

[50]Possanza presents a thorough analysis of Germanicus' treatment of the catasterism myth of Virgo-Iustitia including the influence of Vergil. In the most popular version of this classical story, the goddess Dike inhabited the earth happily interacting with mortal beings in the Golden Age; conditions began to deteriorate in the Silver Age and even more so in the Bronze Age. Dike disliked the Bronze Race, and so she swiftly "flew away over the mountains, leaving the people thunderstruck" in order to take her place in the heavens, becoming catasterized as the beautiful constellation of Virgo.

[51]The constellation Virgo/Dike is discussed in more detail in Appendix C.

important publication from Bellandi et al. (2001) analyze the Latin translations of Germanicus and Avienus examining parallel passages and literary contest of their different versions.[52] Each of the translators interpreted the Phaenomena through their own goals and preferences; Avienus reads the poem through the 'lens of Neoplatonism' (Gee 2013: 21).

Avienus' version of the poem was utilized to expand scanty sections of the other authors, or to flesh out skimpy passages; for example, a few of Avienus' verses appear in the Leiden *Aratea,* to replace the missing lines of Germanicus. His poem is "the most mature expression of a living, evolving translation that spanned six centuries" (Poochigian 2010: xxvii) But Avienus' poem was perhaps too lengthy to be copied by those interested in possessing a Latin copy of the *Phaenomena.* Ultimately the translations of either Germanicus or Cicero were a more practical choice when considering the expense of parchment, the space required if including constellation pictures and the costly process of manuscript production in general. Therefore, the reception of Avienus' *Aratea* was not as celebrated, survival of his manuscripts was not as numerous, and commentaries on his work are much fewer, identifying him as the least successful of the three Latin Arateans. But his extant poetic writings represent a final radiant and complete flowering of Latin literature at the close of the Classical Era.

Reception and Influence of the *Phaenomena* in the Roman World

The *Phaenomena's* reception in the Roman literary world reflected its popularity among Greek intellectuals, as the educated Latin world admired and appropriated the art, architecture, and intellectual accomplishments of Greek society. Greek mythology was adopted and altered to suit Roman tastes; Hellenistic literature was imitated; classical artworks, especially their naturalistic sculptures, were replicated by Roman artists, and the 'philosophical way of life' was the idealized lifestyle (Zanker 1988: 337). Of particular importance for a deeper understanding of the reception and influence of the Latin versions of the *Phaenomena* and their prominent position in Roman poetic and philosophical circles are two influential books: a recent study by Emma Gee, *Aratus and the Astronomical Tradition* (2013) and an authoritative investigation of Germanicus' translation published by Mark Possanza, *Translating the Heavens*. In an earlier book, Gee presents specific

[52]In a Bryn Mawr Classical Review of the Bellandi, et al., study, K. Volk analyses their comparison of Germanicus and Avienus. Their work focuses strictly on this one small aspect of the poem. Her main criticism is that they write on their chosen author separately and do not actually compare the two author's approaches. It remains for the readers to interpret their singular discussions for themselves.

examples that detail how the *Phaenomena* of Aratus influenced the poetry of Ovid (43 BCE–CE 17) in both his *Fasti*[53] and *Metamorphoses* (Gee 2000: 4–7).[54] Gee provides a detailed treatment of the poem and its reception, a significant addition to the scholarly literature.

The astronomical poetry of Aratus attracted much consideration among Roman authors as witnessed in particular by his influence on the writings of Lucretius and Virgil (Mair 1969: 197) The poem's success in imperial Rome was further assured by the reputation and exceptional abilities of his Latin translators, Cicero, and Germanicus, both highly respected in Roman society. For a general understanding of the cultural and artistic developments taking place at the time the *Phaenomena* was translated by Germanicus, Paul Zanker's informative study, *The Power of Images in the Age of Augustus*, provides essential background. He also discusses the progressive changes within Roman society as a result of Augustus' effective policy of peace and prosperity. Zanker examines the "complex interrelationship of the establishment of monarchy, the transformation of society, and the creation of a whole new method of visual communication" (Zanker 1998: 3).

Cicero's *Aratea* stirred great interest in the works of Aratus due to his poem of 89 BCE; his philosophical writings in the years 60 and 50 BCE carry on this particular attraction. The popularity of Aratus in Rome can also be attested by the Cinna epigram and the *Ephemeris* of Varro of Atax, who based his poem on the *Phaenomena* (Gee 2013: 62). In addition Possanza evaluates the complexity and sophistication of the *Aratea* by Germanicus and discusses influences from contemporary and historic writers implanted in his poem. These and additional studies by classical scholars furnish vital background material for this time period; their analyses enable researchers to achieve a sharper perception of the Roman literary society in which Cicero, and Germanicus lived and wrote. The survival of this entire body of ancient descriptions of the constellations and heavenly motions from Greek and Roman poets presents contemporary scholars with an enormous amount of historical material to analyze, compare and mine for deeper meaning. Since the central focus of this study concerns the history of astronomy and the transmission of astronomical knowledge, an appreciation of the prominence of that science in

---

[53]The *Fasti* is examined more closely in the next section. Ovid's *Metamorphoses* is an epic poem in fifteen books, his only surviving poetry in hexameters, composed in the years immediately preceding his exile—a collection of imaginative tales from Greek and Near Eastern myths and legends, each alluding to a supernatural shape-shifting, change of form—from animals to humans, animate to inanimate.

[54]Gee states that Ovid was indebted to Aratus and the integrated universe he presented. "The unified cosmos of Aratus' Stoic poem is fragmented across Ovid's Roman calendar, and at the same time tied in with the shifting universe of the *Metamorphoses*. To my mind, Ovid's fragmentation of a poetic text which, as we shall see, carried political importance in Rome, and assimilation of it to a more anarchic world, runs counter to the Julian and Augustan impulse towards unity, particularly as manifest in the calendar reform of 46 BCE." She also affirms along with F. Bömer that the three works: the *Fasti*, *the Metamorphoses*, and the Germanicus' *Aratea,* were written around the same time, and all deal with the causes of time, of natural phenomena, and the stars.

Roman culture will also help bring a better understanding of attraction of celestial phenomena.

Commentaries on the astronomy and poetry of the *Phaenomena* were plentiful in the Roman world, just as they had been in the Hellenistic world; authors of twenty-seven ancient commentaries are known, and in fact commentaries continue to be written to the present day. The number of books and journal articles generated by the ancient poem of Aratus and the works of his translators increases rather than lessens with time. When you compare the thick editions of scholia and commentaries compiled by Kidd, Maass, Martin, Possanza, Gee and numerous other authors with the slim volume of the 1154-line poem, one can appreciate the enormous output of scholarship and the printed exegesis inspired by the original and its three major Latin translations. Avienus' *Aratea* appears to be the last Roman version of a 400-year tradition of attempts to outdo Cicero's translation.

The *Phaenomena* was admired and commented upon in the Roman World, not only for its ability to capture astronomical and mythological details in a straightforward format, but also for the poetic artistry of the original and the Arateans. This poem provides a unique combination of natural 'science' incorporated into a rhythmic textual format. When speaking of the poem's popularity, one needs to differentiate between its more mundane function as a utilitarian book of astronomy, memorized for its cues in identifying the constellations, necessary for travel over land and sea, and its qualities as a work of verbal artistry fully appreciated only by those with the education and sophistication to understand its illusions to the classical predecessors (Possanza 2004: 87). Familiarity with the rhythmic celestial cycles provided one with the assurance of accurate time and direction throughout the dark nights on land or sea. Because astronomical knowledge, its reception and transmission are the central focus of this investigation, it is helpful to determine the amount of public interest in astronomy and establish its place in Roman culture.

Astronomy, Astrology, and Catasterism in Imperial Rome

During the reign of Emperor Augustus (27 BCE—CE 14), interest in public divination and religious sacrifices, characteristic of the previous Republic Period, began to decline while interest in astronomical and astrological subject matter increased dramatically (Green 2004: 146).[55] Understandably Rome's victory over Cleopatra's Egypt spurred an interest in Egyptian customs and in the accumulated knowledge of that mysterious, inscrutable culture, particularly the 'science' of astrology and development of individual horoscopes instigated by Hellenistic astrologers, mainly in Alexandria (Tester 1987: 19). Astrology in a more general

[55]According to Green, "The Julian family, in particular, manipulated astronomical and astrological readings as a means of legitimizing their own power and curbing that of others; such activity was particularly fervent in the latter years of Augustus and the early years of Tiberius."

sense, that heavenly motions portend earthly events or destinies for royals or governments, such as wars and plagues, dates back thousands of years but forecasting heavenly influences based on natal horoscopes can be traced to Hellenistic astrologers. Ancient Egypt produced no astrological hypothesis of its own; the carved ceiling zodiac of Dendera and other astrological artworks were not created in ancient Egypt but date to the first- and second-century CE. Certainly astrological concepts were greatly influenced by traditional Egyptian and Babylonian beliefs, but all known writings related to horoscopic astrology defining a specific moment in time were composed by Hellenistic Greeks during the Ptolemaic era beginning in the third-century BCE.

In addition to embracing Grecian art, literature and way of life, Roman leaders also appropriated the long-held Egyptian belief that after death, the divine pharaoh set sail in his solar boat to join the cosmic gods in his designated place in the heavens, becoming a star in the afterlife, his (or her) catasterism or astral apotheosis. Roman astronomer-priests or augurs combed the night skies tracking the course of the planets and watching for unusual celestial phenomena, predictors of events to come. This reading of portents was not merely superstition or folklore; it was part of the state religion of Rome; the planets were gods. An unexpected but fortuitous occurrence in the sky provided a source of propaganda and reinforced this newly adopted Roman belief.

During Julius Caesar's funeral games, the Games of Venus Genetrix, commissioned by Augustus in 44 BCE, a bright comet miraculously appeared in the sky over Rome. Augustus later recorded in his autobiography that this auspicious comet had been visible all over the world for 7 days and its appearance had been construed as an indisputable sign of Julius Caesar's apotheosis (Zanker 1998: 35).[56] According to an account by Pliny the Elder in his *Naturalis Historia* (Book 2, Chap. 23),[57] the comet appeared in the northern reaches of the sky, an area that had already been associated with Caesar's ancestors, the Julian dynasty (Gee 2000: 157–192).[58] The auspicious comet provided visual proof that Caesar's immortal soul was traversing the vast sky, making its way to his eternal resting place and

[56]By promoting the deification of Caesar, Augustus as his adopted son (actually the son of Julius Caesar's sister Julia) could promote his own power and glorification by referring to himself as *divi filius*. Tracing one's ancestry back to the Greek gods or heroes was a sacred tradition among Roman aristocracy. Since Julius Caesar considered himself descended from Venus and Aeneas, Augustus and Germanicus could also claim the same divine heritage and protection. Vergil had written the *Aeneid* at Augustus' request, which closely connected the mythology of Venus and the Fall of Troy with the founding of Rome, creating a preordained triumph that glorified the role of the Julian dynasty.

[57]Pliny classified ten different types of comets, but he was more an encyclopedist than a scientist, collecting truths, half-truths, myths, and outright nonsense (Yeomans 1991: 11).

[58]Gee presents an in-depth analysis of this concept.

Fig. 1.11 Roman Coin Augustus placing a star on the statue of Julius Caesar, Ancient Rome, c. first century CE, Public Domain

adding further verification of his divine status and celestial apotheosis. The comet's sudden appearance, while the assassinated ruler's funeral games were in progress, heralded a powerful proclamation to the entire Roman society that Caesar had assumed his rightful position among those exalted figures of antiquity that had previously ascended to the heavens, including the powerful pharaohs of the distant past and the gods, goddesses and heroes of mythology.[59]

After the games Augustus placed a star on the head of the statue of Caesar in the Roman Forum; and soon the *sidus Iulium* appeared on many other imperial statues. Coins were repeatedly struck engraved with the name of Julius and the image of a star. In the past Julius Caesar himself had shown interest in the heavens by issuing coins imprinted with the star of Venus. Caesar's family felt it important to claim their relationship with Aeneas, the legendary founder of Rome. Augustus felt it important to claim his close relationship with Julius Caesar, and he exploited that familial and celestial propaganda to intensify his favorable political advantage (Figs. 1.11 and 1.12).

> Hic, Auguste, tuum genitali corpore numen
> attonitas inter gentis patriamque pautentem
> in caelum tulit et maternis reddidit astris.[60]

---

[59]Some contemporary scholars currently question the authenticity of these early reports of Caesar's comet.

[60]"Augustus, you were carried into the sky on the body of this sign, under which you were born, and returned to your mother stars." (Gain's translation—68) In general, the astrological interpretation of a comet's significance varied depending on the region of the sky in which it was observed, more often its arrival was thought to portend negative consequences, frequently foretelling the defeat or death of a ruler as evidenced by the Bayeux Tapestry's depiction of the Comet of 1066 (Halley's Comet).

Fig. 1.12 Roman Coin Divine Caesar with Star, Ancient Rome, c. first century CE, Public Domain

Another example of this increasing fascination with the heavens can be observed in the gigantic Horologium complex commissioned by Augustus in the first decade BCE with its exceptional components, impressive astronomical orientations, and monumental gilded bronze lettering. Constructed in the Campus Martius, the field of Mars, the Horologium functioned as an enormous public sundial, the largest ever built, designed to operate as both a clock and a calendar (Zanker 1988: 144)[61] (Fig. 1.13).

Next to the Horologium, Augustus commissioned an enormous circular ancestral mausoleum, where the remains of family members were placed and where his own ashes would eventually rest. His astronomical complex included a 90-foot-tall red granite obelisk, seized from Heliopolis by Augustus, the first obelisk transported from Egypt. The shadow cast by the obelisk acted as a gnomon on a carefully constructed grid indicating the time of day and month of the year. Just to the east of the Horologium stood the renowned *Ara Pacis,* the *Altar of Augustan Peace* commissioned by the Senate to honor Emperor Augustus the year before his death. The elaborate sculptural program covering the inner and outer walls of the open air temple celebrates the peace and prosperity brought about by the civic programs of Augustus, the successful reign memorialized by Germanicus in his famous poem. Historians propose that young Germanicus appears as one of the imperial children in the religious procession portrayed on the outside wall of the *Ara Pacis* (Fig. 1.14).

Following the ancient tradition of aligning sacred architecture with the motions of the heavens, contemporaries reported that the Egyptian obelisk in the

[61]Zanker reports that the mausoleum was 260' round x 120' tall. A network of markers within the Horologium grid was inscribed in both Greek and Latin noting the time of day as well as the time of year.

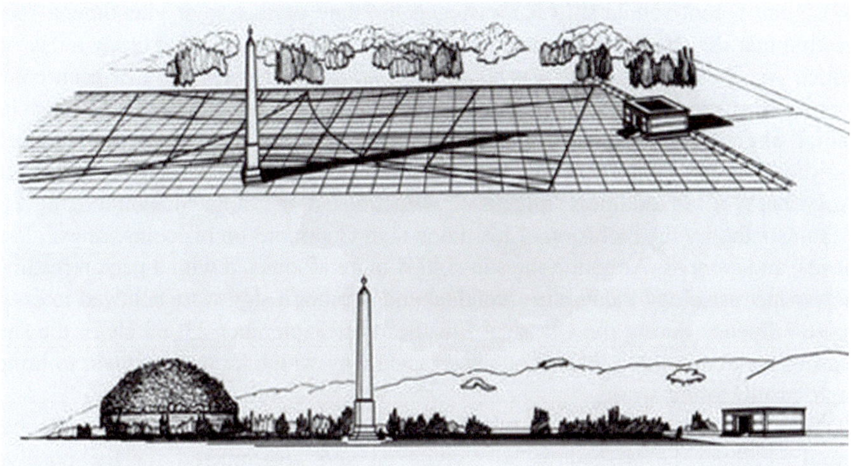

Fig. 1.13 Horologium, Mausoleum and Ara Pacis of Augustus in Campus Martius, first century CE

Fig. 1.14 Procession sculpture on Ara Pacis, Germanicus is thought to be one of the children, Public Domain

Horologium had been precisely oriented to cast a long shadow that reached into the temple's doorway and across the center of the marble altar of the *Ara Pacis* on the anniversary of Augustus' birth, September 23, the autumn equinox (Gleadow 1968:

64),[62] but recent scholarship has disputed the date of that solar phenomena suggesting that the shadow was cast into the altar on other dates, particularly a day on which Apollo was honored. Several decades after the Horologium had been completed, Pliny the Elder wrote that the massive clock had lost its accuracy because its initial alignment no longer synchronized with the heavens, but the admired Horologium complex exemplified the cosmic resonance and peaceful reign of the Augustan era. In addition Augustus' attraction to astrological symbolism was demonstrated by the inclusion of his moon sign Capricorn on his coins, rings, glass beads, and cameos. Although the sun sign is more associated with a person reading newspaper astrology today, the ascendant and the moon sign were believed to exert more influence during the Classical Era. Born on September 23, 62 BCE, the sun sign of Augustus was Libra, the balance scales, by which he was destined to bring equilibrium to the world.

Why did Augustus prefer to appropriate his moon sign Capricorn as his personal motif? Perhaps he wished to be recognized and remembered for the positive characteristics inherited by those under the sign of Capricorn, especially strong moral authority. Another significant aspect of Capricorn refers to its synchronicity with the winter solstice; the birth of the new celestial year could symbolize the birth of Augustus' new era of Roman peace and prosperity. The *Gemma Augustae* (Gem of Augustus), a large onyx cameo, depicts the emperor at his apotheosis assuming the pose and identity of Jupiter, being crowned with a victorious wreath, his sacred eagle stands at his feet (thought carved during the reign of Caligula—CE 37–41). The personification of Rome sits next to him; their feet trample enemy shields. Above the emperor, the sea-goat Capricorn appears in a roundel. Some historians have proposed that the young man standing next to Roma may be Germanicus (Figs. 1.15 and 1.16).

Another surviving image of Augustus also pays homage to his divine heritage. Shortly after the death of Augustus, a life-size marble statue of the emperor was commissioned, possibly by Emperor Tiberius for his mother; the statue is now called *Augustus of Primaporta,* because it was discovered in his wife Livia's family villa at Primaporta, near Rome. A youthful Augustus stands proudly clad in military armor, although the statue was commissioned after his death. To acknowledge and honor the divine ancestry of Augustus, an image of Cupid, son of the goddess Venus born of the sea, sits astride a dolphin at the statue's base. This powerful royal image signifies the claim of the Julian family of descent from Venus through her human son Aeneas, and promotes visual enhancement of Roman imperial propaganda (Fig. 1.17).

---

[62] According to Gleadow, the date of Augustus' birth is debated, but was most probably September 22, 64 BCE. The Sun had just moved into Libra, the Moon was in Capricorn strengthened by an opposition to Jupiter and making a favorable aspect to Mercury, both in their exaltation. At that time the sign of the Moon was considered more important in the horoscope than the sun sign. Gleadow says that Augustus' dawn birth added the impact of a Capricorn ascendant. This is impossible for if Augustus was born at dawn, his ascendant would also be Libra. Jupiter and Saturn were both positioned at the mid-heaven, another favorable aspect in astrology.

Fig. 1.15 Gemma Augusta Onyx, early first century CE, Kunsthistorisches Museum, Vienna, Public Domain

Fig. 1.16 Roman Coin Augustus and Capricorn with Large Cornucopia, first century CE, Public Domain

Modeling Caesar's example, Augustus allied himself with Venus and the heavens in a similar way. Invigorated by his auspicious natal horoscope, Augustus believed his role as "savior of the state" was "in his stars" (Zanker 1998: 48).[63] In his astronomical poem, Germanicus does not refer to his honored ancestor Julius Caesar directly, but lines 558–60 refer to the death and deification of Augustus. Germanicus incorporated his allusions to the divine status of Augustus first by

[63]Starting in 30 BCE, Augustus' birthday was celebrated officially as a day of good omens.

Fig. 1.17 Augustus of Prima Porta, first century CE, Vatican Museums, Public Domain

replacing Aratus' hymn to Zeus with his ode to Augustus, and again when he refers to the occasion of Capricorn, Augustus' adopted emblem, transporting Augustus to his destined place in the celestial sphere (Gee 2000: 157).[64]

Germanicus and Ovid's *Fasti*

Further evidence for evaluating the prestige and influence of Germanicus and his translation of the *Phaenomena* appears in the poetic calendar, the *Fasti*, composed by the illustrious Roman poet, Ovid. (43BCE–CE17) Ovid created the *Fasti* in

[64]Gee cites Tamsyn Barton for providing this connection.

keeping with the contemporary concern with astronomy, time and the calendar. This interest was stimulated in part by Julius Caesar's undertaking to reform the calendar, a task completed by Augustus, and the installation of official calendars in public places.[65] Ovid's *Fasti* ranks close in popularity to his *Metamorphoses*, a hexameter epic poem in fifteen books, the most celebrated and influential of Ovid's works. The *Fasti* was composed about the same time as Germanicus' *Aratea*, probably begun after he completed the *Metamorphoses,* and mostly finished by CE 8; possibly he worked on his two projects simultaneously.

With his six-book *Fasti*, Ovid produced a calendar in elegiac couplets devoted to the Roman year and composed in chronological order with one book dedicated to each of the first six months. Particular days of the month are noted by recording pertinent historical, religious, and astronomical traditions. The *Fasti* considers the official Roman holidays, games, and sacrifices; the rising and setting of the stars; mythical stories explaining the origin of the constellations; his verses celebrate the Julian monarchy as well. Ovid describes the mythology, festivals, and commemorative events of Rome; and then synchronizes the stories and occasions with the phenomena of the cosmos, making *Fasti* one of the most valued records of Roman religious practices and cyclical celebrations.

Ovid was about 50 years old, at the time he began writing the *Fasti* and at the height of his brilliant and successful career. In CE 8, the first six books were in a 'reasonably finished form' when Augustus banished him to Tomis on the Black Sea, a culturally barren city at the far edge of the Roman Empire (Green 2004: 17).[66] At Tomis Ovid made revisions to the *Fasti* that related to Germanicus. Originally Ovid had dedicated his poem to Augustus; but after the emperor's death in AD 14, Ovid revised his poem to honor instead the popular leader, Germanicus, neglecting the unpopular Tiberius who reigned at that time (Fantham 1998: 3).[67] In the following lines from *Fasti's Proem*, Ovid both invokes and imitates Germanicus, just as Germanicus had invoked his 'father' to accept his translation of Aratus' *Phaenomena*.

> The seasons distributed throughout the Latin year, their causes,
> and the constellations which set beneath the earth and rise,
> I shall celebrate. Caesar Germanicus, welcome this work
> with a tranquil countenance and guide the voyage of my timorous craft.

---

[65]In 46 BCE Julius Caesar introduced his new calendar, developed by the astronomer, Sosigenes, whose framework continued in use until the Gregorian calendar reforms in 1582. One of the Roman calendar's main accomplishments was to separate the solar year from any connection to the lunar year.

[66]Green has published a recent commentary on Book I of the *Fasti*, which includes many references to Germanicus. Some historians suggest that the final six books of the *Fasti* were in some fragmentary form, but were either lost or destroyed; others insist they were never written, for not a trace of the later six books survives.

[67]Most probably, Ovid replaced Augustus to gain favor from Germanicus whom everyone expected to be the next emperor, succeeding Tiberius, and who might possibly allow Ovid to return to Rome. Ovid extended the revision to other sections, especially in Book 1, in order to integrate his new dedicatee significantly into the poem, referring to him as a poet in various places, and calling him the 'learned prince'.

> Do not turn from a trifling honor but come, lend a god-like presence
> to this service here offered you. You will recognize the sacred rites
> extracted from ancient chronicles, and by what distinction each day is known.
> There too, you will discover the festivals of your own family;
> often must your father and your grandfather be read about;
>
> the honors which they bear and which decorate the red-lettered calendar,
> you too shall carry off with your brother Drusus.................................
> .... And we know too, when inspiration has moved you to poetry,
> how powerfully the stream of your genius flows. If it is permissible by law
> human and divine, guide as a poet the reins of a poet, so that under your
> auspices the whole year may proceed with favor (Herbert-Brown 1994:174).

The first part of Ovid's *Proem* was addressed to Germanicus as a distinguished member of Julius Caesar's dynasty and the second part to the 'genius' of Germanicus as an orator and poet (Herbert-Brown 1994: 177).[68] The astral material of the *Fasti* may have been partly motivated by the literary interests of Germanicus, but the astronomy in the *Fasti* is definitely not based on either Aratus or Germanicus, in fact it is 'wildly inaccurate' (Fantham 1998: 36). No actual mention of the *Phaenomena* can be found in the *Fasti*; however, Ovid certainly knew the work of Aratus and even translated the Greek poem himself, for fragments of it survive, even though he does not refer to it directly (Green 2004: 145–48).[69] But in Ovid's rendition of the catasterism myths, the presence of Aratus can be felt (Possanza 2004: 3).

Many Roman writers thought the calendar a theme worthy of poetry, illustration, and scholarly investigation of its origins. Not only Ovid, but other Roman authors, (such as Varro, Filocalus, and later Macrobius), historians, grammarians, and specialists studied and wrote about the Roman calendar and the yearly religious festivals. A surviving example of an illustrated calendar, the famous *Calendar of 354* by Filocalus, will be discussed in Chap. 2.

The *Fasti* remained quite popular beyond the Roman period and, like the *Phaenomena*, provided a means for the transmittal of astronomical concepts.[70] Like

[68]Both Herbert-Brown and Gee question whether Germanicus' talent as a poet was significant enough to warrant such a compliment from the 'urbane, sophisticated' Ovid, since the *Phaenomena* is the only extant evidence of his poetic abilities.

[69]Green discusses the possible influence of Aratus and states that Ovid was probably aware of Germanicus' astronomical interests though not of his translation of the *Phaenomena*, and that "any allusion between the two is probably (happily) coincidental.".

[70]The *Fasti* has been studied by historians and literary experts as a source of specific information on Roman ceremonies and ritual practices, and it has attracted new attention both as a complex work of elegiac poetry and a record of religious beliefs during the Augustan Age. Green states that the *Fasti* has undergone a revolution in the past 25 years and scholars "have uncovered a work which is both rich in literary heritage and a powerful witness and critic of late-Augustan and early-Tiberian imperial Rome" (ix).

the *Aratea*, Ovid's poem was appreciated and enjoyed throughout the Middle Ages, as more than 170 manuscripts of the *Fasti* survive, but none of these extant elegiac works are illustrated. (Fantham 1998: 50).[71] Was the *Fasti* ever illustrated? Pächt remarks that it is inconceivable that Ovid's works were not improved with illustrations, but surprisingly not a trace of a picture has ever been discovered (Pächt 1966: 33).[72]

Another theme that will be explored in this study is how the lofty subject matter of this celebrated poem of Aratus seemed to resonate with the royal courts of Europe through history. The *Phaenomena* was originally commissioned by a Hellenistic monarch, and then was cleverly appropriated by the Romans during the early Empire period. Astronomical symbolism was an obvious association that appealed to the exalted ideals of Augustus and the Julio-Claudian dynasty. Emperor Charlemagne and the Carolingian dynasty sanctioned a conscious desire to re-establish the glory of the Roman Empire and to associate Carolingian emperors with the stars in the heavens, as the Roman emperors and many other monarchs had done.[73] The popularity of Aratus' poem in its Latin translations was renewed in medieval libraries. Astronomy-loving Carolingians copied all three Latin versions of the poem many times and also adopted classical astronomical imagery in their decorative programs. This theme of royal designation of astronomic motifs and iconography continues among the courts of European nobility and then bursts forth again in the courts of Italian Renaissance aristocrats.

Text History of the *Phaenomena*

Relevant to its appreciation among Aratus's contemporaries, Kidd accepts that there were several copies of the *Phaenomena* in Alexandria with at least one copy held in the great Library. Hipparchus' corrections and criticisms of Aratus' astronomy were available in the second-century BCE. Text editions of the *Phaenomena* date from the late first-century BCE, at which time the grammarian Theon of Alexandria is credited with writing an edition that became the standard in late antiquity and the foundation for medieval manuscript traditions (Kidd 1997: 49).

---

[71]Among the most important medieval manuscripts containing the *Fasti* are: Vat. Reg. 1709-10th century; Vat. Lat. 3262-11th century; Brussels, BR.5369-11th century; and Oxford Bodl. Auct. F.4.25-15th century. I examined the two Vatican copies on microfilm, and neither copy was illustrated.

[72]Pächt writes that in antiquity, only certain works were illustrated and in addition, that the Middle Ages had a passive attitude toward the illustrations of classical texts right up to the Gothic period. Illustrations were readily copied when available, but if there were no pictures in the earlier model, usually none were added.

[73]See P.E. Dutton, "Of Carolingian Kings and Their Stars" in *Charlemagne's Mustache and other Cultural Clusters of a Dark Age*.

As enumerated by Kidd, 50 copies of Aratus' Greek poem still survive and are held in various libraries (Kidd 1997: 57–68). Eight versions on papyri dating from the second-century BCE to the fourth-century CE and fourteen early Greek manuscripts are those which he utilized for the basis of his 1997 edition. Text scholars, Ernst Maass and Jean Martin, each published large volumes based on extant Greek scrolls and manuscripts that contributed to the accurate reading and interpretation of the poem and its *scholia*. Variations of Aratus' text attributed to Eratosthenes began to appear during the Alexandrian Age which became very popular in the Greco-Roman world. For the best explanation of this complex textual history see Martin, 1956. A possible stemma for the varying Germanicus manuscripts was proposed by nineteenth-century specialists who divided the early texts into two different families. Every line, every word of Aratus' *Phaenomena* has been analyzed by textual specialists for deeper meaning, references and subtle nuances.

The *Phaenomena* was even translated into Arabic; a version was commissioned in the ninth century by Abbāsid Caliph Al-Ma'mun (786–833) for his famous House of Wisdom in Bagdad which will be discussed in Chap. 4. Another translation of the poem was cited by Al-Biruni (973–1048) who is considered one of the most knowledgeable scholars of the medieval Islamic era; he was especially well-learned in astronomy, physics, mathematics, and natural sciences also discussed in Chap. 4. The popularity of the *Phaenomena* of Aratus and the Latin *Aratea* manuscripts spread throughout many areas of the medieval world.

As mentioned previously, this monograph does not discuss the surviving Greek copies of Aratus' original poem since they were not illustrated, but focuses on the surviving Latin *Aratea* manuscripts abundantly illustrated during the Middle Ages. The study includes but is not limited to the merits of the text, the poetic content or literary qualities of the *Aratea* but instead considers the history and characteristics of the component parts of the extant manuscripts and their survival patterns. This monograph will delineate factual data on the *Aratea* manuscripts, looking at physical condition, content, production, readership, and especially the format and iconography of the constellation illustrations.

The earliest surviving illuminated manuscripts of the *Aratea* date to the ninth century; these will be described and compared in a later chapter. A variety of scholia became incorporated into the Germanicus' *Aratea*. By the end of the second century, the Greek collection of constellation myths and positions was translated into Latin and added to the text of Germanicus which is now referred to as *Scholia Basileensia*. This scholia adds more details to the mythology and reduces the amount of poetic verses in medieval Germanicus manuscripts.

Many fifteenth-century-illustrated manuscripts of Germanicus' *Aratea* are extant as well. By the start of the sixteenth century, handwritten and hand-painted manuscripts were no longer being produced except for rare instances. But the *Aratea* still lived on and circulated even more extensively among educated readers in a less expensive, more available form. The *Aratea* of Germanicus, one of the first

Fig. 1.18 Cover of *Syntagma Arateorum* Printed Edition by Hugo Grotius 1600, Public Domain

books printed in Italy, was published in Bologna in 1474. In 1488, it was joined with the *Aratea* of Cicero and printed in Venice by Antonio de Strata, and then again in Venice in 1499 as part of a collection of astronomical works published by the Aldine Press. In 1600, Hugo Grotius, the Dutch humanist, published a Latin edition in his *Syntagma Arateorum,* in which he adapted the painted constellation illustrations from the ninth-century Leiden *Aratea* of Germanicus, replacing them with woodcut engravings by Jacob de Gheyn (Figs. 1.18 and 1.19).

Evidence that the ancient astronomical poem is still read and valued today, can be demonstrated by a selection of the large number of current translations, interpretations and studies of Aratus's *Phaenomena* that were listed earlier. In addition to the three widely read Latin translations, another translation of Aratus' work circulated throughout the Middle Ages called the *Aratus Latinus*.

Fig. 1.19 *Syntagma Arateorum* Zodiacal Constellation Aries encircled by large equinoctial ring, Public Domain

The Aratus Latinus

The *Aratus Latinus* survives as a later version of the Greek *Phaenomena*, probably translated in the first half of the eighth century by an anonymous author in France, most likely at Corbie, France. A later hand, dated to the second half of the eighth century, made revisions to the original copy. This manuscript includes a commentary on the poem and associated texts. Maass (Comm.175—306) has produced the only printed edition of this translation based on two extant manuscripts, one held in Brussels (BR 10698) and another in Paris (BNF Lat.7886), which is not illustrated. When compared with Aratus' original, this Latin version is incomplete with 85 lines missing; Maass describes this version as 'barbarous' and often 'nonsensical' (Kidd 1997: 52–53). Poochigian calls it "clumsy and often mere gibberish" (Poochigian 2010: xxvii).

The translator of the *Aratus Latinus* lacked the skill and sophistication of the other Arateans; his version appears stilted and choppy for the scribe has simply exchanged Latin vocabulary for Greek. The author of the *Aratus Latinus* was not adequately skilled in Greek and misunderstood parts of the original poem; obviously the deficient expertise of this medieval translator would result in a poor translation compromised by a wrong division of words and mistaken interpretations. Greek words unknown to the translator were sometimes just eliminated, which may explain the many missing lines in this eighth-century version. But in spite of its many deficiencies, the *Aratus Latinus* has been useful in tracing the transmission of the Greek poem in the Latin West (Kidd 1997: 54). On occasion, the term *Aratus Latinus* has been misused by academics as referring to the whole Latin tradition of *Aratea* manuscripts. The *Aratus Latinus* refers to only one particular badly translated version that was rarely seen or read; only four manuscripts survive. The original *Aratus Latinus*, in addition to the *Aratea*, contained the biography of Aratus (*Arati genus*), several different introductions, and a commentary (Lott 1983: 148). The Scholia Sangermanensia (so named because the original manuscript that was simplified was held at St Germain), which was added to the text of some *Aratea* manuscripts is actually a modification of the seventh-century *Aratus Latinus* translation.

Misinterpretations of the *Phaenomena* and the Arateans

The lengthy and convoluted history of the *Phaenomena* has allowed many misperceptions concerning its significance and function to accumulate in art history. Because of the complicated circumstances in the transmission of this particular astronomical knowledge even meticulous scholars have serious misunderstandings concerning the textual material of Aratus' *Phaenomena* and the *Aratea*. Without a thorough examination of the text editions of Aratus' poem in conjunction with the text editions of the three different *Aratea* versions and the *Aratus Latinus*, it becomes difficult to be assured each interpretation is correct. Scholarly literature referring to the *Aratea* is currently replete with misinterpretations of these astronomical poems. Often historians or references use Aratus' Greek poem and the Arateans' Latin poems interchangeably but they are in reality quite different, created by different poets, at distinct time periods, living in dissimilar cultures, with individual histories, criticisms and interpretations. But through time the various texts have been intermingled with each other and with other writers, Hyginus in particular. Each of the poems has been mistakenly referred to as 'a treatise', 'an astronomical textbook', 'a book of astrology', and 'a computistical miscellany'.

A common mistake in describing the *Phaenomena* and the *Aratea* has been their original and medieval functions. The *Aratea* were not treatises or essays filled with technical, scientific or mathematical information; they are astronomical poems relevant to their specific time in history. The *Aratea* were not designed as astronomical textbooks; they were originally sophisticated poems describing the relative positions of the constellations and their myths along with celestial motions, *Weather Signs*, philosophy and contemporary issues. Since the *Phaenomena* was the most famous astronomical text, and there were limited descriptions of the stars available, it may have been used as a text for educational instructions, but there is no evidence to support that claim. On the other hand, during the medieval period the *Aratea* had many competitive sources for learning the basics of astronomy as understood at that time. By the thirteenth century, the writings of Sacrobosco had become the standard astronomical text; before that, other choices existed in monastic and cathedral schools, such as Pliny the Elder, Macrobius, Bede, and Isidore of Seville. Monastic and convent libraries often included copies of huge medieval 'encyclopedias' that contained the accumulated knowledge collected to date, such as the *Hortus deliciarum* by Abbess Herrad of Landsberg and the *Liber Floridus* by the cannon Lambert of St. Omer, both produced in the twelfth century. These texts and illustrations will be discussed in Chap. 3.

To further clarify the writings of Aratus and the Arateans, their poems do not include or make reference to astrology, the popular pseudoscience. Ancient Greek scholars such as Aristarchus of Samos, Anaxagoras, Eratosthenes, and Thales of Miletus were highly advanced in the study of astronomy, but astrology with its precisely designed horoscopic charts, twelve individual houses, ascendant and mid-heaven was not formulated until the second-century BCE to the second-century CE by well-known Hellenistic astrologers, Paulus Alexandrinus, Dorotheus of Sidon, Vittius Valens, and especially Claudius Ptolemy. The *Tetrabiblos* of Ptolemy, the most prominent astronomer and astrologer of his era, was composed in Alexandria in the second-century CE and firmly established the foundation of the individual Western astrological traditions. The poems of Aratus and the Arateans contain only astronomical and weather information; they have no correlation with astrology, the belief that movements of the stars, planets, sun, and moon determine and predict events in human lives.

Another often-repeated error states that these astronomical poems, written centuries before the spread of the Christian religion, were necessary for religious practices and consulted by medieval monks and priests in order to determine the date of Easter, the calendric process known as computus. The process of fixing the date of Easter, the first Sunday after the first full moon after the spring equinox, was not determined by direct astronomical observations of the vernal equinox and the full moon but by accessing the Julian calendar currently in use. Nor did astrologers look to the night sky when calculating an individual's horoscope. Both monks and astrologers consulted a star chart, an ephemeris, listing the positions of the sun, moon, planets and zodiacal constellations, and then adjusted those positions for the subject's location on earth.

When monks or priests planned the yearly Church calendar, Easter Sunday had to be established far in advance in order to establish the dates of Ash Wednesday and the season of Lent. Christian clerics were required to understand computistical studies; each religious community was required to determine their date independently; as communication between distant dioceses was often challenging or nonexistent. The Christian Church considered it most important that all dioceses of Europe celebrated the religious holiday on the correct date. Considering that Aratus wrote his poem 300 years before the birth of Christ, clearly the *Phaenomena* was not studied to establish Easter's proper date; the poem does not hold that type of data. Cicero and Germanicus also wrote their astronomical poems in advance of Easter's occurrence; therefore their translations had no connection with the rituals of Christianity. Although Avienus did live during the Early Christian Era, he was not Christian and made no mention of Christianity or its tenets in his version of the poem.

Throughout the Medieval Era and into the Italian Renaissance, the transmission of the *Aratea*, rather than becoming more simplified, actually became even more complicated. The poetic verses of astronomical and mythological descriptions were intermingled with the writings of other authors, such as medieval encyclopedists and specialists. Through the centuries, the constellations which lacked mythological descriptions in the *Aratea* were expanded by adding information from other writers such as Hyginus, ps.-Bede, ps.-Eratosthenes, and the *scholia Sangermanensia* and *scholia Strozziani*. The illustrations and celestial diagrams in the *Aratea* developed different characteristics and iconography reflecting the cultural changes and the civil developments in the various regions of Europe. This investigation of illustrated *Aratea* manuscripts will explain and clarify some of the misinformation that currently abounds in the historical literature and bring more transparency to a vast and murky topic.

References

Alexander, J. J. G., ed. (1994) *The Painted Page: Italian Renaissance Book Illumination, 1450–1550*, Munich: N Y: Prestel.

Aratus, Kidd, D. (1997) ed. *Aratus: Phaenomena:* Cambridge: Cambridge University Press.

Bellandi, F., E. Berti, and M. Ciappi (2001) *Iustissima Virgo: Il mito della Vergine in Germanico e in Avieno*, Pisa: Giardini.

Blume, D., M. Haffner, and W. Metzger. (2012) *Sternbilder des Mittelalters: Der gemalte Himmel zwischen Wissenschaft und Phantasie*. Part One: 800–1200. Berlin: Akademie Verlag.

Buescu, V. (1941) *Cicéron. Les Aratea*. Bucharest, repr. Hildesheim: Georg Olms 1966.

Dekker, E. (2013) *Illustrating the Phaenomena: Celestial Cartography in Antiquity and the Middle Ages,* Oxford: Oxford University Press.

Dekker, E. (2010) 'The Provenance of the Stars in the Leiden *Aratea* Picture Book", *Journal of the Warburg and Courtauld Institutes* 73, pp. 1–37.

De Santillana, G. and H. Von Dechend (1977) *Hamlet's Mill: an Essay Investigating the Origins of Human Knowledge and its Transmission through Myth,* Boston: D. R. Godin, Publisher, Inc.

Dicks, D. (1985) *Early Greek Astronomy to Aristotle,* Ithaca, NY: Cornell University Press.

Dodwell, C.R. (1971) *Painting in Europe, 800–1200*, Harmondsworth: Pelican History.

Dutton, P. E. (2004) *Charlemagne's Moustache and Other Cultural Clusters of a Dark Age,* New York: Palgrave Macmillan.

Erren, M. (1967) *Die Phainomena des Aratos von Soloi, undetersuchungen zum sach-und sinnverstandnis*, Wiesbaden; Franz Steiner Verlag GMBH.

Fantham, E. (1998) ed. *Ovid, Fasti Book IV,* United Kingdom: Cambridge University Press.

Faber, R. (1993) *The Apostle & the Poet: Paul & Aratus*, Spindle Works, http://spindleworks.com.

Gain, D.B. (1976) *The Aratus ascribed to Germanicus Caesar*, London.

Gee, E. (2000) *Ovid, Aratus and Augustus: Astronomy in ovid's Fasti*, Cambridge: Cambridge University Press.

Gee, E. (2001) "Cicero's Astronomy", *Classical Quarterly*, 51.2 pp. 520–536.

Gee, E. (2007) "Quintus Cicero's Astronomy", *Classical Quarterly* 57, pp. 565–585.

Gee, E. (2013) *Aratus and the Astronomical Tradition*, Oxford: Oxford University Press.

Germanicus. *Aratea, Kommentarzum Aratus des Germanicus, MS. Voss. Lat. Q.79* Bibliotheek der Rijksuniversiteit Leiden, Faksimile Edition and Commentary, 2 volumes, B. Bischoff, B. Eastwood, T.A.-P. Klein, (1989) F. Mütherich and P.F.J. Obbema, Lucerne: Verlag.
Godwin, J. (2002) *The Pagan Dream of the Renaissance*, Grand Rapids, MI: Phanes Press.
Green, S. (2004) *Ovid, Fasti I, A Commentary*, Leiden: Brill.
Gleadow, R. (1968) *The Origin of the Zodiac*, London: Trinity Press.
Hard, R. (2015) *Eratosthenes and Hyginus, Constellation Myths with Aratus' Phaenomena*, Oxford: Oxford University Press.
Herbert-Brown, G. (1994) *Ovid and the Fasti: An Historical Study*, UK: Clarendon Press.
Hornblower, S. and A. Spawforth, eds. (1998) *Oxford Companion to Classical Civilization*, Oxford: Oxford University Press.
Hunter, R. (1995) "Written in the Stars: Poetry and Philosophy in the *phaenomena* of Aratus." *Arachnion* 2, 1–34. Online at: www.cisi.Unito.it//archne/num2.hunter.html.
Lippincott, K. (2016) *The Saxl Project*, http://www.kristenlippincott.com/the-saxl-project.
Lombardo, S. (1976) *Aratus' Phaenomena: an introduction and translation*, PhD dissertation, University of Texas, Austin.
Mair, G. R., and A.W. Mair (1969) *Hymns and Epigrams. Lycophron: Alexandra, Aratus: Phaenomena*, trans. G. R. Mair, Loeb Classic Library, Cambridge: Harvard University Press.
Martin, J. (1998) *Aratos, Phénomènes*, Paris: Les Belles Lettres.
North, J. (2008) *Cosmos: an Illustrated History of Astronomy and Cosmology*, Chicago: University of Chicago Press.
Pächt, O. and J. J. G. Alexander (1966–73) *Illuminated Manuscripts in the Bodleian Library*, Oxford: Clarendon Press.
Poochigian, A. (2010) *Aratus Phaenomena: Translated with an introduction and notes*, Maryland: John Hopkins University Press.
Possanza, M. (2004) *Translating the Heavens: Aratus, Germanicus, and the Poetics of Latin Translation*, NY: Lang Classical Studies.
Saxl, F. and H. Meier (1966) *Verzeichnis astrologischer und mythologischer illllustrierter Handschriften des latinischen mittelalters*, 4 parts: *In römischen Bibliotheken* (1915) 2. *Die Handschriften des National-Bibliotek in Wein* (1925) 3. two vols. *Manuscripts in English Libraries* ed. H. Bober (1953) London. 4. *Astrological Manuscripts in Italian Libraries* ed. Patrick McGurk, London: The Warburg Institute.
Seznec, J. (1953) *The Survival of the Pagan God*, trans. by Barbara F. Sessions, NJ: Princeton University Press.
Soubiran, J. (1972) *Cicéron: Aratea Fragments Poètiques*, Paris: Les Belles Lettres.
Tester, S. J. (1987) *A History of Western Astrology*, NY: Ballantine Books.
Thomas, R. (1991) *Literacy and Orality in Ancient Greece*, Cambridge: Cambridge University Press.
Weitzman, K. (1959) *Ancient Book Illumination*. Cambridge, MA: Harvard University Press.
Yeomans, D. (1991) *Comets: a Chronological History of Observation, Science, Myth and Folklore*, NY: Wiley Science Editions.
Zanker, P. (1988) *The Power of Images in the Age of Augustus*, trans. by A. Shapiro. Ann Arbor: University of Michigan Press.

Chapter 2
Possible Classical Models for *Aratea* Illustrations

Abstract Many volumes have been written on the transmission of astronomical knowledge from ancient Greece through the Middle Ages to the time of Copernicus, Kepler, and Newton but much less has been written about the celestial images that often accompanied that astronomical literature. This chapter examines the type of astronomical and astrological images that decorated Roman public buildings, temples, and images that appeared on celestial globes and in the painted pages of Late Antique manuscripts. Foremost examples of these early models will be discussed in this chapter. From the early Middle Ages through the Renaissance, the 12 signs of the zodiac became a significant iconographic symbol appearing in unusual places, carved on church facades, and painted in religious and calendric manuscripts, eventually drawn in the stunning Books of Hours cherished by wealthy patrons, and even decorating Jewish synagogues. The Early Christian era witnessed the transition of recording literature and illustrations on ancient papyrus scrolls that were rolled and stacked on library shelves to the new parchment codex that was trimmed and tightly bound between heavy covers to protect the treated animal skins and preserve the precious text and carefully painted illuminations. The pictorial transition from scroll to codex is discussed as well and the earliest surviving illustrated manuscripts are described.

The transmission of the text of Aratus' *Phaenomena* and of the *Aratea* by its Latin translators can be easily traced since the poems have been thoroughly studied by text specialists in classical literature for centuries, almost since they were first written, but this has not been true for the transmission of the astronomical images found in *Aratea* manuscripts. In antiquity only a very limited number of texts contained illustrations or any type of décor, particularly when the papyrus roll was the main material for preserving written information (Pächt 1984: 22). Illustrated 'books' did exist in that era but they were a slender exception; this lack of illustrative history creates immense difficulties in learning how much, if anything, the classical and Late Antique codices inherited from earlier papyrus rolls.

Since the surviving copies of the *Phaenomena* in Greek are not illustrated, and there is no evidence that the original poem ever contained pictures, we must

M. Dolan, *Astronomical Knowledge Transmission Through Illustrated Aratea Manuscripts*, Historical & Cultural Astronomy,
DOI 10.1007/978-3-319-56784-6_2

question where and when the images, so prominent in the surviving medieval *Aratea* manuscripts, were added to the text. Potential sources for the medieval illuminations of the Sun, Moon, planets, and constellations can be found in numerous artistic media: sculptures, wall frescoes, public monuments, celestial globes, floor mosaics, decorative arts, metal work, coins, or even Greek vase painting. Among the most obvious sources were scientific, astronomical, and astrological illustrations that do survive in early literary works on both papyrus scroll fragments and parchment manuscripts and particularly in calendars.

In ancient Rome, the calendar, filled with the dates of recurrent Roman holidays and religious festivals, served as a vital element of their culture, a part of what it meant to be a Roman citizen, for both pagans and early Christians alike (Blackburn, 1992: 669). In order for the calendar to be effective in proclaiming the events of the periodic seasons, the dates must be accurate and aligned to the harmonious motions of the stars and planets. By the first century BCE, the Roman calendar was no longer synchronized with celestial motions, a problem that occurs quickly over time if not continually revised. Because the existing calendar was no longer compatible with the heavens, Julius Caesar ordered a complete modification in 46 BCE. The mathematical and astronomical calculations of astronomer Sosigenes, who was assigned the difficult project, provided a more accurate, updated calendar. The Julian calendar, one of Rome's great successes, was utilized throughout the Middle Ages (McClusky 1998: 24). Calendars were often posted in public places to keep citizens informed, sometimes even carved in stone, called *fasti*; they also appeared in written form, such as Ovid's *Fasti*, (discussed in Chap. 1) and were sometimes decorated with pertinent pictures.[1] The choice of illustrations and the iconography of prominent artworks embody the ideals and beliefs of the society that produced it; the astronomical images utilized during the Roman Era tell their own story.

Codex-Calendar of 354

When searching for possible prototypes for the illustrations found in the earliest surviving copies of the *Aratea*, a most likely candidate would be the *Codex-Calendar of 354* the only calendar that survives from the time of the Roman Empire; it has been definitely dated to the fourth century (Weitzmann 1977: 79). Although the original *Codex-Calendar* is no longer extant, the work is known in its entirety and available for study through later copies[2](Salzman 1990: xxi). This lavishly illustrated Roman compendium for the year CE 354 is a likely model for images of the zodiac, planets, and constellations in medieval astronomical codices,

[1]To view images drawn in a late fifteenth-century printed and highly annotated copy of Ovid's *Fasti* in three volumes: see University of Leeds Library, Special Collections, The Brotherton Ovid Digital Resource.

[2]Another important source for information concerning the *Calendar of 354* is a study published in 1953 by Henri Stern, *Le Calendrier de 354: etude sur son texte et ses illustrations.*

and in particular, the Leiden *Aratea* of Germanicus, one of the most, if not the most, important surviving *Aratea* manuscript. The original *Calendar of 354* did not survive beyond the ninth century but a complete and accurate Carolingian copy was discovered during the Renaissance in the scholars' search for ancient documents (Salzman 1990: 70).[3] This ninth-century codex along with its decorative program was copied copiously in the sixteenth and seventeenth centuries, though none of these are entirely intact, six illustrated copies survive in part (Salzman 1990: 4).[4] Unfortunately the intermediary Carolingian copy has been lost, but the Roman prototype has been duplicated so reliably that the intervention of the ninth-century copyist can hardly be noticed. No elements suggestive of the art of the early Middle Ages can be found in the *Calendar of 354*, not even in the decorative features; the Carolingian artists showed a thorough understanding of the Classical style and reproduced those elements faithfully (Nordenfalk 1988: 8). Fortunately that same faithfulness to Classical exemplars can be detected in the ninth-century copies of the *Aratea* manuscripts.

The *Calendar of 354* was only a small section of a larger manuscript of more than a dozen varied texts that had been assembled into a single volume, appropriately called the *Codex-Calendar of 354*. Much more than a basic calendar, this was a collection of chronological material such as histories, lists of festivals, and important dates throughout the year. This deluxe version of the official Roman calendar had been prepared as a gift for a Christian aristocrat named Valentinus and composed by the "most famous calligrapher of the century" Furius Dionysius Filocalus, also a Christian. (Salzman 1990: 3). This elaborate gift included colorful illustrations of the planetary gods and the constellations which employ Classical iconography; the illustrations are possibly the work of Filocalus as well. In the calendar of Filocalus, both pagan and Christian elements are included side by side without a problem; for example, it contains chronicles of the world since the time of Adam from the Christian Bible, along with chronicles of Roman history from the time of Romulus, with traditional blessings of their pagan gods and goddesses. "Its

[3]This manuscript tradition was first clarified by the modern scholar, T. Mommsen, who established that all later copies were derived from the lost ninth-century copy, the Luxemburgensis, a copy of the fourth-century original. The text and illustrations were described by the seventeenth-century scholar and collector of antiquities Nicholas-Claude Fabri de Peiresc in a letter which has been published by Mommsen. Luxemburgensis disappeared at Peiresc's death in 1637.

[4]Salzman names the surviving copies, fragments, and descriptions. The Vatican Library holds Romanus 1 (Barb. Lat. 2154) and Romanus 2 (Vat. Lat. 9135). A description by Nicholas-Claude Fabri de Peiresc can also be found in the Vatican Library. Along with these partial manuscripts, small portions of the *Calendar* are held in various other libraries. Salzman's lengthy study presents an iconographic analysis and comparison of each picture. Salzman has reconstructed the *Codex-Calendar of 354* from the various surviving portions and has published a thorough study of its Roman context, explaining the layout of the *Calendar* in precise detail. After the *Calendar's* opening dedication page decorated with a calligraphic design held by angels, the *Codex* begins with full-page images of female personifications of the four major cities of the Roman Empire—Rome, Alexandria, Constantinople and Trier—along with their identifying iconography. These impressive illuminations were followed by a list of the apotheosized Roman emperors from Augustus to Constantine including the current emperor, Constantius.

Fig. 2.1 December folio, *Calendar of 354*, Seventeenth Century Reproduction

invocations illustrations, historical and festal notices—secular, Christian and imperial—attest to the blending of these elements at a high social and intellectual level" (Weitzmann 1977: 79).

The initial sections are followed by illustrations of the 'seven planets' including their legends (Jupiter and Venus are now missing) and an illustrated calendar page for each of the 12 months decorated with the appropriate sign of the zodiac in a small medallion. A full-page miniature faces each calendar page depicting a single figure shown in an activity applicable to that month and accompanied by descriptive verses; next follows the series of lists, feast days, and civic dates. Bringing the book to a close are the historical chronicles of Rome complete with a king list from the time of Romulus and a description of the city itself (Fig. 2.1).

The layout of each calendar folio is exactly the same format seen in later devotional Books of Hours and in many medieval manuscripts containing calendars. The page lists the columns of letters applying the typical Roman system of time reckoning divided into Kalends, Nones, and Ides. Each day of the month is noted by the important event or anniversary celebrated on that date, even indicating the inauspicious Egyptian days (*dies aegyptiaci*).[5] Red ink was always employed to specify the most significant dates, known as the 'red letter days'.

The appropriate zodiacal sign was pictured for each month, enclosed in a small circle, a motif that persists in calendars of medieval psalters and sacramentaries. Since this manuscript comprised much more than a typical calendar, *Codex-Calendar of 354* could be described more accurately as an almanac. Similar to the numerous almanacs produced in the seventeenth and eighteenth centuries, the Roman calendar included lists of important dates as well as astrological lore concerning the planetary week, the zodiac and helpful tidbits of information or predictions. The 'seven planets' with their negative and positive qualities were described along with the attributes of the days and the hours over which they presided according to astrological theory. As in most astrological interpretations, Saturn and Mars represented mostly negative forces; Mercury, Moon, and Sun signify neutral influences; the more auspicious planets Jupiter and Venus were considered bearers of beneficial aspects (Blackburn, 1991: 570).[6] In ancient Rome, producing a calendar for a private individual and including astrological data was not unusual, as classical sources confirm similar examples of private calendars; their appearance affirms the mounting popular interest in the prophetic characteristics of natal astrology (Salzman 1990: 9).[7]

The *Calendar of 354* encompasses an immeasurable source of information from that crucial period of transition after the imperial administration had relocated to Byzantium in the East and the long-established pagan religion of the Western Roman Empire was converting and adapting to the new Christian doctrines. In addition these significant writings offer insight into the treatment of both text and illustrations during the conversion from the papyrus scroll to the parchment codex. The *Calendar* is recognized as the earliest dated manuscript with full-page illuminations in Western art, comprising not just one picture cycle, but several sets of pictures drawing from different religious and secular traditions (Salzman 1990: 33).

The superior artistic quality of the miniatures in the calendar attests to the complexity of their sources. The *Calendar* manifests both the 'new' format displaying full-page illuminations created for the codex as well as retaining pictorial elements that were commonly found in papyrus scrolls. Its miniatures exhibit

[5]In most medieval calendars, each month contained at least two Egypt Days which were considered bad luck days, a time when certain activities were to be avoided, beginning a journey, planting crops, especially medical treatments, and bloodletting.

[6]The *Oxford Companion to the Year* gives a full account of the astrological interpretations of the planets and their influence on the days of the week especially for medical conditions and treatments.

[7]Salzman cites writings of Petronius, Juvenal and other literary references that support the conjunction of calendars with the practice of astrology.

ornamental complexity, as the full-sized figures are embellished with architectural devices, columns and curtains (Nordenfalk 1988: 8).[8] The format of the zodiacal images, placed in small medallions that illustrate the text, is still in what Weitzman calls the 'papyrus tradition' (Weitzman, 1971: 106).[9] The signs of the zodiac have not advance to full-page miniatures, as we will see in the Leiden *Aratea*, but were treated in the established manner and inserted into the text pages.[10]

The illustrations of the *Calendar of 354* are especially important when studying *Aratea* manuscripts because these astronomical images can be accurately dated. Also they are the earliest calendric examples on parchment that survive and they hold traces of images that may have appeared in papyrus scrolls, although it is impossible to determine if papyrus calendars were illustrated, as none survive. The codex format of the *Calendar of 354* coincides with the development of parchment manuscripts, most likely a fourth-century innovation, and so could not have been produced before the end of the third century.

Thiele, Nordenfalk, and others have compared some of the drawings of the Leiden *Aratea* to those in the *Calendar of 354*, and Salzman continues this *comparanda*, which will be discussed in greater detail in Chapter Seven. Some of the illuminations in the Leiden *Aratea* may have been modeled after those in the *Calendar of 354*, but there are other possible sources for the artwork as well. A small number of illustrated manuscripts survive from antiquity that can also be examined as potential models for the picture cycles in the ninth-century *Aratea* manuscripts. The calendar tradition passed on poplar and basic astronomical data, keeping alive essential classical iconography, but offering no real scientific information. Celestial globes are another potential source for astronomical iconography; these do retain more technical details, depicting the sphere of the heavens and the relative positions of the constellations.

Aratus and the Art of Celestial Globes

The concept of constructing an actual model of the sky, picturing the location of the constellations including the Milky Way and the celestial coordinates, and demonstrating how the stars move as a fixed unit around a central axis, appears to have originated in Greek antiquity. As mentioned previously, Cicero recorded the belief that the Greek astronomer Eudoxus had created a star globe that accurately depicted

[8]Nordenfalk states that we know from Egyptian papyri that it was already customary to frame tabular texts with ornamental colonnettes in scrolls of the Hellenistic epoch and this format continued in calendar illustrations remaining popular until the Renaissance.

[9]Weitzmann discusses many possible models for the illustrations of the *Codex-Calendar of 354*.

[10]In addition to their value as evidence of the transition in artistic styles underway during the Early Christian period, these calendar miniatures are historically important, since four pictures show pagan religious festivals which Romans continued to celebrate into the fourth-century Christian era.

Fig. 2.2 Farnese Atlas, Second Century CE, Naples, National Archaeological Museum, Public Domain

the formations of the constellations (Mair 1969: 196).[11] Careful analysis has shown that Aratus used the globe of Eudoxus along with his treatise on astronomy as guides for describing the positions and appearances of the constellations (Thiele 1898).[12] The Farnese Atlas, now at the National Archaeological Museum in Naples, bears the weight of an extant celestial globe, carved in stone, a direct descendent of the globe of Eudoxus. In the early 16th century the famous globe was acquired by Alessandro Cardinal Farnese, reflecting the fact that the orb had been long exhibited in the Villa Farnese (Figs. 2.2 and 2.3).

The Farnese Globe is a second-century CE Roman copy of a Hellenistic sculpture depicting Atlas bent on one knee, weary from upholding the immense

[11]Mair 196, Cicero *De re publ.* 14 and *De orat.i.*15.

[12]A nineteenth-century study by Georg Thiele, *Antike Himmelsbild*, 1898 concluded that the constellations as described by Aratus were not based on an earlier version of the Farnese Globe. A connection between the two would be impossibility since they differ greatly in their placement of the constellations; in addition Aratus was not aware of the colures which are included on the Farnese Globe. But the placement of the constellations on the Farnese orb do demonstrate an "absolutely certain basis for the relationship of the sphere to Hipparchus." Thiele logically asks would not an artist who wanted to display Atlas supporting a celestial sphere choose the most famous and widely known version as a model.

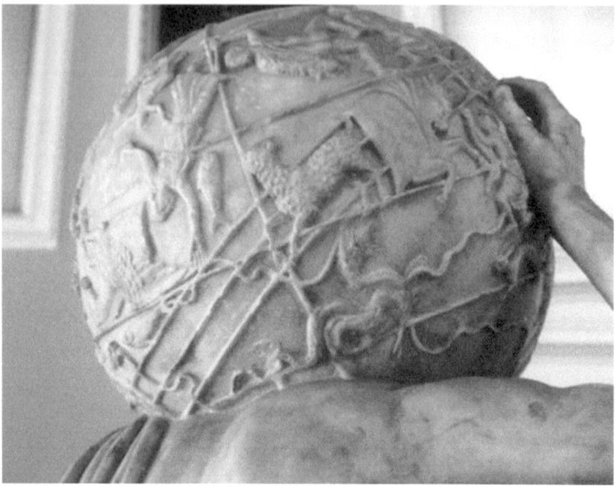

Fig. 2.3 Farnese Atlas detail, Second Century CE, Naples, National Archaeological Museum, Public Domain

weight of the universe on his shoulders (Tester 1987: 106).[13] This sculpture (although partially restored) is the oldest surviving representation of both Atlas and the celestial sphere; little is known about exactly where or when it was carved. The slightly damaged globe and figure, missing its face, arms, and legs, were rediscovered at Rome in the sixteenth century. The impressive Titan statue now stands 7-feet tall, supporting the heavenly orb that measures 65 cm in diameter. Carved on Atlas' marble sphere are remarkably accurate depictions of 41 constellations, precisely positioned against a grid of reference circles including the celestial equator, the tropics, the colures, the ecliptic, the Arctic and Antarctic Circles. The bas relief images stand out sharply from the background sky but depict no individual stars. Various experts have presented different conclusions on which astronomer's calculations the Hellenistic or Roman sculptor consulted when mapping the constellations. A comprehensive study was undertaken by Bradley Schaefer in 2005 which conducted a detailed comparison of the positions of these constellations with the known works of Aratus (based on Eudoxus), Hipparchus, Ptolemy and Ps.-Eratosthenes.[14] Celestial globes could help one in learning the relative positions of the constellations but no writings on their actual usage are extant from Antiquity. But since there were many Aratean texts and their scholia available for Greek–Roman education, it was possible to learn basic astronomy from a descriptive tradition instead. The oldest examples of books on the use of the celestial globe are found in the Middle East. A work entitled, *Book on the Sphere and its Use* by Habash al-Hasib, which is dated to the ninth century, is the earliest yet found (Dekker 2013: 282). Much was written about the celestial globe as a useful devise

[13]Tester states that Thiele determined the date, based on the Farnese's artistic styles related to the Hadrian epoch, surely before 150 CE.

[14]The study was published in the *Journal for the History of Astronomy*; B. E. Schaefer's work aligns to that, the Epoch of the Constellations on the Farnese Atlas and their Origin in Hipparchus's Lost Catalogue, of Thiele.

for learning basic astronomy, but useless for making any type of observations or accurate readings.

As we have seen, Aratus' poem made a huge impact both on the literary world and on the advancement of astronomical knowledge. During the Classical Age, the popularity of the *Phaenomena* tended to "freeze the development of the constellations as well as to define the basic properties of the group". The stellar positions imbedded in the *Phaenomena* carry a wealth of information and allow astronomers even today to draw informed conclusions. The same holds true for the constellation positions on the Farnese Sphere. In addition to other telling evidence, the placement of the summer solstice at the start of Cancer next to the head of Pollux rather than next to the start of Leo indicates that the source for the sphere's astronomy was dated after the era of Aratus. Also the constellation descriptions by Aratus are verbal, and would have been more difficult to transfer to a globe than actually copying from a celestial sphere or a star chart. Comparing the extant ancient sources with the Farnese Atlas produced an almost perfect agreement with the constellation descriptions of Hipparchus; other early sources show serious differences. Schaefer's comprehensive study establishes a strong case for identifying the original observer and demonstrates that the Farnese Globe must be a depiction of the lost star catalog of Hipparchus not Aratus or Eudoxus.

Depending on the model consulted, the constellation figures on a celestial sphere could be drawn from two opposite perspectives. In the first mode, viewers were expected to imagine themselves in a geocentric position within the heavenly globe or on the earth, looking upward at a frontal view of the mythical characters. The other perspective reverses the position of each image, as if the viewer is outside the celestial sphere looking down in a god-like manner onto the backs of the figures. Already at this early date, the mythological human figures on the Farnese sphere were given the god-like perspective, drawn facing inward, displaying the backside of the constellation figures.[15] The Farnese Globe shows most human constellation figures from the rear, some clothed, some naked. Another indication of the outside view is the depiction of Taurus; when seen from earth the half-bull faces to the left; on the Farnese the half-bull faces right as would appear from outside the universe. This practice of depicting some of the constellation figures from outside the stars became an established convention, found in manuscript illustrations, in the fabulous ceiling frescoes of Renaissance palaces, and still appears on modern celestial globes.[16]

Few celestial globes survive from before the sixteenth century, but images of the constellations can be viewed engraved on astronomical instruments, invented in ancient Greece but improved and advanced by Arab astronomers, particularly the astrolabe. The archetypal mythology of the constellations applied not only to its

[15]For a more thorough discussion of this earth view versus god view concept, see Thiele and Stevens, "The Figure of the Earth in Isidore's *De natura rerum*", *Isis*, 71 (1980): 268–277.

[16]See Dekker, *Illustrating the Phaenomena*, 2012 for a full description of the placement and exacting details of the constellations on the Farnese Atlas Globe, pp. 84-102.

primary function as a technical device for organizing the sky, for it also provided easily remembered stories as mnemonic devices. As we have seen, the iconography of the constellations was appropriated from the depictions in the sky for political propaganda by the Caesars, but astronomical or astrological symbolism was also taken over by religious cults and even by Christianity.

Astronomical Art in the Cult of Mithra

Mithraism, one of the most successful mystery cults that pervaded the Roman Empire, must not be overlooked when investigating the strong interest in astronomy and astrology during the Classical Age, its importance in daily life of Roman men, and as a source for inspiring the production astronomical artworks. An extremely rich iconographic program, concentrated on cosmological imagery, has been preserved in Mithraic temple sites, but as a consequence of the utmost secrecy required of its members, the interpretation and deciphering of that imagery has been questionable. A plethora of cosmically inspired statuary: the Dioscuri, Aion, Phanes, and the unnamed lion-headed God whose body is inscribed with the signs of the zodiac, have been excavated in many Mithraic temples. The dominant artwork embraced by Mithraism, an enigmatic cult image called the Tauroctony, seems to have been an essential element of their devotional activities. An image of the God Mithras slaying a large bull called the Tauroctony, as a wall fresco or a carved statue, has been discovered in almost every Mithraeum to date (Ulansey 1989: 6).

The Mithraic-centered religion began to spread throughout the Roman Empire during the first-century BCE and attained its peak in the third-century CE, but was effectively wiped out by Christianity by the beginning of the fifth century (Ulansey 1989: 4). Fortunately hundreds, perhaps thousands, of Mithraic temples, decorated with cryptic astronomical frescoes and sculptures, survive as witnesses to the mystery cult's success. Temples devoted to Mithras have been uncovered throughout all areas of the vast Roman Empire even the frontiers, especially where Roman soldiers were stationed. Mithraic ruins have been discovered from as distant locales as Hadrian's Wall, western Germany, Dura Europus, and scattered across North Africa. In the port city of Ostia as many as 14 Mithraea temples have been excavated. One of the largest Mithraic temples, built around 200 CE was discovered in the nineteenth century beneath the Church of San Clemente in Rome (Figs. 2.4 and 2.5).

In addition to military devotees, wealthy merchants and educated laymen were initiated into this all-male secretive and exclusive cult as well. Undoubtedly their utilization of caves and underground locations for their preferred places of worship improved their chances for survival. The interior of the Mithraea were mostly dark, lighted only by candles, as opposed to Sun temples open to the heavens, the stars, and the constantly changing sky. Most Mithraic temples were relatively small; many had adequate space for only a dozen or two devotees to assemble. Ceilings of the caves or subterranean temples were designed to simulate the night sky. Research

Fig. 2.4 Mithraic temple ruins under San Clemente Church, Rome, Second Century CE

Fig. 2.5 Mithraic temple ruins Campania, Santa Maria Capua, Vitere, Second Century CE

has suggested that some of these sanctuaries were actually used as observatories and that holes were drilled in the walls and ceiling for specific astronomical purposes (Ulansey 1989: 17). The majority of the Mithraea are designed on an east–west axis with a single entrance from the west leading the cult devotees and participants to the focal point of the sanctuary, the carved or painted relief of the

Fig. 2.6 Tauroctony Fresco, Mithraeum, Dura Europus, Syria, 168–171 CE

god Mithras 'Slaying the Bull'. The cult icon consists of an image of Mithras astride a massive bull; also depicted are a snake, a dog, a raven, a cup, a scorpion, and occasionally a lion. The dog and snake are pictured striving for the sacred blood flowing from the knife wound of the sacrificial bull, the scorpion appears near the bull's genitals and the signs of the zodiac usually surround the whole symbolic scene. The revered figure Mithras always appears clad in Eastern pajama-like garb adorned with a wind-blown cape decorated with stars; his exotic costume was completed with a floppy Phrygian cap.[17] The youthful god turns his head away from the distressed bull and gazing skyward, he plunges his knife deep into the bull's neck (Fig. 2.6).

The abundance of cosmological references in the artworks and artifacts excavated from the ruins of Mithraic temples provides proof that the cult of Mithras sustained a strong astronomical focus. Religions with an astral base, such as Mithraism, and the spiritualizing philosophies of Hermetism and Gnosticism were in effect attempts to assist the devotee to avoid or become liberated after death from a fate directed by the stars. Fortunately, the large quantity of iconographic evidence provides clues into the cult's beliefs that are missing in early literary works. The dearth of contemporary written documentation leaves a vacuum in evidential material and has hampered progress in divulging Mithraic mysteries since adepts of the cult were compelled by sacred vows of secrecy not to reveal their beliefs, activities or religious rituals. Contemporary references are generally filtered through early Christian sources which tend to present biased or derogatory accounts of the mysterious pagan religion. Astral religions such as Mithraism, spiritualizing

[17]See also, F. Saxl, Mithras I and Mithras II, *Saxl Lectures*, (1957) 13–44 for more information on Mithraic iconography.

Fig. 2.7 Tauroctony scene, unknown provenance

philosophies such as Gnosticism and Hermetism, and the development of a body of astral magic were all attempts to help the individual to escape or become liberated after death from his fate decreed by the stars.

The enigmatic iconography of this ritual Tauroctony artwork has been subjected to astronomical and astrological interpretations. Mithraic scholar Ulansey has constructed a very convincing argument that every figure poised on the Mithraic icon represents a heavenly constellation and Mithras himself personifies the constellation of Perseus. Moreover, all of these pictured constellations are aligned between the zodiacal signs of Taurus and Scorpio. Taurus the bull, Canis Minor the dog, Hydra the snake, Crater the cup, Corvus the raven, and Scorpio the scorpion all fall along the celestial equator; and in the identical order usually seen on the icon.[18] The bull slaying scene depicted in every Mithraic temple is actually a celestial map (Fig. 2.7).

Also included within this Mithraic pantheon are two attendants, male torch-bearers Cautes and Cautopates, attired as smaller versions of Mithras with chiasmic-crossed legs. Both carry long torches, the torch of Cautes points up; the torch of Cautopates points down. The iconography of these two companion figures suggests an association with the spring (Cautes) and fall (Cautopates) equinoxes when the duration of day and night becomes approximately equal, and the sun crosses the celestial equator indicating the change of season. Other possible associations of the torchbearers that have been suggested by Mithraic specialists are: the brightest stars of Taurus (Aldebaran and Antares), a symbol of light (the rising and

[18]Ulansey credits a 19th century German scholar K.B. Stark for first proposing an astronomical explanation of Mithras which was dismissed by the scholar of ancient religions Franz Cumont, who assumed the cult had Iranian roots, and therefore that explanation was ignored until the late 1970s. See Ulansey for a detailed discussion of his astronomical theories.

setting sun), or the formation of a trio (the thrice-great Mithras), or the Neoplatonic theories of life, death and the reentering of the soul into a new existence.

The cosmic tableau of the Mithraeum has been gradually divulging its mysteries and the debated reinterpretation of the cult iconography, if accurate, gives even more credence to the influence of the heavens in the daily life of those living in Roman Empire. Artistic remnants preserved from Mithraic rituals provide an excellent example of how astronomical and astrological information could be encoded into the iconography of artworks. Astronomical artworks, statuary, and relevant images utilized in the cult of Mithras, have been excavated throughout the widespread territory of Roman occupation and are currently being studied in consideration of their celestial implications. The artistic adornments of Mithraic temples provided abundant representations of contemporary astronomical iconography, especially zodiacal depictions, available for adaptation by early manuscript illustrators. The signs of zodiac originated in Mesopotamia early in the first-millennium BCE, but they were not used in Greece before the fifth-century BCE. (North 2008: 70) Dating from the Classical Era, the signs of the zodiac became a convenient and significant artistic symbol easily recognized by varied cultures through many time periods, even in most unsuspected places.

Zodiacal Mosaics in the Roman World

In addition to the previously mentioned exemplars, another possible model for astronomical imagery can be found in mosaic artworks. Mosaic floors were ubiquitous in the ancient world and have been uncovered in most archeological sites throughout the Roman Empire. Since mosaic images formed from stone, marble and glass tesserae are long-lived compared to the more fragile materials, such as papyrus scrolls and wall frescoes, they provide an important record of iconographic forms surviving from Late Antiquity. Mosaicists of that time relied upon the familiar and valued designs depicting historical, religious, and mythological images and stories. Popular themes for wall and floor mosaics included the calendar, the signs of the zodiac, the planetary gods and goddesses, mythical heroes, female personifications of the four seasons and the nine muses; these appeared in a wide variety of artistic creations in both public and private artworks (Parrish 1984: 15).

The themes bearing the most significance for this examination of astronomical art are the mosaic reproductions of Helios in his quadriga, Luna in her ox cart, the signs of the zodiac, and the heavenly constellations. A frequent companion to the four seasons in North African floor mosaics is the well-known image of Aion, whose attribute is the elliptical band of the zodiacal signs (Parrish 1984: 46).[19] In some areas iconic images of

[19]The god of eternity appeared in both imperial and private Roman monuments, sometimes as youthful but often as an elderly man. Aion with his zodiacal circle is found in connection with Mithraic cult sites as well.

Fig. 2.8 Helios in Quadriga surrounded by Zodiac, unknown provenance

Helios/Apollo and Luna appear with the zodiac and the four seasons. Many remarkable examples of zodiacal floor mosaics are extant from the Roman World, but the most extraordinary images are those discovered in early Jewish synagogues, preserved in various conditions. In the words of Hershel Shanks, editor of Biblical Archaeology Review, "What's a Greek god doing in an ancient synagogue?" (Fig. 2.8).

Zodiacal Art in Ancient Jewish Synagogues

During the Roman-Byzantine period, the villages or towns which had a sizeable Jewish community also had at least one synagogue and possibly even more, fulfilling the need for places for worship after the destruction of the Second Temple by the Roman army under Emperor Titus in 70 CE. In ancient Palestine alone, over 100 synagogues have been uncovered in archeological investigations dating from CE 100 to 600 (Levine 1998: 139). These religious assembly buildings were designed in a variety of styles and the type of décor found within them reflected the current popular tastes of the larger Roman-dominated society, but were deliberately and consistently adapted to meet the beliefs and ideals of the Galilean Jewish communities. From the age of Constantine onward, the cult of Holy Places established by Helena, mother of Constantine, in Palestine led to the construction of huge numbers of Christian churches as well (Dunbabin 1999: 192). They too were decorated with elaborate floor mosaics.

Similar to most Roman and Byzantine public and private buildings of that era, the floors of the synagogues that have been excavated were customarily decked with artistically designed mosaic pavement; it seems that the Jewish prohibition of figural representations had been relaxed, at least in some areas (Dunbabin 1999: 188). But surprisingly, the Jewish religious images and symbols in the mosaics often appear alongside traditional Roman pagan symbols. In addition to the expected Jewish icons, the menorah, ram's horn and tabernacle, Roman images are found in conjunction with religious emblems: the eagle, symbol of imperial Rome, naked cupids holding Greco–Roman wreaths, and winemakers pressing grapes in a Bacchic scene.

Most unexpectedly, in some of these ancient synagogue floor mosaics, a central and dominant position has been devoted to astronomical or astrological imagery. Examples of these zodiacal mosaic floors have been discovered at six archeological sites in ancient Palestine dating from the fourth to the sixth century: Beth Alpha, Hammath-Tiberius, Husifa, Khirbet Susiya, Na'aran, and the highly sophisticated floor mosaic of Sepphoris discovered in 1993. The seventh synagogue was found at Yafia, but the site is too damaged to be sure of its mosaics. The basically similar oblong compositions display a zodiacal wheel in the center where images of the twelve signs of the zodiac are encased in pie-shaped wedges; the zodiac is framed by emblematic as well as historical symbols of the Temple and the Tabernacle in the desert. Bianca Kühnel has investigated these anomalies and characterizes the remarkable floor mosaics as "one of the most authentic creations of Jewish art" (Kühnel 2000: 43).

Jewish religious principles and in particular the Second Commandment forbid polytheism and the depiction of images of God or Yahweh as well as other living beings or animals; followers of Islam of course adhere to these same restrictions. Christianity in the East also struggled with the display and worship of sacred images attested by iconoclasm in the eighth and ninth centuries. These iconoclastic views were not so tightly adhered to in Jewish society during the Late Antique and Early Byzantine Eras, as demonstrated by the extensive synagogue frescoes discovered at the Roman military town of Dura Europus. This synagogue is one of the earliest to survive; the construction was dated by an inscription to 244 CE (Fig. 2.9). (See Weitzmann, K. and Kessler, H. (1990)

But the discovery of mosaic images of the Greek and Roman Sun God, Apollo/Helios/Sol Invictus, and representations of the pagan mythological world, combined with scenes from Jewish sacred scriptures was initially quite shocking. These surviving zodiacal mosaics are relatively consistent in the images and symbolic pattern included, as well as in the symbols that they exclude. Although the image of Apollo standing in his *quadriga* dominates the central position of these zodiacal floors, Apollo's female counterpart Luna or Selene was either omitted entirely or reduced to an insignificant crescent moon in these heavenly scenes. Surprisingly the Moon Goddess, whose traditional iconography portrays her steering a small cart pulled by two oxen or cattle and holding a flaming torch, does not appear in any of the synagogue zodiacal mosaics. The customary symbols of the Sun and Moon as seen in later astronomical manuscripts are usually pictured side by side, or one above the other in the center of the circle displaying the twelve signs of the zodiac.

Fig. 2.9 Synagogue frescoes, Dura Europus, Syria, Third Century CE, Public Domain

The artistic diagram of the zodiacal sphere was certainly not an original creation by Jewish synagogue artists, but was envisioned much earlier in Roman artworks. (Kühnel 2000: 42) The appearance of astronomical symbols and the signs of the zodiac in decorative arts, mosaics, and on gold and silver coins was well established in Roman times especially among aristocrats; even Emperor Constantine, governing in the growing Christian world, was still using astrological symbolism on his coins (Fine 2005: 90). But zodiacal mosaics were adopted for symbolic synagogue décor from non-Jewish sources and were maintained for centuries, even after the Christian world had mostly terminated their use of this symbolism in architectural contexts until it was re-worked during the Carolingian and then Romanesque Periods. Religious historians are still offering explanations and justifications for this ostensibly sacrilegious iconography that decorated many ancient Near Eastern synagogues.

The Zodiacal Mosaic at Beth Alpha Synagogue

Discovered in 1928, the Beth Alpha floor mosaic remains the best preserved and most highly publicized of the surviving synagogue zodiacs and is pictured in many art publications.[20] The Beth Alpha synagogue had been destroyed in an earthquake,

[20]For example, the Beth Alpha zodiacal mosaic is featured and discussed in Snyder's *Medieval Art*, p. 6 and in Stokstad, *Art History*, Fourth Edition Volume 1, p. 221.

Fig. 2.10 Zodiacal Mosaic, Beth Alpha Synagogue, Israel, Early Sixth Century, Public Domain

but fortunately for historians, the important zodiacal mosaic survived, carefully protected by a layer of fallen plaster. As a result, the central floor mosaic carpet is still in an excellent state of preservation, while the mosaics in the eastern aisle, narthex and courtyard were mostly destroyed. Information about the local artisans who laid the floor, Marianos and his son Haninah, was preserved in two mosaic inscriptions, written in both Greek and Aramaic, along with the names of various donors and the date of the Beth Alpha synagogue's dedication, during the reign of Byzantine Emperor Justin I, in the early sixth century (Negev 1990: 55) (Fig. 2.10).

The Beth Alpha floor mosaic is divided into three separate panels; each surrounded by an ornate border, and all are designed to be viewed from the north. The first section shows Abraham preparing to sacrifice Isaac; the panel closest to the pulpit and apse holds a depiction of a tabernacle flanked on each side by a menorah, a ram's horn and other symbolic Jewish articles. The religious objects in these two divisions are scattered about to fill the neutral ground, a technique common in the design of late Roman mosaic pavements. The mosaicists used 21 different colors of stone and glass tesserae.

Our main interest focuses on the square central panel containing the zodiacal diagram divided into 12 equal sections that hold its appropriate constellation figure identified by an inscription in Hebrew. Each constellation is pictured in its recognizable form, but presented in an unclassical manner, more awkward and amateurish in its rendering. The center of the circle holds an image of Apollo/Helios standing in his *quadriga* and driving his four horses; each of the four horses had been individualized in a slightly different way. The bright rays of light encircling Apollo's head form a golden crown, only his head and neck are pictured; on the sun god's left is a tiny crescent moon, stars fill the background, most of which have fallen to the bottom of the circle, indicating that the starry night is being eclipsed by the rising sun. The signs of the zodiac are read in the usual counterclockwise direction, but begin with Libra in the ascendant position, which is unusual. More often its counterpart or opposite sign, Aries, the sign of spring and the vernal equinox holds that primary position.

In the four corners of the square that encloses the zodiac are personifications of the Four Seasons, female half-figures with wings, shown with appropriate attributes, each is inscribed with its Hebrew name (Ness 1990: 5).[21] The artistic workmanship of this multicolored pavement falls far below that of the other mosaic synagogue floors, such as the one at Hammath. The Beth Alpha artisans were more accomplished in their creation of abstract patterning than in their human and animal likenesses, as the braided bands that make up the circles and dividers of the zodiac are very carefully executed. The depictions of the astrological signs, although astrology was condemned by the prophets, became a widely used decorative program appearing in early synagogues. The astronomical mosaic designs at Beth Alpha are not unique but represent a colorful folk art rich in motifs; although the pagan character of the central scene remains striking.

Synagogue at En Gedi

En Gedi, a popular oasis in the Judean desert, had a large Jewish community during the early Byzantine Empire (Ness 1990: 5/12). The ancient En Gedi synagogue was discovered by local farmers in 1966, excavation followed in 1970–1972. The archeologists concluded that the synagogue had been constructed in three phases with the zodiacal inscriptions belonging to the latest phase, which was destroyed by fire around 530 CE (Ness 1990: 5/13). An important inscription was uncovered in the western aisle, one of the longest ones yet found; it contains a list of both the 12 signs of the zodiac and the months in Hebrew, among the other inscriptions (Fine 2005: 198). The circular diagram of the zodiac, with at least six known examples, was not found here; but the zodiacal signs were inscribed, the only known instance of this format. The zodiacal signs are followed by the listing of the 12 Jewish months, another rare appearance among synagogue inscriptions.

[21]Ness describes the Four Seasons, "Spring, or Nisan, holds a shepherd's crook, for example, while Summer, or Tammuz, sits among fruit and grain crops. Fall has a star overhead, as at Na'aran. Winter is marked by red cheeks."

Synagogue at Hammat Tiberias

On the shore of the Sea of Galilee, the Hellenistic city of Tiberias, named for Emperor Tiberius, was renowned in antiquity for its therapeutic hot springs; a famous Roman temple dedicated to Hygiea, goddess of health, was constructed there, as well as several synagogues. A fourth-century synagogue mosaic, discovered in 1947 at Hammat Tiberias, is among the earliest zodiacal floors, dated to the mid-fourth century; its artistic conception and workmanship can be rated far superior than many of the later synagogues and is outstanding even when compared with its contemporary public and private mosaic work. The entire floor of the building was decorated with mosaic patterns of the finest quality of richly colored browns, oranges, and beiges (Negev 1990: 166).[22] Its classical style is its most impressive feature, quite different from the more 'primitive' style of other Jewish mosaics.

Unfortunately a later stone wall was constructed through the circular diagram, obliterating part of the central design along with four constellation figures. The zodiac found at Hammath is similar to the other synagogue floor designs with its inclusion of both pagan and Jewish symbolism, although the entities are kept separate and do not intermingle. This mosaic floor is also divided into three sections each with a distinctive border; the portion closest to the entrance consists of nine sections, each inscribed in Greek. The middle section is decorated with the familiar zodiacal pattern, and the third part depicts an Ark or Tabernacle, flanked by a menorah and ram's horn on either side (Negev 1990: 167) (Fig. 2.11).

The center of the diagram has a splendid, but partially damaged Apollo, attired in Imperial Roman style with a cloak decorated with a large circular pin. The sun god's youthful face is beardless and encircling his light-tinted hair, a shining halo emits seven rays of light. His right hand is raised as if giving a blessing and his left hand holds a whip and an orb with two circles intersecting.[23] The chariot in which Apollo rides has been destroyed, but seems to emerge from the sea, visible in the lower part of the inner circle; the sky behind him includes only one large star on his left and a pale crescent moon to his right. The 12 signs of the zodiac are symbolized in their usual Hellenistic portrayals, even including nude, uncircumcised male figures for Libra and Aquarius and every month has its name inscribed in Hebrew (Ness 1990: 5/10).[24]

---

[22]Negev states that they were probably created by foreign artists, while local artisans, a father and son team, executed the designs on the floors at Beth Alpha and Beth Shean which are of much lesser quality.

[23]One author calls the globe a shield, but this could not be correct, as a shield is never displayed in the palm of the hand.

[24]Ness suggests that perhaps the mosaic was made by gentile workmen.

Fig. 2.11 Zodiacal Floor Mosaic at Hammath-Tiberias Synagogue, Fourth Century

Synagogue at Husifa or Ishya

The Husifa zodiacal floor mosaic was discovered in 1930 in the ancient synagogue located on Mt. Carmel not far from Haifa. The synagogue was oriented toward the holy city of Jerusalem, but the building had been destroyed by fire, perhaps during the reign of Justinian (Ness 1990: 5/6). Since it lies beneath modern buildings, only half of this ancient sacred site has been excavated but a similar astronomical mosaic carpet was revealed there. Originally the entire synagogue floor was covered with outstanding mosaics that are now badly damaged, but due to the ravages of time, not iconoclasts. In the nave, traces of a dedicatory inscription still survive, in addition to the geometric carpet and the traditional zodiacal diagram. The zodiacal panel consists of two circles inscribed on a square; the smallest of the zodiacs found so far (Ness 1990: 5/7). The inner circle is almost entirely destroyed and none of the central figure survives.

Fragments of the five zodiacal signs from Sagittarius to Aries can be identified; they were designed in a parallel manner to those at Beth Alpha. Sagittarius at both of these sites is depicted as a nude human figure, not a centaur, holding a bow, and covered with a cloak, similar to the iconic lion skin held by Hercules. Only the horns remain of the sign of Capricorn, but Aquarius survives mostly intact, not the expected male figure pouring water, instead a lone jar from which a stream of water flows forth. The double fish representing Pisces have been damaged considerably, only the fin and tail can be recognized; merely the tip of its tail and hind hooves remain of Aries the ram. The Husifa zodiacal mosaic has not been positively dated, but its similarity to the Beth Alpha and Na'aran mosaics allow it to be securely placed in the fifth or sixth century as well (Ness 1990: 5/7).

Synagogue at Khirbet Susyia

Traces of a zodiacal diagram, along with a scene depicting Daniel in the Lions' Den, were discovered at Khirbet Susyia, an ancient synagogue near Hebron in southern Palestine. The site was surveyed in the nineteenth century and excavated in 1971–1972, revealing a complex history (Ness 1990: 5/10). Actively used from the fifth to the eighth or ninth centuries, only traces survive from the various levels of the floor mosaics in this synagogue, but there is enough evidence available to suggest a zodiacal circle design. The Biblical scenes and the zodiacal wheel from the fifth century were removed in the sixth century and replaced by a more neutral and less controversial floral and geometric carpet of cruder quality (Fine 2005: 95).[25] Part of a zodiacal wheel can be distinguished, thus it is 'probably reasonable' to add this site to the list of astrological mosaics (Ness 1990: 5/11). It is quite possible that this mosaic design may have been destroyed by iconoclasts, as was the mosaic at Na'aran.

Synagogue at Na'aran Yafiha

In 1918 a German bomb exploded in Na'aran, 2½ miles north of Jericho, accidentally exposing the mosaic pavement inside an ancient synagogue that had survived from antiquity in excellent condition, the first zodiacal floor design discovered. The date of the mosaic's production is still debated by archeologists, most likely within the fifth and sixth centuries (Negev 1990: 273). The mosaic floor had been decorated with many types of birds and animals that had been deliberately and carefully removed in the eighth or ninth centuries; the destruction had not been carried out by anti-Jewish fervor, as expected, but by Jewish iconoclasts during a time of strict religious orthodoxy (Fine 2005: 84).[26]

Even though seriously vandalized, enough of the floor remains to permit an accurate reconstruction. The zodiacal diagram appears in a large square panel containing two concentric circles with the usual depiction of Apollo in the center, driving his *quadriga*. Here, as was the case with the other living creatures in the

---

[25]Fine reports evidence of other iconoclastic activity at Susiya, as images of animals and human figure were removed from the screens surrounding the main bema as well.

[26]Fine says that after the removal of the sacrilegious images, the religious life of the Jewish community seemed to continue as it had in the past.

Na'aran mosaic, the figure of Helios and the faces of his four horses have been gouged out leaving only the wheels of the chariot, the rays of light emitted from his halo, his whip, and his cloak decorated with stars. But utmost care had been taken not to damage the names of the months as they were written in Hebrew, the sacred language.[27] The 12 zodiacal signs were badly damaged; but from what remains, their iconography seems comparable to other synagogue mosaics. The symbolism depicted in the Na'aran mosaics matches that found in other synagogue floors; the workmanship appears distinctive, but not outstanding.

In addition to the zodiacal pattern, the floor was decorated with the image of a Torah Shrine flanked by two menorots; nearby another mosaic scene depicting the prophet Daniel between two lions had also been ruined. The most astonishing feature of this archeological site lies in the scale of the vandalism that the zodiacal mosaic suffered and the careful selectiveness of that destruction. The zealots attacked only animal and human forms, especially their faces. Plant life, Jewish symbols, and Hebrew inscriptions were left unharmed, implying that the vandals were Jews, prompted by religious motivation (Ness 1990: 5/4). Although each synagogue's zodiacal mosaic has its own distinctive features, the most intriguing ancient floor must be the one excavated at the celebrated ancient city of Sepphoris.

Synagogue at Sepphoris

The Sepphoris Synagogue's zodiacal mosaic floor was excavated in 1993; the structure was built in the first half of the fifth century and destroyed at the end of the Byzantine Period, mid-fifteenth century.[28] Recent excavations at Sepphoris have detected more than 40 mosaic floors found in both public and private structures, rating the city as one of the most important sites in the ancient world for its wealth and for its variety of mosaic decorations surviving from the third to the sixth centuries (Weiss and Netzer 1998: 10). The Sepphoris floor mosaics are noted not only for their exceptional artistry and richness, but for their sensitive iconography and classical style. This synagogue mosaic ranks as the most sophisticated of the seven discovered in its size, layout, and extensive pictorial design, and comprises the most complex decorative program preserved in a Palestinian synagogue. Its arrangement of 14 separate panels has been compared to the extensive mural designs uncovered within the ancient synagogue of Dura Europa (Weiss and Netzer

---

[27]Sacred writings in Hebrew could not be destroyed, but if no longer usable, were kept in the *genizah* found in every synagogue.

[28]Located in Lower Galilee, Sepphoris has a rich history going back more than 2000 years, rebuilt by Herod Antipas in the early first century and ruled alternately by Romans, Jews and Christians. The city of around 30,000 people was quite cosmopolitan and became a renowned center for learning.

1998: 15).[29] Because of its key location, design, and dating, this mosaic is crucial for the interpretation of the whole group of ancient Jewish Zodiacs (Kühnel 2000: 31).[30]

This mosaic was divided into seven horizontal sections of unequal dimensions; the other synagogue floor mosaics are all divided into three panels, with the signs of the zodiac central in the design. Like the other synagogues, a scene is devoted to Abraham's Sacrifice of Isaac, and another area depicts the Tabernacle and Temple at Jerusalem, as anticipated, the familiar zodiacal diagram was centrally placed. The *quadriga* of Helios with its four horses lacks the prominent figure of Apollo holding the reins, not destroyed by fervent iconoclasts, but purposely replaced by a brilliant sun disc, emitting ten light rays reaching out in every direction (Kühnel 2000: 39).[31] The rider-less chariot faces the viewers, wavy blue lines representing water fill the lower part of the circle; the mosaic portrays the dawn sun rising from the sea.[32] Next to the fiery orb, created with white tesserae, sits a slightly smaller crescent moon, as if glowing, tucked inside its full face, shaded a deep brown; a single large star burns beside the Moon.

The signs of the zodiac are represented by their customary symbols and in some months as youthful figures performing the labors of the month as well. Each sign and the name of each month are labeled in Hebrew; those sections not damaged include a large singular star. Only 4 months are totally intact, Libra, Scorpio, Sagittarius, and Pisces; the others have varied amounts of damage; but at least one detail of each sign can be positively identified. Aries is placed in the expected ascendant position and the remaining signs are read counterclockwise, as in astrological charts.

The iconography of Gemini, or the Dioscuri, at Sepphoris is especially significant; the two youthful brothers, Castor and Pollux, are depicted frontally, naked and embracing; but do not carry their battle-ready spear and shield. The young man on the left grasps a club, and the one on the right holds a stringed instrument; these symbols indicate the iconography belonging to different sets of mythical twins, Amphion and his brother Zethus, who were frequently interchanged with Castor

---

[29]This is not to suggest that the makers of the Sepphoris mosaic were directly influenced by the wall paintings at Dura Europus, but instead, that both groups of craftsmen drew from a tradition that was common in the Roman East. If this is correct, it is possible that more examples arranged in this manner may still be found in Israel. This mosaic forms a link between the third-century frescos of Dura Europus and late Byzantine and medieval illuminated manuscripts. In addition, the scenes of this mosaic may provide indirect evidence for the existence of illustrated manuscripts of Biblical material in the Jewish community, which may have been models for the mosaic at Sepphoris.

[30]Kühnel calls this mosaic 'the until-now missing link', as it is intermediate between Hammat-Tiberias and Beth Alpha.

[31]Kühnel remarks that "according to an ancient Roman tradition, the four horses of Sol's quadriga were linked to the four seasons and also the times of day: *Huic quoque quadrigam scribunt, illam ab caussam, quod aut quadripartitis temporum varietatibus anni circulum paeragat; aut quod quadrifido limite diei metiatur spatium* (Fulgentius, *Myth.* 1.11).

[32]Neptune is often shown in a similar manner riding in a chariot that rises from the sea; perhaps their iconography has been conflated.

and Pollux.[33] Overall, the most amazing feature of the Sepphoris floor mosaic is the absence of Apollo since every other depiction of this zodiacal diagram included his image; but at Sepphoris his *quadriga* is empty except for the solar rays. The deletion of the anthropomorphic sun god in this single instance raises many challenging questions.

Interpretations of Zodiacal Artwork in Jewish Synagogues

The mystifying inquiry as to what these zodiacal diagrams pictured on synagogue floor mosaics actually meant to Jewish worshippers has spurred scholars to seek plausible explanations. These images of the 4 seasons, the 12 months, the pagan Sun God, and the signs of the zodiac remind the viewer of both the rhythmic pattern of nature and earth's interaction with the eternal cycle of the heavens. In ancient cosmological theories, the microcosm of the earthly sphere was considered a reflection of the macrocosm of the celestial sphere. Religious scholars suggest that the powerful figure of the Sun God Apollo was intended to stimulate the audience to imagine god, not in human terms but in celestial terms, as a reminder of god's omnipotence, that the symbolic sun god represents the invisible force that drives the universe, its stars, seasons, and human labors. But that still does not answer the unrelenting question, why was the celebrated pagan image of the Greek god Apollo, encircled with the 12 constellations of the zodiac, adopted for the central position in at least six major Jewish synagogues?

Depictions of the four seasons and the signs of the zodiac with Helios and Luna in the center were familiar images taken over from artworks popular in the Hellenistic era and Late Antiquity. But the combination of these particular themes, minus Luna, with scenes of Abraham and Isaac and symbols of the Tabernacle and Ark has been discovered only in ancient synagogues. The earliest example of the three-part mosaic occurs in Hammat Tiberias, fourth-century CE; Sepphoris in the fifth century; Beth Alpha, Khirbet Susyia, Na'arn and probably Husifa date to the sixth century. Appearing in synagogues repeatedly for over 300 years implies a consistency of purpose and a traditional significance for this pictorial pattern. Consider how many more of these floor mosaics may yet be uncovered, or may have been destroyed over the centuries.

Some historians consider the appearance of the zodiacal signs in synagogues in the same manner as they interpret the sculpted zodiacs that appear on the west facade of many Romanesque and Gothic cathedrals. Those Church officials who commissioned the use of these multipurpose astrological symbols on religious structures relate this theme to the calendrical cycle of time, the yearly cycle of the

[33]This switch in the iconography of the twin figures of Gemini will be discussed in more detail in Appendix C.

four seasons and the formation of the heavens by the divine creator, with no suggestion of astrological implications.

In the same way, the idea that the image of the zodiac in a Jewish religious setting had astrological undertones has been thoroughly rejected by religious historians. But a study by Lawrence Ness states an opposing view; he investigated the manner in which the Jews of the Hellenized world "adopted and adapted astrology" (Ness 1990: 4–16).[34] Ness provides a summary of modern research on the topic of astrology and Jewish art, and submits that astrology was part of Hellenistic science and philosophy that Jewish theologians could not ignore. He states that astrology was seen as an emblem of the power of YHWH in managing the world and caring for his chosen people and also that artistic illustrations of the zodiac in the ancient synagogues were symbolic of his power and concern for his creation (Ness 1990: 4–16).

The zodiacal floor mosaics have been a topic of considerable research and scholarly essays among religious historians and archeologists; their symbolism and function within Palestinian synagogues and Jewish culture is still being debated. Herbert Kessler stresses that in this early period, there is no doubt that "Christians and Jews were in conversation with one another."[35] He doubts that there was one simple pictorial tradition, but suggests that there was cross-fertilization between the two cultures. Other scholars interpreted the mosaics as 'expressions of Hellenistic mysticism', as 'meaningless decorative elements', or having 'a cosmological interpretation' or 'an imperial one', 'part of a Jewish eschatological program', 'a Jewish calendar with a practical purpose', or even an interest in astrological theory (Kühnel 1990: 36).[36]

In *From Synagogue to Church*, John Wilkinson has written a critical analysis of the sizes, proportions, and layouts of the early synagogues, using the zodiacal mosaics as crucial components for his study. Often the zodiac sits in the center of the mosaic floor and Wilkinson compares the center of the synagogue with Jerusalem as the center of religion and the world, and relates this theory to the synagogue zodiacs. Suggestions that the zodiacal scene represents the calendar have been rejected since the facts dispute this possibility. At Na'aran the zodiac signs circle the image clockwise, but the seasons are arranged counterclockwise; and at Beth Alpha the zodiac signs do not match the seasons. To a viewer who wanted to

[34]Ness has published his dissertation, *Astrology and Judaism in Late Antiquity,* in a book format, *Written in the Stars*. The Jewish floor mosaics are examined as an important piece of evidence in the development of his thesis.

[35]Quoted from a symposium paper delivered by Kessler on September 14, 1997, at an international conference held in conjunction with the Sepphoris Mosaic Exhibition, *Revealing an Ancient Message: A Synagogue Mosaic from Sepphoris*, in New York; the report on the symposium was published in the Jewish Heritage Report, Vol. I, Nos. 3–4/Winter 1997–98.

[36]Kühnel writes that "texts characterize the Jewish attitude to astrology as clearly differentiated: a line was drawn between cosmic phenomena and the pagan gods associated with them in Hellenistic thought and art. As long as astrology did not lead Jews to worship the gods of the planets, it was not seen as an active danger".

use these designs as guides to the calendar, such visual errors could not be tolerated (Wilkinson 2002: 91). The three themes: Abraham and his son Isaac, the zodiacal circle and symbols of the Tabernacle and Temple at Jerusalem are represented as a thematic unit in these distinctive mosaic floors. An explanation of their symbolism as a group has not been definitively accepted by religious historians. Because of their impressive numbers and key locations, the mosaics need to be considered as part of official dogma (Kühnel 2000: 36).

The artwork created for the ancient Near East synagogues has set in stone the ideas and beliefs of those who worshipped there. The fact that the zodiacal symbolism was at one time accepted as suitable decor, and then later no longer permitted invites great interest. Some pertinent questions are: would these carefully executed, expensive mosaic floors, bearing traditional sacred images, have been walked upon? If the floor designs had been covered with a carpet, the images would not be visible and its symbolism lost. How would the images on the floor interact within the religious services or ritual activities taking place in the synagogue? Why would the Jewish community display a Hellenistic zodiac in their synagogue, especially since they used a lunar rather than a solar calendar, and why would they adopt a prominent Greek god as a metaphor for the omnipotence of Yahweh, who was never visualized? Did the celestial design hold some type of astronomical knowledge? Was the objective directed to a messianic or eschatological meaning? Could the religious or cosmological design have been a teaching device?

The significance of the zodiac and its function in these ancient Jewish rituals raises intriguing questions beyond the scope of this study. But the format, style, and design of these zodiacal scenes is an integral piece of the puzzle when tracing the transmittal of astronomical iconography. Although the Jewish synagogue mosaics were in all likelihood not a direct source of images for the illustrators of *Aratea* manuscripts, the designers of the imagery for both types of artworks must have been influenced by preexisting exemplars and drew from the same accepted astronomical and mythological traditions prevalent in the Greco–Roman world. The similarities could be based on direct influence or on comparable interpretations. "However there can be no doubt that the Jewish reworking of an ancient scheme eased the way for mediaeval Christian representations" (Kühnel 2000: 43). The illustrations of the constellations in *Aratea* manuscripts are not comparable to the Jewish mosaics since the Aratean images treat each constellation in a singular manner. They do not separate the zodiacal signs in a specific arrangement or pay any particular attention to their appearance among the nearby star patterns.

Zodiacal Mosaics in Early Christianity

Although mosaics were commonly used to decorate floors and walls in domestic and religious buildings, zodiacal imagery seems to have had only an incidental role in Early Christian art. Historian Steven Fine concludes that this subject matter was 'deliberately excluded' (Fine 2005: 199). Since the zodiac was so closely associated with Jewish synagogues through the seventh century, perhaps the Christians in the East consciously chose not to include this motif as part of their iconographic programs. With few exceptions, the earliest surviving zodiacal representations in Christian art date to the ninth century, and were modeled on the early astronomical and astrological manuscripts that were being copied in the Carolingian court schools. One Christian mosaic from the sixth century appears somewhat similar to the synagogue zodiacal diagrams and this lone example emphasizes the Christian reluctance in respect to this subject (Kühnel 2000: 43).

This floor mosaic from the Monastery of the Lady Mary in Bet She'an shows a 12-part zodiacal circle filled with large human images representing the months. The 12 standing males are each attired in Roman casual garb with attributes that refer more to the Labors of the Month with no direct indication of any familiar constellations of the zodiac, although the male worker of September holds a shaft of wheat, always ascribed to Virgo and the star Spica, and the figure representing April holds a ram. Interestingly May, the month of Gemini, has one young man holding a lyre, the iconography of Amphion, who is sometimes substituted for either Castor or Pollux. In the central circle of this mosaic are busts of a feminine-like Helios, eyes looking heavenward and honored with rays and holding a huge torch, and a serene Selene crowned with a crescent moon, holding a scepter.[37] No trace of their chariots can be detected. The lack of zodiacal imagery in the Early Christian art and architecture is in direct contrast to the Middle Ages when the zodiac readily appeared in tympanum sculptures, stained glass windows, tapestries and of course, illuminated manuscripts. Some of the earliest depictions of the zodiac in Carolingian manuscripts appear surprisingly similar to the early Jewish designs. A significant exception to this general statement is the mosaic image of Christ discovered in Rome.

Christ as Apollo

The iconographic image of Apollo/Helios directing his *quadriga* as it was transported across the heavens by four powerful horses was borrowed by Early Christian artists for one of the earliest depictions of Christ. Excavations at the Necropolis beneath Saint Peter's Basilica in Rome revealed an evidential mosaic in the vault of Mausoleum M of the Julii in the Vatican Grottoes which dates to the mid-third century, now called *Christ as Helios* (Fig. 2.12).[38]

Apollo, the pagan sun god, has become conflated with Jesus Christ, the Christian Son of God. The partially damaged figure was placed within a golden background and surrounded by a common motif of curling vines with lacy leaves, instead of the

[37]Possibly the Sun God had taken on the appearance of Mary, in a monastery dedicated to her.

[38]In Snyder's *Medieval Art*, this mosaic is pictured and briefly discussed on page 25, first edition, and page 6 in the second edition.

Fig. 2.12 Christ as Helios Mosaic, Tomb M, Vatican Grottoes, St Peter's Basilica, Rome, Third Century, Public Domain

expected signs of the zodiac. Just one wheel of the chariot survives, and two white horses face toward the center of the image, the damaged area most probably held the other two horses. The image of Christ-Helios was rendered as blond, youthful, and beardless, dressed in a striped garment with a cape blown by the wind. A glowing aura emits rays of light around his head that form a cross that extends from the halo rather than within the halo, which later becomes the iconographic crossed nimbus.

Similar to the images of Mithras, Christ's eyes are directed skyward, not toward the viewer; his left hand holds a globe, symbol of his power and protection that continues through illustrations of the Middle Ages and Renaissance.[39] The dazzling Sun-God icon has had a pervasive life history; developed as the ideal male form in the Hellenistic world, taken over by the Romans, adopted and adapted to conform to the Jewish religious culture, and then embraced by Early Christian artists as one of the accepted models for the portrayal of Christ. Astronomical images in artwork of the Latin West do not survive in abundance until the Carolingian renaissance of the ninth century, but there are examples to explore in both scroll and codex illustrations.

[39]This portrayal of Christ has a strong resemblance to some images of Mithras, especially the eyes.

Late Antique and Early Christian Scroll Illustrations

Papyrus scrolls enhanced with illustrations have been discovered in tombs and in sarcophagi dating as far back as Old Kingdom Egypt, 3000 BCE; many of these ancient hieroglyphic writings include verses, spells, and incantations from the *Book of the Dead.* As in all early cultures, astronomy played a crucial role in Egypt in determining the yearly cycles, the time of the annual flood of the Nile, creating a calendar and predicting the time for religious festivals. Though the Egyptians practiced astronomy and astrology, the iconography of the familiar constellations seems to date to classical Greece and was not derived directly from Egyptian sources. Ancient Babylonian relief sculpture and cuneiform tablets often include astronomical imagery and the signs of the zodiac, but as far as presently known, the picture cycle containing the entire 44 to 48 constellations on which this study focuses, does not appear until Hellenistic times.

Survival of manuscripts from the late antique period is rare, only a small number are extant: some Biblical, some scientific and some literary. Therefore, the history and transmission of the *Aratea* manuscripts from the time they were originally written until the ninth century cannot be written; the question of exactly when the poems were first decorated with appropriate pictures cannot be answered. The earliest dated papyrus fragments of the *Phaenomena* were written in Greek and were not illustrated; no scrolls of the *Aratea* survive from the time of the Roman Empire. But other papyrus fragments from this early time period demonstrate that the text of scrolls was often decorated with pictures. According to Weitzmann, there was a vast tradition of book illumination in classical times, it was said that the Library of Alexandria held 700,000 scrolls; many of these would have included pictures (Weitzmann 1971: 9).

We can assume that many natural science texts were clarified with descriptive pictures, instructive charts, and diagrams; the precise, realistically painted Vienna Dioscurides which discusses the medicinal qualities and healing power of various plants is an extant example. Pliny the Elder writes (N.H. XXV.IV.8) that the herbals of Crateuas, Dionysius. and Metrodorus were 'most attractively' illustrated with the plants and herbs that they discussed (Weitzmann 1971: 9). In fact, one of the earliest extant papyrus rolls in Greek is adorned with astronomical images. That papyrus scroll, 2 meters in length and containing an astronomical treatise, is now held by the Louvre. The text columns of this roll are explained with diagrams and drawings of the heavenly bodies. (Fig. 1-3 pictured in Chap. 1).

These simplified sketches with no background embellishment or defining frame are positioned between the columns of text; Weitzmann coins the term 'papyrus-style' illustrations for these 'scaled-down' images.[40] He demonstrates that these papyrus-style pictures continued as models for centuries, even after more

[40]The Utrecht Psalter survives as a perfect example of this earlier 'papyrus style'.

advanced modes of illustrating had appeared.[41] In his study of the development of illustrations from the papyrus scroll to the parchment codex, Weitzmann uses representations found in astronomical texts, especially the *Aratea* manuscripts, to exemplify his discussion (Weitzmann 1971:96). He selected astronomical treatises since that is the earliest topic in Greek papyrus scrolls to be illustrated, and in addition, astronomical imagery was available for his study in a great number of medieval manuscripts.

Illustrations in Epic Poetry Codices

In order to create an ancient "book" in classical times, numerous sections of papyrus were glued together and then tightly rolled; some extensive scrolls stretched as long as 22 feet, tags were added listing the various contents (Thomas 1992: 8). Literacy of course was the primary requirement, but reading through an entire papyrus scroll entailed a specific set of skills. While holding the scroll steady, the reader proceeded to unwind with one hand and rewind with the other, which required a special posture and coordination. The script, written on brittle and flaking papyrus, had no punctuation, paragraphs, or word divisions making the entire process less than easy.

By the end of the fourth century, the transference of classical and Early Christian literature from the unwieldy papyrus scroll to the parchment codex format was basically accomplished (Pächt 1984: 15). This transition from papyrus to parchment changed illustrative art even more dramatically than it modified the art of writing. Radical changes in the artwork took place when the pictures began to be separated from the text and then eventually enlarged to fill the entire folio. This advancement was possible since vellum was a much smoother, receptive medium for paint than fragile papyrus scrolls. Once the sketchy images were no longer crammed into tiny spaces between columns of text, miniature painting could become more expansive, varied, and colorful by borrowing from the art of frescoes, panels, or mosaics, in addition to copying the drawings that decorated obsolete scrolls. Only a relatively small number of manuscript texts were illustrated or decorated in any way during Late Antiquity and the Early Christian Eras (Pächt 1984: 33).[42] But among the

[41]Weitzmann writes that "A manuscript in Munich (Staats, cod. 210) written in the year 818 in Salzburg shows exactly the same principle of illustration, except that the simple diagrams with partly Egyptian symbols were replaced by images essentially based on Greek mythology, images which surely were not invented for the *Phaenomena* of Aratus-although this whole group of texts is known as *Aratea*-but most likely are from the *Katasterismoi* ascribed to Eratosthenes of Cyrene." He does not say exactly, but he implies that this manuscript is an *Aratea*; it actually contains the writings of Hyginus. He also writes that the accompanying verses in the Leiden manuscript are taken from Avienus, when in fact the verses are mainly those of Germanicus with small sections of Avienus added.

[42]Pächt finds it quite extraordinary that Ovid's *Metamorphoses* inspired no pictorial representations until the time of the Renaissance.

earliest extant illustrated manuscripts we find examples of epic poetry, one of the most revered and studied topics of those times.

Although there is no indication whether or not they were illustrated, written texts of the epic poems of Homer, the *Iliad* and the *Odyssey,* were known from references to have circulated throughout the Greek world by the fifth-century BCE. In the fourth-century BCE, the Homeric texts were attacked by Plato and studied by Aristotle and other contemporary Greeks (Young 2003: 29–63). Dramatic narratives derived from Homeric verses were among the artistic scenes appearing on Grecian black-figure and red-figure vases, kraters, amphorae, plates, and even sarcophagi as early as the fifth-century BCE. The earliest fragments of the Homeric text, written on papyrus in the third-century BCE, were found in Egypt, preserved fortunately by the parched desert climate. Huge numbers of Homeric manuscripts survive from the medieval period, but not surprisingly, manuscripts of Virgil's *Aeneid* are close to Homer in survival numbers (Weitzmann 1977: 13).

The earliest illustrated manuscripts of classical poetry still surviving are not copies of Aratus or the *Aratea*; only three examples of manuscripts with painted miniatures are extant from Late Antiquity, one by Homer and two by Virgil. The single extant antique Homeric manuscript, which originally contained the whole *Iliad*, survives now in myriad pieces, cut into 58 fragmentary miniatures, probably removed from the codex before the thirteenth century. Textual excerpts from the *Iliad* appear on the reverse of the paintings in Greek uncial script. This incomplete parchment manuscript, known as the *Ambrosian Iliad* or *Ilias Picta*, (Milan Biblioteca Ambrosiana, Cod. F.205 Inf.) has been dated to the late fifth or early sixth century.

Along with the Ambrosian *Iliad,* two manuscripts containing the *Aeneid* of Virgil (90–19 BCE),[43] the *Vergilius Vaticanus* and the *Vergilius Romanus*, are the only copies of illustrated classical literature that survive from antiquity. Begun in 29 BCE, the *Aeneid* is considered one of the supreme literary achievements of the Latin West, even referred to as the most important single work in Western literature. Virgil worked on his heroic poem made up of twelve books that 'sings' the adventures of Aeneas, the legendary founder of Rome, until his untimely death ten years later, when the poem was essentially finished except for minor editing. Especially encouraged by Augustus to produce his poem, Virgil dedicated his epic poetry to the celebrated Roman emperor, just as Germanicus did when creating his translation of the *Phaenomena* in the second decade CE.[44]

[43]The early spelling of Virgil used an "e", while the modern spelling prefers an "i"; thus the two different versions of his name.

[44]The basis of the accepted *Aeneid* text used currently is derived from three manuscripts of Virgil's works from the fourth and fifth centuries CE; remarkably, these early manuscripts are relatively free from errors and mostly agree, which demonstrates that the scribes who were copying the text had high-quality exemplars.

The two earliest illustrated manuscripts of the *Aeneid* are both housed in the Vatican Library; *Vergilius Vaticanus,* Vatican City, Biblioteca Apostolica, Cod. Vat. lat. 3225, which dates to the early fifth century, and *Vergilius Romanus*, Vatican City, Biblioteca Apostolica, Cod. Vat. lat. 3867, dating to the later fifth century, commonly called the Vatican Vergil and the Roman Vergil. Like the ninth-century *Aratea* manuscripts, the two Late Antique Virgil manuscripts are totally different in size, style, and iconography, and in spite of this huge divergence, (The foremost authority on illustrated Virgil manuscripts is D. Wright) experts now agree that they were both created in Rome, as it was the only possible production center for manuscripts of such high quality in art, parchment, and calligraphy. (The Vatican Virgil is now digitalized and available online.)

The text of both codices is incomplete containing fragments of the *Aeneid* and the *Georgics,* which are also decorated with miniatures of seasonal occupations and animal folklore. The illusionistic miniatures in the Vatican Vergil and the Roman Vergil are frequently set apart from the text by bright red frames, which have been dated without exception to the early fourth century. The poetic text is positioned either above or below the distinctive framing. This particular decorative framing feature also reappears in the Leiden *Aratea*, modeled on similar Late Antique exemplars. Reaffirming this decisive era, Nordenfalk states, "The first efflorescence of the art of illumination took place about the time when the culture of classical antiquity was beginning to decline," (Nordenfalk 1988: 7).

Both the Vatican Vergil and the Roman Vergil are written in rustic capitol script, the style used on most monument engravings, which has no separation between words and rarely any punctuation throughout the manuscript, adding to the burden of paleographers. But reading the text aloud eliminates the problem of running the words together; which illustrates an important fact about ancient Greek and Latin literature, it was always experience aurally. Even wealthy Romans fortunate enough to possess a copy of a literary work would enjoy the long-established custom of reciting or singing it aloud. The written literary culture of Greece and Rome inherited a wealth of oral traditions passed down through endless thousands of years, "since most of human history has existed with orality" (Lundwall 2015: 50). Before literary texts were abundantly illustrated, oral recitations of the poems to enlighten families and entertain friends were readily appreciated. But the addition of numerous miniatures to the poetic verses changes the focus of the interaction between reader and manuscript; in order to examine the paintings carefully and compare the scenes to the text requires that the reader have a more private, solitary relationship with the book (Fig. 2.13).

The Vatican Vergil was created during a transitional period in Roman history when the long-admired classical culture was being revived in an outmoded pagan society while at the same time the Christian Church was imposing organized efforts to suppress pagan religions. This innovative period introduced a new type of book with greatly expanded emphasis on descriptive illustrations designed to accompany

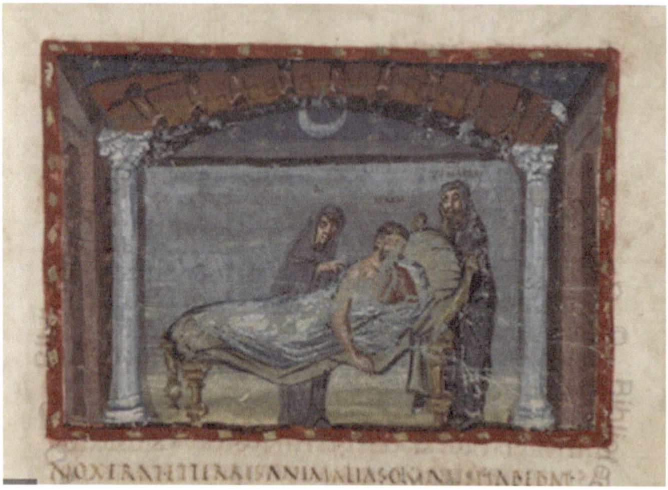

Fig. 2.13 Aeneas Abed, Vatican Vergil, Biblioteca Apostolica, Cod. Vat. lat. 3225, Fifth Century, Public Domain

Fig. 2.14 Shepherds, Roman Vergil, Biblioteca Apostolica, Cod. Vat. lat. 3867, Rome, Fifth Century, Public Domain

the poetic text. The Vatican Vergil contains 50 original miniatures on 76 folios, including 10 of the 12 books of the *Aeneid* and 2 of the 4 books of the *Georgics*. If the manuscript had held the complete collection of Virgil's poetry, the number of folios would increase to 440 and the number of illustrations to 280. The entire text was copied by the same scribe and the illustrations are placed within the text column, except for a few larger miniatures that cover the entire folio. The pictorial program includes dramatic scenes depicting Aeneas' adventures, detailed landscapes, and architectural details (Fig. 2.14). (A facsimile of the Vatican Vergil was produced in 1984, see D. Wright)

In comparison to the Vatican Vergil, the Roman Vergil is a huge codex with 410 folios that have been abundantly illustrated with large-scale miniatures.[45] The text of the Roman Vergil text is written in rustic capitals with no division between words and containing 18 lines on each folio; the manuscript has only 19 surviving illustrations; some but not all miniatures are framed. Analysis of the miniatures in the Roman Vergil depicts distinctive changes from the traditional classical paintings and the start of a new less-realistic depiction of narrative scenes. The human form becomes abstracted and flattened and the naturalistic representation of space is eluded. This unrealistic style develops to a greater degree a few centuries later in the medieval period especially in Hiberno-Saxon artworks where human figures take on the appearance of cardboard cut-outs.

As the Christian tenets take hold, emphasis was directed away from the physical world and the temptations of the body and instead centered on the inner world of prayer and thoughts of the heavenly rewards waiting in the afterlife. These societal conditions began after Emperor Constantine moved the Roman administration center to Byzantium beginning in 324 CE. Interactions with Eastern book production influenced artists in the West; the Virgil manuscripts are witnesses to the general climate and reflect current developments in the shifting governmental and religious ideals.

An additional extant fragmentary manuscript of Virgil's writings from the Late Classical Period, modest but important, is the *Vergilius Augusteus*, also dated to the crucial fourth century. This partial manuscript retains only seven leaves, four of which are in the Vatican Library, MS 3256, and the other three in the Staatsbibliothek zu Berlin, Lat. Fol. 416. These disconnected folios hold textual sections of Virgil's *Georgics*; the eighth leaf contained fragments of the *Aeneid* but is mostly lost. Unlike the rustic capital text of the other two Virgil manuscripts, this text is written in Roman square capitals. The insignificant number of poetic verses, 284 out of almost 13,000 in the original, reduces its influence in the textual transmission of the poem. The importance of the *Vergilius Augusteus* lies not in its text but in its decoration. These seven folios of Virgil's poetry are not illustrated with full scale miniatures, but instead are decorated with enlarged initial letters at the top of each folio. The decorated initials are not aligned to divisions of the text but appear at the start of the poetry verse that appears at the top of the folio. These 15 decorated initials are the earliest examples to survive and denote the start of letter development which eventually lead to the vast variety of innumerable decorated initials that appeared in illuminated manuscripts through the Middle Ages.

[45]Walker calls the Vatican Vergil the "most impressive example we have of Roman book illumination at the moment of transition from classical to medieval practices".

Produced during the transition from antique to Byzantine art, these early Virgil manuscripts reflect influences from Hellenistic and Roman artistic styles. Their influence on the artistic production of medieval manuscripts is impressive in many ways; in addition to the introduction of decorated initials and large framed illustrations, the first folio of the Aeneid text is the earliest known decorated '*incipit*' page.[46] The artistic approaches and iconography that appear in the Virgil manuscripts are a consideration in the study of *Aratea* manuscripts as they both provide evidence on the transmission of artistic motifs. Like the *Aratea* manuscript tradition, the Homeric and Virgilian poetic texts were studied and copied in enormous numbers during the Carolingian Era and strong interest revived again during the Italian Renaissance. Attesting the reception and continuing fascination held by these codices, all three Late Antique Virgil manuscripts have been reproduced in print and digital facsimiles.

Looking back to examples of Late Classical artwork provides evidence that illustrations in ninth-century codices were not created in a vacuum, but drew on the vast variety of traditional art communicated through scrolls, wall paintings, model books, and other artistic media. The astronomical picture cycles that appear in *Aratea* manuscripts were adapted for the codex format from archetypal images of the constellations already well-established in the Classical world. Mythological figures, such as Herakles, Perseus, Andromeda, and Castor and Pollux, had already become commonly recognizable through their specific iconography transmitted from ancient art, especially Greek vase paintings, and Roman relief carvings, statuary, frescos, mosaics, globes, metalwork, and coinage treasured by noble families.

Manuscripts of the *Aratea* which survive from the ninth century differ greatly from each other in their arrangement and style; therefore artists must have had a variety of choices from which to procure their mythological imagery. In addition to the fore-mentioned artistic sources, textual models carrying early astronomical imagery could most certainly be found in the works of other astronomical writers as well, in particular, Pliny the Elder, Hyginus, Martianus Cappella, and Macrobius. Their treatises survive in later copies that were definitely illustrated; and as in the case of the *Aratea*, it is not known for certain if they were illustrated originally, or at what point pictures first adorned these astronomical writings. That Christian patrons and scribes took advantage of the new book form is a fact, but one of the manuscripts chosen to be copied and illuminated by Christian patrons and clergy was an astronomical poem honoring the pagan gods and pagan philosophical concepts that conflicted greatly with accepted Christian dogma is rather curious. Recognizing this exceptional status gives further credence to the esteem with which the ancient writing of Aratus was held, as the poem from the classical culture was readily accepted into the Christian literary world. For more than 2000 years, the *Phaenomena* and its Latin translations have been continuously analyzed by experts

[46]The incipit is the first words of the script used as an identifying label; incipits were often written in different colors or in a different script from the general text.

and enjoyed by interested readers. The illustrations of the *Aratea* have been admired since at least the ninth century when created by the Carolingian artists.

This section of the inquiry has examined some of the many possible sources for the astronomical imagery found in *Aratea* manuscripts. In summary, common exemplars for astronomical imagery may have been: calendars (*Codex-Calendar of 354*), celestial globes (the Farnese Atlas), the cult of Mithras (the Tauroctony), zodiacal mosaics (Jewish synagogues), and Late Antique illustrated codices. In order to better understand the role of the text and images of the *Aratea* in medieval society, Chap. 3 will investigate the manuscript traditions of other principal authors whose 'scientific' texts were instrumental in transmitting 'popular' astronomical knowledge through the Middle Ages.

References

Aratus. *Phaenomena.* trans. G. R. Mair, (1969) Loeb Classic Library, Cambridge: Harvard University Press.

Dekker, E. (2013) *Illustrating the Phenomena: Celestial Cartography in Antiquity and the Middle Ages*, Oxford: Oxford University Press.

Dunbabin, K. (1999) *Mosaics of the Greek and Roman World*, Cambridge UK: Cambridge University Press.

Fine, S. (2005) *Art and Judaism in the Greco-Roman World*, NY: Cambridge Univ. Press.

Kühnel, B. (2000) "The Synagogue Floor Mosaic in Sepphoris: between Paganism and Christianity", *From Dura to Sepphoris: Studies in Jewish Art and Society in Late Antiquity*, eds. L. I. Levine and Z. Weiss, RI: pp. 31–43.

Levine, L. I. (1998) *Judaism and Hellenism in Antiquity: Conflict or Confluence?* Seattle: University of Washington Press.

Lundwall, J. K. (2015) *Mythos and Cosmos: Mind and Meaning in the Oral Age*, Pleasant Grove: Utah: C & L Press, Academic Publishing.

Mair, G. R., A.W. Mair (1969) *Hymns and Epigrams. Lycophron: Alexandra, Aratus: Phaenomena.* trans. G. R. Mair, Loeb Classic Library, Cambridge: Harvard University Press.

McCluskey, S. C. (1998) Astronomies and Cultures in Early Medieval Europe, Cambridge: Cambridge University Press.

Ness, L. J. (1990) *Astrology and Judaism in Late Antiquity*, Diss. University of Miami, Ohio Online at: www.smoe.org.

Negev, A. (1990) *The Archaeological Encyclopedia of the Holy Land*, Third Edition, New York: Prentice Hall.

Nordenfalk, C. (1988) *Early Medieval Book Illumination*, New York: Rizzoli International Publications, Inc.

Pächt, O. (1984) *Book Illumination in the Middle Ages: An Introduction*, London: Harvey Miller Publishers.

Parrish, D. (1984) *Season Mosaics of Roman North Africa*, Rome: G. Bretschneider Editore.

Salzman, M. R. (1990) *On Roman Time: The Codex-Calendar of 354 and the Rhythms of Urban Life in Late Antiquity*, California: University of California Press.

Schaefer, B. E. (2005) "The Epoch of the Constellations on the Farnese Atlas and their Origin in Hipparchus's Lost Catalogue', Journal for the History of Astronomy, v35 n2k, pp. 167-196.

Thiele, G. (1898) *Antike Himmelsbilder*, Berlin: trans. Hugh Thurston

Thomas, R. (1992) Literacy and Orality in Ancient Greece, Cambridge: Cambridge Universitiy Press.

Ulansey. (1989) The Origins of the Mithraid Mysteries: Cosmology and Salvation in the ANcient World, NY: Oxford University Press.

Weiss, Z. and E. Netzer. (1998) *Promise and Redemption: A Synagogue Mosaic from Sepphoris*, Jerusalem; The Israel Museum.

Weitzman, K. (1970) *Illustrations in Roll and Codex: a Study of the Origin and Method of Text Illustration*, Princeton: Princeton University Press.

Weitzman, K. (1971) *Studies in Classical and Byzantine Manuscript Illumination*. Chicago: University of Chicago Press.

Weitzman, K. (1977) *Late Antique and Early Christian Book Illumination*, NY: George Braziller, Inc.

Weitzmann, K. and Kessler, H. (1990) *The Frescoes of the Dura Synagogue and Christian Art*, Dumbarton Oaks Research Library and Collection.

Wilkinson, J. (2002) *From Synagogue to Church: the Traditional Design*, UK: Routledge Curzon.

Wright, D. (1993) *The Vatican Vergil: a Masterpiece of Late Antique Art*, University of CA Press.

Wright, D. (2001) *The Roman Vergil and the Origins of Medieval Book Designs*, University of Toronto Press.

Wright, D. H. (1984) *Vergilius Vaticanus: vollständige Faksimile-Ausgabe im Original format des Codex Vaticanus Latinus 3225 der Biblioteca Apostolica Vaticana*. Graz, Austria: Akademische Druck- u. Verlagsanstalt.

Young, P. H. (2003) *The Printed Homer: A 3,000 Year Publishing and Translation History of the Illiad and the Odyssey*, North Carolina: McFarland & Company Inc.

Chapter 3
Astronomical Knowledge Transmission Through Illustrated *Aratea* Manuscript

I know that I am mortal by nature and ephemeral, but when I trace at my pleasure the windings to and fro of the heavenly bodies, I no longer touch earth with my feet. I stand in the presence of Zeus himself and take my fill of ambrosia

Ptolemy, Almagest.

Abstract This chapter presents an overview of the principal authors and manuscript traditions well known for transmitting much of the existing astronomical information through the Middle Ages, and provides an account of surviving numbers of their medieval manuscripts. These astronomical texts were comprised of basic descriptions of astronomical principles, celestial motions, and the relative positions of the constellations; many of those produced after the twelfth century included astrological information as well. Most of these astronomical manuscript traditions included a complete constellation cycle including illustrations, number of stars and mythological information; after the writings of Bede, computus and particular astronomical calculations became important for Christian calendars and liturgical events. Most of the advanced mathematical astronomy achieved through the efforts of early Greek and Babylonian astronomers was unknown in the Latin West except for brief references by other authors; that knowledge was thought to be lost forever. Authors and encyclopedists toiled to preserve the few basic astronomical principles they possessed. Astrology had been practiced regularly during the flourishing centuries of the Roman Empire and was still available in later years, but passed into disfavor as Christianity gradually spread throughout the Western countries until the twelfth century. After the Arabic to Latin translation period, astrology returned in a new guise and often became intimately intertwined with astronomy.

In the early Middle Ages, astronomical knowledge was transmitted throughout the Latin West by many sources other than the Aratean poetry of Cicero, Germanicus and Avienus; most of these other widely read astronomical treatises were illustrated with cosmological diagrams and many contained the full cycle of constellation pictures. This section investigates the writings of these key authors individually in order to distinguish their various works in style and content from the *Aratea*, discussing the

M. Dolan, *Astronomical Knowledge Transmission Through Illustrated Aratea Manuscripts*, Historical & Cultural Astronomy,
DOI 10.1007/978-3-319-56784-6_3

other well-known astronomical traditions, in order to determine which authors are likely to include the constellation cycle and which manuscripts were not usually illustrated. The most influential sources of astronomical information that will be discussed are: Pliny the Elder, Hyginus, Martianus Capella, Macrobius, Isidore of Seville and Bede. Many ninth-century manuscripts of the *Aratea* are smaller in dimensions and in volume than these astronomical writings, limited to the short Latin poem and the cycle of 44 or so constellation illuminations. In contrast, the Aratean poems from later centuries of the medieval period were often expanded with addition writings and scholia and were frequently included in miscellanies, anthologies and 'scientific' compilations; in most instances, the Latin *Aratea* poem was bound in the same manuscript with astronomical writings of these other popular authors.

For comparison, this section presents a brief overview of the type of astronomical knowledge included in later illuminated manuscripts, such as the twelfth-century encyclopedic compilation of Lambert of St-Omer, *Liber Floridus*, considering the illustrations of the constellation cycle in particular, and how the text relates to the illustrations. By explaining and comparing the text attributed to each author, and defining the picture cycle that accompanies each, a better understanding of the various manuscript traditions can be reached.

Authors of Astronomical Texts Accompanied with Constellation Illustrations from the First to the Twelfth Century

When investigating the transmission of astronomical knowledge, art historians and historians of science are limited unfortunately to those texts that have survived; so much has perished that accuracy is questionable, the lack of most early Greek and Hellenistic astronomical texts, known only from early references, seems regrettable. Thanks to Carolingian copyists, the *Aratea* manuscripts survive, mostly intact. Even though easily read today, the early Greek and Babylonian texts that do survive are relatively silent as to how they were understood; how they were used in teaching; how they were disseminated; and what role the long tradition of oral transmission played in passing on scientific knowledge. We do know that the Greeks created a satisfactory picture of the physical world and explained the phenomena of earth and sky through mathematical and philosophical principles. By the time of Eudoxus, Greek science had separated itself from creation myths and early legends and had moved on to a more rational basis (North 2008: 68). In earlier times, much more information had to be committed to memory, since learners lacked the vast amount of resources we are fortunate to have readily available today. Although ancient Greece was considered a literate society, it was in many ways an oral society as well; poetry was performed and sung before an audience, not read in silence by one lone person. "Far more was heard or spoken, rather than written and read" (Thomas 1992: 3). In this study, orality and the art of memory appear as a small but not inconsequential component that must be considered in the transmission of technical knowledge, as astronomical features, especially the signs of the zodiac were often used as a mnemonic tool.[1] Informative illustrations too act as an instrumental aid in the retention of accurate details, and beyond their aesthetic value, may perform an invaluable mnemonic function as well.

Astronomical Writings of Claudius Ptolemy

Without question the most important astronomer until the time of Copernicus, Brahe, Galileo, and Kepler, was Claudius Ptolemy of Alexandria (fl. 2nd century CE), mainly through his *Almagest*, which has been firmly dated to the reign of Emperor Antoninus Pius (r. 137–161 CE), 'one of the five good emperors'. The last stellar observation recorded in this book is February 2, 141 (Toomey 1998: 1). His work in 13 books presents a total description of mathematical astronomy, in which he develops the necessary trigonometrical theory. During the period up to and including the time of Kepler in the seventeenth century, virtually all accurate astronomical texts, written in Arabic, Hebrew, Byzantine Greek, or Latin, were strongly influenced by the *Almagest*, which was the fundamental key to understanding other authors. Contemporary commentaries show that in antiquity the *Almagest* was the standard textbook for advanced students in Alexandria and probably at Athens and Antioch also (Toomey 1998: 2). But Ptolemy's writings actually did not become a direct factor in the transmission of scientific knowledge in the Latin West until the late-twelfth century. Although revered among scientists as an exceptional Hellenistic scholar of antiquity during most of the Middle Ages, he was little more than an ancient name, sometimes even mistaken for a Ptolemaic king of Egypt. Few details are known concerning the life of Ptolemy other than he lived and worked in Alexandria, the cultural and scientific epicenter of his age, and he was appointed the head of the great library there.

Medieval astronomers were cognizant of Ptolemy's great work from commentaries and *scholia* from other scientific authors, his actual treatises had been lost to the West from the early medieval period. But by the year 800, Ptolemy's scientific writings were translated from the original Greek into Arabic and became available to Persian and other Arabic-speaking astronomers. Ptolemy himself built his theories and understanding of the cosmos on a collection of new and ancient techniques, integrating, discarding, and modifying astronomical data to create an intelligible whole. After the documents had been translated, Arabic astronomers began working with these early writings and further refined, corrected, and improved upon Ptolemy's mathematical and astronomical systems. During the

[1]See Francis Yates, *The Art of Memory*; Mary Carruthers, *The Book of Memory*; and Carruthers and Ziolkowski, *The Medieval Craft of Memory*. This research on the art of memory and the mnemonic uses of the constellations will be discussed in Chap. 8.

twelfth century, Ptolemy's astronomical texts were recovered from Arabic libraries chiefly in Spain and Sicily and translated into Latin by teams of Christian, Islamic, and Jewish scholars.

Ptolemy borrowed an older technique that included an epicycle on a deferent-carrying circle to which he added the equant. His work was technically challenging and still is; his theories were spread mainly by commentaries and less difficult renditions by adept scholars, especially Sacrobosco in the thirteenth century. Ptolemy reconciled observations of heavenly movements with Aristotle's principle of uniform, circular motion by applying certain mechanisms: the eccentric, the epicycle, and the deferent. He not only transmitted the knowledge accumulated by the ancient Greek scientists, he organized and developed the received material into geometrical models, establishing mathematically based systems from which positions of the sun, moon, and planets could be accurately defined and predicted. Ptolemy described the stars in 48 constellation figures and listed 1022 'fixed stars' and their magnitudes.[2] For each star group, he gave its relative location, brightness, and color; its longitude; its latitude; and direction. The constellations were described from the view of an observer on earth, looking at the inside of a globe and facing him (Toomey 1998:15). Ptolemy put together past and current theories and his own observations in his treatise which he called the *Mathematical Compilation.* Like Aratus' *Phaenomena*, Ptolemy's book became instantly popular in Alexandrian circles attracting many commentaries. The title of his masterful treatise evolved to *The Megiste Syntaxis* (*The Greatest Compilation*), which became transliterated in the Arab world as *al-Majisti* and eventually became known in the West as the *Almagest*. After the *Almagest* had been translated from Arabic into Latin in the twelfth century, it became a standard astronomical textbook throughout the later Middle Ages, but in a less technical version.

Ptolemy's second major composition, written after 147 CE, is his four-book treatise on the art of astrology *Quadrapartitum,* or *Tetrabiblos*, which later astronomers had apprehensions toward attributing to him, not wanting to accept the fact that the exceptional astronomer was also an active astrologer. Ptolemy considered astronomy and astrology as a complementary pair. Book One of *Tetrabiblos* describes the technical aspects of astrology, Book Two discusses the influences of the heavens on the earth in general, and Books Three and Four explain the astrological theories and how they influence human lives and events. *Tetrabiblos* provides details on the variety of powers thought to emanate from the heavenly bodies and how these influences affect the earth and even more so, its human inhabitants. One aspect of Ptolemy's astrological theory was the idea that humans living in different areas and climates have different temperaments due to the distinct planetary influences and that people move through seven stages of life as influenced by the sun, moon, and five planets (Whitney 2004: 171).

[2]Hipparchus always defined the individual stars as a bright star or half-bright star. Ptolemy was the first to organize the stars into six magnitudes—first level as brightest and then lessening.

In his scientific writings, Ptolemy was not interested in the poetry and myths of the constellations of the heavens, such as appeared in the *Phaenomena* of Aratus and the Latin Arateans, but instead was entirely concerned with the harmonious movements of the stars, the complicated spheres and epicycles of the planets, theoretical laws, and mathematical aspects of astronomy, introducing what is now called spherical trigonometry. Ptolemy's Greek texts were not known to be illustrated since his writings were concerned with relative motions and descriptions of the stars' intrinsic qualities, not the descriptions of Greek gods, heroes, and legends. His treatises were often accompanied with complex geometrical diagrams that helped explain his cosmological concepts; but these are quite different from the colorful and easily understood diagrammatic *rotae* attributed to Macrobius or Isidore of Seville. Most of their explanatory diagrams were derived ultimately from the writings of Pliny the Elder not from Ptolemy. The type of astronomical knowledge contained in most early illustrated manuscripts is considered basic 'cultural' or 'popular' astronomy rather than the sophisticated technical and mathematical astronomy of Ptolemy.

Although most of Ptolemy's works were received from Arabic sources, a medieval translation of Ptolemy's writings directly from Greek into Latin was made by William of Moerbeke (1215/35–c.1286), a Flemish Dominican priest, classical scholar translated philosophical, medical, scientific, and astronomical texts. His version by-passed the Latin translations made through the Arabic texts. It is thought that Thomas Aquinas asked Moerbeke to undertake the work to determine whether the Arabic sources had distorted or misunderstood the original Greek. Moerbeke's translations were a major scholarly achievement and quite influential during the late Middle Ages, and are still of substantial value to current scholars. For example, his Latin translation of the *Tetrabiblos* of Ptolemy has recently been translated into English for the first time.[3]

Manuscripts of Pliny the Elder

Most astronomical information that was available in the early Middle Ages was derived from Roman rather than Greek sources. Although not as well known for his astronomical writings as for his works on architecture, Vitruvius (1st century BCE) discusses the topic in his *Ten Books on Architecture*, where he wrote about the constellations, planets, phases of the moon and astrology. A much more important Roman source for astronomy was Pliny the Elder (CE 23/79) in his *Naturalis Historia*, but the number of medieval scholars who knew Pliny writings firsthand

[3]See *Ptolemy's Tetrabiblos in the Translation of William of Moerbeke*: Claudii Ptolemaei Liber Iudicialium by Gudrun Vullemin-Diem; Carlos F. Steel; and Pieter de Leemans, which provides an edition of Moebeke's Latin version which is now in the Digby collection, Bodleian Library. It was previously owned by Thomas Allen, the famous mathematician and astrologer to Robert Dudley first earl of Leicester.

was very small (Laistner 1941: 37).[4] Pliny's *Natural History* itself is not an original creation, as most of his material was gathered from earlier writers; yet his work contains a myriad of facts (Lindberg 1992: 141).[5] His *Natural History* is quite broad in scope, an encyclopedia in 37 books; the astronomical material was not treated as a separate subject, but discussed in different sections throughout the book.

In the first part of Book Two, Pliny discusses cosmology and the motions of the sun, moon, planets and stars in addition to some more technical aspects, such as the ecliptic, eclipses, and the sphericity of the earth (McCluskey 1998: 16). Pliny also distinguishes between stars and planets and supplies the periods of their orbits and estimates of their distances from earth. His writings are Stoic and influenced by astrological theories; his cosmic system relies on forces, maintaining order along with the basic patterns of celestial events (Eastwood 1986: 197). In Book Six, Pliny discusses geographical issues, the size and shape of the earth, the concept of latitude and the fluctuating length of daylight. Most of Book 18 is devoted to a general discussion of time, the calendar, an explanation of practical astronomy, and related meteorological considerations (McCluskey 1993: 162).[6] In a macro/microcosmic scheme, Pliny associates the motions of the heavens in relationship to human needs, following the course set by Hesiod, Aratus and Virgil, but he provides more specific scientific details.[7]

Research has shown that the *Naturalis Historia* of Pliny was known and consulted by medieval scholars such as Alcuin, Helperic, and Lupus of Ferrières at the premier centers of learning at Aachen and Auxerre. Quotes from Pliny are cited in letters that survive between Charlemagne and Alcuin, and his work is mentioned again when listed in the contents of the library of Louis the Pious (McCluskey 1993: 163). These facts, plus the large number of surviving manuscripts and diagrams attributed to Pliny, demonstrate that he was one of the strongest influences on

[4]Laistner claims that Pliny's writings were quoted and commented upon secondhand by numerous other authors.

[5]In the preface to *Natural History*, Pliny tells that he and his assistants searched 2000 volumes by 100 different authors and drew 20,000 facts from them. Through his extensive work, he then produced a 'storehouse of interesting and entertaining information'.

[6]McCluskey states that the astronomy related by Pliny is 'stellar'. Although Pliny knew that features of astronomy varied from place to place because of the sphericity of the earth, he refrains from introducing mathematical concepts.

[7]Large sections of Pliny survive from the eighth and ninth centuries in six different manuscripts containing diagrams (Eastwood 1993:162). Only one manuscript contains the complete text of Pliny (Paris BNF lat. 6795); it definitely dates to the ninth century. A manuscript dating to the first half of the ninth century, possibly from Lorsch, is held by the Morgan Library (MS. M.871); it contains about one-half of Pliny's *Naturalis Historia*, ending before Book 18. An eighth-century manuscript of Pliny at Leiden University Library (UB MS. Voss.F.4) survives in part, beginning after the astronomical section of Book Two. A fourth manuscript of Pliny's encyclopedic work, made in northern France around 800, is incomplete and has been divided between two different collections (Vat. Lat. 3861 and Leiden Voss. F.61). Another Leiden manuscript (UB MS. Voss. Lat. Q.69) preserves the fifth Pliny segment, but only in excerpts, about seventy from the early books, including the important Book Two. The sixth extant Pliny manuscript was produced in Corbie and dates from the early ninth century (Paris lat. 6796).

Carolingian planetary theory. Pliny's writings were still being read and copied in the twelfth century, for excerpts of Pliny's astronomy and one of his diagrams appear in the great encyclopedic compilation, *Hortus Deliciarum* of Herrad of Hohenbourg (Eastwood 1990).[8]

Manuscripts of Marcus Manilius

Another Roman poet who owes much to the authorship of Aratus and the Arateans is Marcus Manilius (f. 1st c. CE),[9] author of *Astronomica*, a didactic astrological poem composed between CE 9–15, during the final years of Augustus' rule or the early reign of Tiberius. From references in his poem, it is fairly certain that Manilius survived Augustus at least by a few years (Garrod 1911: lxi). Although a contemporary of Tiberius and Germanicus, the exact dates of Manilius' birth and death are unknown, in fact almost nothing is known about his life beyond his extant writings and that he was associated with the Roman imperial circle (Garrod 1911: lxi).[10] His fame was long-lasting for Albrecht Durer included an image of him, along with Aratus, Ptolemy and al-Sufi, on his map of the heavens (Figs. 3.1 and 3.2).

What survives of *Astronomica,* a massive Stoic poem written in hexameter verse, is contained in five books, with 800–900 verses in each (Tester 1987: 30).[11] Today it may seem strange that Manilius would compose an instructional book of astronomical and astrological theory in poetic form; but poetry was the customary format of most philosophical treatises at that time. Manilius lived in the same cultural milieu as Cicero and Germanicus and other important authors who wrote not for commoners, but for the intellectuals and educated elite of Roman literary society, where verse was the accepted didactic medium (Tester 1987: 30).[12] Manilius and other authors of technical material were admired for their ability to put such complicated, conceptual discourse into clever verse, which was what ultimately attracted so many readers (Tester 1987: 30). Text scholars have long commented on the close connections of Manilius with Aratus and noticed the similarities between the structure of the two Stoic poems. Some believe Manilius

[8]See "The Diagram *Sphera Celestis* in the *Hortus deliciarum*: A Confused Amalgam from the Astronomies of Pliny and Martianus Capella." The diagram appears on p. 93, plate 5, of Rosalie Green et al., *Hortus Deliciarum/Herrad of Hohenbourg,* Warburg Institute, 1979. See also Green and Volk, (2011) Forgotten Stars: Rediscovering Manilius' Astronomica.

[9]Manilius' image is taken from Dürer's celestial map of 1515.

[10]Manilius seems to have a high regard for the sign of Gemini, "he tells that under Gemini are *born poets and astrologers.*" Perhaps he himself was a Gemini.

[11]Even his name has been questioned; prompting discussion and debate among scholars, but the name of Marcus Manilius is accepted by most historians today.

[12]Another classic example of didactic poetry from an earlier date is the *De rerum natura* of Lucretius, which Manilius often imitated.

Fig. 3.1 Celestial Map, Albrecht Durer, drawing, 1515 Manilius bottom left, Aratus upper left, Ptolemy upper right, al-Sufi lower right. Public Domain

had Aratus in mind when he made significant changes in his poetry.[13] In spite of his literary achievements, Manilius was not quoted by his contemporaries though his writings were surely read. Due to the astronomical advancements in the eighteenth and nineteenth century, astrology was reduced to a pseudoscience, an embarrassment to the scientific community; as a result, Manilius' astrological poem has been mostly ignored by contemporary scholars also.

Astronomica contains five distinct poetic books; each one holds eight to nine hundred verses and contains the earliest text surviving on astrology. Book I presents an overview of the constellations; Book II explains the signs of the zodiac,

[13]A comparison study of particular segments of the poems, the Proem and the Dike catasterism was published in 2007 by Josephe-Henriette Abry, *Manilius, and Aratus: Two Stoic Poets*, Leeds International Classical Studies 6.01 available online.

Fig. 3.2 Celestial Map Detail, Aratus, Albrecht Durer, drawing, 1515, Public Domain

instructions on how to calculate a horoscope, and the astrological interpretation of their relative positions. Book III expounds upon the "houses", the areas of life governed by the 12 sections of the heavens; this chapter also covers the ascendant and the crucial question of life expectancy. Book IV describes the qualities of specific degrees of the 360° circle of the zodiac; lastly is Book V which discusses the "paranatellonta", the constellations which lie just beyond the zodiacal signs but according to astrological beliefs are still able to cast influences on human lives.

Most astronomical material in Book I of *Astronomica* was derived from the informative Latin verses of the *Aratea*, since they were thought to be the only astronomical text that had been translated into Latin and available at that time.[14] Manilius' astrological didactic poem paraphrases the *Phaenomena* 19-558 as he describes the constellations and structure of the universe (Gee 2013: 5). Book I introduces elementary astronomical principles: the *sphaera*, the basic description of the heavens, including the circles of the tropics and equator, the horizon, meridian and the Milky Way. The verses also praise the order and reliability of the

[14]As mentioned in Chap. 1, many Roman authors had translated the Greek poem of Aratus into Latin. Ptolemy did not write his authoritative astronomical treatises until the second century CE, which remained the 'bible' of astronomy until early modern times.

movements of the heavenly bodies and conclude with a lengthy discourse on comets, particularly their significance as omens of future events, furnishing specific examples from the past. The "popular" astronomy in Book I is relatively easy to follow, students of Latin poetry could understand the basic theories he introduced.[15] For instance, Manilius wrote in 12 lines of verse a description of the 12 signs of the zodiac, perhaps to be memorized as a mnemonic exercise.

> First Aries shining in his golden fleece
> Wonders to see the back of Taurus rise,
> Taurus who call, with lowered head, the Twins,
> Whom Cancer follows; Leo follows him,
> Then Virgo; Libra next, day equalling night,
> Draws on the Scorpion with its blazing star,
> Whose tail the Half-horse aims at with his bow,
> Ever about to loose his arrow swift.
> Then comes the narrow curve of Capricorn,
> And after him Aquarius pours from his urn
> Waters the following Fishes greedily use,
> Which Aries touches, last of all the signs (263–274) (Tester 1987: 31).[16]

Although it is thought that Manilius drew his astronomical material from one of the many *Aratea* versions available in Roman intellectual circles, all ultimately derived from the *Phaenomena* (Garrod 1911: 3).[17] Manilius' astrological material had to be adapted from other sources, most probably Hellenistic and Egyptian (Tester 1987: 31). That is certain because Aratus' poem contains no astrology, since the 'science of astrology' was still in its formative stage when Aratus was writing; but the *Weather Signs* at the end of the poem brim with accounts of animals' behavior and natural phenomena that act as signs and predictors of forthcoming natural events. When Manilius was composing his astrological verses during the early first century CE, there were numerous astrological treatises being written and already in circulation in both Greek and Latin that he may have consulted, such as Dorotheus of Sidon and Vettius Valens. One of the most valued and sophisticated sources of early astrology were produced in Ptolemaic Egypt; the writings purportedly by Nechepso and Petosiris, an Egyptian pharaoh and his priest, included a textbook of horoscopic astrology attributed to Hermes (Tester 1987: 20). In addition to borrowing the astrology of Hellenistic experts and the astronomy of Aratus, Manilius adopted the Stoic philosophy presented by Aratus when he recognizes the 'all-controlling Reason' for keeping the heavens in motion and the constellations following their appointed courses.

[15]See *M. Manilii Astronomica*, Ed. and trans. G. P. Gould.

[16]"…the position of the zodiac itself is not described until 400 lines later," Tester's translation.

[17]Though not mentioning him by name, Manilius refers to Aratus when he says, "there are certain of the poets who have described to us the diverse configuration of the stars, and have assigned to their proper class and cause all the constellations that glide at large in the spreading heaven." II 22–26.

The longest and most complicated segment of the five-part *Astronomica* is Book Two, for the astronomical material becomes much more difficult and somewhat daunting for the average reader. Manilius' versification of the scientific terminology concerning triangles, quadrangles, hexagons, and dodecatemories formed by the signs of the zodiac becomes incomprehensible, 'a poetic nightmare' (Garrod 1911: vi).[18] His didactic poetic verses that have survived mostly intact are among the earliest sources of Hellenistic astrological theory. But critics imply that perhaps he had more interest in creating an extraordinary work of art than in faithfully reporting the techniques of astrological practices.

For example unlike most astrological logicians, Manilius does not include the significance or symbolism of the 'seven' wandering planets. This seems quite unusual since the transits of the planets through the zodiacal signs became one of the key components of astrological theory. Perhaps, he was following the lead of Aratus in this respect. Setting a precedent for this aversion, Aratus too evaded a long discourse on the motions of the planets, but he blamed that omission on the erratic, unpredictable behavior of the planets in respect to their celestial orbits. Like Pliny, Manilius used astrology to link the heavenly and divine macrocosm with the earthly and human microcosm, affirming the ancient Hermetic proverb, "As above, so below."

By the first century CE, astrology had taken a strong hold on Roman society; religion, literature, and popular thought were thoroughly infiltrated with the prevailing and accepted astrological doctrines (Garrod 1911: lxvii).[19] Manilius was more philosophical in his attitudes toward astrology, not promoting these esoteric principles for special benefits or advanced knowledge. The principles of the astrological doctrine advocated by Manilius were not based on the superstition or personal advancement, but on the oneness of a divine intelligence that pervades all the phenomena of nature. "God hath descended into man, and in man's seeking of him seeks himself." Manilius II, (Garrod 1922: lxxi) Manilius' Stoic philosophy states, or more correctly sings, that Fate is immutable and astrology exists as a means of understanding the natural order, but not for predicting or changing future events. The Stoic tenet that each person is born with a certain fate became incorporated into the astrological view, where the planets became the agents of the irrevocable fate in human destiny. Like Aratus and the Arateans, Manilius expresses his philosophical ideas in compelling poetic form.

---

[18]"...we have yet to learn how to deal with a poet who versifies the multiplication tables and sets to the measure of Vergil the reasoning of Euclid." Book II also discusses the 12 signs of the zodiac defining their various relationships; the hostility of oppositions, aspects of signs that can be squares, triangles, and conjunctions. At the start of Book II, Manilius invokes the memory of Homer and Hesiod, speaking of the historical, mythological, and astronomical references found in their poetry.

[19]Garrod writes that when reading the work of an astrologer, "we looked for a quack or an imbecile," but "the most striking impressions that we receive from the reading of the *Astronomica* are undoubtedly those of a rare purity of mind in its author and a singular freedom from superstition. ...Manilius is the peer of Lucretius."

For I will sing of the silent strong workings of nature,
and of god interfused in earth and sky and sea
controlling in impartial covenant the mighty mass of things.
I will sing of how the whole universe is alive with harmony
upon this side and that, and is moved by the motions of Reason,
one spirit indwelling in all its parts,
shedding its dews upon the round world,
swiftly speeding through all things,
and fashioning the body of things to breathing life (Garrod 1911: lxxi).

Throughout his lengthy book, Manilius refers to additional astrological material that cannot presently be found; therefore specialists assume that large portions of his writings have been lost or destroyed. While his poetic account of astrology holds much technical data, the verses show little evidence that Manilius was actually an active astrologer himself. Although the astrological concept of 'houses' or *templa*, connecting a dominant aspect of human life—family, health, career, marriage—with each sign of the zodiac, had developed over centuries, the term makes its first appearance in *Astronomica*. (The method of assigning zodiacal signs and planets to various "houses" had been well established by Hellenistic astrologers in Ptolemy's time, but he almost completely ignored them in his *Tetrabiblos*). In spite of its vast amount of astrological material, one could not use Manilius' poem as a textbook or learn to cast a horoscopic chart by reading it, for his objective was not didactic but artistic. The lengthy astronomical/astrological (the two disciplines were considered as one and the same by many until the seventeenth century) poem was copied over and over throughout the medieval period, admired for its richness and complexity and thankfully it survives almost intact.

The poetic text of Manilius depends primarily on just three medieval manuscripts (Reeve 1983: 235),[20] even though 22 manuscripts of *Astronomica* are still extant (Garrod 1911: xv). These 22 manuscripts have been grouped into three families: the Belgian (six), Italian (six), and Hungarian (ten). The three most important survivals are two from the Belgian family and one Italian manuscript from which the other five Italian copies derive; all six Italian copies date to the fifteenth century. The earliest *Astronomica*, Brussels, MS.10012, was produced in the eleventh century; not surprisingly, all but four of the surviving manuscripts date to the fifteenth century when interest in astrology expanded significantly in the intellectual centers of Europe. None of the surviving manuscripts of Manilius from the fifteenth century are included in the Blume et al. catalog *Sternbilder des Mittelalters*. From early references, scholars also know of the existence of four lost manuscripts of *Astronomica*.

[20]In "Manilius", published in *Texts and Transmissions*, his more recent text analysis, Reeve states that the three primary *Astronomia* manuscripts are Leipzig 1465, Brussels 10012, and Madrid 3678. He recommends the Garrod edition and the excellent Loeb edition of G. P. Gould, 1977.

Text scholars do not mention whether any of these manuscripts are illustrated with pictures or useful diagrams. *Astronomica* concentrates only on the 12 constellations of the zodiac; the numerous additional constellations that fill the night sky are not included in astrological systems; thus they are not mentioned. Therefore, the inclusion of a pictorial cycle in *Astronomica*, often found in other astronomical treatises would not be expected. Although not contained in the astrological poem of Manilius, outstanding illustrations of the constellations are found in astronomical manuscripts of another Roman author named Hyginus.

Illustrated Manuscripts of Gaius Julius Hyginus

Another contemporary of Germanicus and the other first-century poets is the writer of astronomical mythology, Gaius Julius Hyginus (64 BCE—CE17), a native of Spain or perhaps Alexandria, and a freedman of Emperor Augustus, by whom he was appointed superintendent of the Palatine library (Suetonius, *De grammaticism*: 20). It has not been firmly established that the man Suetonius describes is actually Hyginus the author. Scholars debate whether the learned man who oversaw the Palatine Library was the same person who wrote the *Fabulae* and *De astronomia*; they also disagree on whether the two texts were actually written by the same author. Hyginus was known as a voluminous writer; his works included topographical and biological treatises, on such varied topics as bee-keeping and commentaries on Helvius Cinna, a poet of the late Roman Republic, and on the poems of Virgil; all of which have been lost. The avid interest within Roman society on the topics of apotheosis, astronomy, astrology, and celestial imagery may have inspired Hyginus to create his astronomical handbook. Hyginus most probably gathered material for his handbook during the time between the translations of Cicero and Germanicus.

In addition to *Poetica astronomica* more commonly known as *De astronomia*, Hyginus produced another surviving prose composition named *Fabulae*, a huge compilation of Greek and Hellenistic mythology and astronomical lore. The *Fabulae* includes about 300 very brief and simply written mythological legends and celestial genealogies.[21] Presented in an unembellished format, the *Fabulae* myths taught the fundamental details that were required knowledge that every educated Roman in the early Empire needed to know concerning Greek mythology. Although the writings of Hyginus have been criticized for not remaining close to the original versions and for including Roman-influenced additions, more recent interpretations modify that claim for "myths constantly change in order to maintain their function within a culture, so must the works that collect these myths change"

[21]For an overview of the text, see M. D. Reeve, "Hyginus" in *Text and Transmission: A Survey of Latin Classics*.

(Fletcher 2013: 134).[22] Translating prose or poetry to another language requires the process of making material from one language and culture significant in another. Translation often allows for and determines a work's survival (Fletcher 2013: 164).

The *De astronomia* of Hyginus comprises an elementary treatise on astronomy with some technical data, but the work is not original, as most of its mythological content had simply been translated from the *Catasterisms* of Ps-Eratosthenes. His prose writing has been criticized for its numerous errors, but in spite of that, his astronomical work became a highly popular and influential sourcebook from the time it was written. The composition is divided into four sections plus an introduction; Book I includes the general definitions of the sphere, the zodiac and the earth. Book II deals with folklore of the stars, a compilation of catasteristic myths; in Book III, Hyginus names and illustrates 42 constellations, indicating the number of stars in each group. Book IV describes the circles of the sphere, day and night, the signs of the zodiac and the planets.[23] Although Hyginus' descriptions of the constellations are more extensive than the Ps-Eratosthenes, they are closely linked; Hyginus mentions Eratosthenes twice in Book III but 21 times throughout the entire work. The lists of the stars in *De astronomia* deviate from Eratosthenes and from the Aratean sequence but follow in the same order as in Ptolemy's *Almagest* which supports the idea of a second-century compilation suggested by some myth experts, more evidence for the misidentification of the author. Hyginus' allocation of the constellation positions indicates that, in addition to his main literary sources, he could have also used a celestial globe for assistance in his writing (Fig. 3.3).

Since portions of the astronomical handbook of Hyginus are often found bound with, integrated with, and sometimes mistaken for, *Aratea* manuscripts, its illustrations are integral to this discussion. In fact the text of Hyginus' *De astronomia* can be easily recognized since his descriptions include the locations and number of stars in each constellation, which are not mentioned in the *Aratea*. Scholia on Cicero's version of the *Phaenomena* derive from a text of Hyginus (Reeve 1983: 189).

Just as Aratean poems attained great popularity in the ninth century, Hyginus' prose writings appealed to the courts of the Carolingian rulers since the earliest surviving manuscripts of *De astronomia* date to that pivotal period. At least 11 copies of Hyginus' astronomical handbook survive from the ninth century. Like the *Aratea*, numerous manuscripts of *De astronomia* also survive from the fifteenth century. Hyginus became a favorite with humanists and booklovers, possibly because his work made the pure classical tradition of astronomical knowledge accessible without reference to Arabic or medieval intermediaries.

By the fifteenth century, astrology had reached the apex of its intellectual position and influence, attracting many noble and wealthy patrons who sought to

---

[22]Fletcher, *Hyginus' Fabulae: Toward a Roman Mythology.*

[23]For a detailed discussion of Hyginus' text and examples of manuscript illustrations see Lippincott, The Saxl Project—an ongoing project to be completed in stages which plans to provide complete digital access to over 250 medieval and Renaissance illustrated astronomical and astrological manuscripts.

Fig. 3.3 Perseus woodcut, Hyginus *de Astronomia*, printed book, Hungary, Sixteenth Century, Public domain

commission books that featured elegant illuminations with cosmological themes. Appealing to their refined tastes, the readers residing within wealthy Renaissance courts also enjoyed the work of Hyginus for its classical allusions, attractive illustrations, and mythic tales of ancient Greece combined with contemporary astronomical 'science' (Saxl and Meier 1953, vol. III, I, liii).[24] The first printed and

[24]See also Wormald and Giles, *A Descriptive Catalogue of the Additions in the Fitzwilliam Museum*, Cambridge, 1982, pp.192–94. At least two copies of *De astronomia* were associated with the Gonzaga court in Mantua: the Fitzwilliam Museum MS 260, c. 1475 and the formerly Dyson Perrins MS 92.

illustrated copy of *De astronomia* was published in 1482 by Erhard Ratdolt in Venice and titled *Clarissimi uiri Hyginii Poeticon astronomicon opus utilissimum.*

New York Public Library, Spencer Collection MS 28

A small but exquisite manuscript of Hyginus' *De astronomia*, held by the New York Public Library, was displayed in an exhibition there, entitled *The Splendor of the Word* (Alexander 2005).[25] This Latin astronomical treatise on eighty folios, created c. 1465–70, is decorated with 38 illustrations of the constellations drawn on paper by two or perhaps three different artists (Alexander 1994: 120).[26] While the elegant humanist script has been recognized as that of Francesco Buzzaccarini, the miniatures for the most part are attributed to Giovanni Vendramin and demonstrate that the manuscript was connected with Paduan intellectuals in the 1470's.[27] Three of the constellation illuminations are attributed to another artist, the Douce Master, who also painted a breviary written in 1473 (Canova 2005: 339). The second illuminator painted in a manner similar to the first artist, imitating his technique, so there would be no dissonance among the drawings when the patron was viewing his manuscript. The celestial miniatures are placed between sections of text, some constellations are framed with a red tasseled cord; while on other folios, the illustrations reach into the written words. The drawings show the characteristic quasi-antique influence commonly preferred in the Italian Renaissance, and the figures appear in their traditional anthropomorphic or zoomorphic forms. The Spencer Collection Hyginus *De astronomia* is included and pictured in the *Sternbilder des Mittelalters* catalog of Blume et al., along with twenty additional thirteenth to fifteenth century illustrated Hyginus manuscripts (Figs. 3.4 and 3.5).

When compared to the enormous volume of scholarship centered upon Aratus and the *Aratea*, the numerous illustrated manuscripts of Hyginus have held a minor position in art historical literature; until recently they had been little studied neither individually nor as a manuscript tradition. Text scholar Reeve declares, "Has any classical text been so ill served by recent scholarship as this?" (Reeve 1983: 187) Myth specialist Fletcher adds, "…of the major writers of myths—none has received such scorn as Hyginus" (Fletcher 2013: 133).

The Hyginus manuscripts often contain the full cycle of constellation pictures; the artworks in most copies share a very high quality in their artistic interpretations, yet simple and unembellished. During the first and second centuries CE, large

[25]The exhibition catalog, *The Splendor of the Word*, was written by J.J.G. Alexander, J. H. Marrow, and L. F. Sandler, with contributions by other authors. The exhibition ran from October 2005 until February 2006. The article discussing the Hyginus manuscript was written by G. M. Canova. See also DeRicci, II, p. 1341, no. 28 and J.J.G. Alexander, *The Painted Page*, 120.

[26]Alexander asserts that there are five different illustrators represented in this Hyginus manuscript.

[27]Padua was one of the main centers for the study of astrology in all Europe and, along with Florence, was one of the main centers of humanism as well.

Fig. 3.4 Eridanus, Hyginus *de Astronomia*, Fifteenth Century, New York Public Library, Public Domain

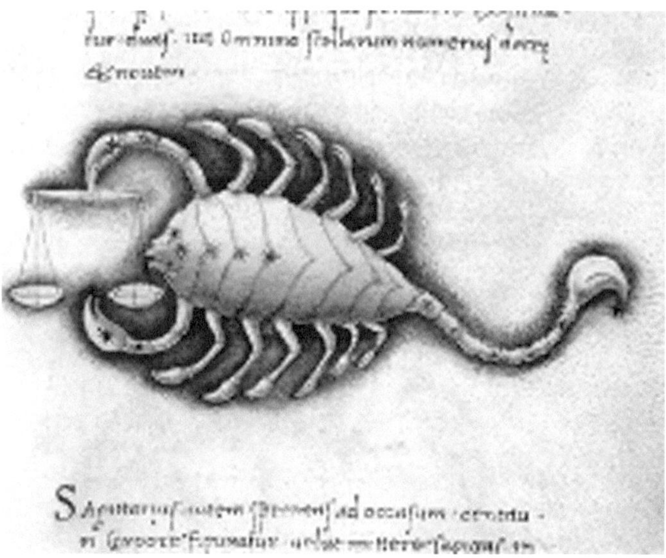

Fig. 3.5 Scorpio holding Libra, Hyginus *de Astronomia*, New York Public Library, Public Domain

numbers of astrological treatises were written and circulated within intellectual circles. These were accompanied by astrological charts or horoscopes, but not decorated with pictures. After the burst of activity among writers of astronomical discourses in the first- and second-century CE, there are no outstanding authors whose astronomical works were illustrated until the appearance of Martianus Capella and Macrobius in the fifth century. If there were others, they were not readily copied, nor did they circulate in the Latin West.

Illustrated Manuscripts of Martianus Capella

During the Middle Ages, in addition to Pliny the Elder and the *Aratea*, the three major authorities mined for astronomical information were Boethius (ca. 480–524),[28] Martianus Capella (fl.ca. 410–439) and Macrobius (ca. 360–422). When comparing the writing style of these three Late Antique authors to the verse format preferred by Aratus, the *Aratea* translators, and Manilius, poetry is no longer the desired arrangement for 'scientific' information. Instead all three of these Latin prose writers present their astronomical knowledge, not in a technical treatise, but rather in an allegorical form. Their respective works, *The Consolation of Philosophy*, *De nuptiis mercurii et philologiae*, and *Commentum in somnium Scipionis*, intersperse their 'scientific' or astronomical information with an entertaining narrative, perhaps as a mnemonic devise to assist the reader in retaining the more technical aspects (McCluskey 1993: 122–23).[29] The allegorical stories of these three authors related their cosmological viewpoints which had mainly been derived from the works of Plato and the Neoplatonists, but Plato's *Timaeus* was the only original Greek philosophical work available to the Latin world in the early medieval period. The works of all three of these authors were widely read throughout the Middle Ages; they were designed for advanced students and educated layman, but none of them contained challenging mathematical astronomy (Tester 1987: 115).[30]

Martianus Capellais thought to have composed *The Marriage of Mercury and Philology* in his old age, probably the first third of the fifth century; (Laistner 1957: 40) although scholarly debate on the exact dating continues. This narrative work is

---

[28]Boethius was an important writer in the early 6th century, both a Christian and a Roman patriot. His *Arithmetic* was almost the only source of arithmetical knowledge for the early Middle Ages.

[29]Boethius wrote his greatest work, the *Consolation*, while in prison awaiting his fate on the charge of conspiracy for which he was executed in 524. Lady Philosophy appears and reminds him that earthly fame and suffering are insignificant compared to the vastness of the cosmos. His writings have Christian overtones, but since they were not illustrated, they are not part of this study.

[30]Another Roman author of astronomical data was Calcidius, a late fourth-century astronomer, although his influence was slight, since his writings were almost unknown until the twelfth century. Calcidius translated the cosmological section of Plato's *Timaeus* and wrote a long Neoplatonist commentary on it. His commentary discusses the sphericity of the earth and the universe, in addition to the imaginary circles that surround the cosmos, i.e., the equator, the tropics, the Arctic and Antarctic circles, the zodiac, the horizon and the meridian.

divided into nine books and is essentially a treatise summarizing the seven liberal arts; but converting the scholarly details into a mythological story and embedding the teachings into myth created a more enticing narrative. Most of the astronomical information is found in Book VIII of *De Nuptiis*.[31] The allegory focuses on a wedding where Mercury, the traditional bearer of learning and the inventor of letters, introduces the seven beautiful handmaidens to his bride, Philology, the daughter of Wisdom; each of the seven maidens symbolizes one of the seven liberal arts. First are the "trivial" three, the Trivium of grammar, rhetoric and dialectic, and then the more advanced four, the Quadrivium of the sciences which were geometry, arithmetic, astronomy, and music. Then each of the 'Arts' in turn presents her discipline to the assembled gods. In her hands Astronomy holds a measuring instrument and a book containing the predestined courses of the stars and planets (Fig. 3.6).

Although hidden beneath the shadowy veil of allegory, the account given by Martianus Capella was replete with general astronomical information, and despite the fact that the data was not presented in a scholarly or methodical manner, the treatise was used as an advanced text in Carolingian 'schools'. His careful description of planetary theory and the planetary spheres is clearer than any of the other early writers, simple to follow, mainly derived from Pliny the Elder. Extensive commentaries on the work of Martianus Capella were written by Carolingian scholar, John Scotus Eriugena (815–877) in 852.[32]

The Carolingian educators were quite fascinated with the writings of Martianus Capella, and copied his allegory of the Seven Liberal Arts numerous times. Seventeen manuscripts of the complete *De Nuptiis* and four manuscripts of Book VIII survive from the ninth century, many of which are illustrated or glossed. In the twelfth-century revival, it was popular as both a textbook and a literary source. As evidence of its great popularity, there are 241 manuscripts of the *De Nuptiis* extant (Bowersock 1999: 566). The graceful handmaidens imagined by Martianus Capella were immortalized in the later Middle Ages by sculptors in artistic personifications of the Seven Liberal Arts on the west facade of Chartres Cathedral and in illuminated manuscripts, such as the *Hortus deliciarum* created by the twelfth-century abbess Herrad of Landsberg.[33]

Studies of the astronomy of Martianus Capella by Bruce Eastwood in *The Revival of Planetary Astronomy in Carolingian and Post-Carolingian Europe* credit him with the survival of heliocentric theories significant in the history of

[31]See L. Lenaz, D.R. Shanzer, and W.H. Stahl (1952) for commentaries on *Martianus Capella's De Nuptiis Philologiae et Mercurii Liber 1.*

[32]For a full discussion of the Martianus Capella commentaries see McCluskey (1993) 171-73.

[33]For detailed information on the personification of the Seven Liberal Arts in literature and sculpture, see Emile Mâle, *L'art religieux de XIII siècle en France*, Paris, 1948. See also Rosalie Green et al., *Hortus Deliciarum/Herrad of Hohenbourg,* Warburg Institute, 1979, *Seven Liberal Arts* miniature appears on f. 32, pg. 104, Plate 18.

Fig. 3.6 Allegory of Music, Seven Liberal Arts, Martianus Capella *De nuptiis mercurii et philologiae*, fifteenth century image, Public Domain

science.[34] When discussing the astronomy of the Latin writers of late antiquity, Eastwood remarks that Martianus alone offers a heliocentric orbit for the planets Mercury and Venus (Lindberg 1992: 146).[35] This theory appears in diagrammatic form in Carolingian manuscripts, proving that the heliocentric hypothesis of Greek science still flourished in part during the fifth century CE. This important information will be discussed in more detail in Chap. 7. Taken as a whole, Martianus' work is a competent summary of the Seven Liberal Arts, and his allegorical presentations exerted a strong appeal for both students and other authors for a 1000 years. Martianus' personification of the Seven Liberal Arts became the standard representation preserved in medieval art and architecture.

[34]See Eastwood and Graßhoff for a thorough explanation of astronomical diagrams included in Roman works on astronomy which could be found in medieval libraries. They identify four chief sources for Roman astronomy in the Middles Ages—Pliny, Macrobius, Calcidius, and Martianus Capella—their writings on astronomy reentered the broad European intellectual world through the Carolingian scholars and schools. Eastwood and Graßhoff provide a descriptive and critical analysis of the Carolingian copies with references to the original writings; they identify and interpret the textual and visual elements with commendable accuracy.

[35]This feature is also reported in other studies such as Lindberg (1992). He mentions that Copernicus cites Martianus Capella in support of his heliocentric system.

Illustrated Manuscripts of Macrobius Ambrosius Theodosius

Although a near contemporary of Martianus Capella, almost nothing is known for certain about the life of Macrobius Ambrosius Theodosius, other than he is credited with three works preserved from the Late Classical period. Of the three treatises, the most eagerly read, *Commentum in somnium Scipionis*, has survived intact; his much longer work, *Saturnalia,* is incomplete; and his treatise *On the Differences and Similarities of the Greek and Latin Verb* is now lost, but a medieval abridgement of it survives. (Stahl 1990:3) The *Commentary on the Dream of Scipio* by Macrobius achieved enormous popularity throughout the Middle Ages, and was considered one of the basic source books for medieval science. This tradition is important for its role in the transmission of astronomical information, but not for illustrations.

Macrobius, born circa 360 CE, was a Roman author and Neoplatonist philosopher who flourished during the reigns of emperors Honorius and Arcadius (CE 395–423) (Stahl 1990: 4).[36] It is thought that he is the Macrobius who was mentioned in the Codex Theodosianus as a praetorian prefect of Spain in 399–400, proconsul of Africa in 410, lord chamberlain in 422 (Stahl 1990: 6).[37]

The *Commentary* of Macrobius is based on the brief but well-known *Dream of Scipio* narrated by Cicero in the sixth and last book of his *De Republica*, written between 54 and 51 BCE. In Macrobius' *Dream*, Scipio appears to his adopted grandson and describes to him the life of the good after death and presents the nature of the cosmos from the Stoic point of view. He added a dramatic form to his basic astronomical descriptions. Cicero's purpose was not to teach astronomical science, but to use cosmology to explain the value and immortality of the human soul. "Macrobius's literary style of exposition was less colorful that Martianus's, and it owed more to the style of Cicero and Plato" (North 2008: 244). Macrobius uses the *Dream* to elaborate upon the doctrines of Neoplatonist philosophy that he had studied extensively (Laistner 1941: 42). Macrobius" volumes were an important source of Platonism in the Latin West, 'praised for its succinctness and lucidity'

[36]Few personal facts can be found in Macrobius' writings; he mentions a dearly loved son Eustachius to whom he dedicated his major works, and he discloses that he was 'born under another sky,' not a native of Italy. He states the various positions of contemporary researchers as to the birthplace of Macrobius. Most scholars place him in Africa, a few suggest Greece, but he writes in Latin. Furthermore, Macrobius' treatises display a greater familiarity with Latin than with Greek authors and he frequently mistranslates Greek words.

[37]Stahl presents a full account of the reasoning of scholars on whether Macrobius was Christian and why Christian dogma would not enter into the material of the ancient Greek authors under discussion. Scholars still debate his identity since the holders of high office at that particular time were required to be Christian. The extant writings of Macrobius fail to make a single reference to any doctrines of Christianity, but recent scholarship sees little conflict between this fact and whether or not he actually was a Christian. There is no reason to expect references to Christianity since his subject matter in *Commentary* concerns the philosophical tenets of Neoplatonism and *Saturnalia* pertains to pagan antiquities and classical literature.

and for making the basic doctrines of Neoplatonism intelligible to his readers (Stahl 1990: 10).[38] The first four chapters of Book I are a basic introduction to his philosophy; Chaps. 4–14 discuss the core of his work on the origin and descent of souls, reinforcing the belief of Plato and Cicero and citing the opinions of 21 philosophers in support of the immortality of the soul after death. From Chap. 14 to the end of Book II the cosmographical and astronomical discourse is found, comprising nearly half of the *Commentary*, and often these sections were bound separately. Numerous manuscripts survive that contain only this section. Historians of science consider Macrobius one of the most prominent authorities on astronomy and geography during the Middle Ages (Stahl 1952: 24–39) (Fig. 3.7).[39]

Macrobius' *Commentary* was widely read throughout the medieval era as his subject matter was of enormous interest. In addition to the astronomical information, he had compiled information on the interpretation of dreams, the sacred Pythagorean number symbolism, the nature of the soul, the descent of the soul from its celestial origin into human bodies, and Neoplatonic philosophy. Macrobius' writings encompassed the motions of the celestial spheres and the order of the planets, the method used in distinguishing the signs of the zodiac, proof that the Earth is in the center of the universe, numerical ratios of planetary distances, and the origin of the harmony of the spheres. An examination of Macrobius' work shows that he elaborates on the attributes of the planets, their spheres, and their appearance in each sign of the zodiac, but barely mentions any of the other stellar constellations by name. He provides no listing of the known constellations, and therefore illustrations of the pictorial cycle found in manuscripts of the *Aratea* are not expected, nor are they found.

Even though figural representations of the constellations are missing, explanatory diagrams are plentiful. Manuscripts of the *Commentary* include numerous types of diagrams, both celestial and terrestrial. The cosmological schema include depictions of the entire Earth-centered Universe, the causes of eclipses of the Sun and Moon, and diagrams showing the position of the planets of the solar system. The terrestrial diagrams depict the zones of the Earth marked with five climatic zones: torrid at the equator, bound by temperate zones above and below, and finally frigid zones at the extremities, and world maps, and even rain patterns are plentiful. Some depict the habitable world of the northern hemisphere and the uninhabited world of the southern hemisphere. Maps of Macrobius are oriented with North at the uppermost area, unlike the majority of medieval *mappae mundi* with their

[38]*Scipio's Dream* by Cicero is an imitation of the Vision of Er, the closing episode in Plato's *Republic;* Macrobius is credited with its survival. According to the philosophy that Macrobius discusses, souls descend from the sky at the intersection of the Milky Way and the zodiac. As individual souls pass through each of the planetary spheres, they acquire a specific attribute which they are to experience and learn from during their life on earth.

[39]Stahl explains the latest scholarship on the various sources from which Macrobius drew. Scholars have now shown that, like other Latin compilers, Macrobius borrowed extensively from contemporary writers; but cited classical authors as their sources. It is generally agreed that the writings of Macrobius were not original, and that he borrowed extensively from Porphyry.

Fig. 3.7 Zonal Diagram, Macrobius medieval manuscript, *Commentum in somnium Scipionis*, Public Domain

Christian orientation; these picture East and the Garden of Eden at the very top and honor Jerusalem as the center of the world. These various types of celestial diagrams and world maps found in Macrobian manuscripts were later reproduced in the works of Bede, Lambert of St-Omer and other medieval encyclopedists.

The *Commentary on the Dream of Scipio* by Macrobius was very influential among early medieval writers such as Isidore, Bede and Johannes Scottus Erigena; and even more so later as he is quoted as a classical authority by Adelard of Bath, Bernard Silvester, John of Salisbury, William of Conches, and Vincent of Beauvais, all of whom drew heavily from his work. Even Albertus Magnus, Thomas Aquinas, and Chaucer often cited the writings of Macrobius. When looking at extant secular manuscripts from the medieval period, copies of the *Commentary* are among the most numerous, numbering in the hundreds (Stahl 1952: 60).[40] Like Aratus, Macrobius has been recognized by modern astronomers as they have named a crater on the Moon in his honor.

Illustrated Manuscripts of Isidore of Seville

As the years of Late Antiquity gave way to the early Middle Ages, the more substantial scientific texts of the classical authors had all been lost. After a period of great advances in the discovery and formulation of astronomical knowledge through Hellenistic times, 'science' had declined to an elementary level. During the seventh century, Germanic invasions and the dissolution of the Roman Empire had severely disrupted education and scientific inquiry in the Latin West. The tenets of Christianity had taken hold and redirected the central concepts of astronomy. Natural phenomena of the celestial and terrestrial worlds were no longer associated with mythical gods or with vague philosophic forces, but instead were orchestrated and controlled by one omnipotent Ruler overseeing His creation. The Stoic philosophy of Aratus, which emphasizes the phenomena of the heavens as signs from Zeus to assist humanity, was no longer appropriate in the cosmological outlook of Christianity. In spite of these great changes, the astronomical poem of Aratus persisted, surviving in its Latin forms, although its transmission through this early period cannot be accurately traced except for brief appearances. In his writings, Isidore of Seville looked back to the manuscripts of Germanicus and compiled some of his own astronomical information from the scholia of Germanicus' *Aratea* (Laistner 1941: 124).

Isidore of Seville

The first encyclopedist of the Middle Ages, Isidore of Seville (ca.570–636) is considered the most learned man of his era: an archbishop, theologian, and compiler of ancient knowledge. When only 30-years old, Isidore succeeded his brother Leander as bishop of Seville; he remained in that position until his death in 636. His entire life was devoted to scholarship, which extended to every known field of study; his desire for secular and scientific knowledge was as strong as his religious concerns. The extent of Isidore's knowledge and the diversity of topics that he included in his written works supply proof of the richness of the library held at

[40]Lists of surviving manuscripts of Macrobius can be found in M. Manitius, *Handschriften antiker Autoren in mittelalterlichen Bibliothekskatalogen* (Leipzig 1935), pp. 227–32 and in *Macrobius Opera quae supersunt*, Ed, L von Jan, Leipzig, 1848–52.

Seville at that early date (Saxl 1957a, b: 228–9).[41] In addition to his copious religious duties as bishop, presiding twice over church councils, in Seville (619) and Toledo (633), Isidore managed to produce numerous writings embracing a huge range of subjects both secular and religious: *De natura rerum*, *Synonyma*, *Sententiae*, a history of the Goths, and his masterwork reaching almost 1000 pages in twenty books, the *Etymologiae* (McCluskey 1998: 124).[42] In the *Etymologies* (or *Origins*), Isidore explains the meaning of a word by discussing its derivation (Dreyer 1953: 220). Isidore's method in culling and writing this vast amount of material was to adapt and condense what he had read, concentrating on the information he thought most essential to preserve. Despite the many deficiencies of his volumes, Isidore rendered a great effort to preserve and transmit the tattered remnants of ancient knowledge for later readers by passing on the rudiments of Greek and Roman scientific thought, especially astronomy; as a faithful churchman he amplified that early knowledge by relating his Christian commentary.

Isidore has sometimes been berated by historians for his inaccuracies and lack of originality, because, other than his diagrams, most of his writings are simply compilations of earlier classical or patristic authors. Some of his comments show a faulty understanding of scientific concepts well known in the Greek world (Whitney 2004: 165). He was known for his productivity and eloquence but contributed little in the introduction of new discoveries, creative ideas, or innovative interpretations, either to current theological doctrine or to secular material. But it is necessary to remember that the introduction of originality or creativity was never the main goal of scholarship during this early era. The intention of most early scholars was to explain, clarify, and disseminate what was already known, that 'sacred' knowledge passed down from the Golden Age of antiquity or from Biblical times, and to protect that corpus of received knowledge against error or change. "Isidore preserved Roman cosmology which formed the basis of early medieval cosmological thinking and was the main pre-Carolingian channel for the transmission of ancient knowledge and concepts" (Kühnel 2003: 93). His compilations became convenient, standard reference sources made available for study; *De Natura Rerum* and *Etymologiae* could be found in every sizeable monastic library; Isidore's anthologies were consulted and copied profusely by students and by later generations assuring that science would remain an important part of monastic education (Laistner 1941:124).[43] For this study, the only Isidorian writings of concern are his

---

[41]Saxl, "Illustrated Encyclopedias 1." The model for Isidore's encyclopedic work goes back to Latin models, not Greek, as the Greeks had no inclination for composing encyclopedias. Varro wrote 41 books on the antiquities of human and divine things, but Isidore's library did not contain a copy.

[42]The Visigothic King Sisebut asked Isidore to explain the causes of natural phenomena, such as the lunar and solar eclipses seen in Spain. Following the royal request, Isidore wrote a treatise *On the Nature of Things*.

[43]A long list of writers from the seventh to the tenth century, who adapted excerpts from Isidore, could be assembled. A complete list of extant manuscripts produced from Isidore's time until the mid-ninth century can be found in C.H. Beeson, *Isidorstudien* (Munich, 1913).

compilations of astronomical information that are illustrated with constellation figures, found in both *De natura rerum* and *Etymologiae*.

Isidore's *De Natura Rerum*

De natura rerum is the shorter of his two encyclopedic works; both include secular and religious material. Isidore was the first medieval writer to develop a method of bringing text and illuminations together without using the common classical and Early Christian technique of adding mythological personifications (Kühnel 2003: 74). Isidore describes celestial phenomena: the world axis, the poles and the ecliptic and was one of the first authors to distinguish between *astronomia* and *astrologica*. He clarifies the obvious distinctions between astronomy and astrology, which he describes as prophecy by the stars, scrupulously rejecting the practice as baseless superstition (Tester 1987: 19). These condemnations and bitter attacks against astrology by Christian churchmen actually performed the opposite effect by keeping the subject alive and arousing the laymen's curiosity and desire to access the forbidden doctrines.

Isidore's writings are augmented with numerous descriptive rotae, and in fact many of the large variety of diagrammatic forms were considered original innovations and credited to the author himself. *De natura rerum*, known also as *Liber rotarum,* presents a breakthrough in scientific book illumination because of its addition of many explanatory circular diagrams. *De natura rerum* became an instant success, as the various monastic scriptoria began to reproduce his composition almost immediately after it was completed. The first recension (or revised edition) is dated 613; in addition there are 17 pre-Carolingian manuscripts of *De natura rerum* extant, (Kühnel 2003: 95) and seven manuscripts from Fleury and Cambrai alone dating to the seventh century, all in the Library of Fleury. In addition, 33 manuscripts of *De natura rerum* are known from the ninth century.

Isidore's Etymologiae

The *Etymologiae* begins with the Seven Liberal Arts which were considered from long before Isidore's time to include all necessary human knowledge. In addition to the Trivium and the Quadrivium, Isidore amassed information on medicine, the four elements, the theory of humors, geography, and history; he also brought together in diagrammatic form; the 6 days of Creation, the six world eras, and the six ages of human life.[44] The astronomical texts in *Etymologiae* are not extensive and are

[44]For detailed information on the illustrative tradition of medieval encyclopedias, see Saxl (1957a, b) Illustrated Encyclopedias 1.

scattered throughout the wide-ranging work; most are found in Book II, but his cosmological accounts are merely brief explanations of earlier scholarship (Laistner 1941: 122).[45] Isidore's astronomical material was not the advanced mathematical calculations of Greek 'philosophers', but primarily the popular 'science' found in *De astronomia* of Hyginus and the *scholia* composed on Germanicus' *Aratea* (Laistner 1941: 124). In *Etymologiae* he included the spherical shape of the earth, a discussion of the winds, the zones of the Earth, and the motions of the stars and planets.

Isidore's universe was composed of a primordial substance possessing neither quality nor form; its shape was determined by the four elementary qualities: coldness, dryness, wetness, and hotness. Following, Lucretius and many Greek cosmographers, Isidore regarded these elements as in constant flux between the earth and the solar fire at the center of the universe. He shared the ancient Hermetic theory of each individual human being as a microcosm paralleling the macrocosm The *Etymologies* of Isidore, 'although a confused and ill-conceived blend of classical learning' has been named "one of the most popular books of the entire Middle Ages," as almost every monastic library included a copy.[46] Early manuscripts of Isidore's work survive not only from Spain but from South Germany, Austria, and Northern Italy; in fact the *Etymologiae* was so popular, it survives in more than a thousand medieval manuscripts (Lindberg 1992: 158). The first English edition of the Etymologies was published in 2006, making the purposes and processes of Isidore more available and intelligible to a wider audience.[47]

Was Isidore's encyclopedic work illustrated originally? According to Saxl, probably not, but later versions may have been fully illustrated with both images and with numerous diagrams, the schemata relating to the basic structure of the Earth and explaining the known 'sciences' and astronomical phenomena. Later medieval encyclopedists all look back to Isidore's seventh-century model. Isidore's writings began to circulate beyond Spain during his lifetime and his popularity spread throughout the rest of Europe. "Only the works of Gregory the Great were diffused as rapidly as those of Isidore" (Bischoff 1994: 4). But Isidore's encyclopedic work of the early seventh century was soon eclipsed by the diligence, scholarship and original research of Bede in the early eighth century. Some historians have confused the 'popular' astronomy found in Aratean manuscripts with

[45]Laistner reports that *Etymologiae* had been edited and separated into twenty books by Isidore's friend, Braulio, bishop of Saragossa, who also composed a short biography of Isidore and a chronological compilation of his writings.

[46]Editions of Isidore are Faustinus Arevalo, *Isidori Hispalensis opera omnia*, 7 vols.(Rome, 1797-1803), in J. P. Migne *Patrologia latina*, LXXXI-LXXXIV; W. M. Lindsay, *Isidori Hispalensis Etymologiarum sive Originum libri*, 2 vols. (Oxford, 1911); and Jacques Fontaine, *Isidore de Seville: Traité de la nature* (Bordeaux 1960).

[47]See Barney, S. A., Lewis, W. J., Beach, J. A., and Berghof, O., (2006) *The Etymologies of Isidore of Seville,* Cambridge: Cambridge University Press.

the more exact astronomical science and celestial charts and data necessary for establishing the ecclesiastical calendar, a topic of crucial concern throughout the Middle Ages. The writings of Bede became the basis for organizing and updating the calendar and establishing the procedures of computus.

Illustrated Manuscripts of Bede

Born in the isolated, windswept area of England's northeast coast, Bede (673–735) entered the Benedictine monastery at Wearmouth-Jarrow at age seven where he was educated by Benedict Biscop and Ceolfrid, abbot of Jarrow, both accomplished scholars. Though never leaving his remote abbey which probably had less than 300 separate works in its library, Bede mastered Greek, Latin, mathematics, music, and astronomy and became the most illustrious scholar, teacher, historian, and author of medieval England, honored with the title Venerable. His extensive writings especially on biblical exegesis, martyrology, hymns, and poems confirm his mastery over most areas of traditional Christian learning, including astronomy (North 2008: 236).

Bede's exacting study of astronomical motions led him to undertake his independent scientific research by collecting data on daily, monthly, and yearly celestial patterns, ocean tides and timekeeping from his own observations and experiences. Bede began composing his computistical handbooks shortly after he was ordained a priest around the year 703. In addition to his own scientific analyses, Bede drew from Scripture and from the few earlier and available authors, Pliny the Elder, Macrobius and Isidore of Seville in particular.[48] Specialists have traced his many sources from Egypt, Roman Africa, and Italy, through Spain, Gaul, and Ireland to his monastic library at Jarrow. Bede's use of those early sources for creating and clarifying the European calendar still regulates our modern consciousness of time. When evaluating his writings, scholars have determined a wide range of expertise in Bede's scientific work; some was less than adequate, some was quite precise, and "some stands out as so remarkable that it seems to surpass the boundaries for his time and place and culture" (Stevens 1987: 1). The adoption of his system of organizing time by counting the years successively from the birth of Christ, *Anno Domino* or the 'Christian era', still in use today, became accepted throughout Europe mainly through the erudition and influence of this lone Northumberland monk (Stevens 1987: 169). Bede called his work the "little book about the fleeting and wave-tossed course of time."[49]

---

[48]Bede quotes the second book of Pliny Secundus in Books XXVII, XXX, XXXI, XXXIV, and XXXV of his *Natural History*. (Charles W. Jones edition).

[49]*Noster libellus de uolubili ac fluctiuago temporum lapsu: The Reckoning of time,* ch. 71 (544.91-92).

During the medieval era each religious community throughout the Christian world was expected to determine the correct date of Easter even though the calculations were tedious and difficult to surmount. Of course there was no uniform, universal calendar; each area had its own means of calculating time. Bede's pivotal writings on time, the *computus*, became the standard reference work for the next 1000 years, directed of course toward Christian exegesis, and predicting future dates for the celebration of ecclesiastical events, especially Easter. In spite of the controversies and confusion of *computus*, manuscripts based on time-reckoning survive in enormous numbers (Stevens 1987: intro).[50] The astronomical data dealing with the cycles of time were never entirely satisfactory, and required constant adaption of the reckonings to keep computations in line with the irregular motions of the sun and moon for which Bede's computations and instructions were fundamental. The titles of his three well-known treatises that concentrate mainly on astronomy are *De natura rerum* (703), *De temporibus* (703), and *De temporum ratione* (725); in addition Bede produced more than forty texts relating to history, science, poetry and hagiography. Among his most valuable compositions, *The Ecclesiastical History of the English People,* composed in 731, survives as the earliest account of Anglo-Saxon England ever written, starting with Julius Caesar to his current year.

Bede's *De Temporibus*

On Time was one of Bede's earliest works, written when he was about 30, his first treatise devoted to the calendar, for which he was indebted to earlier Irish models (Wallis 1999: lxv). This work is a focused and concise computistical manual explaining the units of time, the nineteen-year cycle of the moon, the Paschal tables, and constructing a Christian calendar. Bede avoided mixing computus with cosmology, and so does not mention the sun, moon, zodiac, or solstices and equinoxes in this treatise. Therefore illustrations and diagrams would not be required and would have no place in *De temporibus*, only the Easter tables were included with these computistical writings. Although he wrote this work for the monks of Jarrow, it was too difficult for most of them to understand; therefore he wrote *De natura rerum* and then another, much longer treatise on time, *De temporum ratione*.

[50]According to *computus* scholar Wesley Stevens, more than 9000 computistical tracts can be found in Latin manuscripts transcribed before 1600, approximately 1200 of them were written between the eighth and the eleventh centuries. He reports, "Those manuscripts have revealed unexpected evidence that early Christians drew more heavily and more specifically upon Hellenistic mathematics and natural sciences than any lectures, articles, or books had ever recognized much less explained."

Bede's *De Natura Rerum*

Bede produced *De natura rerum* shortly after his *De temporibus*, drawing from ancient sources and heavily from Isidore's work of the same name, except that Bede extended his version by three chapters, and he arranged some of the material in a new way (Bober 1956–57: 19–20). In the same mode as Isidore, Bede begins his treatise with the universe as a whole and then descends to earthly phenomena discussing the atmosphere, oceans, rivers and lands. He employs the traditional model of the four basic elements; fire is associated with the heavens, air refers to the atmosphere, water to the seas, and earth to the land masses (Wallis 1999: lxv). Isidore included a section on time in his *De natura rerum*, but Bede removed that topic since he had already written extensively on the matter of time. Chapter 15 includes quotes from Pliny verbatim concerning the colors of the planets; Chap. 16 is borrowed directly from Pliny's account of the zodiac and the planetary wanderings (Eastwood 1986: 200). The subjects included in this work seem likely areas to have inspired both diagrams and illustrations. But Bober states that was not the case, "Bede's *De natura rerum* was not originally accompanied by illuminations, as attested by the earliest copies which exist in quite a large number of examples; the earliest manuscripts of *De natura rerum* to be accompanied by illuminations date to1066" (Bober 1956–57: 19).

Bede's *De Temporum Ratione*

Composed in 725, *De temporum ratione* followed as a sequel to Bede's other two books on science, divided into four separate parts that contain in total, 71 chapters. This treatise stands as a fusion of calendar science and natural history in which Bede discusses important concepts: the units of time, instructions for calculating the date of Easter, a world chronicle, and even looks to the future of life on earth envisioning the very end of time, the Apocalypse (Kühnel 2003: 100). In this extensive essay, Bede clarified and systematized the collected knowledge in computistics so comprehensively that the *De temporum ratione* became the primary source for the Western European calendar until the early modern era (Stevens 1987: 169). In this and all of his compositions, Bede carefully avoids referring to astrological theories. He speaks quite bluntly as in Book XXIII, "Should anyone be so lazy or slow-witted as to wish to know the course of the Moon without any of the trouble of calculating he must rely on the alphabets which he sees."[51] In Book XIII, Bede states that the stars were created of different sorts of matter that affect the weather; however, he specifically maintains that the stars have no influence upon man's character and have no power to determine human activities.[52]

[51]Wallis translation, *Bede: On the Reckoning of Time*, p. 71.

[52]*De natura rerum* XIII, ed. C. W. Jones (1943), Cambridge: Medieval Academy of America.

This computistical treatise discusses the movements of the heavens, especially the path of the Sun and Moon through the circle of the zodiac. Time reckoning requires familiarity with the division of the sky into the 12 signs of the zodiac along with the concept of 360° as a basic scale for systematizing observations of the positions of the sun and moon. In Book XVI *De signis duodecum mensium*, Bede discusses the months and pairs each with its appropriate sign of the zodiac, but he does not mention any of the other 36 Ptolemaic constellations; and even those 12 are used strictly as fixed mathematical points for determining the 360° course of the sun in its daily circuit through the sky.

The original text of *De temporum ratione* does not call for diagrams and scholars generally agree that it was not accompanied with them until many years later. Bede scholar C.W. Jones collated many of the surviving manuscripts of Bede's computistical compositions and published an edition in 1943. He examined 104 extant manuscripts of the *De temporum ratione* but none of these was written within 70 years of the original composition dated to 725. Jones' study was strictly concerned with the text and glosses, paying almost no attention to illustrations or diagrams. The earliest surviving manuscript to contain illustrative diagrams has been unanimously attributed to the scriptorium of Fleury ca.795 (Paris BNF Lat. MS 5543); astronomical diagrams also appear in (BL Harley MS 3017), another slightly later London manuscript (Kühnel 2003: 69). Illustrative diagrams became particularly popular in the medieval period and were included in many types of manuscripts. Diagrams assisted in encapsulating the necessary information, a useful mnemonic tool; a typical diagram from an eleventh century Bede manuscript, seen below, denotes the four seasons. In England, the spring equinox was celebrated with a feast honoring the Saxon goddess Ēostre. Bede's *De temporum ratione* identifies this season as Easturmonath, the root for both words derives from the direction East, where the sun rises directly on the vernal equinox (Fig. 3.8).[53]

In Kühnel's study of diagrams, she notes that both Bede's *De temporibus* and his *De temporum ratione* "were probably accompanied by illustrations from an early period, since parts of them are already illuminated in the anthology held in Cologne (Dombibliothek, Cod. 83 II, ca. 805) which contains almost the complete text of *De temporum ratione* at the beginning of the ninth century" (Kühnel 2003: 100). These large survival numbers reinforce the enormous importance of Bede's three great works, often incorporated into science-based manuscripts with other astronomical texts.[54] Bede was by far the most widely read and admired scholar of his age.

[53]Bede discusses The Four Seasons, Elements, and Humors in Chap. 35 of *De temporum ratione*.

[54]In Stevens', "Bede's Scientific Achievement", Jarrow Lecture of 1985 in *Cycles of Time*, Appendix I, he states that *De natura rerum* can be found in 135 manuscripts; his *De temporibus*, Books 1–16 survives in 97 manuscripts, Books 17–22 in 25; and *De temporum ratione,* Books 1–65 are extant in 245 copies, Books 66–71 in 41 manuscripts. Appendix II presents a complete list of *De temporum ratione* manuscripts from the eighth and ninth centuries. Unfortunately there is no indication whether any of these are illustrated with constellation figures or explanatory diagrams.

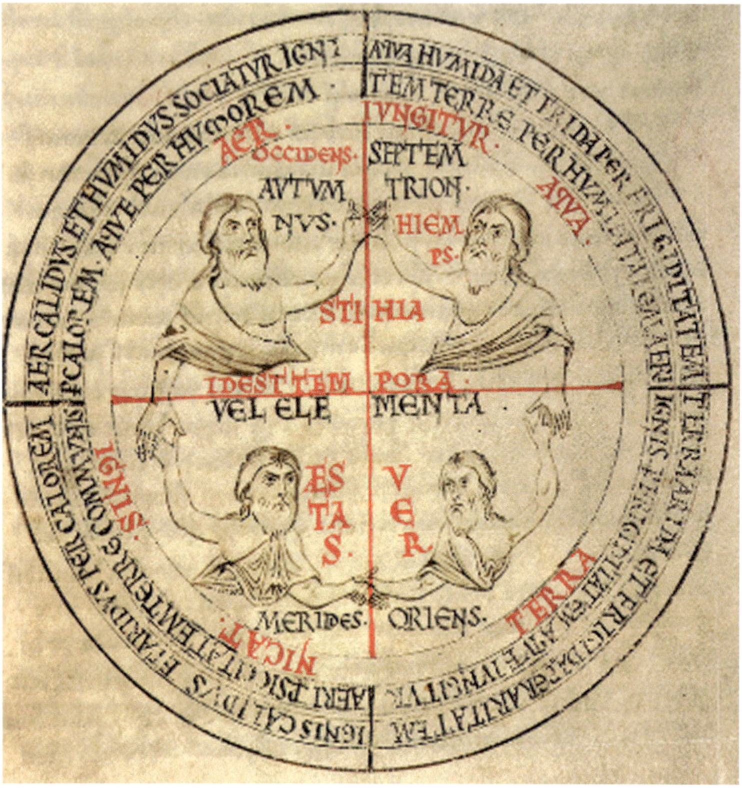

Fig. 3.8 Personification of Four Seasons, Bede manuscript, *De temporum ratione*, early Eleventh Century, Public Domain

De Signis Coeli of Pseudo-Bede

A widely utilized treatise on elementary astronomy called *De signis coeli*, became wrongly credited to Bede during the medieval period, but this work was hardly more than a star catalog.[55] The Pseudo-Bede composition, most popular from the tenth to the thirteenth centuries, had a clear mathematical basis and purpose; its illustrations took the form not of geometric diagrams, but of constellation images. A thirteenth-century manuscript of this work (Bodleian Oxford, MS Laud 644) is illustrated with the typical constellation cycle images and includes a short text accompanying each picture. Forty constellations are pictured on ff. 8r-10v (Saxl and

[55]See Ps-Beda Venerablis, *De Signis coeli* ed. E. Maass, *Commentariorum in Aratum reliquiae*, Berlin, 1898, pp. 582–594. Also cited in Van de Vyver, 'Traductions', Osiris I (1936) and in Saxl and Meier. The entire text is online at www.kristenlippincott.com/assets/Uploads/ps-Bede-de-signis.

Meier 1966: 386–93). Also in this manuscript, fragments of the *revised Aratus Latinus* (lines 582–594) are included on the same folio as the Pseudo-Bede; in addition this Oxford compilation holds partial works of Boethius, Abu Ma'shar, Robert Grosseteste, and Gerard of Cremona's translation of Ptolemy's *Almagest* from Arabic to Latin.

The cycle of the 48 constellation illuminations appear prominently in the encyclopedic tradition first established in the Middle Ages with *Etymologiae* and *De natura rerum* compiled by Isidore, it drew from Classical encyclopedists, especially Pliny the Elder, and continued with the writings of Bede/Pseudo-Bede and was greatly expanded in the ninth century in the writings of Hrabanus Maurus which brought in more religious material.

Astronomical Illustrations of Hrabanus Maurus

During the Carolingian renaissance, the growth of monastic and cathedral schools saw the spread of Anglo-Saxon and Irish scholarship throughout France and Germany. Although Bishop Alcuin of York (735-804) never produced an important scholarly work, he brought his experience and expertise from his post in England to upgrade the educational system on the continent. Alcuin's exceptional student, the eminent scholar Hrabanus Maurus (c.780–856) became Germany's most erudite teacher in the ninth century; both Alcuin and Hrabanus enjoyed and encouraged the study of astronomy. Hrabanus taught at the royal palace in the Carolingian capitol of Aachen, where he composed his highly praised religious discourses. He was an educator at the Benedictine monastery of Fulda as well, and elected abbot there in 822; under his rigorous supervision, Fulda became one of the most respected schools in Europe. Through enormous effort, Hrabanus expanded the library at Fulda into the leading center of learning in the Carolingian Empire for sacred as well as secular texts (Stevens 1987: 178). Eventually Hrabanus was honored for his achievements by being named archbishop of Mainz (Fig. 3.9).

During his retirement at the monastery of Petersburg near Fulda, Hrabanus Maurus completed his all-inclusive work *De rerum naturalis* (842–47). Like Isidore of Seville, he amassed information from a great variety of available sources, rewriting the original material in a more comprehensive and accessible form, making the material understandable and available for students. The 20 volume collection composed by Isidore was amended and enlarged into a massive 22 volume encyclopedic work filled with a vast quantity of information, including a bestiary, decorated with numerous illustrations and Christianized with lengthy spiritual commentaries.[56] Only five illustrated copies of Maurus' encyclopedia are

[56]Hrabanus' sources, beyond Isidore and Bede, included church fathers, Ambrose and Augustine, interspersed with pagan writers Virgil, Priscian, Donatus, Pliny and Hyginus, whom he quotes in Chap. 28.

Fig. 3.9 Hrabanus Maurus and Alcuin before St Martin of Tours, medieval illuminated manuscript, Public Domain

extant, and none date to the ninth century. The earliest copy (Cod. Casinensis 132), dated to 1022–35, is still preserved at Montecassino, where it was initially produced; the manuscript contains 339 miniatures and was written in Beneventan script (Le Berrurier 1978: 2–3).[57]

[57]"Only the oldest and the most recent copies have a full set of miniatures." In addition to the Monte- cassino manuscript, the second fully illustrated copy is Vat. Lib. Cod. Pal. lat. 291, which has been dated to 1425 and contains 338 miniatures. The other three surviving copies are: Vat. Lib. Cod. Reg. lat. 391, dated to the beginning of the fifteenth century, with only the first eight miniatures; Berlin, Preussische Staats. Cod. fol.lat. (930) attributed to Catalonian artists of the late fourteenth century, has 141 miniatures; Paris, BN MS. Lat. 17177, produced around 1200 in western Germany has only four miniatures on two separate folios.

A dissertation by Diane Le Berrurier on the pictorial sources of Maurus' writings examines the sources and pictorial traditions of the surviving illustrated manuscripts of *De rerum naturalis,* concentrating on the mythological and scientific illustrations; monsters, astronomical and botanical illuminations.[58] Her study describes the number and location of extant manuscripts.[59] Stevens provides a fuller listing of 16 extant manuscripts of Hrabanus' *De rerum naturalis,* but does not mention any illuminations.[60]

Most of the astronomical illustrations of *De rerum naturalis* are found in Book Nine, which consists of a prolog entitled, *De mundo et quatuor plagis ejus,* and 29 chapters which discuss meteorology and astronomy, accompanied by lengthy Christian commentaries. The astronomical portions of 22 chapters are illustrated and according to Le Berrurier these were modeled on illustrations of the *Aratea*.[61] She lists the titles of 22 illustrations, of these only: *de coelo*, *de sole*, *de luna*, de Pleiadibus, *de Arcturo*, and *de Orione et Hyadibus*, appear regularly in *Aratea* manuscripts. The other astronomical illustrations: *de Lucifero*, *de vespero*, *de nubius*, *de igne*, *de carbonibus*, *de cineribus* and others are not part of the cycle of constellations that are included in the typical *Aratea* image program. Among the five surviving illustrated copies, the maximum number of constellation illuminations found is only 23, compared to 42 to 48 that appear in most complete constellation pictorial cycles. The miniatures found in the encyclopedic work of Hrabanus were derived from sources, such as illustrated copies of Dioscurides, Solinus, and the *Aratea* of Cicero. The Hrabanus pictures are quite unpretentious in their Italian recension but they provide interesting insights into the Carolingian conceptions of the world and the universe which seem to have been close to those of Pliny and Cicero.

The familiar celestial diagram from antiquity seen in mosaic floors consisting of the signs of the zodiac surrounding the personifications of Sol and Luna found in Cas.132 is similar to the diagram in the fifteenth-century *Aratea* manuscript (Vat. cod. gr.1087). The signs are arranged in the same counterclockwise order with Aries at the mid-heaven or top of the horoscope. The Vatican *Aratea* manuscript is of a much superior quality; the only difference is that the orb of the sun held by Helios in the *Aratea* is omitted in the Maurus copy.

The illustrations of the constellations found in manuscripts of *De rerum naturalis* by Hrabanus Maurus have been compared to those found in *Aratea*

[58]The section on monsters, trees, and plants, I am not qualified to comment upon, but the section on astronomy is full of errors and erroneous suppositions, for instance she refers to Eridanus as Enidris, does not recognize Canis Major and Sirius. She reports that the *Aratea* were written by Aratus, and that "the *Aratea* became very popular at the end of the Roman Empire."

[59]Le Berrurier, 3–5, The earliest of the five surviving copies of Hrabanus' was decorated with miniatures which exhibit characteristics of southern Italian art of the time, but also show influences from an earlier period. This copy was modeled on a Carolingian manuscript as it has classical prototypes; its ultimate source for illustrations would have been ancient models.

[60]Stevens, *Cycles* 190–96.

[61]Le Berrurier, 63–64.

manuscripts, but they actually have little in common. The writings of Hrabanus do include a few of the standard constellation images, but many of the graphic miniatures are actually quite distinct, displaying very different iconography and including some pictures never seen in *Aratea* manuscripts. The two astronomical traditions seem to have utilized very different models.[62]

Le Berrurier proposes that the Hrabanus manuscript of the eleventh century is based on the Vatican Greek manuscript which dates to the fifteenth century, not quite possible. But there were numerous sources that could have provided this popular astronomical motif.[63] Except for a very few miniatures, the majority of astronomical illustrations in the surviving Hrabanus manuscripts are unique to that tradition and do not belong to the constellation cycle seen in manuscripts of the *Aratea*, nor those in the Hyginus or Isidore/Bede traditions. Therefore, the connection or influences between those two manuscript traditions is insignificant. In addition, the astronomical material in Maurus' manuscripts has been heavily Christianized, which is never the case with the *Aratea*. While assigned at the monastery of Fulda, Hrabanus became the teacher and friend of Lupus of Ferrières, who plays a part in the history of *Aratea* manuscripts as he was known to possess a copy.

Letters of Lupus of Ferrières

Scholar and teacher Lupus of Ferrières (805—c.862) has become an important figure in the transmission of *Aratea* manuscripts, for he mentions in one of his surviving letters that he has possession of a Cicero translation. Lupus corresponded regularly with his fellow scholars on specific literary and religious points. Demonstrating that books were not the only means for the transmission and distribution of scientific knowledge, Lupus wrote many letters 'discussing scholarship' which provide "examples of informal and formal methods for the exchange of ideas" (McKitterick 1994: 26). "Astronomy around the Carolingian period was in many ways weak, even by the lights of Roman astronomy in late antiquity, but it was weak for want of texts, not enthusiasm" (North 2008: 245).

Lupus was born during the last decade of Emperor Charlemagne's reign and was educated at the monastery of Ferrières, France where scholastic resources seemed to

[62]A facsimile of Hrabanus Maurus' encyclopedic work *De rerum naturis*, or *De Universo*, was published in 1996. The facsimile was copied from the surviving manuscript produced in 1022–35, Montecassino Archivio dell'Abbazia; the large-format codex contains 530 pages in Beneventan script and illustrated with more than 360 miniatures.

[63]Le Berrurier also suggests that the illustration of Helios in the Cod. Pal. Lat. 291 copy of Maurus was modeled on the iconography of the constellation Auriga. This is definitely not the case since Auriga is always dressed as a worker, not Apollo, the Lord of the Sun; furthermore Auriga is always accompanied by his goat and its kids and usually has a small square farm cart with two horses, not the magnificent chariot and four powerful horses of the Sun God Apollo.

be scarce (Laistner 1957: 252).[64] His family was wealthy, but claimed no connection to political power. Lupus became abbot of Ferrières in 840, but he was never named bishop, was never assigned to a large or dominant monastery, and never received an official court appointment (Noble 1998: 239). In addition to his religious and scholarly duties, Lupus was also obliged to take part in the campaigns of Emperor Charles the Bald, often acting as emissary (Laistner 1957: 253). Empress Judith, second wife of Louis the Pious and mother of Charles the Bald, was very generous toward Lupus, possibly they were related.

Fortunately 127 of Lupus' letters, written between 830 and 862, survive and these contain valuable information about his literary, theological and humanistic studies. In a letter written to Einhard (c.775–840) biographer of Charlemagne, Lupus refers to a list that he had seen of the books possessed by Einhard.[65] Lupus became an excellent scribe and there are at least a dozen extant manuscripts that were revised or annotated by him. (Laistner 1957:254)[66] What is unparalleled in his studies is his eagerness to obtain a second manuscript of a particular work that he already possessed as an exemplar, so that he could compare the two and then improve his own copy.

Lupus was quite familiar with the writings of Cicero, including his translation of Aratus' *Phaenomena*.[67] His letters show that he borrowed a manuscript of Cicero's *Aratea* from the monk Ansbald of Prüm (fl.860–886), so that he might restore his own defective copy of the *Aratea*, in order to create an improved text (Lupus, Epistle 69). Writing in September of 847 to Ansbald, Lupus says, "I shall collate the letters of Cicero which you send me with my own copy so that, if it be possible, I may get an accurate copy from the two. Will you in turn send me Cicero's *In Arato* with this courier of mine, so that the portions which are lacking in this copy, as pointed out by our Eigil, may be supplied from the one which I hope to obtain" (Regenos 1966: 81). Unfortunately, it is not known for certain which manuscripts Lupus was referring to, nor if either manuscript is extant. Saxl proposed that since the Ciceronian *Aratea*, Harley 647, now held by the British Library, was carefully

[64]The poor, obscure monastery of Ferrières was located south of Paris, between Orleans and Sens. Due to the poverty of the monastery, their library was small and its collection limited; therefore, Lupus was sent to Fulda to complete his education. At Fulda Lupus was taught and befriended by Hrabanus Maurus; he also corresponded with Einhard and met other young monks who, like Lupus, became heads of religious houses.

[65]*Lupus of Ferrières, Ep. I, Correspondence*, ed. L.Levillain (Paris 1964); English translation, G. W. Regenos, *The Letters of Lupus of Ferrières* (The Hague 1966), no.1. For more information on Lupus, see also Beeson, C.H. (1930) *Lupus as Scribe and Text Critic*, Cambridge, MA.

[66]While at Fulda, Lupus wrote and illuminated a large manuscript in which five Germanic law codes were brought together. Among the manuscripts that he handled were Valerius Maximus at Berne (No. 366), Aulus Gellius (Vat. Reg. 597) and Ti. Claudius Donatus' commentary on the first six books of the *Aeneid*, and four manuscripts now in Paris: Cicero's *Letters of Symmachus* (Lat. 8623) and Macrobius' *Commentary on the Somnium Scipionis* (Lat. 6370).

[67]In his earliest letter to Einhard (Epistle 1), Lupus seems to imply that even as an adolescent student in Ferrières, he was already a keen Ciceronian. Lupus' own literary style shows the influence of his careful study of Cicero.

corrected; it may have been revised by Lupus himself or one of his pupils (Saxl 1957a, b: 88).[68] Eastwood affirms Saxl's suggestion and agrees that Lupus did make the corrections to Harley 647, which was at Fleury in the tenth century and was taken to England in the late tenth century as part of a monastic education reform (Eastwood 1993: 168).

Throughout the medieval era, *Aratea* manuscripts continued to be copied and were included in many astronomical anthologies. The well-known twelfth-century compendium, *Liber Floridus,* contains an illustrated cycle of the constellations and has sometimes been referred to as an *Aratea*n manuscript, but it actually holds an often-copied astronomical treatise credited to Pseudo-Bede, but may contain bits of the *revised Aratus Latinus*.

Astronomical Illustrations in the *Liber Floridus* of St Omer

Another key entry in the long tradition of encyclopedic works including astronomical information and constellation images was compiled by a medieval scholar named Lambert, canon of the Church of Our Lady at St-Omer.[69] Since his autograph manuscript survives intact, we know with certainty that it was designed with a fully illustrated constellation cycle at its origin. The *Liber Floridus* was transferred to Ghent shortly after its completion in 1121, possibly as early as 1136, first at St Bavo's Abbey, and today housed at the University Library there.[70] "Illustrated autographs of twelfth-century encyclopedias are so rare that this manuscript is now protected by the Flemish Community's *Decree on the acquisition and protection of rare or exceptional movable patrimony*."[71] Lambert felt that this important information from earlier centuries could be lost, and so he culled the 'flowers' or the best of the earlier authors to copy and preserve, writing on cosmographical, biblical and historical material, but his own voice is rarely heard.

---

[68]"The Belief in Stars in the Twelfth Century," *Saxl Lectures*. This revision of the text of Harley 647 is discussed further in Chap. 7.

[69]*Liber Floridus,* Rijksuniversiteit te Ghent Bibliotheek, MS 92. St-Omer is now in northern France, in the twelfth century it was a part of Flanders, governed by Flemish counts. Lambert writes proudly of local history and of Godfrey de Saint-Omer, one of the founders of the Knights Templar.

[70]There were at least eleven medieval copies made of the *Liber Floridus*, with nine still extant in various European libraries and three partial copies; there is also a facsimile edition produced in 1968. The earliest and most accurate copy of the original is Wolfenbüttel, Ms. Gud. Lat. 201 made in the last quarter of the 12th c., the latest copy is dated 1512, The Hague, Royal Library, Ms. 128C 4, Edingen, The *Liber Floridus* consists of 287 numbered folios with illustrations on 64 of its folios; several sections are missing in the original manuscript but survive complete in its many copies.

[71]Quoted from the Ghent University website, *Liber Floridus* and its bibliography can be viewed online.

The *Liber Floridus* is truly comprehensive in its scope of information, ranging from world history, science, genealogy, botany, and religion; scholars have identified almost 100 different sources consulted as reference works for the textual material (Derolez 1998: 29).[72] He included the Apocalypse, which has been lost, but the eschatological nature of the work can still be perceived. Many of the sources that Lambert borrowed from were known to be in the library of St-Omer or at St-Bertin, a nearby monastery (Derolez 1998: 29).[73] Earlier encyclopedic works had been compiled by Isidore of Seville, Bede, and Hrabanus Maurus; Lambert quotes from these authors; other writers were utilized as models as well. Determining whether Lambert is quoting from the original works of the early authors of from medieval commentaries on the authors is very difficult.

Lambert's work was copied extensively because it includes world maps, a variety of informative illustrations on 64 of its folios, and numerous rotae, which gained greatly in popularity during the Middle Ages. A series of cosmological diagrams were inserted to assist in explaining his text and to visually summarize the main concepts necessary to understand the workings of the heavens. Lambert includes the zonal map inspired by Martianus Capella, and others he attributes to Macrobius, eclipse diagrams, the ecliptic, and a wind rotae; he used the images and diagrams effectively and amended them in order to incorporate Christian ideals (Figs. 3.10, 3.11, 3.12 and 3.13).[74]

The full cycle of constellations is illustrated on seven folios in the *Liber Floridus*. The leading folio of this section is devoted to a verbal description of the 12 months and the signs of the zodiac (88v), arranged clockwise in small squares, with Helios in his Chariot encircled in the center with the five planets shown as flowerlike stars and the crescent Luna above and below. The next six folios (89r-91v) portray 42 constellations squeezed amidst the lines of text that explain each one. The title of this section is *De ordine et positione signorum* and the descriptive text is a variant version of Ps.-Bede's *De signis coeli* (Derolez 1998: 100).[75] The constellations are divided into uneven sections, each outlined to distinguish and separate it from the others; crowded six to eight on a page to conserve space. The constellations follow in almost the exact order presented in the *Phaenomena*, starting with the image of Apollo in his quadriga. The cycle begins with Ursa Major and Minor, which are called Helice and Cynosura, (as per Aratus,

[72]The latest full study of the *Liber Floridus* was published by Derolez (1998); his work focuses on textual analysis and its paleography.

[73]Some evidence indicates that Lambert may have been the librarian at St-Omer. He also left marginal notes, corrections and additions in various books that he used. A listing is provided in the Derolez appendix, pp. 194–96. Also listed are articles written concerning the history, illustrations, and maps of the *Liber Floridus*, including an article by Hans Swarzenski, "Comments on the Figural Illustrations" and another by Jessie Poesch "The Beasts from Job in the Liber Floridus Manuscripts," pp. 5–9.

[74]The full listing and a summary of the contents of the Liber Floridus was completed by Delisle (1903).

[75]Derolez credits E. Mass, (1898) *Commentariorum in Aratum reliquiae,* Berlin, pp. 582–592.

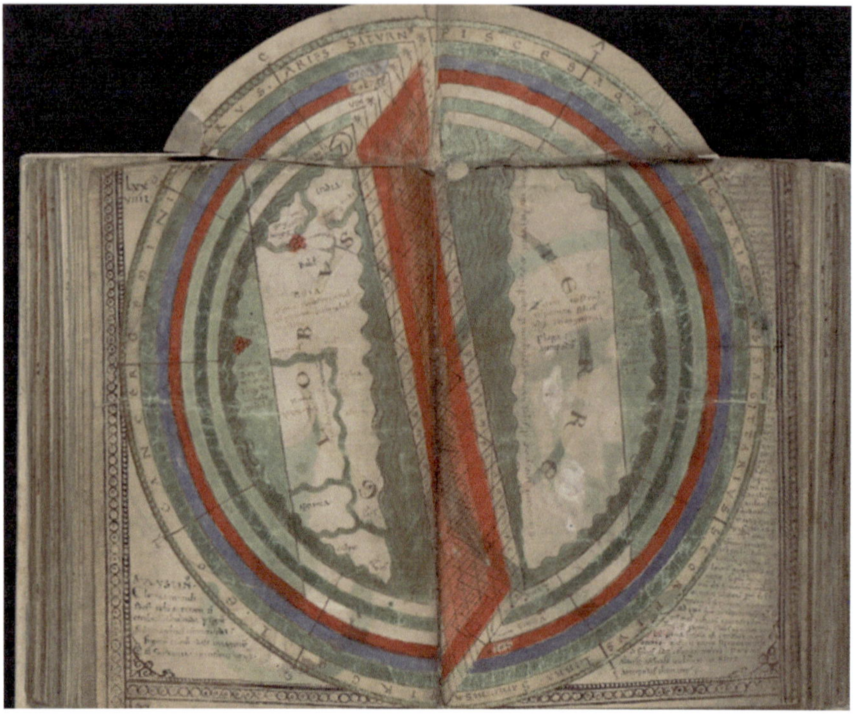

Fig. 3.10 World Map, *Liber Floridus*, Rijksuniversiteit te Ghent Bibliotheek, MS 92, Twelfth Century, Courtesy of Ghent University Library

except in reverse order); Draco is intertwined between them. The constellations illuminations are closely packed to conserve the costly parchment. The constellations end with a repeat of Corvus, Crater, and then Anticanis or Procyon (Figs. 3.14 and 3.15).

The constellation drawings of the *Liber Floridus* were most probably created by Lambert himself, since they are simply drawn and rather amateurish, and directly interwoven with the text which is written entirely in his own hand. Derolez states that Lambert writes in an old-fashioned type of late Caroline minuscule (Derolez 1998: 21). The constellation figures employ the iconographic traditions established in antiquity with their typical motifs and classical drapery style, many nude with cloaks, but his images are thought to be copied from a ninth or tenth century exemplar, not a Late Antique manuscript. Only two mythological heroes, Hercules and Ophiuchus, are depicted naked and from the back; the god-like view usually depicted on a celestial globe.

Lambert of Saint-Omer most probably observed an *Aratea* manuscript since the Leiden *Aratea* was known to have been held in the nearby Saint-Bertin library during Lambert's lifetime. Two copies of the Leiden *Aratea* were produced at

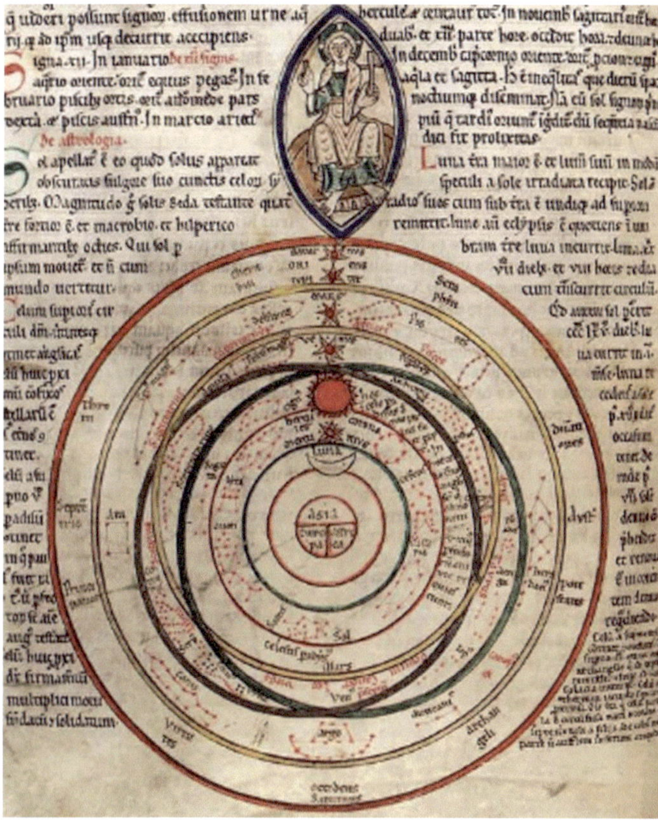

Fig. 3.11 Celestial Map, T-O Map Earth, Courses of the Planets, Christ above the Fixed Stars, *Liber Floridus*, Rijksuniversiteit te Ghent Bibliotheek, MS 92, Twelfth Century, Courtesy of Ghent University Library

Saint-Bertin during the tenth century.[76] The Leiden *Aratea* was not Lambert's model, since only one of his constellation illustrations has a close resemblance to Lambert's drawings (Derolez 1998: 31).[77] Lambert does include text excerpts from an Aratus commentary on folio 94r (Derolez 1998: 100).[78] Derolez states, "The

[76]The two copies are Boulogne-sur-Mer BM MS 188 and Bern MS 88; they are discussed in Appendix B.

[77]"Gerard Isaac Lieftinck has shown that Lambert's depictions of the constellations do not derive from one of the famous Aratus manuscripts of Saint-Bertin, as the latter descend from a Late Roman model, whilst Lambert's constellations belong to the Isidorian tradition." The 'Aratus' manuscripts held at Saint-Bertin are not Aratus poems but actually copies of the *Aratea* of Germanicus.

[78]Here Derolez credits *Excerptum de astrologia Arati*, ed. Maass n. 167, p. 309, for the source of the commentary.

Fig. 3.12 Eclipse Diagram f. *Liber Floridus*, Rijksuniversiteit te Ghent Bibliotheek, MS 92, Twelfth Century, Courtesy of Ghent University Library

Constellations are obviously copied from an Aratus manuscript, but strangely enough not from the Aratus codex in Saint-Bertin abbey" (Derolez 1998: 100).[79] Lambert's text is entirely different from the poetic form of the early *Aratea* manuscripts, as he lists the number of stars in each constellation and tells where they are positioned; his illuminations were probably modeled on the same source as his text. His grouping of many illustrations on a page interspersed between the paragraphs of prose description is similar to the Ps.-Bede design. As demonstrated, *Aratean* manuscripts were not the only source for the cycle of constellation pictures; the stellar images could be found in illustrated manuscripts of Hyginus, Isidore and Ps.-Bede.

In spite of its limitations, Derolez calls the *Liber Floridus* "a work of great originality and exceptional richness", and "…the best example of a Romanesque encyclopedic work" (Derolez 1998: 182). This type of work was used as a didactic resource in monastic, convent and cathedral schools, and then later in universities. Following, Lambert's compendium of contemporary knowledge, the scholastic

[79]Derolez continues, "One really wonders how Lambert ever had access to another copy of these rare illustrations." This study shows that the Aratean illustrations were not rare at all, but many copies dating from the Carolingian era and onward were available in monastic libraries in the twelfth century.

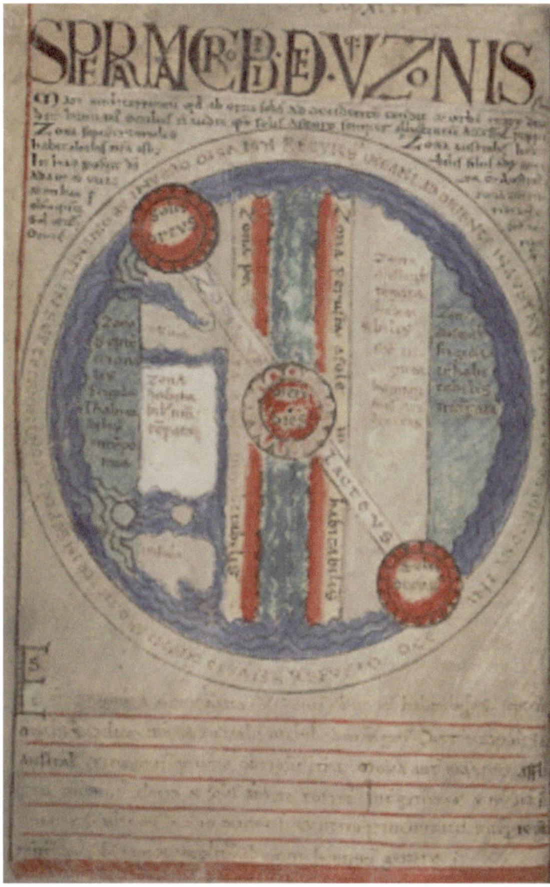

Fig. 3.13 Zonal Diagram of Macrobius f. *Liber Floridus*, Rijksuniversiteit te Ghent Bibliotheek, MS 92, Twelfth Century, Courtesy of Ghent University Library

encyclopedia emerged. But the encyclopedic tradition, in which all the world's knowledge could be compiled or condensed into one large volume, changed drastically during the twelfth century as the quantity and quality of scientific and philosophic writings exploded in an enormous efflorescence thanks to the numerous resources uncovered in Arabic libraries.

This overview of the most widely copied medieval astronomical manuscript traditions is far from comprehensive; it merely sets forth the primary candidates responsible along with its accompanying illustrations and diagrams, for the transmission of astronomical knowledge. The compilations of Macrobius, Martianus Capella, and Isidore served to transmit the remnants of ancient science without achieving any growth in originality or creativity. The writings of Bede show the participation and addition of his observations of natural phenomena, but his astronomical research was mainly directed toward the study of computus.

This chapter investigated the various transmission streams for 'popular' astronomical knowledge that was, along with the *Aratea* manuscripts, available before

Fig. 3.14 Ursa Major, Ursa Minor and Draco, *Liber Floridus*, Twelfth Century, Rijksuniversiteit te Ghent Bibliotheek, MS 92, Courtesy of Ghent University Library

the twelfth century. Each author and their writings were examined for content and for extant numbers of their treatises and to determine if they were illustrated. By looking at the circulation and surviving numbers of manuscripts of these popular authors of astronomical matters, it was possible to gauge the influence of the various manuscript traditions.

Looking back at Pliny, large numbers of his manuscripts survive and his writings include reliable astronomical information, but do not seem to contain the constellation cycle, although his works were noted for their many explanatory diagrams. The writings of Macrobius and Martianus Capella were quite popular for transmitting elementary information on astronomy, but few are illustrated. More often the illustrations are centered on personifications of the seven liberal arts, not the constellation cycle. Their writings are supplemented mainly with cosmological diagrams, celestial, and terrestrial maps. Enormous numbers of Isidore's manuscripts survive, many accompanied by diagrams, but they warrant a separate study as well to determine if or how they were illustrated. Astronomical manuscript traditions, other than the *Aratea*, that were abundantly illustrated at that early date were limited to Hyginus and Ps.-Bede. The Ps-Bede compositions, titled *De signis coeli*, are 'watered down' versions of Venerable Bede, but they are often illustrated with the full constellation cycle. The textual material is quite abbreviated; usually just a sentence or two, and the sketchy illustrations are crammed amidst the text, five or six to a folio.

Fig. 3.15 Lyra to Hydra, Corvus and Crater, *Liber Floridus*, Twelfth Century, Rijksuniversiteit te Ghent Bibliotheek, MS 92, Courtesy of Ghent University Library

Hyginus' *De astronomia* was a very important illustrated manuscript tradition. Without much directed effort, it was possible to accumulate a list of 55 surviving Hyginus manuscripts having wide distribution, with the greatest concentration appearing in Italy where there are at least 18, and in England which holds 15. Most of the Hyginus manuscripts are illuminated; many are watercolors that have an elegant, sophisticated appearance. The clothing is more traditional with soft, muted coloring compared to the bolder artwork in typical Aratean miniatures where the human figures, more often than not, appear cloaked in classical nakedness. Many *Aratea* manuscripts are bound with the writings of Hyginus; some even include a picture cycle for each author.

References

Alexander, J. J. G. ed. (1994). *The Painted Page: Italian Renaissance Book Illumination, 1450–1550*, Munich: N Y: Prestel.

Alexander, J. J. G., J. H. Marrow and L. F. Sandler (2005) *The Splendor of the Word*. London: Harvey Miller Publishers.

Bede, *Bedae, Opera De Temporibus*, ed. C.W. Jones, (1943) Cambridge: Medieval Academy of America.

Bede, *Bede: The Reckoning of Time*, (1999) translated, with introduction, notes and commentary by F. Wallis, Liverpool: Liverpool University Press.
Bischoff, B. (1994) *Manuscripts and Libraries in the Age of Charlemagne*, Ed. and trans. by M. Gorman, New York: Cambridge University Press.
Blume, D., M. Haffner, and W. Metzger. (2012) *Sternbilder des Mittelalters: Der gemalte Himmel zwischen Wissenschaft und Phantasie*. Part One: 800–1200. Berlin: Akademie Verlag.
Bober, H. (1956–57) "Illustrated medieval school-book of Bede's '*De natura rerum*',". *Journal of the Walters Art Gallery*, XIX–XX, pp. 77–91.
Bowersock, G. W., et al (1999) *Late Antiquity; A Guide to the Postclassical World*. Cambridge: Harvard University Press.
Canova, G. M (2005) *Splendor of the Word*, London: Harvey Miller Publishers, pp. 337–339.
Carruthers, M. (1990) *The Book of Memory: A Study of Memory in Medieval Culture*, Cambridge: UK, Cambridge University Press.
Carruthers, M. and Jan M. Ziolkowski, Eds. (2002) *The Medieval Craft of Memory: An Anthology of Texts and Pictures*, Philadelphia: University of Pennsylvania Press.
Derolez, A. (1998) *The Autograph Manuscript of the Liber Floridus, A Key to the Encyclopedia of Lambert of Saint-Omer*, Tvrnholti: Typographi Brepols Editores Pontifich.
Dreyer, J. L. E. (1953) *The History of Astronomy from Thales to Kepler*, New York: Dover Publications.
Eastwood, B. S. (1993) "The Astronomies of Pliny, Martianus Capella and Isidore of Seville in the Carolingian World." *Science in Western and Eastern Carolingian Times*. Eds. P. L. Butzer and D. Lohrmann, pp. 161–180.
Eastwood, B. S. (1990) "The Diagram *Spera Celestis* in the *Hortus deliciarum*: A Confused Amalgam from the Astronomies of Pliny and Martianus Capella", *Speculum,* Vol. 65, No. 4, pp. 1088–94.
Eastwood, B. S. (1986) "Plinian Astronomy in the Middle Ages and Renaissance" in *Science in the Early Roman Empire, Pliny the Elder, his Sources and influence*, R. French and F. Greenaway eds. London, pp. 197–251.
Eastwood, B. S. and G. Graßhoff (2004) "Planetary Diagrams for Roman Astronomy in Medieval Europe, c. 800–1500", *American Philosophical Society* 94, part 3, Philadelphia: American Philosophical Society.
Fletcher, K. F. B. (2013) "Hyginus' fabulae: Toward a Roman Mythology", in *Writing myth*: *Mythography in the ancient world*, Trzaskoma, S. and Smith R. S. eds. pp. 133–164.
Garrod. H. W. ed. (1911) Manilius. *Manili Astronomicon, Liber II*, Oxford: W. Typographeo Academico.
Gee, E. (2013) *Aratus and the Astronomical Tradition*, Oxford: Oxford University Press.
Gould, G. P. (1998) ed. and trans. *M. Manilii Astronomica*, Stutgardiae: Teubner.
Green, S. J. and Volk, K. eds. (2011) *Forgotten Stars: Rediscovering Manilius' Astronomica*, Oxford: Oxford University Press.
Hyginus. *Hygin. L'astronomie*, Text établi et traduit par. Le Boeuffle, A. (1983) Paris.
Jones, C. W. (1994) "Bede, the Schools and the Computus", W. M. Stevens, ed. *Variorum*, Ashgate Publishing Limited.
Kühnel, B. (2003) *The End of Time in the Order of Things: Science and Eschatology in Early Medieval Art*, Regensburg: Schnell & Steiner.
Lambert of Saint-Omer. (ca.1120) *Liber Floridus: Codex Autographus Bibliothecae Universitatis Gandavensis*, A. Derolez and I. Strubbe, ed. (1968) Ghent, Bib. Rijksunivsiteit. MS. 92, Story-Scientia.
Le Berrurier, D. O. (1978) *The Pictorial Sources of Mythological and Scientific Illustrations in Hrabanus Maurus' De rerum naturalis*, Diss. University of Chicago.
Lenaz, L. (1975) *Martiani Capellae: De Nuptiis Philologiae et Mercurii Liber Secundus*. Padua: Liviana editrice.
Levillain, L. ed. (1964) *Lupus of Ferrières, Ep. I, Correspondence*, (Paris, "Les Belles Lettres", 1964); English translation.

Lindberg, D. C. ed. (1978) *Science in the Middle Ages*, Chicago: University of Chicago Press.
Lindberg, D. C. (1992) *The Beginnings of Western Science*. Chicago: Chicago University Press.
Lippincott, K. (1957) "The Textual Tradition of the De Astronomia of Hyginus", *The Saxl Project*, Warburg Institute, University of London. Online website.
M. Manilii Astronomica, (1998) Ed. and trans. G. P. Gould, Stutgardiae: Teubner.
Manilius, *Manili Astronomicon, Liber II*, (1911) H.W. Garrod, ed. Oxford: W. Typographeo Academico.
McClusky, S. C. (1998) *Astronomies and Cultures in Early Medieval Europe*, Cambridge University Press.
McCluskey, S. C. (1993) "Astronomies in the Latin West from the Fifth to the Ninth Centuries", in *Science in Western and Eastern Civilization in Carolingian Times* P. Butzer and D. Lohrmann, eds. Basel: pp. 139–160.
McKitterick, R. Ed. (1994) "Script and Book Production" in *Carolingian Culture: Emulation and Innovation*. Cambridge: Cambridge Univ. Press, pp. 221–247.
Noble, Thomas F. X. (1998) "Lupus of Ferrières in His Carolingian Context" in *After Rome's Fall: Narrators and Sources of Early Medieval History*, Alexander C. Murray, ed. Toronto: University of Toronto Press, pp. 232–50.
North, J. (2008) *Cosmos: an Illustrated History of Astronomy and Cosmology*, Chicago: University of Chicago Press.
Ptolemy, *Ptolemy's Almagest* (1998) Translated and annotated by G.J. Toomer, Princeton: Princeton University Press.
Reeve, M. D. (1983) "*Aratea*" in *Texts and Transmission: A Survey of the Latin Classics*, L. D. Reynolds, ed. Oxford: Clarendon Press, pp. 235–38.
Regenos, G. W. (1966) Translation, Introduction and Notes, *The Letters of Lupus of Ferrières*, The Hague: no.1.
Saxl, F. (1957a) "The Belief in Stars in the Twelfth Century", *Lectures*, 2 vols. The Warburg.
Saxl, F. (1957b) "Illustrated Encyclopedias 1", *Lectures*, The Warburg Institute.
Saxl, F. and H. Meier (1966) *Verzeichnis astrologischer und mythologischer illllustrierter Handschriften des latinischen mittelalters*, 4 parts: 1. *In römischen Bibliotheken* (1915) 2. *Die Handschriften des National-Bibliotekin Wein* (1925) 3. Two vols. *Manuscripts in English Libraries* Ed. Harry Bober, London (1953) 4. *Astrological Manuscripts in Italian Libraries* ed. Patrick McGurk, London: The Warburg Institute.
Shanzer, D. R. (1986) *A Philosophical and Literary Commentary on Martianus Capella's DeNuptiis Philologiae et Mercurii Liber*, Berkeley: University of CA Press.
Stahl, W. H. (1952) *Commentary on the Dream of Scipio by Macrobius*, Translation, Introduction and Notes, New York: Columbia University Press.
Stahl, W. H. and R. W. Johnson with E.L. Burge (1971 and 1977) *Martianus Capella and the Seven Liberal Arts*, 2 vols. NY: Columbia University Press.
Stevens, W. M. (1987) *Cycles of Time and Scientific Learning in Medieval Europe*, Aldershot, UK: Variorum collected studies series.
Tester, S. J. (1987) *A History of Western Astrology*, New York: Ballantine Books.
Thomas, R. (1992) *Literacy and Orality inn Ancient Greece*, Cambridge: Cambridge University Press.
Whitney, E. (2004) *Medieval Science and Technology*, Westport: Greenwood Press.
Yates, F. (1966) *The Art of Memory*, London: Routledge.

Chapter 4
The Transmission of Astronomy and Astrology in the Medieval Islamic World

Abstract This section examines the development and great advances in astronomical and astrological knowledge in Eastern cultures, beginning with the initial establishment of Islam in the eighth century, when Muslim armies had taken over much of the collapsed Roman Empire. The discussion continues through the twelfth-century—over 500 years of progress in Arabic astronomy and in its disreputable twin—astrology. At the start of their long translation period, Islamic scholars interacted with surviving classical traditions from Europe and Byzantium, and also absorbed the astronomical knowledge accumulated by Indian, Persian, and Babylonian specialists. The utilization of paper as an inexpensive substitute for costly parchment allowed Arabic authors to create and collect enormous numbers of manuscripts which was cost prohibitive in the Latin West. Precisely illustrated manuscripts and accurate astronomical charts played a great role in popularizing and passing on the work of Muslim scholars. The foremost scientists in the Islamic world credited with establishing and transmitting celestial knowledge will be discussed. This chapter also discusses the illustrated manuscript traditions of the Arabic authors along with their illustrations, particularly Abu Ma'shar and Al-Sufi. In spite of the long distance traveled, Arabic manuscripts and artifacts were instrumental in handing down Eastern-influenced pictures of the constellations, supplementing the separate traditions that had come through Rome and Byzantium. Both pictorial traditions have left traces in the modern world, particularly in the astrological domain.

The research of contemporary scholars has established that Western mathematical astronomy derives from at least two different early cultures, Greek and Babylonian, which merged during the Hellenistic Era (Neugebauer 1983: 130). The relationships of Greek, Babylonian, and Indian astronomy can be studied with more meaningful results now that the translation and understanding of cuneiform writing have greatly improved (Pingree 1963: 229). The study of ancient Mesopotamian knowledge has become more effective with the digitizing of texts and the creation of cuneiform dictionaries by the universities of Oxford, Chicago, UCLA, and others. The practice of correlating civic, administrative, or military activities with the positions of the

M. Dolan, *Astronomical Knowledge Transmission Through Illustrated Aratea Manuscripts*, Historical & Cultural Astronomy,
DOI 10.1007/978-3-319-56784-6_4

stars and planets can be traced to the age of the Babylonians and probably even much earlier. Babylonian scholars devised many of the fundamental astronomical and mathematical concepts and systems we still use today.

Their astrological practices can be documented now that huge numbers of cuneiform tablets created by Babylonian scribes have been made available. Most early societies—including Egyptian, Persian, Indian, Chinese, and Mesoamerican—developed traditions associating kingship, warfare, religion, and ceremonial rites with the positions and diurnal movements of celestial bodies.[1] The Vedic religion, written in Sanskrit, was among the earliest to be recorded in literary form, composed over a 500- or 600-year period, no later than the eleventh-century BCE. Through these early texts, specialists have shown the relationships developed between the cosmic and the divine (North 2008: 171). But creating personal horoscopes for individuals or for specific occasions, which corresponded with a snapshot of the sky at that specific moment, was a concept formulated by Hellenistic astrologers; a concept received by each advanced society with different attitudes and varying degrees of acceptance. In general, horoscopic or natal charts directed to a specific person or event date to the late second-century BCE.[2]

Astrology plays a huge role in the history of astronomy, many early Hellenistic and Roman astronomical calculations and studies were not undertaken for purely scientific purposes but for ascertaining astrological interpretations and predictions. The general concern with the physical influences of the heavens on human lives has a long history; it was diminished but not entirely discontinued in the early medieval period. Astrology as we understand it, forecasting future actions and purposeful events in human lives based on the movements and angular positions of the stars and planets "became a skilled and public art in the early centuries B.C." This assemblage of astrological criteria of houses, zodiacal signs, paranatellonta, and lunar stations was developed by Hellenistic astronomers working mainly in Alexandria, the epicenter of Greek astrology (Tester 1987: 2).

Since the end of the seventeenth century, astrology has been viewed by scientists as an inconsequential by-product in the history of astronomy, but until that time, the two disciplines were often resolutely intermingled. In his scientific paper, "The Study of Wretched Subjects," Neugebauer states: "The only hope of obtaining a few glimpses of the astronomical methods of the time of Hipparchus rests in the painstaking investigation of wretched writers like Vettius Valens or Paulus Alexandrinus" (Neugebauer 1951: 111). The astrological portion of history, "the

[1]For example, the Chinese rituals and certain gods were associated with the four directions, the ancient Maya only called for warfare when Venus was visible (Sivin 1986). "Chinese archaeoastronomy: between two worlds" in *World Archaeoastronomy,* pp. 55–64; (Krupp 1986). "The cosmic temples of old Beijing", in *World Archaeoastronomy*; Milbrath (1999) *Star Gods of the Maya: Astronomy in Art, Folklore, and Calendars.*

[2]The earliest known horoscope was cast for the year 410 BCE (A. Sachs, Babylonian horoscopes, J. of Cuneiform Studies 6 (1952, p. 49–75) as cited by Neugebauer 1969: 197). "The earliest Greek horoscope known at this time is that of the coronation by Pompey of Antiochus I of Commagene in 62 BCE. Horoscopes on papyrus or in Greek literature start at the beginning of this era.".

superstitious flotsam" of scientific progress practiced by the 'lunatic fringe', was not considered worthy of serious study by many in the academic world until more open-minded researchers realized its influential status throughout the history of science. Like it or not, ancient astrology has now been established as an acceptable area of academic research, evidence of its impact and relevance during certain historic time periods is compelling. Astrology's place in cultural history has been suitably rectified, but still studied with considerable caution. In the later Middle Ages, astrological manuscripts have inspired creative and informative illuminations.

The establishment of Hellenistic astrology was a complicated and multisourced field; the founders built their theoretical and technical doctrines, not on original or imaginative ideas, but upon a combination of ancient Babylonian traditions and on Stoic, Middle Platonic and Neo-Pythagorean foundations. In the late centuries BCE and early centuries CE, interest in astronomical principles advanced immensely because of the powerful attraction for learning the secrets of astrology and the ability to predict or anticipate the future. By the time, Cicero and then Germanicus were translating their versions of the *Phaenomena*, astrology had come of age within the Roman Empire; but since Aratus had no model of astrology from Greek astronomers, astrological theories and practices play no significant role in the long history of the *Phaenomena* nor in the *Aratea* manuscripts. The *Aratea* texts were amended in various ways but the patrons or scribes had never introduced astrological dogma or practices.

During the Classical and Late Classical periods opinions varied, some natural philosophers argued against astrology while simultaneously accepting certain astral theories. Academic skeptics like Cicero, and Epicureans like Lucretius, "with their atheistic materialism and their desire to rid man of all superstitious fears," were bound to oppose astrology (Tester 1987: 50). In spite of these mixed opinions on the validity of astrology, most Roman emperors eagerly engaged court astrologers to attain worldly advice. We saw that Emperor Augustus endorsed the inclusion of his astrological sign of Capricorn on artworks and carved gems. Emperor Tiberius of course was deeply involved with both astronomy and astrology; he became greatly influenced by his astrologer friend Thrasyllus, a grammarian and self-professed Pythagorean who studied in Rhodes.[3] During Tiberius' reign, astrology had reached its apogee throughout the Roman Empire.

By the fifteenth century, a copy of Cicero's *Aratea* with scholia, was included in a compendium of astronomical and astrological texts, Vat. Reg.1324, which showed that readers considered Cicero's poem an adjunct to the study of astrology. "Cicero himself would have been horrified at the incorporation of the *Aratea* into the discipline of astrology, which he describes at *De Divinatione* 2.90 as *deliratio incredibili.*" (Gee 2001: 523). Germanicus' attitude toward astrology does not seem

[3]Roman imperial astrologer Balbilius (d. c.79 CE), probably the son of Thrasyllus, acted as court astrologer to the Emperors Claudius, Nero and Vespasian. Balbilius was known for his astrological method of determining a subject's "length of life from starter to destroyer" (Granstan 2012).

to be recorded, but as a decisive military leader, defending distant borders, having the necessary time or patience to analyze astrological charts seems most unlikely. Mastering mathematical astronomy and determining planetary positions and their resultant interpretation requires far different skills than translating poetic verses.

During the early medieval centuries, the distinction between magic and faith, between superstition and mythology, between witchcraft and medical practice, between fairy tales and lives of the saints was not well established. Symbolic truths intermingled with fantasy, geomancy, alchemy and magic; "astrology hovered on the fringes of all this" (Tester 1987: 145). In the later Middle Ages 'scientific astrology' was reintroduced to the West fortified by newly discovered scholarly texts by respected Hellenistic astronomers and further supported by Arabic treatises written by wise men of the East, reminiscent of the Magi who long ago were said to read the stars and follow them to Bethlehem to worship the Christ Child.

In "The Transmission of Astrology in the Early Middle Ages," Valerie Flint states that previously, those recounting the history of astrology's transmittal assumed that the subject was virtually unknown in the West until the ninth and tenth centuries, but she challenges that view (Flint 1990).[4] Others thought that in the clock-less ages before the twelfth century, only a faint memory of a lost and disreputable art lingered in the minds of those interested in astronomy. Historians also stated that after the twelfth century astrology was "rehabilitated as a discipline appropriate to Christians," mainly through the Islamic intermediaries. Again Flint disagrees, arguing that the science of astrology was known and practiced throughout Europe in the early Middle Ages, but not by scientists or scholars, but by non-Christian leaders and 'magicians'. Those who practiced this 'underground' astrology kept interest alive, but added nothing to its written principles and left no manuscripts for verification.

> During the later Middle Ages, in fact until the end of the seventeenth century, astrology was intimately associated with astronomy; the terms were often used interchangeably. Astrology survived from its Hellenistic beginnings for a thousand years with very few changes, but its practice was frowned upon by both church and state (Tester 1987: 94).

Often criticized as contradicting the notion of free will, predictive or 'judicial' astrology, when acting on human lives, was controversial and seldom reinforced by empirical evidence. Since the practitioner of astrology believed in a combination of fate, chance, and free will, the astrologer was sometimes accused of conjuring a close connection with demonic magic. At the same time, the scientific study of the apparent physical influences of the stars and planets on the earth was regarded as a recognized part of natural philosophy, especially when involving medical procedures and predictions. Medical practitioners believed that each sign of the zodiac influenced a different part of the human body and ruled the effectiveness of

[4]Tester states emphatically that there was no astrology in Western Europe from the early sixth century to the late twelfth century. To attest to this he comments that there was no known mention or condemnation of astrological practices in the writings of church fathers or in sermons, but warnings of the evils of demons, superstitions and necromancy were common (1987: 200).

medicine and disease. Medicinal plants and herbs were collected only when the aligned planet was visible, otherwise the cures would not be successful. The line between occult and scientific astrology, which predicted the probable actions of physical bodies and the course of life, was often blurred.

Astrology, though practiced, did not actually reach its zenith in the Latin West until the twelfth century. The scientific aspects of astrology acquired a new intellectual framework in the West with the translation in the 1130s of two of the most important astrological texts: Ptolemy's *Tetrabiblos* and Abu Ma'shar's *Greater Introduction to the Science of Astrology*. The dynamic understanding and deference to astrological principles, as presented by Islamic astrologers, was reinforced by a multitude of seemingly scientific evidence which was consistent with the current understanding of astronomy. The undeniable superiority of the incorruptible heavens over the corruptible sublunary sphere was an adequate justification for accepting the influence of the stars and planets over terrestrial events and human activities. An enormous number of manuscripts in both Latin and Arabic were produced discussing and analyzing these principles; illustrations and diagrams assisted the reader in understanding and committing these ideas to memory.

Arabic Astronomical and Astrological Manuscripts

> For more than a thousand years the Muslim East and the Christian West, notwithstanding the differences in matters of creed, ideology and social traditions and the intervening opposition of defenders of orthodoxy on both sides, cultivated science in mutual contact
> Paul Kunitzsch.[5]

A dominant factor in the transmission of astronomical and astrological knowledge in the later Middle Ages can be attributed to the contribution of scientists and astronomers from the Islamic World through Arabic manuscripts, along with their diagrams and illustrations. Science arose in Arabic-speaking countries shortly after Islam was established. Leaders of the early Islamic empire, inspired by the Qur'an's dictum to study all of God's works, appointed scholars to collect Persian, Sanskrit and especially Greek texts in order to translate them into Arabic for study and application. "One of the most important events in the history of science was the transmission of the Greek heritage to the Islamic World," (Goldstein 1986: 82). Under the sponsorship of enlightened rulers, intellectual centers, mainly in the cities of present-day Iraq, especially Baghdad, were established to encourage the collection and scrutiny of these early scientific texts. "Unprecedented numbers of astronomers with well-deserved and lasting reputations" were working and writing scientific texts in Muslim countries (North 2008: 199). The first astronomical works

[5]Quoted from *Scientific Contacts and Influences between the Islamic World and Europe: the Case for Astronomy*—online article at Muslim Heritage website.

of a mathematical nature written in Arabic were sets of tables called a *zij*[6]; these initial writings came from the area which is now Afghanistan and the Sind area of Pakistan. Through their usage over time, Islamic astronomers made many improvements on the details of the *zij*.

Several centuries of scholarship and experimentation, based on the received knowledge, brought impressive advancements in all areas of science, medicine and philosophy; these remarkable developments achieved in the Middle East were completely unknown in the Latin West. Early medieval Europe no longer had access to the writings and knowledge detailed by the ancient Greeks, but Muslim scientists searched out these ancient works, then translated and copied those they discovered. As a result they possessed Arabic translations of virtually the entire corpus of Greek science. Because of their great concern with celestial phenomena, medieval Islamic astronomers and astrologers studied and commented upon their accumulated astronomical knowledge for more than 500 years before their accomplishments were realized by their counterparts in the West.[7] Arabic scholars in the Near East, Northern Africa, and Andalusia worked and reworked astronomical principles, observations, and mathematics, from which they produced a significant number of original writings. In addition to their own cultural and mythological heritage, Arabic authors assimilated the corpus of astronomical knowledge from Greek, Byzantine, Babylonian, and Indian sources.

Recent research has questioned to what degree and in what ways Hellenism remained a strong influence in that its essential writings had endured years and years of interpretation and change. These primary works were rarely used exactly as written, but as raw material borrowed and manipulated for another's use or advantage. Large sections of the original scientific writings of early cultures, Syriac, Persian, and especially Greek had been "translated, copied, amended, and adopted" for almost 500 years before they were adopted by Arabic scholars. Montgomery asks, "To what extent was this material still specifically Hellenistic?" (Montgomery 2000: 91). Without the original works for verification, the later translations may always be questioned by specialists.

The enlightened writings of Arabic scientists had a profound effect on the advancement of knowledge as their achievements were especially noteworthy not only in astronomy, but also in mathematics, chemistry, optics, and medicine. Through the utilization of astronomical instruments and construction of

[6]A *zij* was a compilation, including a new type of table in which celestial times and positions of Sun, Moon, planets, and stars were tabulated. The *zij* also explained how to use the tables and how to solve astronomical problems. A *zij* also contained mathematical aids for their calculation, canons, and commentaries. Since each *zij* was specific to a certain time and place, many were produced throughout the Arab World. Their translation along with the use of the astrolabe provided European astronomers the means to make their own observations and calculations. *Zijes* were influential in introducing the use of Arabic numerals to Europe, and in accurate star mapping.

[7]I am using the term Arabic is its broadest sense, not limited to those of Arab heritage which would restrict the discussion to inhabitants of Arabia—present-day Saudi Arabia and southern Syria. Arabic science refers to all contributors under the rule of the Abbāsids, whose official language was Arabic.

observatories, libraries, and learning centers, the understanding of science and technology in the Arab World far exceeded what was known in the Latin West. The sophistication of astronomical knowledge and accurate star descriptions held in Arabic libraries can be credited to the fortunate circumstances that led to their possession of Ptolemy's great works. Since the time of Ptolemy (fl. CE130), the principal center of astronomical studies had been Alexandria, particularly at the Neo-Platonic Museion, which at its peak had collected an estimated 700,000 papyrus rolls encompassing every field of knowledge. Although greatly diminished and then mostly burned in 642, those ancient scrolls and codices that survived were transferred to Antioch in the early eighth century, by which time the prevailing Islamic leaders had taken control of the Middle East (Butzer 1993: 465). Not only scientific writings of Greek specialists had been preserved, but also astronomical instruments from Greek traditions such as the celestial globe and the astrolabe were known and used.

Arabic Translation Movement

After Muslim rulers conquered vast areas of the former Roman Empire, considerable treasures and resources came into their possession, including many regional libraries stacked with treatises on science, medicine, philosophy, and literature in Greek, Aramaic, and Latin. Important centers of learning that had survived the decline of the Roman Empire still existed in the great cities of Asia Minor and North Africa; they too came under Persian rule, invoking an intellectual awakening in the Arab World. Without delay, these early texts were studied and translated into Arabic throughout the ninth and tenth centuries, inspiring original contributions. During the first decades of the ninth century, the *Phaenomena* of Aratus was translated from Greek into Arabic, perhaps by a Jewish astronomer Abu 'Uthmān Sahl b. Habi̇̄b who was the court astronomer of a general under caliph al-Ma'mun (Dekker 2013: 279). Unfortunately no early copies survive in Arabic and it is not known if it was illustrated. Many translations of Ptolemy's writings were available as well. Even before the translation movement of the Islamic era, astronomical and medical texts had been translated from India to Pahlavi in the Sasanian Empire and from Greek to Syriac in the Byzantine, Sasanian, and early Umayyad empires (al-Khalili 2011: 39). Many scholars tend to give precedence to the Greco-Arabic translation events, but several historians, particularly David Pingree and George Saliba, have studied and emphasis the influence of Indian and Persian contributions. At the time of the ninth-century translations, Islamic astronomy had already been well established. The combined knowledge of Eastern and Western traditions in astronomy and astrology opened new areas in which Arabic scholars had made exciting progress in written works, observations, stellar charts, and the utilization of astronomical instruments.

The Abbāsid Caliphate, which ruled the Islamic Empire from 750 to 1258, created one of the most influential and longest lived Muslim dynasties, reaching toward China and India in the East and toward Byzantium in the West; they profited from adopting advanced ideas and practices from these enlightened cultures. After conquering Damascus and basically eliminating the Umayyad Caliphate, one of the most effective changes carried out by the Abbāsids was to relocate the capital of the Islamic empire; the administrative center of the Abbāsid Caliphate was transferred to a new city much further east, rising on the banks of the Tigris River.

The second Abbāsid caliph, al-Mansūr (714–775), who secured the dynasty's claim to power, became the chief architect of the early Abbāsid Empire and initiated the creation of a new capital city with a unique, symbolic design. In 762, al-Mansūr relocated the power base of his empire from Syria, where it had been based throughout the Umayyad period, first to Kufa and then to his newly founded city on the Tigris. Intended to be an impressive capital, Baghdad (also called Madinet as-Salam or City of Peace) was created in a circular pattern representing the circle as the most ideal form, an example of the radial-concentric scheme with no beginning or end (Kostoff 1991: 13). Surrounded by 80-foot-high defensive walls, the caliph's luxurious palace dominated the central position of the geometric ground plan for within the center lies eternity.

The Abbāsids were of Persian background where astrology and astrological medical practices had strong roots. By the mid-eighth century, Hellenistic astrology had ascended to such prominence in the Islamic empire that al-Mansūr directed his three best astrologers to calculate the exact date and time that would insure an auspicious beginning for the Empire's new administrative center (al-Khalili 2011: 40).[8] Baghdad rapidly became the largest and wealthiest city in the world outside of China; the cultural and political center of the Arabic world, where art, poetry, science, and astronomy flourished profusely.[9] The huge expanse of the Abbāsid Caliphate territory allowed contact and shared borders with many distant empires; therefore, along with Greek and Roman texts, scholars also had access to the collected knowledge of Persia, Egypt, India, China, and Byzantium which, beginning in the mid-eighth century, motivated the Arabic translation movement. By the mid-ninth century, a new tradition of original scientific and philosophical scholarship encouraged an even stronger desire for more translations of earlier works. Many of these early books were illustrated, including astronomical, historical fables, and literary works.[10] The initial appearance of illustrated books in Arabic cannot be definitely verified since no illustrated manuscript survives before

[8]The three astrologers were an Arab, al-Fazāri, a Persian Nawbakht and a Jew, Mashā allah, and also Umar b. Farrukhn al-Abar; they agreed on the date July 30, 762 for the setting of the first stone in the founding of Baghdad. Al- Fazāri, is thought to be the first Muslim to build an astrolabe and he wrote on the use of the armillary sphere. Al- Fazāri was also involved with translating astronomical texts from Sanskrit into Arabic, although there were several scholars with that same name. See also al-Fazari, Encyclopedia of Islam, Third Edition (2016) by Julio Samsó.

[9]For additional references on the many Islamic astronomers and scholars see, Goldstein (1985); Lindberg (1992); Kunitzsch (1989a, b); and Pingree (1989).

[10]See Hoffman (2000) "Beginnings of Illustrated Arabic Books: An Intersection between Art and Scholarship."

the year 1000, when it was already well established; the earliest representative art survives mostly on loose single pages.

While the Islamic world maintained a relatively intact corpus of ancient knowledge, medieval Europe, in stark contrast, had lost access to the Hellenistic texts and to most received knowledge of Greek science. While in the East, intellectual activities and teaching were encouraged and supported by many powerful Abbāsid caliphs; two of the most influential early caliphs recognized for sponsoring Arabic scholars were the grandson of al-Mansūr, Hārūn al-Rashīd (763–809), and his son al-Ma'mūn (786–833), during whose reign Baghdad almost certainly reached its apogee. Hārūn al-Rashīd, the fifth caliph of the Abbāsid dynasty, brought enormous prosperity to his circular capital city, which had quickly expanded far beyond its original walls; activities of his court inspired *The Thousand and One Nights* tales of Shahrazad.

Research into the details of the political climate during the rule of Hārūn al-Rashīd has led Saliba to assert that the ability to rise in the reigning court was aided by specialization in advanced sciences. This drive for knowledge would help those involved gain an edge in the struggles for powerful positions. This competition concentrated attention on translating the more advanced sciences (Saliba 2007: 65). These conditions and others need more investigation to provide an alternative motive for the Arabic translation movement. Saliba is working to verify this alternate reality rather than the traditional one in the development of Islamic thought.

> It is no wonder that the names of philosophers and scientists who were not even contemporaries and who all produced their knowledge before the third century of the Christian era (Plato, Aristotle, Galen, Ptolemy and Diophantus among others) became household names in ninth-century Baghdad (Saliba 2007: 66).

Known as a generous patron of the arts and sciences and famous for the wealth and luxury of his royal court, Hārūn al-Rashīd sent Arabic agents to Byzantium and to the distant courts of India and China in search of important manuscripts to copy and to translate and also requesting contact with scholars (Lindberg 1992: 168). Having heard of Charlemagne's vast kingdom in the West, Hārūn sent envoys bearing gifts to the emperor's court at far-off Aachen. A biographer of Charlemagne, Notker the Stammerer, (c.840–912) relates the tale of diplomats sent by the Persians during the reign of Caliph Hārūn al-Rashīd. "These same Persian envoys brought the Emperor an elephant, some monkeys, balsam, nard, unguents of various sorts, spices, scents and a wide variety of medicaments."[11] The reverse was true as well, for Hārūn's court became an attraction drawing many visitors who wished to observe his successful programs and effective practices. Notker also wrote of the representatives from his Frankish kingdom that traveled to Baghdad bearing gifts in order to pay homage and to consult with Caliph al-Rashīd; who in return sent even more elaborate gifts to Charlemagne, including a much-admired

[11]Notker, the Stammerer, Thorpe L. ed. (1969) *Two Lives of Charlemagne*, 143–45, written originally 883/4.

Fig. 4.1 *Caliph Hārūn al-Rashīd Receives the Delegation of Charlemagne in Baghdad,* 786 by Julian Köchert, 1864. Oil on canvas Public domain; courtesy of Wikimedia

mechanical water clock. From 797 onward, a frequent exchange of delegations took place between Hārūn al-Rashīd and Charlemagne; they accepted and admired each other as the most powerful rulers of their respective empires. Even though these contacts flourished, none of the scientific knowledge under study in the Islamic world seems to have reached Europe or the scholarly court of Charlemagne (Edson and Savage-Smith 2004: 29) (Fig. 4.1).

The House of Wisdom

The House of Wisdom *(Bayt al-Hikma)* was a distinct institution begun under Caliph Hārūn al-Rashīd and then promoted to its zenith by his son, al-Ma'mūn, who became the seventh Abbāsid caliph after a civil war with his older brother, al-Amin. This 'proto-research institute' was thought to include a library, observatory instruments, and translation rooms where noted scientists studied, taught, and composed texts (Butzer 1993: 466–67).[12] Since no evidence of the physical location has been found today, some scholars question the exaggerated claims of past historians and even its actual existence. Whether the House of Wisdom existed as an actual building or was strictly an intellectual idea, impressive scholarship and original research was carried out under its aegis. Following the preliminary

---

[12]Among the astronomers and mathematicians who were thought to work at the House of Wisdom were al-Jawhar, and al-Farghan who wrote an exposition of the *Almagest* which was translated in 1137 by John of Seville and Gerard of Cremona. The first philosopher of Islam, Abu Yūsuf Ya'qūb ibn 'Isḥāq aṣ-Ṣabbāḥ al-Kindī (801–873) also worked in Iraq at this time.

objectives of his father, al-Ma'mūn expanded the extraordinary center for the study of the sciences, especially astronomy; as a result, the House of Wisdom, by the mid-ninth century, was said to be the world's largest book depository. Scientific knowledge advanced greatly during the reigns of Hārūn al-Rashīd and al-Ma'mūn through the work of numerous specialists in Baghdad. Both Muslim and non-Muslim scholars were able to study and translate the accumulated knowledge of scientific history in one specific location and into one language—Arabic. The learned class thrived not only in Baghdad but within all the major cities of the Islamic empire; competent scholars were drawn from many diverse populations (Goldstein 1986: 80) (Figs. 4.2 and 4.3).

Beginning in al-Rashīd's reign, an important series of translations of the most desired astronomical work, Ptolemy's *Almagest,* were undertaken for the first time, texts were copied from Greek and Syriac into Arabic, but no volumes of these initial works are extant. The earliest surviving translation of the *Almagest* was produced circa 830 by al-Hajjaj (c.807–c.874), during the rule of al-Ma'mūn. Al-Hajjaj was a well-known Arabic translator who traveled extensively throughout the Near East in order to collect scientific texts (Goldstein 1986: 84). Unfortunately hardly any significant Arabic texts survive from the first two centuries of the Islamic period. But despite the paucity of extant texts, experts are able to reconstruct the

Fig. 4.2 House of Wisdom, Baghdad Public Domain

Fig. 4.3 Al-Ma'mūn Exchange of Information, Public Pomain

works produced during this prolific period by referring to the reports of the Muslim scholar al-Biruni (973–1048) and other later authors (Goldstein 1986: 91). By seeking out scientific texts from India, Persia, Egypt, and Greece, diligent Arabic scholars synthesized their diverse sources into a universal body of knowledge which they nurtured and expanded.[13] Under the aegis of al-Ma'mūn, the earliest known Arabic treatise concerning astrolabes was written, texts of both Euclid and Aristotle were also translated in addition to many others (Fig. 4.4).

Historians have stated that Caliph al-Ma'mūn was well educated; a scholar in his own right, especially interested in astronomy (Hattstein 2000: 90). He built the first astronomical observatories in the Islamic world and commissioned astronomers to make accurate observations and to compare them with the values recorded by early Greek astronomers (al-Khalili 2011: 78). His fascination with ancient Egypt can be easily confirmed, for he is remembered for breaking into the Great Pyramid in 820, the first person to explore its interior rooms and corridors since the Old Kingdom. Records indicate at least one important scientific expedition took place under the auspices of this academy using groups of scientists working together to determine the length of a terrestrial degree, an unprecedented investigation for this time period (Goldstein 1986: 85). Under the rule of al-Ma'mūn, diligent Islamic astronomers improved methods for measuring and calculating the motions of celestial bodies and created physical models of the universe which imitated the orbits of the planets. The accuracy of observations achieved in the ninth century by these pre-telescopic Islamic astronomers was surprisingly precise, within a few minutes of arc.[14]

[13]See also Hill (1993). "Science and Technology in 9th Century Baghdad" in *Science in Western and Eastern Civilization in Carolingian Times*, and (Kunitzsch 2004). *Stars and Numbers: Astronomy and Mathematics in the Medieval Arab and Western Worlds.*

[14]Their precisely recorded cosmological positions were not exceeded until the observations of Tycho Brahe in the sixteenth century, who was able to make slight improvements on the Arabic annotations. The noted Danish astronomer made the most detailed and accurate astronomical, stellar and planetary observations possible with the naked eye (Goldstein 1986: 87).

Fig. 4.4 Embassy of John the Grammarian in 829, Emperor Theophilos and Abbasid Caliph Al-Ma'mun. Wikimedia Commons, the free media repository

The Abbasid caliphs provided essential support for the translation process that occurred during the early establishment of Islam, a process that lasted two centuries, creating hundreds of scientific manuscripts that drew from four major languages: Syriac, Greek, Pahlavi, and Sanskrit. "The secular knowledge introduced in this manner eventually came to affect writing and thought in almost every conceivable discipline…" (Montgomery 2000: 93). The Arabic translation movement has been studied and transcribed by Western historians through the nineteenth and twentieth centuries, but the focus of scholarship since the start of the twenty-first century has expanded into new directions after gaining awareness of previously unrealized attitudes.

Current scholarship has progressed significantly since recognizing an inherent bias in many earlier writings that described the Golden Age of Islam and its role in the transmission of scientific knowledge. Recent articles have pointed out the misunderstandings in previous accounts of the role and accomplishments of the Arabic translation movement. Arabic authors were respected merely for the preservation and safekeeping of early scientific knowledge; their acknowledged role was seen as rescuing and passing on the Greek heritage to its rightful owners in the Latin West. These ethnocentric assumptions by earlier historians imply that the true home of Greek knowledge was Western Europe and that the temporary possession of that knowledge by Arabic intermediaries had little consequence other than transferring that rich legacy (Montgomery 2000: 132). The studied opinions of scholars, such as Edward Said author of the controversial-but-ground-breaking book *Orientalism*, have pointed out that subjective Orientalist viewpoint, explaining how history written by non-Muslim about Muslims often degrades the original contributions and portrays the culture in unrealistic ways.[15] Although Said wrote of historical attitudes instituted by colonialism in the nineteenth and twentieth centuries, those same underlying methodologies may have influenced earlier studies of the Arabic translation movement.

> *To investigate Orientalism is also to propose intellectual ways* for handling the methodological problems that history has brought forward, so to speak, in its subject matter, the Orient (Said 1978: 110).

Astronomy, Astrology, and Islamic Religion

Through the centuries, Muslim cities such as Cairo, Isfahan, and most prominently Baghdad attracted a multitude of leading scientists who endeavored to transcribe early scientific manuscripts from many cultures in order to study their contents and write commentaries on them. The caliphs invited scholars with expertise not only in astronomy, but also in mathematics, medicine and architecture; numerous original works were encouraged and composed under their patronage.[16] Throughout their history, astronomy was a particular attraction for Arabic speakers in order to provide directions, especially during dark nights at sea and on long caravan treks through the blistering deserts, as they often had to transverse the trackless expanse at night.

Arabic astronomers were active in studying the phenomena of the heavens, in making more accurate observations with their improved and refined instruments, and in methods for recording eclipses, comets and stellar positions. This body of knowledge was established partly because of specific obligations and requirements of the Islamic religion. Just as Christianity required astronomical readings for determining religious holidays, for orienting churches toward the east and for determining proper times for the monastic rounds of daily prayer, similar needs confronted the Muslim world. Three of the Five Pillars of Islam require astronomically precise detailed information: the orientation of obligatory prayers and sacred architecture, fasting times during the month of Ramadan, and the one-time pilgrimage to the sacred city of Mecca.

[15]See also Sabra (1987). Sabra argued against the theories of Pierre Duhem, and demonstrated that Islamic societies did not merely receive and protect Greek science passively but actively appropriated and modified it. The inherent Orientalism prevalent in art history is another topic currently under revision.

[16]Arabic scholars had made great advances in mathematics by applying the concept of zero and substituting the cumbersome Roman numerals with Indian/Arabic numbers. They achieved huge progress in astronomy by improving their measuring instruments, keeping accurate charts, developing technical terminology, applying mathematics, and naming individual stars. They enriched scientific knowledge with many original and creative ideas, discoveries, and inventions.

Accurate calendars and directions were vital for all Islamic religious observances, especially for the required visit to Mecca for every able Muslim. Mosques must be oriented toward the Ka'aba, the sacred shrine within the holy city. Determining a precise alignment between a mosque and Mecca encouraged the development of mathematical and spherical geography and accurate calculations of latitude and longitude. The *qibla* problem (the holy niche directed toward Mecca) had to be addressed every time a new mosque was built anywhere in the widespread Islamic world. Sustained by the rapid promulgation of the Islamic religion, astronomers and mathematicians were required to determine accurate hours for the call to prayer, the moment of sunrise and sunset for fasting, and to track the phases of the moon that marked the start of a new lunar month. Many mosques engaged a full-time astronomer, called a *muqqawit*, to determine the hours of prayer and consult lunar calendars to fix the dates for Ramadan and other religious occasions. All of these essential religious requirements motivated increasingly sophisticated and advanced astronomical knowledge and the creation of an improved calendrical system.

The attitudes of medieval Muslim religious figures to astrological practices were similar to the attitudes of medieval Christian clergymen. The number of theologians, jurists, and philosophers who wrote anti-astrological texts indicates that the practice was controversial and not fully accepted as a scientific or ethical activity. Many Muslims thought that to suggest that influences, other than from God, could determine human actions was anathema to the strict doctrines of Islam. Even though religious authorities mostly rejected astrology's fundamental tenets, described as taking positions that belong only to God, astrological systems, without doubt, became a significant part of the Islamic tradition. New astrological procedures were borrowed from the highly developed cultures of Egypt, India, and Persia; their traditions were integrated with Hellenistic principles. Among the Jews in the eastern regions, astrology was publicly discouraged by the orthodox leaders; but in the early years of Islam, it was practiced by many of both faiths; Jewish astrological writers became well respected. A few of the best known were Sahl ibn Bishr, Rabban al-Tabari and one of the most influential astrologers of all time, Māshā'allāh (North 2008: 188) (Figs. 4.5 and 4.6).

Abu Ma'shar, al-Biruni (976–1048), and al-Din al-Tusi (1201–1274) were just a few of the well-known Muslim astronomers who wrote widely read astrological texts. Detailed manuscripts, explaining the complicated astrological concepts, were produced and then copied in huge numbers; many were enhanced with elegant calligraphy and imaginative illustrations. The raw materials used to record and consolidate this rapidly accumulating knowledge were a supplemental factor in preserving these written volumes within a traditionally oral culture. An unexpected factor in the transmission of astronomical and astrological data can be attributed to the impact of paper and paper production. Paper was not commonly used in the Latin West until the implementation of the printing press in the mid-fifteenth century, but paper was used extensively and inexpensively throughout the Arabic World.

Fig. 4.5 Islamic Astronomers

Fig. 4.6 Al-Biruni Commemorative Stamp

Arabic Manuscripts and the Production of Paper

When the Islamic religion began to take control politically and philosophically in the century after the death of the Prophet in 632, Arabs were mainly an unlettered people in an oral society. The script used at that time was elementary and undeveloped, but by the time the Abbāsid Caliphate became established in Baghdad in the late eighth century, the Arabic script had undergone effective reformation. The developing art of calligraphy brought new emphasis on the written word. The major factor inspiring this artistic rendering of text was the sacred nature of the Qur'an.

Among Islamic devotees, that holy text is believed a direct revelation from God; therefore the words must be written perfectly with respect and elegance worthy of its divine origin. Arabic calligraphy itself became an artistic manifestation of deep spirituality (Gaur 1984: 165). In the Islamic world, a talented scribe was highly regarded, especially one considered skillful enough to copy the Qur'an; calligraphers were honored for being the instrument of transmitting God's word and being aligned with God's will. The three dominant religions—Christianity, Judaism, and Islam—are all religions of the Book; sacred writings—new or old—were always treated with extreme care and devotion.

The majority of medieval manuscripts produced in the Latin West were written on prepared animal skin—sheep, goat, or the most common, calf. Exceptional volumes, commissioned for kings and cardinals, often used intrauterine skins, the softest, whitest, and most pliable parchment available. Parchment or vellum was quite expensive as the process required a lengthy, labor-intensive, and malodorous ordeal to prepare the animal skins for the reception of the calligraphy of scribes and the paints of artists. Creating an entire Bible would require killing a huge number of animals, often well over a hundred. The costly materials needed to produce an illuminated manuscript limited their acquisition to the most prominent members of society. Texts for students and monastic libraries were more cheaply made but still beyond the reach of most readers. A less-costly material was not introduced to European scribes until the end of the first millennium. Paper was invented by the Chinese in the second- or first-century BCE; but the Muslim world was most instrumental in developing its variety of possibilities. The Latin West adopted the use of paper, not directly from the Chinese, but from Muslims (or Moors) in Andalusia where they began to imitate the Arabic practice begun in the East of using cheap and durable paper, but only for government and commercial records initially. The first European paper factory was built in Xativa, Spain (near Valencia) in the twelfth century, which produced paper for the dissemination of Arabic texts before and after they were translated into Latin.

In Jewish communities during the Talmudic period, strict rules governed the writing of the Torah scroll. The Hebrew scribe was not allowed to write from memory but was always required to copy the text from a written exemplar. Before beginning work on sacred scripture, a scribe was expected to take a ritual bath and prepare himself mentally, so that he would be worthy to write the name of God in a proper state of devotion and ritual purity (Gaur 1984: 154). Christian scribes had similar obligations of silence and respect when copying the words of Holy Scripture and the writings of honored disciples and Church Fathers. The fundamental importance of writing in Islamic cultures is attributed to the dominant position of the Qur'an. The most sacred writings of all three religions were copied on parchment until Islamic scribes adopted the use of paper for all their written documents, even the Qur'an.

Muslims first encountered paper in the early eighth century after conquering major cities in central Asia. The products they had previously used for writing, scarce Egyptian papyrus and expensive parchment folios, were quickly replaced by Chinese-style paper, which could be produced cheaply in almost unlimited amounts, using a variety of easily accessed raw materials. Artistic Muslim craftsmen soon turned papermaking into a sophisticated art form, creating reams of various qualities, colors, and textures. By the late eighth century, paper production had spread to many other Islamic cities; ultimately the caliph ordered that paper was to be used for all writing. In the ninth and tenth centuries, paper making throughout Islamic territories became "a remarkable historical and technological achievement that transformed society in its wake" (Bloom 2001: 47). During the ensuing centuries, paper quality became finer and whiter but calligraphers often preferred paper with tinting; they felt it was easier on the eyes. The preferred shades were reddish yellow, reddish orange, lime green, pistachio, and buff. Through time Arabic manuscripts increased dramatically in size and number produced. After paper was established as the accepted material, new stylish and elegant scripts were developed and deemed acceptable for copying the Qur'an. The papyrus-producing industry, which had thrived in the Nile Delta of Egypt for over 4000 years, no longer prospered.

In China, conditions were ideal for production of vast amounts of paper; its warm and humid semitropical regions were perfect for growing the plant fibers needed for manufacturing paper. While in the country's many arid areas, the papermakers used textile waste, especially linen and cotton rags; luxurious paper was made from silk as early as 650 CE. Chinese papermaking quickly spread to many other Asian countries. In the eighth century, the Chinese invented printing which resulted in great advances in the process of papermaking; this in turn created an increased demand for paper (Bloom 2001: 73). Handled carefully, paper can survive almost as long as papyrus or parchment. One of the most important ancient astronomical documents, surviving on paper, was discovered in a Chinese treasure cave sealed for a 1000 years.

The World's Oldest Surviving Star Map

In 1909, the Hungarian-British archeologist Sir Aurel Stein, after many years of study and preparation, began an exploration of the Silk Road along the easternmost areas of central Asia. Stein had heard accounts of great possibilities for discovering valuable artifacts, historical documents, and art works that survived from early caravans, travelers, and settlers.[17] When undertaking his archeological quest to these vast unexplored and mostly isolated areas, Stein followed the route taken by Buddhist monks, and traders not unlike Marco Polo, throughout the first millennium, the network of age-old caravan trails from the Mediterranean across Central

[17]Stein was Hungarian-born but had adopted British nationality; he worked under the auspices of the Archaeological Survey of India, controlled by Great Britain. He knew of the potential of that specific area from Count Loczy who had been a member of a Hungarian expedition two decades earlier. Stein made three separate expeditions within fifteen years (Whitfield 1999: 3).

Asia to the Chinese capital Chang'an or Xi'an, known today as the Silk Road (Whitfield 1999: 2).

Inspired by the writings of a seventh-century Chinese Buddhist monk named Xuanzang, who had undertaken a pilgrimage seeking spiritual rewards in India, Stein investigated the long-buried remnants of that earlier era.[18] From its origins at Varanasi in northern India, Buddhism spread throughout Southeast Asia conveying its extraordinary art and architecture, and recording its philosophical dogmas utilizing the Chinese techniques of making brushes, ink, and especially paper (Bloom 2001: 38). Buddhist monks created an enormous number of temples, stupas, and monastic dwellings, many were decorated with religious sculptures and exquisitely painted frescoes; most temples were carved into mountain sides or established in natural caves. Most temples and monastic communities held libraries of written scrolls containing Buddhist sutras and canonical texts of Buddhism and other Eastern religions.

Having planned and lead eight major expeditions along Silk Road sites, Stein gained international acclaim for his discoveries at the Buddhist Mogao cave complex of Dunhuang, an important crossroad town on the Silk Road in northwest China. Within a protective cave sealed since the eleventh century, now called the 'library cave', Stein found almost 40,000 manuscripts and printed documents on silk and paper written in many languages, mostly Chinese and Tibetan, and in several different scripts (Whitfield 1999: 5). The writings ranged from fragments dating to the second century BCE containing just a few words to perfectly preserved scrolls reaching 70 feet in length from the seventh century.[19] Hundreds of experts in various fields are still researching these manuscripts, drawings, and artifacts that he found and rescued. Stein discovered many sites in Iran as well which retain their importance in archeological studies (Whitfield 2004: 129).

Included among the mostly Buddhist texts that Stein uncovered at Dunhuang was a manuscript containing a precise star atlas, the earliest yet known from any civilization, composed before 700 CE, but it has much earlier antecedents. (A few older maps do survive but these display only a section of the sky.) These star charts record the entire night sky visible from China onto a well-preserved 13-foot-long paper scroll; the first known pictorial presentation of the classical Chinese constellations. The initial scientific examination of the atlas revealed that it contains 1585 stars arranged in 257 groups, 'chairs' or asterisms; two recognizable con-

[18]Known for his linguistic skills, Stein spoke many languages and wrote over 20 books and nearly 100 articles. From sites along the Silk Road, he uncovered in total over a hundred thousand manuscripts, painting, murals, statues, textiles, potsherds, and items of everyday use which were given to museums worldwide (Whitfield 2004: 13).

[19]Also found at Dunhuang is the oldest surviving printed book in the world—a copy of a Chinese translation of the *Diamond Sutra* dated to 868 (Whitfield 1999: 36). The *Diamond Sutra* is a brief but essential text containing the wisdom of Mahayana Buddhism, now held by the British Library.

stellations appear to represent the Big Dipper and Orion.[20] The stars are carefully depicted in three colors—red, black and white—corresponding to the three ancient schools of positional astronomy (North 2008: 145). Experts studying the charts have suggested that the different colors used on the map indicated those stars observed by three individual Chinese astronomers from more than 1000 years earlier (Bonnet-Bidaud et al. 2009: 39–59). Like the star positions recorded by Aratus, astronomers are dating the actual star cataloging to a far earlier age. These recent analyses raise many fascinating questions (Figs. 4.7 and 4.8).

The detailed accuracy and graphic quality of the Chinese star atlas add to its significance and its important data for assisting modern astronomers. The chart includes both bright and dim stars, all those visible to the naked eye when viewed from north central China, and was probably used as a visual aid for nightly travels. The stars are plotted with remarkable accuracy using a precise mathematical projection system designed to display the curvature of the sky on a flat page of paper (British Library Dunhuang Website). The projection system is quite similar to that used by Gerardus Mercator, the sixteenth-century Flemish cartographer.[21] The place of origin and usage of the star atlas remains unknown, but dissemination of manuscripts of this type would have been carefully controlled for military and governmental purposes only. The celestial chart was probably left at Dunhuang by an official from the early Chinese capital, Chang'an. Many factors contributed to the fortunate survival of the treasure of hand-written and printed texts: the 'library cave' had been tightly sealed from bats and rats and concealed from looters; this section of the Silk Road is exceedingly arid; and the star maps and other documents were recorded on sturdy paper. The stellar identifications and relative positions carefully mapped on the Dunhuang star atlas have become a crucial document in tracing the transmission of astronomical knowledge.

Thanks to China's invention of paper, it had become the essential ingredient necessary for the transmission of religion, literature, and scientific knowledge in the Islamic world. The unlimited use of paper transformed the literary and scientific output of medieval scholars, spurring untapped creativity. The availability of continual supplies of less expensive writing material, produced from rags or plant material, encouraged greater scholarly production in all areas of research. Islamic libraries boasted huge, almost unbelievable numbers, into the hundreds of thousands, of manuscripts written on cheap paper readily available for study or for reading pleasure.

> But paper itself is a powerful medium for the transmission of knowledge and the effects of the diffusion of paper and papermaking … can be seen in the enormous and revolutionary changes to such diverse realms of human activity as literature, mathematics, geography, commerce and the arts in the Islamic lands between the eighth century and the fourteenth (Bloom 2001: 10).

[20]The writings from the 'library cave' are now scattered throughout scholarly institutions and private collections. The International Dunhuang Project, based at the British Library in London, was founded in 1994 to bring the manuscripts back together using computer technology. Many thousands of the manuscripts, and details of their contents, can now be viewed on the Project website.

[21]See also: https://www.bl.uk/collection-items/chinese-star-chart.

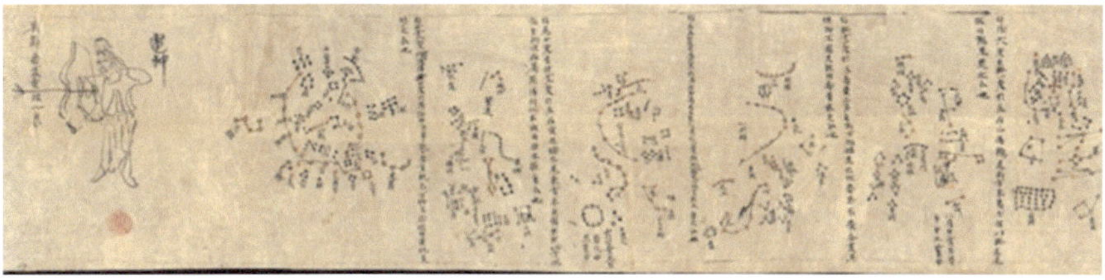

Fig. 4.7 The Dunhuang Star Atlas, last section of Or.8210/S.3326, British Library

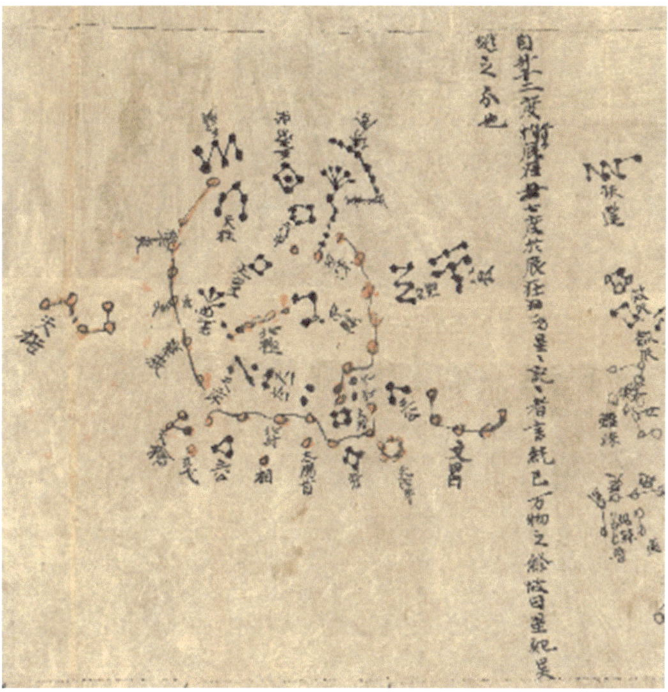

Fig. 4.8 The Dunhuang Star Atlas detail

Artistically designed Arabic calligraphy not only filled uncountable documents, books and scrolls, as visual art, it decorated Islamic architecture, frescoes, and ceramics, in fact anything from mundane serving objects to the most impressive structures, such as the impressive decorative calligraphy on the Taj Mahal. The universal presence of decorative calligraphy suggests that medieval Muslims were more familiar with the art of writing and may have been more literate than other cultures. Assuredly all Muslims could recognize the sacred word Allah (Fig. 4.9).

Fig. 4.9 Taj Mahal, Agra, India, 1632 Calligraphy as Decorative Artwork

Arabic Astronomical Manuscripts

Even before the expansive growth of Islam, the various far-flung Arabic societies already possessed a basic understanding of the heavenly motions vital for orientation in the vast desert areas and for long-distance sailing cultures such as the Phoenicians. Similar to all early cultures, the Arabic-speaking cultures accumulated and encoded astronomical knowledge for calendrical purposes, for determining the seasons, and for planning community and ceremonial events. The folk astronomy and astral myths from earlier periods expanded as Islam spread through the Middle East and North Africa and encountered the scientific heritage of Greece, and also the long-held traditions of India and Persia. Their oral traditions were necessarily recorded in manuscripts to protect, preserve, and improve them. As Islamic science developed in Baghdad, astronomy grew in popularity and the increasing demand for more exact observations and measurements was motivated by an expanding interest in astrology. Technical astronomy was cultivated in Baghdad as well as the mathematics necessary to support it; Greek cosmology, a special aspect of Greek philosophy, was studied as well (North 2008: 191).

As we have seen, the Abbāsid caliphs, until their downfall, were enthusiastic supporters of both astronomy and astrology; research centers were instituted throughout the Islamic world where original and significant advances were made.

The scholarly centers promoted by Caliph al-Ma'mun created a "huge linguistic empire, comparable with those of Greek, Latin, Sanskrit, Chinese, and later English, capable of binding together intellectually large numbers of people of different outlooks and beliefs, so that it became scarcely necessary for them to learn more than a single language to practice their science" (North 2008: 191). Some of the earliest astronomical texts translated into Arabic were from Persian sources, including a copy of Ptolemy (Montgomery 2000: 102). Baghdad ruled as the center of intellectual activity; the first real successor to Alexandria in scientific fields. Improved road systems allowed students to travel great distances and seek learning from individual masters scattered throughout the Islamic empire.

Baghdad and Cairo, both within the geographic range of former Late Antique intellectual centers, became focal points for the production of all varieties of books, including illustrated manuscripts. Illuminated astronomical manuscripts created in Arabic exhibited influences that were deeply rooted in classical iconography; but traditional styles of Islamic artwork introduced many innovations in design and scientific precision. Even though hundreds of Islamic manuscripts copied before the mid-thirteenth century survive, only about three dozen are illustrated, and of these, only three date from before the twelfth century. Manuscripts illustrated for the sheer pleasure that artwork would provide were quite rare in the early Islamic period. Unlike the sacred scriptures of Judaism and Christianity, the Qur'an was never illustrated since the calligraphic text was considered the only feasible representation of God's message. Calligraphy was admired above all the artistic achievements of Islam; calligraphers were treated with the same admiration and respect as were artists of the Italian Renaissance in the West. Unlike the European highly religious practices, the art of book illustration in the Arabic tradition was an entirely secular development, found in both literary and scientific works. Calligraphy was an honorable occupation even for nobles and aristocrats in all areas and during all eras of the Islamic empire (Avrin 1991: 268).

The three earliest illuminated books to survive in Arabic are all scientific texts in which the illustrations are essential for visualizing the text; all three were copied on paper. The use of paper played a major rule in the dissemination of the new learning among levels of society whose knowledge until then had relied entirely on oral transmission. The addition of descriptive images further advanced the reader's understanding; but since so little of the early illustrated book tradition survives, drawing conclusions about the characteristics of illustrated books in the earlier centuries is not possible.

The oldest set of Arabic illustrations has been attributed to the 1009 copy of al-Sufi's *Treatise on the Fixed Stars*, (recent problems with that designation will be discussed further on). The second oldest illustrated manuscript, dated to 1037, is a geographical text originally written some two centuries earlier by the noted mathematician and geographer Muhammad ibn Musa al-Khwārizmī. The third earliest surviving illustrated manuscript is an Arabic translation, dated 1038, of *De materia medica,* which had been originally written a 1000 years earlier by Greek physician and pharmacologist Dioscorides. Its pictures are accurate depictions of medicinal plant life with one figural painting copied from a Greek prototype (Bloom 2001: 163).

The oldest celestial map in the Arabic world (almost contemporary with the Dunhuang star map) was not drawn on parchment or paper but appears as a decorative fresco on the domed ceiling of an early bathhouse at Qusayr ‘Amra. The castle of ‘Amra, near Amman, Jordan, is thought to have been built in the first half of the eighth century. The heavenly panorama dated about the same time as the Dunhuang celestial map was derived not from a planar projection but from a celestial globe. The god-like view of the stars, depicting constellation figures from outside the heavenly sphere, indicates that the model was a celestial globe (Saxl 1932).

Emilie Savage-Smith states: “In painting this early Syrian domed ceiling, the unknown artist… was continuing a well-established pre-Islamic tradition of celestial mapping.” As on a celestial globe, the constellation figures, in spite of its ruinous condition, can be partially determined as appearing within a definite system of coordinates. Through studying photographs, Saxl identified 33 Ptolemaic constellations or partial sections. While the constellation depictions cannot be called scientifically accurate, they are evidence of a thoroughly scientific approach. The zodiac, confined to the zone of the ecliptic, testifies to the artist’s considerable skill in depicting on a hemispherical dome the whole night sky visible from one particular location. A number of technical adjustments were required, and the artist had to erase and redo several parts of his composition; demonstrating that he was interested in accuracy, not just an esthetic effect. “The question of the vault’s model is of interest not only for the history of celestial cartography and globe making but also for the prevailing ideas of transmission of astronomical knowledge” (Dekker 2013: 264). Accuracy is a crucial term that dominates Arabic astronomical art. Islamic astronomers continued to improve their observations of the heavenly objects, devising more refined methods in order to bring their observations in accord with Ptolemy’s model of the world.[22] Dekker includes a comprehensive analysis of the celestial map, constellations, and coordinates of the Qusayr ‘Amra ceiling painting in her study of star maps and celestial globes (Dekker 2013: 260–278).

The Islamic astronomer’s familiarity with Ptolemy’s cataloging system and with Arabic traditions and advancements allowed them to attach names to individual stars within each constellation. Early Greek astronomers had applied customary names to the 48 constellations but did not apply names to specific stars; Arabic astronomers assigned separate names to stars of distinctive brightness, such as Mizar in Ursa Major, Betelgeuse in Orion, or Aldebaran in Taurus, all Arabic stellar names are still in use today.

In contrast to Western images, Arabic scholars considered the visual representations of the constellations as symbols of reality, and made certain that the artists

[22]See Kunitzsch in *Scientific Contacts and Influences between the Islamic World and Europe: the Case for Astronomy* on line article for Muslim Heritage website.

were conscientious in depicting the exact number of stars, positioning each one with absolute accuracy, and even indicating each star's brightness or magnitude.[23] Although very concerned with the placement, size, and appearance of each star in their illustrations, Arabic artists were not particularly interested in retaining the early Greek myths or catasterisms applied to the constellations, or in copying the classical costumes, appearances, or their iconographic aspects. In contrast, the artists' illustrations of the constellations in most *Aratea* manuscripts pay scant attention to the placement of individual stars but do portray the heroic figures and animals in their traditional iconographic stances and attire and include their mythological tales. Unlike Latin astronomical texts which generally treated astrology as separate from astronomy, Arabic writings commonly included astrological principles as an acknowledged part of astronomy. "No Arabic astronomer can be fully understood without a thorough knowledge of astrological concepts" (Neugebauer 1951: 111).

Astrology in the Islamic World

After the fall of Alexandria to Muslim armies in the seventh century, Arabic scholars were introduced to principles of Hellenistic astrology; then in the eighth century, they conveyed those astrological principles and practices to astronomers in Baghdad. The astrological systems of other cultures were studied and absorbed as well, creating an advanced 'science' with historic dimensions. Historians can easily verify that a conflict existed between religion and astrology, and it was conceded from Hellenistic times and forward that the purpose of much astronomical research was in the service of astrology. But this does not indicate that astrology was banned from Islamic society; on the contrary it was, as in later Western cultures, quite popular with many astronomers and the general public.

Initially, horoscopic analyses and astrological readings were limited to the sphere of the royal families, high officials and elite members of society. But astrological theories quickly spread to a wider audience; more and more people were anxious to learn the secrets of the stars and to gain crucial information concerning their lives, futures and even their deaths; even though Ptolemy insists in *Tetrabiblos* that the wise man rules the stars; they do not rule him. The association of astrology with medicine and medical practices came to be the most acceptable and beneficial use of astrological knowledge; medical acumen assured the astrologers' elevated status with the leading figures of Islam as well as the common people.

[23]Ptolemy organized the stars in his *Almagest* by magnitude or brightness graded in a scale of one to six, one indicates the brightest, for instance the brilliant star Sirius is rated a first magnitude; a category six is assigned to the dimmest stars that can be seen with the naked eye. Ptolemy was influenced in this process by Hipparchus' who classified the stars as bright, half-bright, and dark. (Toomey 1998: 323).

> The first 'scientific' discipline to be systematically translated from Pahlavi into Arabic was astrology. One of the earliest texts was the hugely influential five-part astrological work of the prophet Zoroaster, the *Book of Nativities*, which was first translated into Arabic between 747 and 754. Astrology had become a perfectly acceptable branch of knowledge (Pingree 1989: 43).

Another major astrological text to be translated into Arabic in Baghdad came from India in Sanskrit around 770, known in Arabic as the *Sindhind*. Islamic astronomers learned exact methods for determining the positions of the stars and planets from the Indian Sanskrit and Persian Pahlavi manuscripts. They also copied their traditional methods for recording tables of solar, lunar, and stellar positions. For the first time, Arabic astronomers recorded the celestial knowledge of the Bedouins including the passage of the Sun and Moon through the zodiacal signs and symbolism of the lunar mansions. Although various cultural traditions were studied by Arabic astronomers, the Hellenistic Greek astrological writings were certainly the most essential for the establishment of Islamic astrology. The three essential tools of the medieval astrologer were the astrolabe, used to gauge the time by measuring the altitude of the Sun or any visible stellar object; the ephemeris, a table or star chart that lists the positions of astronomical objects in the sky at a particular time; and the dust board, a tablet covered with sand on which calculations could be made and erased. Using all available knowledge, astronomical instruments and their own abilities, Arab astronomers, and cartographers strove for—and frequently achieved—uncanny accuracy in their accomplishments. Even considering the great advances made by current researchers in uncovering the work of medieval Muslim astronomers, many support the view that Islamic scientific tradition was even "richer, more profound and with more complex relations to other cultures than had been thought" (Hogendijk and Sabra 2003: ix). As additional sources slowly become available, the true picture will become even clearer.

Arabic Astronomers and Astrologers

Throughout the ninth and tenth centuries, Islamic researchers compiled an enormous amount of new empirical data, especially in the area of astronomy, astrology, and mathematics. Caliph Al-Mansur followed the models of Persian kings of former centuries who were ardent patrons of astrological and medical practice. Most of the important astrological texts circulating in the Near East were translated into Arabic during his early reign. The primary Hellenistic text, the *Almagest*, was translated into Arabic at least four times between 800 and 890.

> The translations into Arabic took place as an essential ingredient in a broad movement that transformed Islam, generally speaking into an extremely rich, intensely active textual culture – engaged not just in study and teaching, but in writing specifically, in the production of an ever greater number and variety of manuscripts and the accumulation of these works into sometimes massive libraries (Montgomery 2000: 112).

After the extensive translation period, the major area of celestial study among Islamic scientists passed to observational and theoretical astronomy.

Abbāsid Caliph al-Ma'mūn commissioned and supported the work of many scientists; one of the most dynamic was al-Khwārizmī (c.780–863), a mathematician and astronomer, from whom we have the term algorisms. Just after that initial period of intense intellectual activity and Arabic translations, the extensive astrological writings of Abu Ma'shar, whose contribution will be examined later, were produced.[24] Abu Abd Allah Muhammad ibn Jabin al-Battani (c.850–929) was another instrumental Arab astronomer, astrologer, and mathematician. After more than 30 years of meticulous astronomical observations, he created an accurate set of astronomical tables, which, like most Arabic works, were translated into Latin in the twelfth century.

He, like other Islamic scholars, teachers, and translators did not simply absorb the ancient wisdom that was now available; they altered and amended it profoundly to suit their particular moment. During these dynamic centuries, scientific texts did not stand still as originally written; even esteemed works received commentary and alteration. Astronomers worked to explain the factual and observational inaccuracies in Greek astronomy. They would not be considered a serious professional practitioner if they did not attempt to reformulate errors found in Greek astronomy (Saliba 2007: 194). Muslim astrologers also improved and updated earlier Greek and Hellenistic authors. Even Ptolemy stated that astrology could not be considered an exact science; the possible interpretations of horoscopes were extensive and almost infinitely malleable.

Artists began to illustrate descriptions of the historical and fictional texts, as well as the constellations, with informative drawings, and they took advantage of the expanded format of Koran manuscripts to decorate the opening pages not with pictures but with extraordinarily complex geometrical patterns. Pages found at the beginning and the end of a text were often decorated with interlaces and arabesques, which had been popular additions to manuscripts of the sacred writings in the Qu'ran since earliest times (Bloom 2000: 185).

Abu Yusuf Ya'qub ibn Ishaq al-Kindi

The philosophical basis of Arabic astrology was established by Abu Yusuf Ya'qub ibn Ishaq al-Kindi (c.801–873), considered one of the most learned scholars of his time, a leading figure in the House of Wisdom. He lived in Baghdad during the reign of Caliph al-Ma'mun and served as his physician, personal tutor, and overseer of the translations of ancient philosophical and scientific manuscripts, mainly from

[24]In order to make a number of Arabic names and words more accessible to the reader, I have eliminated many diacritics and other accent marks and substituted the more familiar English forms when possible. For example Abu Ma'shar for Ab_ Ma'shar; al-Kindi for Abu Yusuf Ya'qub ibn Ishaq al-Kindi; al-Sufi for Abu al-Sufi for 'Abd al-Rahman al-S_f_; and zij instead of z_j.

Greek to Arabic. A Muslim philosopher, physician, mathematician, and astrologer, al-Kindi was honored as the founder and 'father' of Islamic philosophy, for he succeeded in assimilating Aristotelian and Neoplatonist theories into an Islamic philosophical basis. Aristotle prevailed as al-Kindi's foremost authority in philosophical doctrines, but he followed Neoplatonism also. He established a foundation for astrology as well as other occult topics, magic, and divination (Tester 1987: 158) (Fig. 4.10).[25]

Al-Kindi borrowed from classical ideas the theory of cosmic sympathy that connected the macrocosm and microcosm, like attracts like, as above so below. He thought that the relationship between the changing configurations of the heavens and events in human lives provided evidence of the totality of creation. Submitting to the will of Allah and the teachings of Islam was not incompatible with the fatalism which was inherent in astrology. Many orthodox theologians in both Islam and Christianity argued against astrological beliefs; since God was all-powerful, astrology could only be a delusion similar to magical and demonic thinking. Al-Kindi reasoned that the heavenly bodies are rational entities; their perfect circular motion and influence on earth simply obeys divinely established laws. He wrote extensively on a large variety of subjects including astronomy, astrology, and the astrolabe (Tester 1987: 158).

Many of al-Kindi's works would have been lost except for a single Arabic manuscript, housed in Istanbul, which contains most of al-Kindi's extant philosophical writings.[26] Al-Kindi wrote numerous treatises on these subjects, and his close associate Abu Ma'shar was the greatest figure in Arabic astrology. Both scientists viewed astrology as a rational science, supported by a logical theory of physical causes; they believed that the heavenly spheres were the means by which God exercised providence over the sublunary world. Although the number of Islamic astronomers who wrote of their studies and theories in Arabic includes quite large number, this essay will only discuss the few authors known for their illustrated manuscript tradition.

---

[25]A few of the many treatises written by Al-Kindi are: *The Book of the Judgement of the Stars, On the Stellar Rays, On the Revolutions of the Years, On the Signs of Astronomy as Applied to Medicine.*

[26]This manuscript was edited in Abu Rida 1950 and 1953; several important texts have been edited and translated by Rashed and Jolivet in 1998. This Istanbul manuscript also includes one of the few Arabic copies of al-Kindi's *On the Intellect* to survive; this work also survives in a Latin translation. In the same manuscript, Al-Kindi explains his cosmological theories in two texts, *On the Proximate Agent Cause of Generation and Corruption* and *On the Prostration of the Outermost Sphere*. He also wrote numerous works on meteorology and weather forecasting which apply the same cosmological theories in explaining how heavenly motion produces rain and other meteorological phenomena in the lower world. While these works are influenced by Aristotle, al-Kindi also relies on other Greek sources, especially Ptolemy. A key section of *On Rays* relates that the stars and planets bring about events in the sublunary world by means of rays emitted from the stars and planets to certain points on the earth's surface. Adamson, Peter, "Al-Kindi", *The Stanford Encyclopedia of Philosophy* (Spring 2015 Edition), E. N. Zalta (ed.), http://plato.stanford.edu/archives/spr2015/entries/al-kindi/.

Fig. 4.10 Al-Kindi Commemorative Stamp

Ab Ma'shar: Works and Surviving Arabic Manuscripts

Numerous scholars studied, wrote and taught science in the Middle East during the ninth century, several astronomers standout as the most renowned and influential. One of the earliest was Muhammad Abu Ma'shar al-Balkhi (787–886), "who stood at the crossroads of so many different traditions—the Greek, the Indian, the Iranian and the Syrian" (North 2008: 195). Persian astronomer and astrologer Abu Ma'shar was born near Balkh in Khurasan, (present-day Afghanistan) and during his life traveled throughout the Middle East. Abu Ma'shar (his Latinized name is Albumasar) worked in Baghdad under the Abbāsid Caliphate as court astronomer and astrologer, where experts in both fields were always among their staff[27] (Fig. 4.11).

While consulting with other scholars at the court of al-Ma'mun (813–33), the Neoplatonist philosopher and astrologer al-Kindi persuaded Abu Ma'shar, who was then a religious scholar, to study mathematics in order to better understand philosophy. Through his study of mathematics, he was attracted to astrology where he

[27]For a complete biography see David Pingree, "Abu Ma'shar in Dictionary of Scientific Biography, ed. Charles Gillespie (New York: Charles Scribner's Sons 1970, 1:32–39.

Fig. 4.11 Abu Ma'shar: woodcut, '*Introductorium in Astronomiam*', Erhard Ratdolt 1489

acquired a remarkable reputation as the best by far of all Islamic astrologers, virtually the founder of the topic (Saliba 2007: 36). Utilizing the philosophical arguments of Aristotelian and Platonic writers, Abu Ma'shar attempted to justify the various tenets of astrology. During his long life, Abu Ma'shar produced the seminal Arabic works on astrology that were supported by his thorough understanding of Greek philosophy; although in one of Abu Ma'shar's Arabic treatises, he presents astrology as "the teaching of the Persians" (North 2008: 187). Some of his writings held general information, some were quite specific and some were compendia of other astrologer's ideas; but all of his works became enormously influential in the field of astrology.

His known works are: (1) *The Greater Introduction to Astronomy* (*Introductorium maius*) in eight books; (2) The *Flowers of Astrology (Flores Astrologia)* in nine books; (3) *On the Great Conjunctions* in eight books; (4) On *Revolutions of Nativities* in ten books; (5) *The Thousands*,[28] in eleven books; and also excerpts of various works appeared in his *Shorter Introduction to Astronomy*. In addition, some minor works included two essays on 'editions' and a Book of Tables; all of these extended his collected astrological system.[29] The *Introductorium maius*, also called *Introduction to the Science of the Judgement of the Stars*, asserts his clearly articulated defense of astrology. Abu Ma'shar included a full chapter of his book to opponents of astrology which included skeptics from most sections of society; these defenses indicated the amount of contemporary

[28]See Pingree (1968) the *Thousands of Ab_Ma'shar,* Studies of the Warburg Institute.

[29]Abu-Ma'shar was composing both the *Ysagoge minor* and the *Introductorium maius* during the period when Aristotle's treatises were being translated from Greek into Arabic at the House of Wisdom. Consequently, he absorbed the philosophical views of Aristotle along with his numerous other sources and many Aristotelian concepts found their way into his texts. For more information on this topic see Lemay (1962).

disparity aimed at the predictions of astrologers. His long-winded but pertinent title for Chapter Five avowing its support reads, "Regarding the evidence for the proof of astrology and the rebuttal to anyone who claims that the planets' motion has no power or indication regarding that which comes to be in this world."

Among interested astrologers the Latin West, Abu Ma'shar became the most prominent Islamic writer; the translations of his *Introductorium maius* by John of Seville and Hermann of Carinthia were read and quoted continually until the decline of astrology in the seventeenth century. In the first half of the ninth century, early Islamic astrologers struggled with wide-spread negative opinions of their questionable art in Muslim areas. Between the late ninth and eleventh centuries, Islamic astronomers and astrologers concentrated on analyzing, criticizing, and perfecting the geometrical models of Ptolemy. During the tenth century, Islamic astronomy came under the influence of another important and long-lived astronomer commonly called al-Sufi. He was one of the most successful scientists carrying out this exacting project of correcting Ptolemy. His scientific writings, like Abu Ma'shar's, were fully illustrated with imaginative drawings that influenced Western astronomical manuscripts.

Abd al-Rahman al-Sufi: Life, Works and Illustrated Manuscripts

Abd al-Rahman al-Sufi (Latinized to Azophi) was born in Persia (903–986), near modern Tehran, and probably spent most of his life in that area working under the patronage of the Iranian Buwayhid dynasty, becoming court astronomer in Isfahan.[30] He must have traversed the royal domain with the court for he made observations in different areas (Dekker 2013: 286). He composed numerous scientific treatises, not in his native language Persian but in Arabic, the universal language for scholars among all the wide-spread Islamic countries, just as Latin was universal for the Christian West. Al-Sufi wrote a great variety of astronomical, astrological, and mathematical texts, including a work on the celestial globe and one on the astrolabes. In addition to his scientific writings, reports survive that al-Sufi actually created a celestial globe that was sold in Cairo in 1044 (Kennedy 1997: 48). He also compiled a *zij,* which does not survive, and composed three treatises explaining the construction and use of the astrolabe.[31] But the most exceptional volume by al-Sufi, written circa 964 when in his 60s, is his *Kitab suwar*

[30]An alternate form of al-Sufi's name which goes back to a Spanish-Jewish scholar and astronomer is Azophi. This spelling appears on the famous celestial map of Albrecht Dürer published at Nuremberg in 1515. In the four corners of Dürer's map are imaginative portraits of astronomers from four different eras and cultures, who made contributions to the field, in the lower right corner is Azophi Arabus.

[31]The huge numbers of illustrated astrolabes were another source of astronomical images, see Savage-Smith (2013) and Dekker (2013).

Fig. 4.12 al-Sufi Drawing, Albrecht Durer, Celestial Map Detail, 1515, Public Domain

al-kawakib or *Book of Fixed Stars* modeled in general on Ptolemy's *Almagest*.[32] Al-Sufi's book has been called "the most important treatise on constellation iconography to be produced in the Islamic world" (Savage-Smith 2013: 125). His work became the standard text on Arabic astronomy consulted in all Islamic countries, written originally in Arabic and later translated into Persian and Latin; his fully illustrated manuscript was copied over and over for many centuries (Fig. 4.12).

A large amount of literature had collected in the Islamic world on the subject of the stars; al-Sufi was the first to organize and systematize this huge body of astronomical treatises. In this major work on the constellations, he describes the 48 standard Ptolemaic constellations, discussing not the pertinent Greek myths but each star grouping in general. He then cites the individual stars within the constellation according to the Arabic tradition and states the multiple names for each star or group of stars, adding data on pre-Islamic sky lore. Although adding a few nearby stars to his descriptions, he determined definitive boundaries for each constellation (North 2008: 199). Al-Sufi even mentions the color of eclipses and the reddish color of particular stars. He does not base his accounts strictly on a repetition of Ptolemy's writings; he himself made a series of new observations at

[32]A dissertation by Ihsan Hafez (2010) *Abd al-Rahan al-Sufi and his book of the fixed stars: a journey of* re*discovery*. PhD thesis, James Cook University, includes a study and, for the first time, an English translation of the main sections of the book. There is also a condensed online version.

Isfahan. Al-Sufi calculated a more precise precession constant of one degree in 66 years, discarding Ptolemy's estimate of one degree in 100 years. He updated Ptolemy's stellar longitudes from the year 137 to 964 by adding 12° and 42' to Ptolemy's longitude values to allow for precession, but using his own calculations.

Al-Sufi alsorecorded the coordinates and his own magnitude estimates which frequently deviated from those inPtolemy'swritings and even included the positions of nebulous objects. The *Almagest* named all astronomical objects visible to the naked eye, including nebulous objects but did not include the only galaxy visible from earth. Al-Sufi became the first to identify the Large Magellanic Cloud and to record a description of the Andromeda Galaxy in a written treatise (Savage-Smith 2013: 128).[33] The al-Sufi text called the fuzzy nebula 'a little cloud'; he placed the small cluster near the mouth of a large fish, an early Bedouin constellation formed by the same stars that the Greek astronomers called Andromeda. He drew the 'cloud' as a small clump of dots. Ptolemy mentioned five nebulous objects but not the object now known as the Andromeda Galaxy.

Moreover al-Sufi integrated much relevant material derived from Bedouin folk-astronomy, the constellation system of the *'Anwa*, including the indigenous Arabic names for the stars. Three of the earlier Muslim authors that he drew from are named in his text but none of their writings have survived. His work provided many of the individual star names of Arabic origin currently used today. Al-Sufi consulted both the stellar traditions of early Arabic astronomy and the advanced mathematical discourses of Ptolemy. Through his scholarship, he reconciled and synthesized the two very different traditions, a challenging task since the traditional Greek and Arabic constellations were unrelated and the complicated star patterns usually overlapped.

Al-Sufi then complimented the textual material with not one, but two mirror images of each constellation figure, one as the image would appear looking down on a celestial globe and the other as seen when looking up at the night sky from the earth.[34] The globe view always preceded the sky view for all 48 constellations (Savage-Smith 2013: 134). Lastly he incorporated a chart of the group of stars comprising that constellation, supplying a verbal description of each star's correct location along with its magnitude, latitude, and longitude adjusted to October 1, 964.[35] In his illustrations, al-Sufi indicated the brightness or magnitude of each star

[33]The Andromeda Galaxy is the only galaxy visible to the naked eye and the largest of the Local Group that includes about 45 other 'nearby' galaxies.

[34]The concept of dual interpretations of stellar illustrations was introduced in the discussion of celestial globes and will be mentioned further in Appendix C.

[35]Julio Samsó and Mercè Comes, "Al-S_f_ and Alfonso X" in *Islamic Astronomy in Medieval Spain*, p 69 Al-Sufi makes corrections to Ptolemy's charts to allow for precession, adjusts the stars' magnitudes according to his own observations, and also adds many more stars not included by Ptolemy. His correction for precession equals one degree of longitude for every 66 years, but he actually used 70 years at times.

within the constellation by the size of the red dots; he used his own observations, often criticizing Ptolemy. Many of the common names for stars still used today were passed down ultimately from al-Sufi's meticulous work.

Because the Muslim scribes and artists were more concerned with the positions and descriptions of the constellation stars than with Greek mythology, they often developed new iconography for the figures not consistent with the Western images. The eloquent illustrations of the constellation figures included in Al-Sufi's work were widely imitated, and soon became canonical, eventually influencing the iconography of western constellation images. Through its exacting illustrations, his book established a new standard for accuracy in representations of the constellations and their individual stars. His exemplar for the constellation images could not have been Ptolemy's *Almagest* since that volume contained no illustrations (Savage-Smith 2013: 133). Al-Sufi actually mentions in his text that although he was aware of another illustrated astronomical treatise he did not use it; he copied his constellation images directly from figures engraved on a celestial globe. According to the eleventh-century scientist al-Biruni, al-Sufi explained the process he used. Laying a very thin sheet of paper over a celestial globe, he fitted the paper carefully over the surface of the sphere. After this procedure, he traced the outlines of the

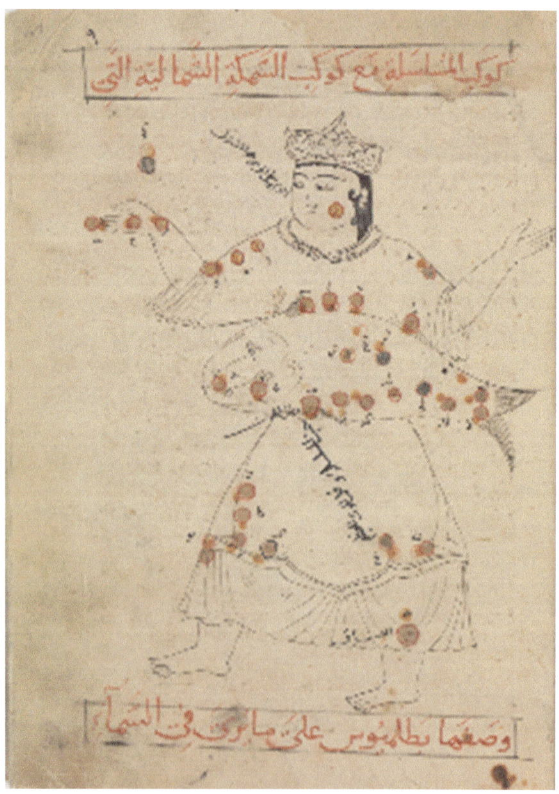

Fig. 4.13 Andromeda, al-Sufi Arabic Manuscript, Fifteenth Century Copy, Public Domain

constellations and positions of the stars on the paper. Al-Biruni later commented that this process "is an [adequate] approximation when the figures are small, but it is far [from adequate] if they are large" (Bloom 2001: 143).

Al-Sufi's monumental work became a veritable handbook for the fixed stars and the constellations, influential for many centuries, among both Arabic and European scholars (Kunitzsch 1989a, b: 57). His detailed volume has been an essential advancement in the history of science and is a verifiable Islamic contribution in knowledge of the fixed stars. Al-Sufi's writings, along with other Arabic astronomers, show that Greek, Persian, and Indian learning experienced a transformation by becoming part of Islamic culture (Montgomery 2000: 133) (Figs. 4.13 and 4.14).

Like the long-lived *Aratea* manuscript tradition, al-Sufi manuscripts also have an extended and complicated history as they were copied over and over for scholars, for caliphs, and for kings. The treatise that survives in Latin manuscripts called, *Books on the Fixed Stars*, also called *Four Books on the Eighth Sphere*, has been credited to al-Sufi, but this Latin version is not a true translation of his original work. This volume actually contains a summary or a compilation of many astronomical texts but al-Sufi's writings furnish the main ingredient. One of these Latin compilations of Arabic writings was produced at the court of King Alfonso X in Toledo, the main contact point from which al-Sufi's work was made known to

Fig. 4.14 Gemini, al-Sufi Arabic Manuscript, undated, Public Domain

scholars in the Latin West (Samsó and Comes 1988). This thirteenth-century astronomical manuscript, produced at the Alfonsine royal court, will be discussed in more detail in Chapter Five. At present, nine manuscripts of the *al-Sufi Latinus* are known (Kunitzsch 1989a, b: 117).[36] These only carry one illustration of the constellation rather than the double appearances of al-Sufi.

Many copies of al-Sufi's original book in Arabic survive in the West and in Muslim countries; 21 Arabic manuscripts are extant dating before the sixteenth century; most of these copies are incomplete, many are missing constellation illustrations. A late fifteenth-century illustrated copy of al-Sufi's *Book of the Fixed Stars* on paper can be seen on display at the Metropolitan Museum in New York.[37] Unfortunately, in every culture the most beautiful images are often removed for personal enjoyment then lost. Some copies of al-Sufi's book were so beautifully illustrated that they are still considered among the masterpieces of Islamic art.

Al-Sufi's Oldest Manuscript Redated—Oxford, Bodleian Library Marsh 144

One of the earliest surviving Arabic manuscripts on any subject is al-Sufi's *Book of the Fixed Stars*, Oxford Marsh 144, bequeathed to the Bodleian Library by Narcissus Marsh in 1713. The astronomical treatise was thought to have been composed and illustrated in 1009 [400 AH], well after al-Sufi's death in 986. This exceptional manuscript on paper, consisting of 419 pages, was written by a scribe who notes in the colophon that he is the son of the author (His son composed a relatively short didactic poem on the stars, but curiously it was not included in this copy.).[38] The codex was produced in Baghdad, the foremost center for book production under the Abbāsid Caliphate; it holds special importance since very few Arabic texts survive from this early period and it is illustrated with elegant double

[36]Kunitzsch lists only eight manuscripts of *al-Sufi Latinus* on p. 68 and includes the illustrations. Current inventories list nine Latin copies. The following can be seen online: the Metropolitan Museum holds a late fifteenth-century and an eighteenth-century copy, both on paper. The Bibliothèque Nationale de France holds a manuscript of *Book of Fixed Stars* that was prepared for Uleg Beg (grandson of Tamerlane) in 1436. St. Petersburg, Institute of Oriental Studies, Russian Academy of Sciences, Ara185 [MS 85 Rosen Catalogue MS C 724] has a sixteenth-century copy, probably made in Istanbul. The Brooklyn Museum owns a single page from a sixteenth-century al-Sufi manuscript, the Constellation of Corvus the Raven, watercolor on paper, but it is not on display.

[37]An Arabic copy of al-Sufi's *Book of the Images of the Fixed Stars* can be found in the collections of the Library of Congress, Washington D.C. This illustrated copy on paper was produced somewhere in south or central Asia, and this too is an exact copy of a manuscript, now lost, prepared for Ulug Beg of Samarkand (present-day Uzbekistan) in 1417 and was rebound circa 1730.

[38]An article by Mora Carey, *Al-Sufi and Son: Ibn al-Sufi's Poem on the Stars and its Prose Parent*, London, can be read online at archnet.org/system/publications/contents/6803/original/DPC3668.

drawings of the constellations in both the earth-bound and god-like views. Because there are few survivals for comparison, we cannot know its exemplar; Marsh 144 has been named "one of most remarkable Arabic medieval scientific manuscripts preserved today." In the early eleventh century, most manuscripts in Arabic were very small about six by three inches; Marsh 144 measures somewhat larger, ten by seven, as illustrated copies were beginning to increase in dimensions (Bloom 2001: 183). A facsimile of the Oxford Marsh 144 manuscript was published in 1986.

Surprisingly, recent studies of Marsh 144 have raised doubts about its long-accepted date of 1009. Questions concerning the colophon and other dubious details have made this copy a subject of exhaustive scrutiny in Oxford astronomical circles. The colophon of Marsh 144 states that it was written in 1009, but Emilie Savage-Smith asserts that it is more likely to have been made not at the beginning of the eleventh century but actually at end of the twelfth century (Savage-Smith 2013: 152). Disagreeing with earlier scholars and the colophon itself, Savage-Smith explains the five separate areas of the manuscript that have been questioned concerning the text and translation, as well as several significant omissions when analyzing the constellation illustrations. She offers several possible solutions for the misdating of Marsh 144. The manuscript has been called a 'semi-fake' as the colophon is now suspected to be a later addition. Despite its questionable dating, the textual material in Marsh 144 remains vital for its astronomy and astrology and particularly for its remarkable artwork which exhibits the Islamic adaptation of iconographic sources that were rooted in classical scientific manuscript traditions.

The unique illustrations of the 48 constellations in Marsh 144 were handled with great care; the stellar figures have been orientalized with characteristics of Sassanid art, painted in the typical casual style. "On the whole, the hybrid style, which was applied in a masterful and original way in the drawings of Marsh 144, includes many features reaching back not only to Western antiquity, but also to an earlier period of Oriental history" (Wellesz 1965: 6). The attire of the human figures is depicted as typical Islamic garb in the medieval period. The iconography of the illustrations was adapted from the classical traditions seen in Late Antique manuscripts, but with some obvious misunderstandings, for example, the drawing of the severed head of Medusa, held by Perseus. When viewing this curious illustration in Marsh 144, Medusa with her snaky hair becomes mistakenly transformed into a demonic male head with wild, uncombed hair; the blood dripping from the severed head of the Gorgon was drawn as a long unkempt beard.[39] Obviously, al-Sufi or the Islamic artist had no interest in the mythological story of the Greek hero. This mistaken identity persists in its present terminology, for the unusual star in Perseus, which is a variable star whose luminosity changes periodically, is still called Algol, meaning 'demon's head' in Arabic *al-Ghul*.[40] Al-Sufi is careful to describe the stars

---

[39]Thanks to the current electronic age, the entire 435 pages of Marsh 144 including the intriguing dual miniatures of the constellations can be viewed online at the Bodleian Digital Image Library.

[40]For examples of an Islamic Perseus see cod. Vat. 8174, cod. Vondob. 5415, and the lapidary of Alfonso X; al-Sufi ms. Paris, Bib. Nat. cod. Arab. 5036; or B L. Cod. Or. 5323.

of Perseus in their relationship to the Milky Way which was not borrowed from Ptolemy for he only defined its northern and southern limits. Within his constellation figures, Al-Sufi differentiated the individual stars by painting them as red dots and labeling them in black; the stars nearby but outside the figure were reversed, represented with black dots and the labels red.

> Thus to avoid confusion and to help a student (his patron?) to bridge the gap between the globe and the sky al-Sufi added two pictures in his book. The result is an extraordinary star atlas composed to suit the student in astronomy in every respect not only by clarifying the distinction between the celestial sphere and its main model, the globe, but also by pointing out objects of special interest in each constellation and further by transferring the names of the stars belonging to older traditional *anwā'* culture into the new mathematical tradition through labeling the indigenous star names (Dekker 2013: 307).

The advanced drawings in Marsh 144 attest to the creation of early astronomical illustrations in Arabic books and demonstrate its broad range and scope as well. After the thirteenth century, Latin translations of al-Sufi began to circulate in Europe. The celestial images complementing the Arabic manuscripts were copied along with the text; these carefully detailed drawings made a significant impact on Western astronomical art.[41]

Al-Sufi's *Book of the Fixed Stars*—MS.2.1998—a Twelfth-Century Copy

In 1998 a previously unknown copy of al-Sufi's *Book of the Fixed Stars*, composed in 1125 [519 AH], was auctioned by Sotheby's in London.[42] This Arabic copy, now thought to be the oldest surviving, was purchased and can now be seen on display at the Museum of Islamic Art in Doha, MS.2.1998. An article on this early copy of al-Sufi's book by Emilie Savage-Smith was published in *God is Beautiful and Loves Beauty: the Object in Islamic Art and Culture* in 2013. Her article entitled, "The Most Authoritative Copy of 'Abd al-Rahman al-Sufi's Tenth-century Guide to the Constellations" analyzes this Arabic paper manuscript of exceptional quality. The scholar/scribe who created this copy noted that he worked on it for three months, aspiring for the most authoritative version possible; obviously he was an astronomer since he stated it was "for his own use" (Savage-Smith 2013: 125). When considering the age and accuracy of the exemplars that the scribe used, this copy, produced in 1125, is without doubt the 'most authoritative copy'. MS.2.1998 includes 179 parchment folios; the descriptions of all 48 constellations survive intact and fully illustrated, adding a pictorial dimension to the textual material.

[41]For a detailed description of the Uranography of al-Sufi see Dekker (2013) pp. 286–307.

[42]A detailed description of the manuscript completed by David A. King, Barbara Brend, and Robert Hillenbrand can be found in Sotheby's auction catalogue (Sotheby's 1998 33–48, lot 34).

The illustrations of the Bedouin constellations that the copyist adds are invaluable since few examples survive of any early nomadic artwork other than in copies of al-Sufi volumes. The illustrations in al-Sufi's treatise and its copies were instrumental in displacing the traditional Bedouin constellation imagery in the Islamic art world and replacing it with the Greek/Ptolemaic system which ultimately came to dominate all astronomical art. Evidence is mounting that this copy of al-Sufi's astronomical masterwork must be the earliest surviving copy. The unadulterated copies of *The Book of the Fixed Stars* are only found in Arabic writings; the Latin versions, such as the Alfonso X's translation and Bibliothèque de l'Arsenal MS 1036, were greatly altered through the translation process.

Al-Sufi's *Book of the Fixed Stars*—Arsenal MS 1036

At some point in the twelfth century, the Arabic al-Sufi manuscript, now the Doha MS 2.1998, appeared in the Latin West. This Doha al-Sufi manuscript, dated to 1175, is a copy of a 1036 CE Arabic version which was another early copy created by an assistant of al-Sufi. The Doha al-Sufi was translated into Latin as the *Liber locis stellarum fixarum*. The earliest extant representative of the whole corpus of Latin al-Sufi manuscripts is held in Paris, Bibliothéque de l'Arsenal MS 1036; this copy was produced in Bologna and dates to the third-quarter of the thirteenth century.[43] Textual evidence shows that this Arsenal manuscript could not have been the archetype of the *Sufi Latinus* tradition, but is one of the later copies (Kunitzsch 1989a, b: 76). There are references to Sicily in several places in the manuscript, mentioning the years of the reign of King William II (r.1166–1189), indicating that perhaps the model for this manuscript was compiled by an unknown scribe in Sicily, probably a member of William's scholarly court circle. This westernized version of al-Sufi retains many specific Islamic characteristics, particularly in the illustrations, and contains a complete star catalog taken from a Latin translation by Gerard of Cremona (Carmody 1956: 94).[44]

Each constellation description found in Arsenal MS 1036 is accompanied by an illustration of exceptional quality but much stiffer than the flowing lines of the Arabic al-Sufi images. When planning the placement of the miniatures, the western composer of MS 1036 was changeable; where al-Sufi provided two drawings for each constellation, this artist chose to include only one, varying between the two opposites, the earth view and the heavenly view. When inspecting the 48

[43]Paris, Bibliothèque de l'Arsenal, MS 1036 is included in Saxl and Meier, 1953, p. XXXII and in *Dix Siècles d'enluminure italienne*, Francois Avril, Bibliothèque Nationale, Paris, 1984, pp. 41–42.

[44]As in other manuscripts based on the work of al-Sufi, this one includes longitudes of stars that are rectified by al-Sufi's empirical observations, as well as the magnitude of each star. Arsenal 1036 also contains John of Seville's Latin translation of *De magnis coniunctionibus* by Abu Ma'shar.

constellation drawings, sixteen are rendered as they appear on a celestial globe and 32 are shown from an earth view. Of the surviving Latin copies, this Paris manuscript is nearest in design to al-Sufi's original work and to Arabic artwork in general as it includes the accurate star positions (Kunitzsch 1989a, b: 69–74). The vivid illustrations are styled after those of Marsh 144, but synthesized and transfigured by its Italian artist to incorporate the Romanesque format popular at that time.

Beginning in the twelfth century when the Arabic texts were initially translated into Latin, the traditional Arabic star names became available to the western world. Until that time, the names of the 48 Ptolemaic constellations were commonly used but not the individual stars. The leading expert on star names in Islamic astronomy, the German historian Paul Kunitzsch, has identified two traditions of star names within the Islamic heritage. The traditional stellar folklore of the Muslim people that he has named 'indigenous-Arabic' is the first; the scientific Islamic tradition, which he designates 'scientific-Arabic', is the second tradition (Kunitzsch 1989a, b: 69–74).

Another surviving Arabic manuscript of al-Sufi's *Book of Fixed Stars* is the thirteenth-century example, British Library MS Or. 5323.[45] This incomplete codex with only 86 folios remaining has been dated between 1260 and 1280 and was probably created in northwestern Iran. The illustrations include double versions of each constellation that are outstanding examples of pen and ink drawings revealing traditional Persian techniques used as models. The popularity of the star charts of al-Sufi can be witnessed by the huge number of copies that survive in Arabic, Persian, and Latin within libraries and museums in the Middle East, Europe, and United States.

The translation of the writings of al-Sufi into Latin, along with the transference of his artwork, became a prime concern of western scholars as soon as they were available.[46] The golden era of Islamic astronomy produced a large number of outstanding scientists and astronomers who created enormous numbers of manuscripts. Arabic science specialist Saliba lists almost 300 Muslim treatises on planetary theories that he has tracked down; only two had been translated into European languages—one into Latin centuries ago and the other into English, in more recent times. Obviously there is much work still to be done.

[45]See Carey (2001) "Painting the Stars in a Century of Change: A Thirteenth-Century Copy of al-Ṣūfī's Treatise on the Fixed Stars, British Library Or. 5323.

[46]Some additional surviving al-Sufi manuscripts in Arabic are: Suleymanie Library, Istanbul, MS Fatih 3422, (1125); Topkapī Sarayī, Istanbul, MS Ahmet III 3493, (1130–31); Bodleian, MS Huntington 212, in poor condition, (1170); Vatican Library, Rome MS Ross. 1033, (1224); Suleymanie Library Istanbul, MS Ayasofya 2595, (1249–50); Bibliothèque Nationale, Paris, MS Arabe Ar 2489, (1266–7); National Library, Cairo, MS Dãr al-Kuftub miqãt 390, (1300); Metropolitan Museum of Art, New York, MS Rogers Fund 1913, (c. 1400); Pertev Pasa Library, Istanbul, MS 375, (16th c.); Kongelige Bibliotek, Copenhagen, MS Copenhagen ms Arabe 83, (1602); Princeton, University Library, MS Garrett 2259, (1607).

Illuminated books became a major art form in Islamic countries much later than in the West, appearing rarely until early in the fourteenth century. After that time, hundreds of volumes of history, poetry, and literature were lavishly illustrated with excellent paintings, mostly on paper. Today, the impressive Islamic artwork in manuscripts is classified as Persian miniature painting. Arabic books and treatises are much more likely to appear in the traditional hand-written format with illustrations drawn by an artist's hand, rather than in printed editions with woodcut illustrations. Due to the cursive form of Arabic scripts and their complicated shapes, the design and casting of individual letters was much more complicated than producing Western scripts. Muslim scribes and patrons strongly resisted printing until the end of the seventeenth century, for calligraphy played a crucial role as a distinctive Islamic art form and a sign of their cultural identity, especially when considering the sacred text of the Qur'an. In contrast to the production of printed Bibles in the West, the first Qur'an printed by Muslims for the Islamic World was not published until 1787 (Bloom 2001: 10).

Throughout the centuries, Baghdad remained one of the most highly populated cities in the world, estimated numbers escalated to a population of one million. Although by the tenth- century the dominant Abbāsid Caliphate was declining in its influence, its capital city continued as the foremost center of scientific studies and education (Kostoff 1991: 37). In 1258, the fearsome Mongol warriors descended on Baghdad; the weakened Islamic military forces offered slight opposition. The famed House of Wisdom, the impressive architecture, observatories, and libraries with their huge collection of books were all totally destroyed. The Islamic Golden Age and the Abbāsid Caliphate came to an inglorious end; headquarters of scientific study, research studies and translation activities had already shifted to Cairo, India and Muslim Spain. Fortunately for Western science, especially astronomy, the overflowing Arabic libraries in Islamic Spain had fallen into the hands of the Christian kings of Leon and Castile.

References

Al-Khalili, J. (2011) *The House of Wisdom*, New York: The Penguin Press.

Avrin, L. (1991) *Scribes, Script and Books: The Book Arts from Antiquity to the Renaissance*, London: The British Library.

Bloom, J. M. (2001) *Paper Before Print: The History and Impact of Paper in the Islamic World*, New Haven: Yale University Press.

Bonnet-Bidaud, J.M., Praderie, F. and Whitfield, S. (2009) "The Dunhuang Chinese Sky: a Comprehensive Study of the Oldest Known Star Atlas", in *Journal of Astronomical History and Heritage* 12.1: pp.39–59.

Butzer, P. L. (1993) "Mathematics in the West and East", in *Science in Western and Eastern Civilization in Carolingian Times*, eds. P.L. Butzer and D. Lohrmann. Basel: Brepols.

Carey, M. (2001) *Painting the Stars in a Century of Change: A Thirteenth-Century Copy of al-Sūfī's Treatise on the Fixed Stars, British Library Or. 5323*—PhD dissertation, SOAS University of London.

Carmody, F. J. (1956) *Arabic Astronomical & Astrological Sciences in Latin Translation*, Berkeley: University of California Press.
Dekker, E. (2013) *Illustrating the Phenomena: Celestial Cartography in Antiquity and the Middle Ages*, Oxford: Oxford University Press.
Edson, E. and E. Savage-Smith (2004) *Medieval Views of the Cosmos*, Bodleian Library: University of Oxford.
Flint, V. (1990) "The Transmission of Astrology in the Early Middle Ages", *Viator, Medieval and Renaissance Studies*, 21, pp. 1–27.
Gaur, A. (1984) *History of Writing*, London: British Library.
Gee, E. (2001) "Cicero's Astronomy", *Classical Quarterly*, 51.2, pp. 520–536.
Goldstein, B. R. (1986) "The Making of Astronomy in Early Islam", *Nuncius: Journal of the History of Science* 1, pp. 79–92.
Granstan, M. (2012) "Balbillus and the method of aphesis", *Greek, Roman, and Byzantine Studies* 52, pp. 587–602. Online at www.Academia.edu.
Hattstein, M. and P. Delius, eds. (2000) *Islam Art and Architecture*, Cologne: Könemann.
Hill, R. (1993) "Science and Technology in 9th Century Baghdad" in *Science in Western and Eastern Civilization in Carolingian Times*, Butzer, P. and Lohrmann, D. eds. Basel: Birkhäuser Verlag pp. 485–502.
Hoffman, E. R. (2000) "Beginnings of Illustrated Arabic Books: An Intersection between Art and Scholarship", *Maquarnas*, Vol. 17, pp. 37–52. www.jstor.org.
Hogendijk, J. P. and A. Sabra, eds. (2003) *The Enterprise of Science in Islam*, Cambridge: The MIT Press.
Kennedy, E. S. (1997) "Al-S_f_ on the Celestial Globe" in *Astronomy and Astrology in the/Medieval Islamic World*, Ashgate, Variorum Collected Studies Series, CS 600, pp. 48–93.
Kostoff, S. (1991) *The City Shaped: Urban Patterns and Meanings Through History*, London: Thames and Hudson.
Krupp, E. C. (1986) "The cosmic temples of old Beijing", in *World Archaeoastronomy*, ed. A. Aveni, Cambridge: Cambridge University Press, pp. 65-75.
Kunitzsch, P. (1989) "Star Catalogues and Star Tables in Mediaeval Astronomy", *The Arabs and the Stars*, UK: Variorum reprints.
Kunitzsch, P. (1989) "The Astronomer Abu 'L-Husayn Al-S_f_ and His Book on the Constellations", *The Arabs and the Stars*, UK: Variorum Collected Studies Series 307, Northampton.
Kunitzsch, P. (2004) *Stars and Numbers: Astronomy and Mathematics in the Medieval Arab and Western Worlds*, Burlington, VT: Ashgate.
Lemay, R. (1962) *Ab_ Ma'shar and Latin Aristotelianism in the Twelfth Century*, Beirut.
Lindberg, D. C. (1992) *The Beginnings of Western Science*, Chicago: Chicago University Press.
Milbrath, S. (1999) *Star Gods of the Maya: Astronomy in Art, Folklore, and Calendars*, Austin: University of Texas.
Montgomery, S. L. (2000) *Science in Translation: Movements of Knowledge through Cultures and Time*, Chicago: University of Chicago Press.
Neugebauer, O. (1951) "The Study of Wretched Subjects", *Isis*, Vol. 42 no. 2; also *The Scientific Enterprises in Antiquity and the Middle Ages*, (2000) Shank, M.H. ed. Chicago: University of Chicago Press.
Neugebauer, O. (1969) *The Exact Sciences in Antiquity*, New York: Dover Publishing.
Neugebauer, O. (1983) *Astronomy and History Selected Essays*, New York.
North, J. (2008) *Cosmos: an Illustrated History of Astronomy and Cosmology*, Chicago: University of Chicago Press.
Pingree, D. (1963) "Astronomy and Astrology in India and Iran", *Isis*, Vol. 54, No. 2, pp. 229–246.
Pingree, D. (1968) *The Thousands of Ab_Ma'shar*, London: Studies of the Warburg Institute.
Pingree, D. (1989) "Classical and Byzantine Astrology in Sasanian Persia", *Dumbarton Oaks Papers*, 43, pp. 227–39.

Sabra, A. I. (2003) *The Enterprise of Science in Islam: New Perspectives*, Cambridge, MA: MIT Press.

Said, E. (1978) *Orientalism*, New York: Random House.

Saliba, G. (1992) "The Role of the Astrologer in Medieval Islamic Society", *Bulletin d'Etudes Orientales* 44 pp. 45–70. Reprinted in *Magic and Divination in Early Islam*, ed. E. Savage-Smith. Ashgate-Variorum, 2004.

Saliba, G. (2002) "Islamic Astronomy in Context: Attacks on Astrology and the Rise of the Hay'a Tradition", *Bulletin of the Royal Institute for Inter-Faith Studies* 4, pp. 25–46.

Saliba, G. (2007) *Islamic Science and the Making of the European Renaissance*, Cambridge, MA: MIT Press.

Samsó, J. and Comes, M (1988) "Al sufi and alfonso X" in *Archives Internationales d'Histoire des Sciences* 38: pp. 67–76. (Reprinted in Samsó, 1994, Chapter XVII)

Savage-Smith, E. (2013) "The Most Authoritative Copy of 'Abd al-Rahman al-Sufi's Tenth-century Guide to the Constellations" in *God is Beautiful and Loves Beauty: the Object in Islamic Art and Culture*, S. Blair and J. Bloom, Eds. New Haven, Yale. University Press, pp. 123–155.

Saxl, F. (1932) "The Zodiac of Qusayr 'Amra" in Creswell, K. A. *Early Muslim Architecture* Tome 1, Oxford: Clarendon.

Sivin, N. (1986) "Chinese archaeoastronomy: between two worlds" in *World Archaeoastronomy*, ed. A. Aveni, Cambridge: Cambridge University Press, pp. 55–64.

Tester, S. J. (1987) *A History of Western Astrology*, New York: Ballantine Books.

Toomey, G. J. (1998) *Ptolemy's Almagest*, Princeton: Princeton University Press.

Wellesz, E. (1965) *An Islamic Book of Constellations*, Bodleian Library, Oxford.

Whitfield, S. (1999) *Life Along the Silk Road*, Berkeley: University of California Press.

Whitfield, S. (2004) *Aurel Stein on the Silk Road*, Serindia Publications in association with the British Museum Press.

Chapter 5
Arabic Astronomy and Illustrative Traditions in the Latin West

Abstract After nearly 400 years of Moorish rule, the thriving Castilian city of Toledo attracted numerous Latin scholars searching for Greek and Arabic scientific manuscripts that they had long heard of and read merely in bits and pieces, but had never seen in their entirety. In the twelfth century, academic specialists were drawn to Arabic sources to search for texts on mathematics, medicine, and astronomy which brought astrology with them. More than one hundred major scientific and philosophical essays were translated from Arabic into Latin and Hebrew in Toledo between 1116 and 1187. After these itinerant scholars translated numerous Arabic works, the innovative writings of Islamic astronomers, physicians, and mathematicians spread quickly throughout Europe, stimulating a great advance in medieval learning and inspiring advances in Latin scholarship. By the end of the twelfth century, many of the essential Islamic astronomical and astrological manuscripts that had become known in the West were copied for many libraries; their unique illustrations too were admired, copied and adopted by Western artists. The production of medieval manuscripts expanded dramatically; the illuminations of constellation cycles advanced in numbers and variety. This chapter will discuss the most important Latin translators and manuscript tradition that followed their efforts, and will expand on those constellation cycles that were illustrated.

Introduction of Arabic Astronomy in the West

By the beginning of the twelfth century, several points of contact had opened between the Arabic East and the Latin West; one of the most important occurred after the Christian forces gained control of the Holy Land following the bloody First Crusade in 1099. As groups of Christian believers began to make pilgrimages to the Biblical sites of the Middle East and noblemen and merchants established

M. Dolan, *Astronomical Knowledge Transmission Through Illustrated Aratea Manuscripts*, Historical & Cultural Astronomy,
DOI 10.1007/978-3-319-56784-6_5

prosperous trading centers in the port cities, Europeans for the first time began to make contact with advanced Islamic civilizations. The kings of Castile and Leon, at the same time, were waging successful campaigns to win back Iberian territory that had been ruled by the Muslim caliphate for four centuries, pushing the Islamic realm further and further south.

In the early Middle Ages, Europe had not completely lost touch with the Greek world or the Persian Empire, but there were few in the West who could read any language beyond Latin. The Iberian Peninsula, like much of post-Roman Europe in the eighth century, had declined from its significant status during the almost 600 years of Roman rule. The Visigoths and Vandals battled over the Hispanic territories; Visigoths took control and finally joined the Christian church in 589. Scientific knowledge had grown very slowly in Hispania; Christian cultures there possessed very little precise astronomical information. This situation changed significantly in the late eleventh- and twelfth century when scholars from Latin Europe, mostly Christian clergymen, set upon former Moorish cities to learn Greek, and more importantly Arabic, in order to access long-lost ancient knowledge as well as newly composed Arabic treatises.

The siege of southern Al-Andalus by Moors and Berbers from North Africa began in 711 and their forces rapidly gained control of the major part of Visigothic Spain. Abd al-Rahman I (731–788), grandson of the last Umayyad caliph (the only family member to escape the fall of his dynasty in Damascus), journeyed to North Africa and was proclaimed emir in 756. After raising an army and conquering much territory in Andalusia the young prince established Cordoba as his capital city and began construction of mosques, palaces, and municipal buildings in the distinctive Islamic style. Under his successors Abd al-Rahman II (792–852) and Abd al-Rahman III (889–961), and especially the bibliophile al-Hakam II (915–976), architecture, culture and the sciences flourished in Islamic Spain.[1] Muslim builders, masons and craftsmen created magnificent structures applying their unique contributions to architecture: calligraphy, garden design, and geometry. They beautified building interiors with graceful horseshoe arches, intricate vaulting and detailed Islamic devices such as muqarnas (stalactite vaults) and decorated their walls with exceptional mosaics and vibrantly patterned ceramic tiles. The tenth-century nun Hrosvita of Gandersheim wrote a complimentary account of Cordoba in 935, based on her conversations with one of the emissaries to Otto's court sent by the caliph Abd

[1]In the tenth century, a contemporary chronicler recounted that Cordoba had 471 mosques, 213,077 working-class houses, 60,300 residents for aristocrats, and 80,455 shops indicating Cordoba's vast size and wealth, possibly exaggerated.

Fig. 5.1 *Muqarnes* Wall Décor, Alhambra, Granada, Spain, Fourteen Century, Public Domain

al-Rahman (just as Charlemagne sent emissaries to Hārūn al-Rashīd's court in Baghdad) (Fig. 5.1).[2]

> The brilliant ornament of the world shone in the west, a noble city newly known for the military prowess that its Hispanic colonizers had brought. Cordoba was its name and it was wealthy and famous and known for its pleasures and resplendent in all things, and especially for its seven streams of wisdom (the Trivium and Quadrivium) and as much for its constant victories.[3]

During the relatively short period of the Andalusian Umayyad caliphate, Cordoba became the most advanced city in Europe. At its peak, Cordoba rivaled Baghdad for its wealth, sophistication, and expansive size. By the end of the tenth century, the royal library, built next to the Great Mosque of Cordoba, was alleged to hold 400,000 Arabic manuscripts, many, many more than any Christian library in Europe which typically held only a few hundred parchment manuscripts and there were many more libraries in that capital city. The vast number of books attributed to Arabic libraries seems exaggerated but a professional Arabic scribe could produce

[2]Hrosvitha, most probably from a noble family, was the earliest-known women poet in Germany; some scholars consider her the first playwright since ancient times. It is believed that before entering Gandersheim Abbey, she spent the major part of her younger years at the court of Otto I, because her writing is similar to others who visited that court. Eight metrical legends of the saints, six plays and two historical epics written by Hrosvitha survive in a manuscript held at the Bayerische Staatsbibliothek, Munich, Germany.

[3]This quotation was taken from Menocal (2002) in *The Ornament of the World*, pg. 32.

one hundred pages a day, which seems incredible, compared to the efforts of a medieval Latin scribe. Examining a page of text in an Arabic book reveals far fewer characters per page and much more of the white sheet than black letters; and also authors and scribes wrote on paper not vellum. (As mentioned in Chap. 4, the first known paper mill in Europe was established in Xàtiva, on the Mediterranean coast of the Iberian Peninsula at least by 1151.) Unbelievably only a single manuscript from that famous Arabic library of Cordoba is known to survive.

As the kings of Leon and Castile expanded their battles, they successfully regained control of much of the Islamic-ruled territories of the Iberian Peninsula, allowing numerous libraries packed with Arabic writings in every discipline to fall into Christian possession. As a result of the eleventh-century *Reconquista,* the prominent Mozarabic city of Toledo was taken over by the successful conquest of the astute and ambitious Christian monarch, King Alfonso VI of Castile in 1085 (Bosch 2000: 26).[4] Toledo remained the most important political and cultural center of Castile until the capital city was reestablished in Madrid in 1560.

The *Reconquista*

Translating texts from Arabic into Latin had already begun at the learning centers of various archbishops in Christian areas and soon numerous Latin scholars were drawn to the Arabic libraries throughout Andalusia in order to access their vast holdings, in particular the long-desired astronomical treatises written in Arabic, Greek and Hebrew. Jewish scholars living peacefully among the Moors in the West and among Muslims in the East were, like Arabic scholars, beneficiaries of the legacy of both classical and Jewish antiquity. In general, the Muslims were a more cultivated and more tolerant society than the Visigoths they displaced, avoiding religious persecution. They did not proselytize and allowed Jewish and Christian residents to practice in their preferred manner, but did assess a special tax. Evidence survives of the Jewish presence in Iberia from the third century until the end of the fifteenth, when the Jews who would not convert to Christianity were expelled by Ferdinand and Isabella.[5] For many centuries Jewish scholars in reconquered Spain participated freely in the intellectual life of the Latin West.[6] Due to the fact that Andalusia had a long, successful tradition of Christian, Muslim, and Jewish cultures

[4]In the seventh century, Toledo had been declared the leading diocese in the country, giving it precedence over other Spanish dioceses. It had also been declared the capital of the Visigothic kingdom, making it the religious and civic center of Visigothic Spain until taken over by the Muslims in the mid-eighth century.

[5]One item that confirms the early Jewish presence is a third-century tombstone, now lost, that commemorated the death of a one-year old girl with an inscription "IVDAEA" (Jewess) (Gambel 1992: 11).

[6]See Goldstein 1986 for further details on medieval Jewish astronomers and scientists.

living and working side by side, *la convivencia* was a time of relative peace, even though Castilian kings now ruled many of the formerly Muslim cities.

After the Muslim world had gathered the large number of ancient texts and then added their newly developed sciences, it was not long before their accumulated knowledge was discovered by Western scholars. The Latin translation movement in the West began in the mid-eleventh century and did not abate until the mid-fourteenth century, during which time scholars from all over Europe were attracted to former Muslim intellectual centers mainly in Spain, but also in Sicily and southern Italy. Sicily had been a part of the Byzantine Empire until 878, when it fell under Muslim control which lasted until the Normans gained possession in 1060. The Norman kings ruled Sicily for three centuries promoting a community like that of al-Andalusia, open to Christian, Jewish and Islamic traders, merchants and scholars. The most revered Norman ruler, Holy Roman Emperor Frederick II (1194–1250), grandson of King Roger II, took control of Sicily at fifteen and then regained his father's and grandfather's title of Holy Roman Emperor in 1120. Frederick II, (called *stupor mundi*, the wonder of the world) sponsored a most distinguished court encouraging science, astronomy and the arts at his palaces in Palermo and Southern Italy.

Throughout that long period, successful Sicilian trading and intellectual centers continued a trilingual culture making it an ideal location for scholarly translating. Relationships had been retained with the Greek East which allowed for the transmittal of important manuscripts and the exchange of both old and new intellectual ideas. In addition, there had also been centuries of connection between Greek scholars and biblical study, so that important religious and philosophical works had been translated and incorporated into the active Christian learning centers (Montgomery 2000: 76).

By the mid-twelfth century, Toledo was thriving as one of Europe's most lively and cosmopolitan learning centers where Latin, Arabic, and Hebrew scholars freely mingled; therefore translation expertise was not difficult to find. Latin researchers, assisted by Arabic and Jewish intermediaries, eagerly began their translations within Arabic libraries that brimmed with coveted knowledge, and worked together translating Arabic texts. Unlike the strong interest in the literature and history of classical antiquity that would be generated during the Italian Renaissance, the twelfth-century Latin translators searched for new scientific and philosophical texts produced by scholars in Arabic, along with the ancient works of Hellenistic, Persian, and Indian authors. These itinerant Latin translators operated under their own directives and ambitions, with no organized support. Unlike the Abbāsid caliphate in Baghdad, there was no facility similar to the House of Wisdom in Toledo, translators received minimal support from Church, state or the developing universities (Montgomery 2000: 142).

The long-sought astronomical texts were the very first to be undertaken by Christian and Jewish translators, revealing the knowledge of Plato, Hipparchus, Aristotle, and Ptolemy, and so many other valuable works thought lost to humanity. Not only did they find that their great classical treatises survived basically intact; but in addition, they had been analyzed, commented upon, and enhanced by Arabic scholars, revealing that their theoretical mastery and their observational techniques were far more highly developed than those of Christian Europe which were completely lacking in experimentation, observation, and innovation. The Latin translators also discovered the production of astronomical instruments by Islamic experts and the compilations of accurate celestial tables, in particular the *Toledan Tables* (or *zij*) which had been compiled around 1069 by a group of Muslim astronomers in Toledo. Throughout history translation movements tend to generate huge advances in the gathering of scientific knowledge.

Introduction of Arabic Astrology to the Latin West

> The twelfth-century Jewish astronomer and philosopher Moses ben Maimon wrote, "Astrology is a disease, not a science".

In the twelfth century, astrology emerged from its centuries of silence in the Latin West and became prominent again in the translation centers of Spain and Sicily. The *sphaera*, the 'scientific' explanation and description of the heavenly bodies and their daily and yearly motions, included the current astrological knowledge along with the purely astronomical; together they provided what was currently known of the universe. Traditional accounts, based on the surviving texts, declare that the 'dark twin' of astronomy, condemned by Church fathers, was basically unknown in the Latin West until the ninth and tenth centuries. Anti-pagan polemics by early Church Fathers, including Augustine, resulted in the gradual suppression of the discipline; in addition, since knowledge of the Greek language had been lost, the capacity to read or access any of the Hellenistic astrological works (all in Greek) was lost as well. Some researchers disagree and argue that text-based accounts of the transmission of learning in the Middle Ages are always inadequate and often distorted.[7] Valerie Flint argues that, in spite of current scholarly opinion, astrology was known and practiced in Europe in the early medieval period, but by non-Christian leaders and "magicians" whose practices left little formal record (Flint 1990: 1–27).

The customary belief that astrology had disappeared entirely between the fall of the Roman Empire and the arrival of Gerbert of Aurillac (946–1003—Pope Sylvester II) is not supported by the evidence from Church councils which often condemned astrologers, or by evidence contained in the writings of Augustine and Isidore of Seville. (Isidore repudiated consultations with astrologers for predictions, but praised its use for medical reasons.) Greek of course was the main language in the Eastern Byzantine Empire and many astrological manuscripts were preserved in imperial and monastic libraries, but Church and secular laws often forbid access to most of them. Although technical astrological texts may have been rare or

[7]See Flint (1990) "The Transmission of Astrology in the Early Middle Ages", *Viator, Medieval and Renaissance Studies 21*.

unattainable in the Latin West during the early Middle Ages, simple astronomical charts and tables could have provided adequate information for the casting of horoscopes and making predictions based on planetary positions. After the Arabic to Latin translation period in the twelfth century, astrological manuscripts became hugely popular as did the practice of casting and interpreting horoscopes. Astrological manuscripts quite often included the complete series of constellation images and exhibited particular favor to the twelve signs of the zodiac and the 'seven planets'. Horoscopic charts, almanacs, and new types of illustrations, such as decans and paranatellonta begin to appear in astrological manuscripts.

Ptolemy—The Impact of Latin Translations in the West

> 'In mathematical astronomy ancient methods prevailed until Newton and his contemporaries opened a fundamentally new age by the introduction of dynamics into the discussion of astronomical phenomena. One can perfectly well understand the "Principia" without much knowledge of earlier astronomy but one cannot read a single chapter in Copernicus or Kepler without a thorough knowledge of Ptolemy's "Almagest" (Neugebauer 1957: 3).

As we have seen, and what cannot be overemphasized, the fundamental source of astronomical knowledge in Arabic manuscripts was Ptolemy's *Almagest*, the most demanding work of classical Greek science.[8] Except for a few rudimentary texts and the *Phaenomena* of Aratus, most of the earlier Greek astronomy has been completely lost; perhaps because its continued existence was considered unimportant at the time. Ptolemy's outstanding work made other astronomical writings obsolete, and "relegated his predecessors to merely historically interesting figures with little chance of survival" (Neugebauer 1957: 55). "His work comprises practically all astronomical achievements that could be reached with the mathematical methods of antiquity," (Neugebauer 1957: 145).

Along with the mathematical aspects of Ptolemy's great work, the *Almagest* comprises a catalog of 1025 stars along with their arrangement into the 48 classical constellations including their latitudes, longitudes, and magnitudes, a comprehensive guide to mathematical astronomy. The *Almagest* had been translated from Greek into Arabic at least four times in the ninth century affirming its widespread interest (Kunitzsch 1974). Although greatly admired in the West, it was through his reputation only; none of Ptolemy's Greek manuscripts had survived into the medieval period; just a few fragments of his great book were known. (The Carolingians missed here, mainly because, unlike the *Phaenomena*, the writings of Ptolemy had not been translated into Latin during the Classical period.)

[8]Ptolemy's *Almagest* was for astronomers what his *Tetrabiblos* or *Quadripartitum* was for astrologers. As mentioned, scholars long doubted that the *Tetrabiblos* was actually written by Ptolemy because they were unwilling to believe that he was a practicing astrologer.

Ptolemy also produced many handbooks containing all that was known about astronomy, astrology, optics, geography, mechanics, musical theory, geometrical constructions using spherical projections, and the structure and size of the universe. His primary goal was to summarize what was already known, but he advanced that current astronomical knowledge as well. His work surpassed all that had gone before and dominated future astronomical studies for over a millennium.

It appears that Ptolemy's early texts were not accompanied with any illustrations beyond technical mathematical and geometric diagrams. What can be examined are the manuscripts of Arabic writers and translators that have survived and were translated into Latin. After the revelation and translation of Arabic treatises, all of this 'new' scientific material became available to scholars in the Latin West for the first time The uncovering of beautifully illustrated Arabic books revealed their proficiency not only in scientific fields but displayed their artistic mastery as well in religious, literary, and mythological texts written in elegant calligraphy. Their creative techniques introduced new elements to Latin readers, displaying a much higher degree of accuracy in the relationship of the pictorial images to the actual appearance of the constellations in the sky.

Of particular interest in this study is the manner in which the illustrations and diagrams that embellished Arabic manuscripts were presented, and how Islamic artwork influenced western astronomical art, especially the *Aratea* illuminations. But it is impossible to separate the artwork from the author, the text, and the manuscript in which the paintings appear as they are symbiotically linked. This section continues the investigation of the transmission and survival of illustrated astronomical manuscripts as the vast corpus of knowledge held by the Arabic world was translated, studied, circulated and then assimilated by the Latin academic world.

Abu Ma'shar—The Impact of Latin Translations in the West

As we have seen, the extensive writings of Abu Ma'shar dominated the field of Islamic astronomy and astrology in Muslim learning centers in the ninth century. Therefore in twelfth-century Spain, along with Ptolemy, Abu Ma'shar was one of the very first Arabic scientists to be translated into Latin, mainly by John of Seville (fl.1133–53) and Hermann of Carinthia (c.1100–1160). For his first attempt, John of Seville translated Abu Ma'shar's *Flores Astrologia*, a collection of brief, useful astrological hints; his Latin translation of the *Greater Introduction* was completed by 1133; this version was the first of Abu Ma'shar's texts to become known in European intellectual centers. John's Latin translation circulated just before Hermann's, which followed shortly after in 1140.

Writings of Abu Ma'shar (or Albumasar) exerted a powerful influence on the development of astrological theory and practice in the West, where his reputation exceeded that in his own country. Most of his treatises were not original creations, but were compiled from the accumulated astrological knowledge of earlier writers, with explanations of the astronomical systems and especially in techniques for preparing and interpreting natal charts and techniques for making predictions. His work introduced a fusion of Hermeticism, Persian chronology, Islamic religious doctrine, Greek science and Mesopotamian astrology. He synthesized information from Hellenistic astrological writers such as Dorotheus and Vettius Valens, and drew from Sanskrit astrological traditions, from Syriac and Neoplatonic philosophy relating to astral influences; he also added theurgy (the interaction with a supernatural power) from al-Kindi and the books of the Harranians.

In addition to his influence on astronomy and astrology, Abu Ma'shar's writings concerning philosophy and natural science made a profound impact on scholars of the twelfth and thirteenth centuries, especially the intellectuals teaching and studying at the 'school' of Chartres. Traces of his astronomical theories can be detected in the work of William of Conches, Bernard and Thierry of Chartres, John of Salisbury and Daniel Morley. Abu Ma'shar's theories were also a consideration in the church's compromised attitude toward astrology in the thirteenth century as defined by Albertus Magnus and Thomas Aquinas (Clark 1979: 27). In his *Greater Introduction,* Abu Ma'shar presents his theory of astrological determinism defending astrology against its detractors. He repeatedly acknowledges divine will as the originator or director of nature, but adds that causal relationships are dependent upon the stars. His astronomical-astrological texts and his mathematical calculations have been included in studies by historians of science and of astrology, as the astronomy of Arabic scholars has motivated a number of current books, articles and conferences.[9]

Manuscripts of Abu Ma'shar's various treatises are extant in both the original Arabic and in Latin translations. Francis Carmody states that almost 50 Arabic manuscripts of his various works have been identified but most of these had not been studied, including 12 copies of *The Greater Introduction* (Carmody 1956: 88). Both John's and Hermann's twelfth-century Latin translations of *The Greater Introduction* survive in significant numbers as well, in diverse states of completeness; some copies are merely fragments. Carmody lists 24 manuscripts with sizeable segments credited to John of Seville's translation and nine copies of Hermann of Carinthia's version (Carmody 1956: 89).[10] Although the Latin versions of *The Greater Introduction* are illustrated, it seems that no illustrations appear in Arabic manuscripts (Clark 1979: 17).

The start of the printing era advanced the readership of medieval astronomical and astrological texts; its popularity continued into the Renaissance and Abu

---

[9]An important source of information focused on the illustrated manuscripts associated with Abu Ma'shar can be found in a dissertation by Vicky Clark written in 1979. More accurately, Clark's dissertation from the University of Pittsburgh focuses on the "Abridged Version of Albumasar's *Greater Introduction* by Georgius Zothorus Zaparus Fendulus" which survives in seven manuscripts. Fendulus and his illuminated manuscripts will be discussed further on.

[10]When tracing John of Seville's 24 Latin translations, there are three surviving from the thirteenth century, four from the fourteenth, five from the fifteenth; the remaining twelve are not dated. Hermann's translations are generally dated earlier, four are from the twelfth century, two are thirteenth century; the remaining three are not dated. Herman's copy of Abu Ma'shar is half the length of John of Seville's because he removed much of the wordiness of the Arabic.

Ma'shar was by far the preferred author. His astrological text *Introductorium Maius,* printed by Erhard Ratdolt at Augsburg in 1489, was the translation of Hermann of Carinthia even though medieval scholars preferred the version by John of Seville. Research has shown John's translation was preferred over Hermann's in the proportion of four or five to one (more than 40 manuscripts of John's compared to ten of Hermann's). Unfortunately the printed text of Ratdolt has numerous errors and omissions; in addition the printed edition "is larded through and through with interpolations by medieval scribes or students" (Lemay 1987: 73). This printed version that circulated in the late fifteenth and sixteenth century did not provide its readers with an accurate account of Abu Ma'shar's writings.

However, Lemay, Clark, and Saliba all stress that a major problem of working in this field, whether it is the history of medicine or astronomy or any other area, is the sheer quantity of manuscripts in Arabic that remain to be translated, read, edited, and analyzed. The number is tremendously large, especially from the tenth to the thirteenth centuries; so far only a very few of these texts have been closely examined. To give some idea of the numbers, Savage-Smith cites the Bodleian Library collection alone, which possesses a very wide-ranging amount in this area; it holds roughly 500 Arabic treatises dating from the ninth century to the most recent written in the late eighteenth century; very few of these have been studied.[11] As historians now realize, a great deal of this Eastern scientific work never even reached Europe due to the fact that the texts were never translated into Latin. The study of the medieval Arabic texts and accompanying illustrations and their influence has not kept pace with the influx of new information and more accurate resultant understanding and interpretations of these works. "Text-critical and philological spade work on a vast scale must be performed before we may hope to understand medieval astrology on its own, historically ascertainable terms" (Lemay 1987: 73).

Translations of Arabic Astronomical Texts by Latin Scholars

Translations by Adelard of Bath

The wandering scholar and humanist Adelard of Bath (c.1080–1152), through his travels and translations, brought a large amount of astronomical knowledge from the Islamic world to the awareness of Latin scholars. Because of his original writings and Arabic translations, Adelard became a central figure in the

[11] *A New Catalogue of Arabic Manuscripts in the Bodleian Library, University of Oxford Volume 1: Medicine* was published in 2013 by E. Savage-Smith. The Library's Arabic collection includes some of the oldest and most important scientific and medical manuscripts that survive. In this catalogue, she describes 377 medical manuscripts representing 242 different treatises written between the ninth and seventeenth centuries.

development of early European thought. Born in Bath, western England, Adelard journeyed abroad to study, visiting France, the Middle East, southern Italy, and Sicily. No evidence exists that Adelard spent time in Spain, but many scholars conclude that he must have visited there to have had access to the numerous Arabic texts which he studied and translated (Fig. 5.2).

Adelard (not to be confused with his more famous Breton contemporary, Benedictine monk Peter Abelard (1079–1142) gained fame through both his innovative treatises in mathematics and astronomy, and for translating numerous Greek and Arabic scientific works. After his stays in Syria and Turkey, Adelard returned to Europe as an enthusiastic promoter not only of Arabic knowledge but also the Arabic tradition of rational scientific inquiry. Where Adelard learned Arabic is not known for certain, nor where he did his years of translating. The few details known about his life have been gathered from occasional personal comments inserted into his own writings, from his diverse interests and from a few local archives.[12] Active as tutor to British royals, his most illustrious student was King Henry II to whom he dedicated his notable treatise on the astrolabe.[13]

Adelard initiated his translating profession with a very challenging treatise, *Centiloquium Ptolemei*, a hundred aphorisms summarizing the astrological theories of Ptolemy (Cochrane 1994: 87).[14] Adelard's version of *Isagoge minor* or *Shorter Introduction to Astronomy* by Abu Ma'shar, completed around 1120, has been evaluated as a much better translation than his Ptolemy; only two of these manuscripts survive (Carmody 1956: 98).[15] Adelard's third translation from Arabic to Latin was the *Liber prestigiorum Thebidis*, a volume on the theory of images by Thābit ibn Qurra. His Arabic to Latin translation of Euclid's *Elements* became a standard mathematical teaching text for centuries (Cochrane 1994: 85).[16] Adelard's Latin version of the Persian mathematician al-Khwarizmi's *Zij,* or astronomical tables, was the first introduction of Arabic notations for numbers and zero. This text is the earliest Arabic astronomical treatise of any size to survive. Today al-Khwarizmi's work is known only through Adelard's Latin translation (North 2008: 197). The pioneering work of al-Khwarizmi on numerals and arithmetical procedures became known in the West as 'algorithms'.

---

[12]Adelard was known to study at the cathedral 'schools' of Tours and Laon. Tours had a reputation for scholarship, especially under the leadership of Bishop Berengar (d.1088), who had studied with the great scholar Bishop Fulbert of Chartres.

[13]A manuscript of Adelard's treatise on the astrolabe still survives, *De opere astrolapsus*, Cambridge, Fitzwilliam Museum, McClean MS 165; see also Tester p. 43 for more information on this manuscript.

[14]This Latin translation was not completed, but broke off at proposition 39, and is considered poorly executed for Adelard's lack of understanding of both Arabic and astronomy. Lemay has shown that the *Centiloquium* was not actually written by Ptolemy, but is a forgery by Ahmed ibn Yusuf, an astrologer from Cairo, c.920.

[15]The two surviving manuscripts are both in England: London BL Sloane 2030 and Oxford, Digby 68. Only the Oxford manuscript is complete.

[16]Cochrane credits R. Lemay for revising the order of Adelard's translations.

Fig. 5.2 Medieval Manuscript Illustration Adelard of Bath, Public Domain

Written about 1150, Adelard's original treatise, *De opere astrolapsus*, opens with a dedication to his former student, the future King Henry II. By this time, the English intellectual was already reaching the end of his career which made him one of the principal scholars credited with establishing Arabic astronomy in England (Burnett 1997: 38). The gap in knowledge of Arabic studies among European astronomers was especially apparent in their knowledge of the astrolabe. Although the astrolabe was a Hellenistic invention, Islamic astronomers added more features and advanced its usages. Adelard was the first to translate an Arabic treatise on the use of an astrolabe, but the West lagged in understanding the extent of its potential. The astrolabe was the most important observational instrument available before the invention of the telescope; it could be used not only for observation of the stars and planets, but for finding the time, latitudes, or heights and distances. It was also helpful in determining astrological tables, for constructing horoscopes, calculating the mundane houses and dividing the natal chart into its 12 astrological houses. In manuscript illuminations, a person holding an astrolabe became an iconographic symbol for an astronomer.

Adelard also wrote a treatise using the Platonic dialogue technique of questions and answers called *Quaestiones naturales* which explains his motivation and methods for teaching scientific principles. The 76 scientific discussions with his skeptical nephew presented a method to promote the Greek philosophy and Arabic science that he had studied in Islamic lands. Adelard underscores the advantages of using reason and experimental data rather than relying on past authority.[17] Astrology too was included in his translations as an accepted part of natural science; his treatise had no Isidorian distinction between *astronomia* and *astrologia* (Tester 1987: 148). There is little published on whether any of his original astronomical writings or Arabic translations had been illustrated. Among contemporary historians, Adelard's translations have made a lesser impact in the history of science than Hermann of Carinthia or Gerard of Cremona; today scholars recognize Hermann as the more proficient translator and a highly respected scholar.

Translations by Hermann of Carinthia

Hermann of Carinthia, sometimes called Herman Dalmatin (ca.1100–1160), known as a philosopher, astronomer, astrologer, and mathematician, has been most widely recognized for his numerous translations of Arabic manuscripts into Latin.[18] In his writings, Hermann states that he was born in 'central Istria', but others say he was actually born on the island of Korcula, today part of Croatia. Educated at a Benedictine monastic school in Istria, Hermann continued his studies in France, chiefly at Chartres Cathedral during the scholarly era of Bishops Bernard and Thierry. During his stay in France, Hermann became aware of the ancient classical astronomical and philosophical texts that were beginning to circulate in European learning centers, thanks to Arabic sources.

After becoming efficient in Arabic, Hermann ventured into a wide range of scientific and philosophical material, translating essential writings of Arabic authors as well as early treatises passed down from the Greeks. The list of works translated by Hermann is most impressive; although he lacks the originality of Adelard of Bath, he mastered a considerable area of the newly discovered mathematics and astronomy. Because he was delving into complicated linguistic territory with no guidelines or antecedents, his translations appear confused at times, but among the transmitters of Greek and Arabic science to the Latin West, his work remains highly esteemed and is considered very influential in the development of medieval European astronomy. Hermann of Carinthia has been noted as "one of the most interesting of mid-century students of Arabic science. His cosmogony, now available to a large number of students, shows a breadth of scientific learning far beyond contemporaries like Thierry of Chartres and William of Conches" (Eastwood 1984: 913).

---

[17]See Adelard of Bath, *Conversations with his Nephew: On the Same and the Different Questions on* Natural *Science and On Birds*, (2006) Cambridge Medieval Classics, Charles Burnett ed. Cambridge University Press.

[18]Not to be confused with an earlier Hermann, who also wrote on astronomical subjects, especially the astrolabe. This Hermann is called Contractus, the lame monk of Reichenau (1013–54).

Sometime before 1160, Hermann traveled to Sicily where he translated Ptolemy's *Almagest* (Kunitzsch 2004: 116).[19] He even dedicated the section of Ptolemy on the *Planisphaerium* to Thierry of Chartres. While a pupil of Thierry, Hermann became acquainted with Adelard's translation of *Introductorium minor* by Abu Ma'shar; later on in his career, he translated the longer version which was of more significance to Latin scholars. His translation of *Introductorium maius* later became the source of the first printed edition of Abu Ma'shar (Lemay 1962: 3–9).[20]

In addition to his many Latin translations of different Arabic authors, Hermann produced his own treatise, *De essentiis,* which was completed in 1143 and again, this was addressed to Thierry of Chartres.[21] Hermann's philosophical treatise demonstrated his knowledge of the Platonic tradition he was exposed to at the 'school of Chartres', as well as his familiarity with literature, the patristic fathers, and Arabic science. Hermann was interested in presenting a cosmology derived from Arabic texts he studied, but even though his mathematical astronomy is elementary, he offered a number of detailed deliberations (Eastwood 1984: 913). *De essentiis* included theories of Aristotle, in particular his writings on natural philosophy. The understanding of Aristotle, until the time many more of his writings were translated, was derived from the interpretations of Aristotle published by Abu Ma'shar, not directly from Aristotle's own words. In this major philosophical work, Hermann describes the physical effects of the motions of the stars and planets and he explains the coming-to-be from the first m*otus*, which is God's creation. He continues through the order of causes until reaching the final triumph of creation, which is man.

Even though he is promoting astrological theories, he, as a clergyman, does not exceed the acceptable limitations of Christian tenets. The prominent thesis of Richard Lemay states that the Arabic astrology of Abu Ma'shar was the major source and stimulus for Aristotelianism prior to the translations of his physical and metaphysical works. These scientific ideas were not conceived directly from Aristotle; they were mostly derived from Hermann's translation of Abu Ma'shar. The foremost translators of the Latin West were most anxious to interpret the astrological theories of Abu Ma'shar as well as his astronomy. From the diffusion of their Latin translations, it is possible to develop a history of the evolution of the various concepts of the powers and potentials of celestial images and the resultant field of astronomical art and artistic imagination.[22]

---

[19]Only one complete manuscript of Hermann's translation survives (Vat. lat. 2056, 13th c.) and also two fragmentary manuscripts.

[20]Hermann's translation of Abu Ma'shar was used by Gerard of Cremona in his public lectures on astrology in Toledo. The full title of Hermann's version, *Introductorium in astronomia Albumasaris Abalachi octo continens libros partials,* was published in 1489 by Erhard Ratdolt.

[21]See Herman of Carinthia, *De essentiis*, ed. and trans. Burnett (1982).

[22]One of Hermann's working associates was a younger English scholar, Robert of Chester (fl.1140–1150). While in Ebro, Spain, Hermann and Robert were commissioned by Peter the Venerable to translate the Qur'an into Latin so that he could understand the religion; they finished the translation in 1143. Robert's translation of al-Khwarizmi's algebraic treatise, made in Segovia in 1145, introduced the first example of algebra into Europe.

Translations by Gerard of Cremona

The third twelfth-century translator who also succeeded in providing Arabic astronomical knowledge to the Latin West and by far the most prolific of the three was Gerard of Cremona, northern Italy (c.1114–1187). Around 1144 Gerard arrived in Toledo, now becoming a hotbed of scholarly work, in search of an Arabic manuscript of Ptolemy's *Almagest,* since he could not find a Latin translation in Italy. The *Almagest* had been translated from Greek to Arabic for the first time in 800 and improved copies followed quickly after that (North 2008: 119). In Arabic libraries, Gerard found a copious source of long-sought scientific manuscripts, and additionally discovered many new to him. In order to retrieve and absorb the *Almagest*, Gerard learned Arabic; the *Almagest,* of course, became his first Latin translation (Haskins 1927: 104).[23] Clearly fascinated with his discoveries, Gerard remained in Toledo for over 40 years and did the majority of his translating there; he was appointed to the position of *diácono* at the Cathedral of Toledo in 1162. This advantageous position as deacon supplied him room and board and allowed him full access to the scholarly works he wanted to translate; he held the title for the rest of his life.

Gerard completed over 80 translations from a variety of scientific disciplines, all from Arabic, including nine works by Galen, many treatises of Aristotle (*Physics*, *On the Heavens, Posterior Analytics,* and *On Generation and Corruption*) written originally in Greek and books such as, Avicenna's *Canon* and al Kindi's *On Optics,* written originally in Arabic.[24] Gerard is also credited with creating Latin copies of many additional Arabic texts on astronomy, alchemy, philosophy, and mathematics; in fact more Arabic science passed through his hands than any other Latin translator (Bosch 2000: 37).[25] By 1150 Toledo was thriving as the center of translating activity from Arabic into Latin; "this 'industry' in Toledo was so successful that it replaced or drove underground much of what went before" (Burnett 2003: 42). Under the guidance of Gerard, there was an organized program of revising and retranslating the treatises thought to be the most essential in astronomy and astrology. Rather than translating classical scientific texts from Greek to Latin, the Toledan translators preferred the Arabic versions which were successful in becoming recognized within the scholarly community (Burnett 2003: 42). He and his team produced translations from most areas of science and from many of the important Islamic authors, such as al-Khwarizmi, al Razi, Ibn al-Haytham, Thabit

[23]Gerard was not the first to translate the *Almagest* into Latin. It had been translated from Greek into Latin in Sicily under the aegis of Henricus Aristippus about 1160. See Kunitzsch (1989) for the transmission of the Ptolemaic star catalogue.

[24]Gerard's translation of al-Kindi survives in thirteen manuscripts; see Francis J. Carmody, *Arabic Astronomical and Astrological Sciences in Latin Translation, a critical bibliography* (1956). Robert of Chester also translated works by al-Kindi; ten manuscripts are extant. It is interesting to note that ten manuscripts of these two translators are held at Oxford.

[25]In addition to these works, Gerard also translated other Arabic authors; al-Gazzali, al-Fargani, and al-Farabí. He most probably did not translate alone but worked with a team of assistants.

ibn Qurra, and he also edited the Toledan Tables for Latin readers, which despite their shortcomings, were still the most accurate compilation of astronomical data in Europe at that time.

Gerard's Latin translation of the *Almagest* became the standard version used throughout Western Europe until the fifteenth century when Greek texts of classical works again became available (Haskins 1927: 120) About 50 Latin manuscripts of Gerard's translation of the *Almagest* are extant, of these 35 are complete, most were not illustrated (Kunitzsch 1989: 117). The first printed version of Gerard's Latin translation of the *Almagest* (from Arabic) was published in Venice in 1515. One of the most unusual and little studied developments of celestial imagery are the manuscripts of Fendulus and the appearance of the *paranatellonta* in the Latin West.

Georgius Zothorus Zaparus Fendulus

An abbreviated version of the astrological work of Abu Ma'shar also survives from the twelfth century. A mysterious medieval author Georgius Zothorus Zaparus Fendulus whose background remains unknown (perhaps an alias for he has never been identified) is credited with writing an abridged and fully illustrated version of the Latin translation of Abu Ma'shar's most popular work.[26] The second translation of I*ntroductorium maius in astronomiam* was completed by Hermann of Carinthia in Toledo about 1140 (the first was by John of Seville). This abridged version and interpretation of Abu Ma'shar was written sometime after 1140, and was inspired by Hermann's work, commonly called the *Greater Introduction* or *Liber astrologiae.* In the incipit of the prolog of the abridged treatise, Fendulus states that he translated the text from Persian to Latin; when in reality, Hermann's translation of 1140 was the version used for his abridgement (Clark 1979:24–32).[27] Fendulus' condensed version of the *Greater Introduction* describes mainly signs of the zodiac, planets, and their astrological influences; then these textual descriptions were transformed into the unique visual depictions that appeared initially in the abridged versions. The earliest and most luxurious surviving copy was made in southern Italy; the words '*atque palatem*' appear after the name of Fendulus which may point

[26]It is quite probable that the alias was assumed by Hermann himself; as a Christian monk it would be acceptable to translate the Arabic codex, but to write an original more-easily understandable astrological book may not have been suitable.

[27]All seven manuscripts are discussed by Clark. For more information on BN lat. 7330 see *Manuscrits Enluminés de la Bibliothèque Nationale, Manuscrits D'Origine Italienne II, XIII Siècle*, F. Avril et M-T. Gousset, Paris: 1984; and also *Liber Astrologiae*, M-T. Gousset and J-P. Verdet and A.G. Avilés, *Imágenes de los decanos en el Liber astrologiae de Fendulus*, Paris, Bibliotheque Nationale, ms. Lat. 7330.

to the court of Holy Roman Emperor Frederick II during the second quarter of the thirteenth century (Clark 1979: 24).[28]

This manuscript tradition is discussed in *Sternbilder des Mittelalters*; some of the miniatures are pictured in color and many more in black and white, including a full-page image of Fendulus clad in the garb of a Muslim potentate.[29] An earlier study of the abridged Abu Ma'shar in a thirteenth-century manuscript (Paris BN Ms. lat. 7330) was published in 1989 by M-T. Gousset and J.-P.Verdet; their scholarship explains the astrological sources and the textual basis of the unusual astrological iconography, distinguished by its beauty and novelty. This abbreviated manuscript *Liber Astrologiae* discusses and illustrates the various aspects of the paranatellonta that rise with the zodiacal signs and the 36 decans that divide each astrological house into three parts.

The astrological decans were part of a reference system that developed independently in early Egypt. The decans represent stars or star configurations that are comprised of 36 segments of the ecliptic that divide each zodiacal sign into 10 degrees. The decans make their appearance very early in drawings and texts inside coffin lids of the Tenth Dynasty (c.2100 BCE) and are shown on the tomb ceiling of Pharaoh Seti I (1318–1304 BCE). The ancient usage of decans appears to have provided the initial basis for the division of the day into 24 hours. In the carved stone ceiling from the Hathor Temple in Egypt, the 36 decans surround the zodiacal circle (Fig. 5.3).

The concept of the *paranatellonta* was not a new addition to astrological practices in the twelfth century nor was it derived from Arabic astronomical works. The earliest text to survive that discussed the *paranatellonta* and their function in detail is found in the didactic poem *Astronomica* by Manilius written in the early first-century CE, cited in Chap. 3. Manilius composed a text about stars and decans that rise with the signs of the zodiac in Book V. Similar to the zodiacal signs and the decans, the *paranatellonta* were time measurers (Tester 1987: 43). They are especially bright stars that ascended at the same time as signs of the zodiac and designated certain decanal sections of the ecliptic, representing every 10 degrees of the zodiac. They expanded the possible interpretations of astrological indicators and stellar influences. Undoubtedly, lists of the *paranatellonta* were first recorded to allow viewers to tell the time of night when the zodiac constellations were not visible. They were used by astrologers for determining the seasons and for

[28]This regal Fendulus manuscript is Paris BN Latin 7330, dating between 1220 and 1240. According to Saxl, this manuscript is not the original by Fendulus, but a copy of a twelfth-century model. Saxl detects at least two models and feels "that the constellations betray an earlier source while the planets reflect a later trend initiated by Michael Scot." In addition to Clark's study of the Fendulus manuscripts, BN Lat. 7330 is included in Saxl and Meier, 1953, II, pp. LII–LXVIII, Figs. 31, 33; and in *Dix Siècles d'enluminure italienne*, Francois Avril, Bibliothèque Nationale, Paris, 1984, pp. 51–52.

[29]See Blume et al. (2016) for their essay on Fendulus, pp. 20–22, and for illuminations from this copy and from other surviving copies: London Sloane 3983: NY Pierpont Morgan MS 785: Paris MS. Lat 7331: Paris MS Lat 7344.

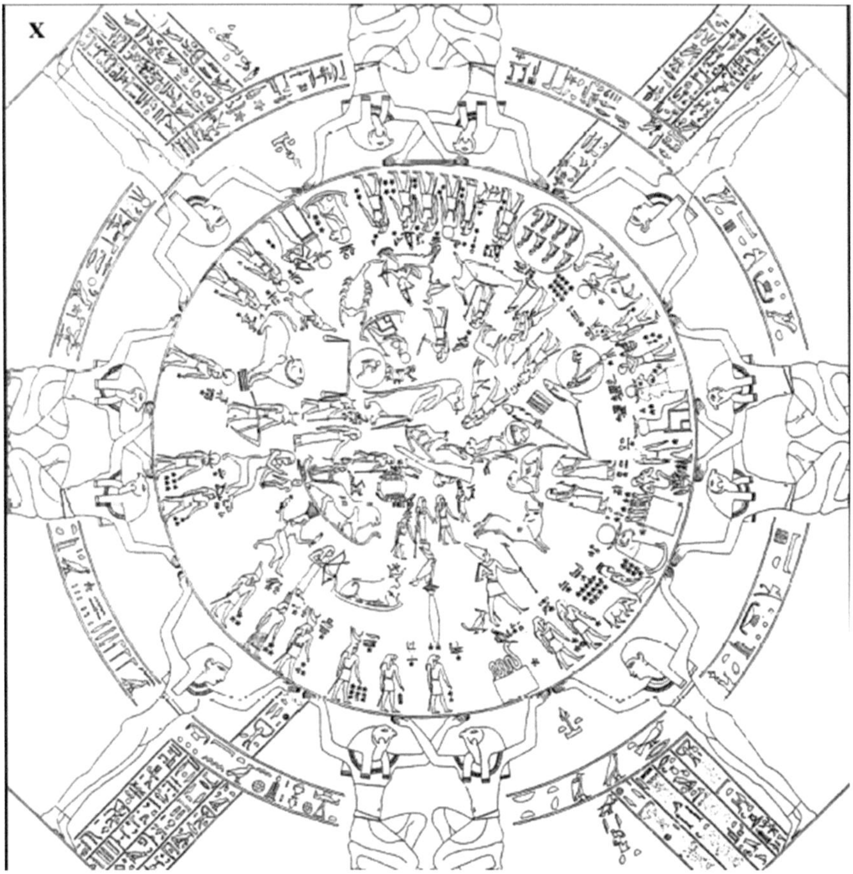

Fig. 5.3 Celestial Ceiling Map Drawing, Hathor Temple, Dendera, c. First Century BCE, Ptolemaic Egypt, Public Domain

determining zodiacal prognoses. Although noted by astrologers in the first century, they were probably used for astrological predictions even earlier. Eudoxus was known to have made a list of them in the early third-century BCE and Hipparchus made a more accurate list of *paranatellonta* a century later.

Illustrations of the *paranatellonta* appear mainly in Fendulus manuscripts and allude to an expansion of astrological theories and predictions created by combining the astrological systems of India, Egypt, Persia, and Greece. These Fendulus manuscripts are essentially elaborate picture books which begin with two full-page portraits: one of Abu Ma'shar and one of Fendulus himself. The brief introductory text describes the 12 signs of the zodiac. The secondary illustrations are faithful to the text and show a portion of a leg, an arm, or an animal. Because of the inclusion of the concept of decans and *paranatellonta*, Fendulus manuscripts seem to be modeled on a completely different astronomical tradition than the *Aratea* or al-Sufi manuscripts. Few medieval Latin astrologers included the influences of the *paranatellonta* in their natal horoscopes. These strange illustrations of various body parts and partial animals lived on until the later Middle Ages; seven copies of Fendulus' *Liber Astrologiae* or *Greater Introduction* are still extant. The Abu-Ma'shar and Fendulus manuscripts were quite influential in reactivating astrological practices in the Latin West.

In direct contrast to the progress and creativity of Arabic astronomers, those in the West were content to study the traditional body of 'scientific' knowledge, not directly from Greek sources, but as redefined by authors in Late Antiquity, making no original contributions or experimentation, and no changes or improvements to what had already been written. Aside from computus studies, the same few popular, but elementary, astronomical manuscript traditions including their basic illustrations and diagrams were copied and recopied with little modification until Islamic contact, when the entire field was inspired and revitalized. This interaction of the Latin world with the Arabic world changed forever the outmoded western tradition of merely protecting and preserving ancient knowledge. These initial encounters or meeting places of Eastern and Western astronomy that began in the eleventh century brought advancements in science and also subtle and fascinating changes to the visual depictions of the stars, planets and mythological characters. Throughout the twelfth century, the gradual exposure of the huge developments attained by Arabic researchers produced a profound impact upon Latin scientists, advancing the development of Western science and astronomical illustrations.

Illustrated Astronomical Texts in the Thirteenth Century

The influx of exciting 'new' information from Greek and Arabic sources and the scientific advances of the twelfth century inspired educational reforms in the Latin West. Latin translators introduced a theoretical and mathematical basis for astronomy derived from Ptolemy to Western scholars replacing the simple descriptions of the seven liberal arts. The discipline of astronomy had accumulated vast amounts of knowledge and more-exact calculating techniques to establish the parameters of the sun and moon, but Christian astronomers, mostly monks and priests were slow to integrate this newly discovered information with their traditional religious doctrines. In the thirteenth century, a new series of treatises, filled with rudimentary knowledge of the technical aspects of astronomy and bearing the generic title of *De sphaera*, circulated in recently established universities at Paris, Oxford, and Bologna. These astronomical texts were supplemented with explanatory diagrams, but did not include the mythological material or the illustrated constellation cycles featured in *Aratea* manuscripts. By far the most popular version of these treatises, copied over and over for European educational centers, was the *De sphaera* treatise composed by Sacrobosco.

Illustrated Manuscripts of Sacrobosco, *Tractus de Sphaera*

The thirteenth-century astronomical treatise *Tractus de sphaera* (or *de Sphaera*) composed by Johannes Sacrobosco (c.1195–1244), an English monk, scholar and astronomer, was one of the most influential and widely used textbooks throughout Europe for almost 500 years, popular until its astronomical information became outmoded at the start of the scientific revolution in the seventeenth century. Little is known for certain about his life or education; John of Sacrobosco also known as John of Holywood, was educated at Oxford and then taught at the University of Paris. (Even these few 'facts' have no reliable documentation.) His *de Sphaera* was published around 1230 and soon after became required reading for all university students. In truth, his treatise was an elementary work explaining the basic celestial phenomena, a simplified version of the Ptolemaic system from the *Almagest*: the rising and setting of the stars and planets, the zodiac, the spherical Earth, the climate, and the Sun's yearly path through the ecliptic. Sacrobosco's treatise was a less mathematical version of Ptolemy with basic spherical geometry and almost no planetary theory. Students of astronomy were no longer concerned with bridesmaids of Martianus Capella or dreams of journeys through space (North 2008: 252).

The *de Sphaera* textbook is organized into four chapters. The first chapter discusses the general structure of the universe, defines a sphere geometrically, and presents viable reasons for asserting that both the universe and the earth are spherical. In the second chapter, Sacrobosco names and defines the circles of the celestial sphere: the equator, the zodiac, the ecliptic, the meridian, the Tropics of Cancer and Capricorn, the solstitial and equinoctial colures, (great circles passing through the celestial poles and the points in the ecliptic corresponding to the solstices and equinoxes) and the Arctic and Antarctic circles. The third chapter discusses the daily rotation of the heavens which causes the rising and setting of the signs of the zodiac, and the diverse climates of the earth. The fourth and final chapter concentrates on the circles and movements of the heavenly bodies and the causes of eclipses.

Although astrological theories were well established and quite popular as an accepted art at this time, Sacrobosco's work avoided all astrological principles and was entirely focused on the astronomical knowledge currently available (Tester 1987: 150). Sacrobosco's surviving texts are most often embellished with astronomical diagrams that helped to clarify his descriptions of solar, lunar, and planetary motions. Hundreds of medieval manuscripts of Sacrobosco's *de Sphaera* survive (McCluskey 1998: 196).[30] By the mid-twelfth century, manuscript production had expanded from lone monks or scriptoria behind thick monastic walls to the domain of lay scribes and artists in dynamic urban workshops producing texts for the new urban schools (Montgomery 2000: 140).

---

[30]See also Thorndike (1949) *Sphere of Sacrobosco and Its Commentators,* and Pedersen (1985) "In Quest of Sacrobosco", Journal *for the History of Astronomy*. An early printed version of Sacrobosco was used by Cuban artist Carlos Mercia as a basis of his art illuminations.

One of the earliest extant copies of Sacrobosco's *Tractus de sphaera*, produced in Paris and dated circa 1262, is held by the New York Public Library (Alexander et al. 2005: 331–2). The astronomical text in NYPL MA 69 is presented in a simplified, clear manner for beginning students; it included *Qualitans* (Quadrant), or A*lgorimus* (Arithmetic) and in addition is elucidated with attractive illuminated diagrams. Sacrobosco's manuscripts contained the elementary information that was though necessary for understanding the workings of the universe. The Morgan Library holds a fifteenth-century humanistic manuscript of Sacrobosco and fragments of another.[31] The Earth-centered diagram demonstrates the theoretical astronomy of Ptolemy with the planets moving on epicycles; the signs of the zodiac are painted around the perimeter (Fig. 5.4).

Along with *de Sphaera*, Sacrobosco is recognized for two additional lengthy compositions: an introduction to the Hindu–Arabic numeral system written circa 1225 called *de Arte numerandi* (or *Algorismus*). Although the Hindu–Arabic numerical system had been introduced to Europe many years earlier, it had not been widely adopted. Sacrobosco's text was the first to bring awareness of Hindu–Arabic numbers into European universities. His other well-known book, *de Anni ratione* (*On reckoning the years*) which was published circa 1235, described the defects of the Julian calendar which through the centuries had accumulated a significant error of 10 days. In this treatise which was his longest and most original, he recommended the corrections for what eventually became the Gregorian calendar three centuries later; *de Anni ratione* also included Sacrobosco's instructions on computus and the ecclesiastical calendar.

In addition to the numerous handwritten medieval copies of *de Sphaera* that survive, there were more than 30 incunabula editions with many survivals. The initial printed version of *de Sphaera* was published in 1472 by Andreas Belfortis, the first printer in Ferrara, the realm of the Este family and a popular gathering place for artists and authors. In 1531, Melanchthon wrote a preface for a new edition of *Sacrobosco* because the book was still commonly used in universities; his comments became widely known (North 2008: 358). More than 200 separate printed editions were published up to the seventeenth century, attesting to Sacrobosco's widespread reputation; some printed versions were illustrated with woodcut illustrations. *De sphaera* was translated from Latin to English by Thorndike in 1949. Sacrobosco's astronomical works were focused on providing an elementary understanding of celestial motions and were not involved with or influenced by the *Aratea* tradition.

[31]*Italian Manuscripts in the Pierpont Morgan Library,* p. 50. The Morgan Library holds a Sacrobosco manuscript M.426, made in Italy, *Sphaera Mundi*, written in Italian in the third quarter of the fifteenth century. M.426 has four astronomical marginal illustrations. Morgan Library also has a fragment of another *De sphaera*, a fifteenth-century copy made in Austria with a colorful diagram of the heavens, M.722, f.18.

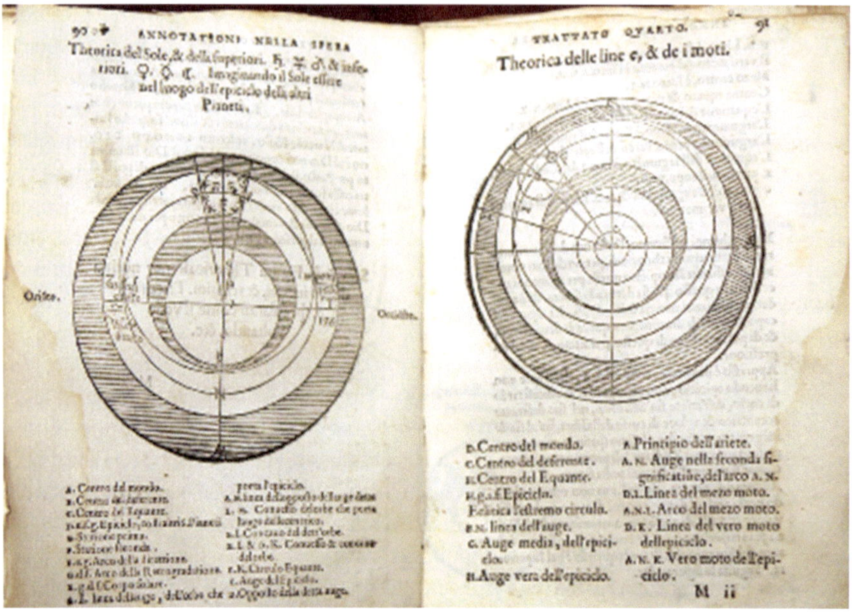

Fig. 5.4 Sphere of Sacrobosco Diagrams, after fifteenth century, print reproduction, Public Domain

Illustrated Astronomical and Astrological Manuscripts at Late Medieval Royal Courts

Progress in the advancement of astronomical knowledge proceeded fitfully through the later Middle Ages. Natural philosophers acknowledged both the advantages and the disadvantages of the Aristotelian and Ptolemaic systems and tried to combine the better features of each system. The masters of the new universities gradually assimilated advanced astronomical knowledge gathered from Eastern sources into the European intellectual tradition. The newly acquired understanding was applied to traditional astronomical problems; European scholars taught, wrote about, and questioned the cosmological implications.

After the writings of Roger Bacon, Albertus Magnus, Robert Grosseteste (1175–1253) and other important thinkers of the thirteenth century had been analyzed and digested, more highly developed scientific texts were produced. The influx of Arabic texts had advanced the complexity of astronomical understanding. Theories of Spanish-Arabic al-Bitrûjî (fl.1180) were especially influential and scholars made extensive use of his astronomical writings in their studies.[32] Aristotle, along with

[32]The important work of al-Bitruji, *de Motibus Celorum* was translated from the Arabic to Latin by Michael Scot in 1217. A critical edition of Scot's translation of this treatise was published in 1952 by F.G. Carmody, University of CA Press.

his Arabic commentators Averroës and Avicenna, was studied, thrown out, and then rewritten to fit Christian ideology. The astrolabe, which was first used in the West in the late eleventh century, allowed for the accurate and automatic calculation of a wide range of astronomical events, including the time of day or night, the computation of the sun's position for any given date, the plotting of the position of the planets against the fixed stars and within the zodiacal constellations. The astronomers' desire for more accurate data was stimulated by the translation of Arabic tables or *zijes*, which allowed for easy calculations of heavenly movements, vital for astrological readings.

Astrology was the uppermost science in the thirteenth and fourteenth centuries, an integral part of learned culture; astrological principles were considered by many as the fundamental natural law; almost to the time of Newton, astrology had dominion over the inferior elementary bodies and many followers gave credence to the ruling influence of the stars (Thorndike 1965: 13). Throughout Europe, astrology had become intimately involved with everyday medical practices. By the end of the sixteenth century, physicians were required to calculate the position of the moon and assure an auspicious time before carrying out complicated medical procedures, especially for surgery or bleeding. Although astrology was a primary element of the medieval worldview in the West, it did not achieve the same level of prominence it held in the Islamic world. Astrological theories incited a potent debate among church, state and university leaders that lasted for centuries. Questions on the orthodoxy and morality of astrological pronouncements brought condemnations and anxieties to many principal voices in the Christian world, but still it is impossible to fully understand later medieval culture without taking astrology into account.

The simplistic type of astronomical information and the constellation illustrations found in *Aratea* manuscripts had no place in this heady company, as thirteenth- and fourteenth-century astronomers were interested in recent discoveries such as the "proper movement of the sun's apogee". Technical, theoretical astronomy with its advanced spherical diagrams and complicated epicycles had supplanted the mythical tales and symbolic stories of the constellations and planets in the writings of Macrobius, Martianus Capella, and Aratus and the fabled images of the *Aratea*. But rather than becoming lost and forgotten in the surge of intellectual advancement, as one would expect, the ancient gods, goddesses, planetary figures, and luminaries aligned with the constellations reemerged in the next century dominating the iconography espoused by the wealthy courts of Europe, especially Italian Renaissance courts.

In the later Middle Ages, royal courts, as always, patronized artists, musicians, scholars, and astronomers, but became especially fascinated by the resurgence of the ancient art of astrology. Astrological theories began to influence the activities and daily life of the courts; astrological and cosmological artistic programs decorated their luxurious regal residences and illuminated manuscripts filled their book shelves. Remarkably, *Aratea* manuscripts not only survived the challenges of time and change throughout the Middle Ages, but actually reached their peak production in the fifteenth century.

Sicily was among the first areas of Europe where the scientific knowledge amassing from Arabic scholars was encouraged and advanced, most importantly by the royal court of Holy Roman Emperor Frederick II (1194–1250). Frederick was one of the most enlightened and respected monarchs of his age; his royal court in Palermo was without doubt the most brilliant in Europe, a center of culture, poetry, natural sciences, and mathematics. He enacted the principal role in all intellectual activities that transpired there and maintained a long-lasting correspondence with Islamic philosophers and scientists. Frederick spoke seven languages including Arabic; he encouraged Muslim scientists, an enlightened policy of Islamic–Christian cooperation begun by his Norman grandfather, King Roger II, who had sponsored the geographer Muhammad al-Idrisi. (His paternal grandfather was German Emperor Frederick Barbarossa.) His court at Palermo, similar to Toledo, became a prominent pedagogic center where Arabic, Jewish, and Christian scholars worked together to translate and reproduce essential scientific texts (Haskins 1922: 244).[33] A courtly 'school' emerged that produced a variety of illuminated manuscripts of the highest quality (Fig. 5.5).

The most prominent production from the court scriptorium, entitled *De arte venandi cum avibus*, (*Treatise on Falconry*), was composed by Frederick II himself in 1250 and dedicated to his son Manfred. This highly admired work, *The Art of Hunting with Birds*, was one of the first experimental studies of the natural world and demonstrated Frederick's commitment to direct observation and precise detail.[34] The erudite emperor spent more than 30 years gathering information and actually performed scientific experiments with his falcons to determine specific results. The manuscripts produced in this royal court (both Palermo and Naples have been proposed as the scriptorium's actual location) show a new and spontaneous element with elegant miniatures painted with somber and delicate shades. The astronomical manuscripts associated with medieval scholar Michael Scot are manifestations of this courtly style associated with Frederick II.

[33]Even before Frederick II came to power, his immediate predecessors, not only grandfather Roger II but also uncles, Kings William I (the Bad) and William II (the Good) promoted scientific activity, especially the translation of Greek treatises on mathematics and astronomy, although they had all died before Frederick was born to Constance, daughter of Roger.

[34]*De Arte Venandi cum Avibus of Frederick II* of Hohenstaufen has been translated into English by C.A. Wood and F.M. Fyfe and published by Stanford University Press and has been published in many modern editions. The beautifully illustrated thirteenth-century work with more than 900 species of birds is transmitted in Codex MS pal. lat. 1701 is now housed in the Biblioteca Apostolica Vaticana.

Emperor Frederick was friendly with the famous import trader and mathematician Leonardi Fibonacci, who introduced Europe to the Hindu-Arabic notation system currently in use that replaced the cumbersome Roman numerals; he presented the emperor a copy of his famous *Liber abaci*. Frederick commissioned copies of many philosophical works and donated them to libraries throughout his vast Empire.

Fig. 5.5 Holy Roman Emperor Frederick II in his *Treatise on Falconry*

Michael Scot: Astronomical Illustrations at the Court of Emperor Frederick II

The intriguing and enigmatic Michael Scot (1175–c.1232) was considered the leading intellectual of his time and became the dominate figure among the scholars supported by Frederick II, even though his career at the royal court in Sicily lasted for only about 4 years in the 1220s. Born in the British Isles, clergyman Michael Scot was educated in Oxford and Paris and since he was often referred to as 'Master Michael Scot', it is assumed he had taken a master's degree. He soon moved into papal circles under Honorius III and then Gregory IX; Honorius III praised him as singularly gifted in science among men of learning (Thorndike 1965: 1). In his writings, Michael reported turning down a benefice in Ireland, as he knew no Irish. He was known to be active in Toledo in 1217 and at Bologna in 1220, probably under the patronage of Holy Roman Emperor Frederick II's court there.[35]

Multitalented Master Michael joined the court of the charismatic emperor as a medical doctor, astronomer, alchemist, astrologer, and translator of Arabic treatises. Scot had actually spent much of his scholarly career translating Arabic treatises in Spain, mainly in Toledo, where, like Gerard of Cremona before him, he was

[35]Frederick II was the first European ruler to deliberately found a university, the University of Naples in 1224—now called University of Naples Frederico II, one of the oldest academic institutions in continuous operation.

appointed deacon at the Cathedral of Toledo. Between 1217 and 1220, Michael translated al-Bitrûjî's *De caelo* and *De motibus celorum* (Carmody 1952: 45).[36] At Frederick's court, he translated Aristotle's *De animalibus* which Frederick had already owned. Carmody's edition lists 13 surviving manuscripts of Michael Scott's translation of *De motibus celorum*. Some are accompanied with mathematical diagrams, but since al-Bitrûjî is concerned with theoretical not popular astronomy, there is no description of the constellations and no illustrations in this group of translations (Fig. 5.6).

Michael Scot also translated texts of Abu Ma'shar as well as the philosophical commentary on Aristotle's *De caelo* by Ibn Rushd (Averroës) (Haskins 1922: 283).[37] His translations of Averroës' commentaries were largely responsible for the introduction of Averroism into western Christian Europe (Thorndike 1965: 24). Scot's treatises on astrology were the basis of his literary fame; his scientific writings show his knowledge of medicine, natural philosophy, and music (Haskins 1922: 285). He translated many Arabic manuscripts for Frederick II, even though the emperor was fluent in Arabic and could have read them for himself. In addition to his many translations, Scot composed some original works such as *De divisione philosophia* and *Liber introductorius*, both written before 1228, and both dedicated to Emperor Frederick II. *Liber introductorius,* considered his most ambitious work, was a handbook written for beginners; the early segments discuss the basics tenets of astronomy, then he presents a summary of the existing knowledge of astrology as it was popularly practiced.[38] He says he wrote for student beginners and "those not over-burdened with intelligence." Both Frederick and Michael appeared to be more interested in scientific questions than in astrology, although both accepted astrology as a proven part of contemporary science (Tester 1987: 190). Michael avoided problems with Church officials by saying the stars were only signs and not causes of future events, but he gained a reputation as an adept in the secret arts.

The illustrations devised by M. Scot were an important influence. The mid-twelfth century German ms Vienna ONB 12600 De ordine had departed from classical prototypes. It is not an astrological manuscript but an astronomical manuscript an illustrated catalog of 42 constellations.

---

[36]Al-Bitrûjî wrote in the early thirteenth century in Spain; his works were among the latest by Arabic astronomers to reach Western scholars. His treatises were translated into Latin soon after they were completed. By 1220 Michael Scot's translation became one of the standard source texts consulted by many astronomers; bringing Scot great recognition. In his original writings, al-Bitruji includes metaphysical principles, time, velocity of movements, precession and spherical geometry; he is also known for his interpretation of Aristotelian axioms.

[37]According to Haskins, "Scot's translations were the occasion of unfavorable judgements on the part of Roger Bacon, who declared that Scot did not really know the languages or the sciences and the work was chiefly done by a Jewish translator named Andrew." Michael Scot did not seem to know any Greek.

[38]Illustrated copies of *Liber introductorius* are: Munich, cod. lat. 10268 (15th c.); Oxford, MS. Bodley 266, which is a copy of the Munich version; Paris BN, Nouv. acq. 1401 (ca.1279). Haskins lists nine more manuscripts which have only fragments of the work; just one has illustrations that one is Seville, Colombina, MS. 7.7.1 (15th c.).

Fig. 5.6 Miniature of the Sun or Apollo in Quadriga from Michael Scot *Liber Introductorius*, Public Domain

The writings of Michael Scot on the constellations and on astrology including the unusual manner of his illuminations were widely copied in through the Middle Ages; by the fifteenth century, most manuscripts that discussed the constellations were directed to astrological purposes. This strong interest in astrology created a lengthy decline of *Aratea* manuscript production during the late medieval period. The untraditional images introduced by Michael influenced the mode of illustrations moving away from the classical prototypes. His *Liber introductorius*, a less technical handbook more readily understandable, (See Blume et al. (2016) for further discussion and examples of his creative imagery.) became quite popular in Italy and also the German areas. The illustrations of the constellation cycle appear in one of Michael's surviving manuscript in Munich, and in its copy at Oxford, and a third copy at Paris. Haskins reports "the remarkable figures of the constellations and planets in the Munich and Oxford manuscripts represent an antique tradition which is ascribed by Boll to the scholia of Germanicus" (Haskins 1922: 288).[39] His illuminated manuscripts bring in completely new imagery showing great vigor and inventiveness that influences future depictions of astronomical art. Illustrations from surviving manuscripts of *Liber Introductorius* are available online at the Saxl Project headed by Kirsten Lippincott. Most manuscripts include images of the full constellation cycle as well as astronomical diagrams.The illustrations of Michael Scott manuscripts are discussed and pictured in Sternbilder des Mittelalters;, Blume et al (2016).

An astrologer to the end, a contemporary account reports that Michael predicted he would be killed after being hit by a small stone; he even determined its weight.

[39]Haskins cites Boll's *Sphäera*, p. 441.

To avoid his fate, Michael invented a protective head covering which he wore most of the time; but during Mass, he respectfully removed his special headgear. Michael Scot died c.1236 after being struck in the head by crumbling masonry in the church (Thorndike 1956: 39).[40]

King Alphonso X, the Wise: Astronomical Illustrations in His Court

The observable association of the royal courts of Europe with astronomy and astrology and its symbolism reached a zenith in Spain during the thirteenth century under the patronage of Alfonso X, (1221–1284) king of Castile, León and Galicia for over 30 years, crowned in 1252. Widely renowned as a poet, a devoted patron of music and arts and as a distinguished lawmaker, Alfonso X was honored in his own lifetime with the title, *el Sabio*. Alfonso X went far beyond the usual role of royal patronage, merely supplying funds, for he actively participated in many of the cultural enterprises that he sponsored. He himself was a musician, poet and also a competent scholar, even revising some of the astronomical work of his collaborators (Chabás and Goldstein 2003: 225) More than 400 poems praising the Virgin Mary, known as Cantigas de Santa Maria, were composed or commissioned by Alfonso X. (Figs. 5.7 and 5.8).

Like Charlemagne, Frederick II, and other imperial patrons, Alfonso X gathered to his court at Toledo the foremost intellectuals of his day: mathematicians, physicians, technicians, astronomers, and astrologers (Bosch 2000: 23).[41] When his ancestor Alfonso VI gained control of the famed intellectual center of Toledo, the crown acquired an enormous number of manuscripts as well as a city that boasted a multi-lingual society. In the manner of Frederick II, Alfonso embraced the brightest Arabic, Jewish, and Christian scholars, who worked together and created a collection of writings that were of great significance for various branches of the history of science. Many of the writings that they produced were related to astronomy: including instruments, clocks, tables, cosmography, star catalogs, and of course, astrology (Chabás and Goldstein 2003: 2).[42] Alfonso had many of these treatises translated into Spanish or Castilian, which made him the first to use the vernacular as a scientific language. The Christian, Muslim, and Jewish scholars assembled by Alfonso, drawing from Arabic writings, composed a collection of monographs on astronomy, on various astronomical instruments, and other related texts which were assembled into a large handbook known as *Libros del saber de astrologia* (Kunitzsch 1989: 64).

---

[40]Michael Scot was succeeded in the court of Frederick II by another astrologer named Theodore.

[41]Toledo's association with astrology looks back to antiquity thanks to one of its founding legends. The ancient myths credit the birth of the city to "Ferecio, a Greek astrologer, who came to Spain with Teucer, Ulysses, and Diomedes after the siege of Troy; later he consecrated the city to Heracles. Ferecio taught magic and astrology to the native inhabitants: hence the appellation *arte Toledano*, used to refer to the occult arts practiced in Spain."

[42]"The astronomers worked for a secular court led by the king and they had no known connection to such educational institutions as the recently founded University of Salamanca."

Fig. 5.7 King Alfonso X el Sabio Statue in Madrid, Public Domain

Fig. 5.8 King Alfonso X el Sabio Las Cantigas, his musical compositions, Public Domain

This Alfonsine compilation proposed to be, but is not exactly, a translation of al-Sufi's Book of Fixed Stars.[43] The text summarizes a large percentage of the data taken from al-Sufi, but it has been somewhat Christianized since this sensitive material might be considered questionable; not only because it relied on Muslim sources, but also because it revolved around the dangerous topic of astrology.[44]

Even before the translation and compilation of al-Sufi treatises at the court of Alfonso X, an earlier European tradition had developed around the writings of al-Sufi, called the *Sufi Latinus*. Like the Alfonsine version, these Latin manuscripts are a compilation of disparate astronomical and astrological material, only parts of which are actually related to al-Sufi. The main component of this group of manuscripts is a star catalog, accompanied by a series of drawings depicting each of the 48 constellations taken from al-Sufi's double drawings of each constellation. The catalog of stars found in these manuscripts is essentially Gerard of Cremona's Latin translation of the *Almagest* with the longitudes of the stars revised and brought up to date by al-Sufi. Constellation drawings that accompany the text in the *Sufi Latinus* are meticulous in showing the individual stars precisely as they appear in the night sky, according to both number and placement of each individual star. The constellations are precise but provide no information on their relative positions. The individual stars dominate each constellation and are numbered in order of their brightness; in addition the nearby stars not part of the constellations are accurately depicted as well. The comparative sizes of the stars in the illustrations indicate their magnitude (Fig. 5.9).

Alfonso's court scholars had vast resources at hand, not only Ptolemy's major astronomical treatise, the *Almagest*, which had been translated into Latin in Toledo by Gerard of Cremona in the previous century. A well-known heretical quote credited to King Alfonso is that "if God had consulted him when creating the world, he would have given Him good advice". It was assumed the king was referring to the complicated system of deferents and epicycles put forward by Ptolemy to explain the workings of the heavens. Although this undocumented quotation

[43]Spurred by his interest in astronomy and astrology, Alfonso X sponsored a compilation of available material subsequently called the *Books on the Knowledge of Astronomy* in which al-Sufi's catalog of stars was the fundamental text. A first draft of this Alfonsine version of al-Sufi was not regarded as satisfactory; it was revised by court astronomers in 1276. The prolog of this work maintains that Alfonso X himself made stylistic corrections to the text in Spanish, evidence that king was intimately involved in this revision Surprisingly the text of this compilation was not written in Latin, but in the vernacular or to be more exact in Castilian. Many Arabic texts had been translated into Spanish or Castilian but these were practically unknown among European astronomers.

[44]The revised book departs from its exemplar with the addition of fresh material from the imagination of the Alfonsine author. The digressions rarely have any astronomical importance, but much art historical importance. The additions are mainly concerned with iconography, descriptions of the animals of the constellations, or with other aspects of everyday life. For instance the description of Aquarius which has 45 stars is quite brief, the description of Canis with only two stars, has a very long section on small dogs and their usefulness, talismanic magic and astrology. See Gonzalo Menéndez Pidal, *La España del siglo XIII leida en imágenes*, Madrid, 1986.

Fig. 5.9 Constellation Cepheus, Sufi Latinus Manuscript, Public Domain

accredited to Alfonso is still repeated in history books, a rebuttal was published by B. Goldstein in 1991 after he decided to trace the historical foundation of this blasphemous story.[45] His study produced no thirteenth-century documentation of such a statement by Alfonso and finally put to rest the irreverent myth accredited to the learned king.

The scholars at Alfonso's court enjoyed easy access to the entire corpus of Arabic scientific material that had attained its highest scientific level in Muslim Spain in the eleventh century, including numerous copies of Arabic astronomical and astrological works for references (Chabás and Goldstein 2003: 1).[46] The richness of reliable sources and the encouragement of royal patronage enabled these scholars to produce a large body of astronomical works which were in part Castilian translations of texts by earlier Arabic astronomers, but also included some original and more advanced studies. See also Samsó, J. (1994) "Islamic Astronomy and Medieval Spain"

[45]Goldstein (1991) "The Blasphemy of Alfonso X: History or Myth?"

[46]Alfonso's father, King Ferdinand III, had already conquered much of the Moor's territory, including the important Islamic cities and learning centers, Seville and Córdoba. Alfonso relied on Christian scholars and Jewish intellectuals who were fluent in Arabic to translate their astronomical works into Castilian.

Alfonso X has been recognized and esteemed primarily as the producer of the well-known, widely utilized astronomical tables, which bear his name; as a result his name appears prominently in the history of astronomy. "In contrast, the translations and original works in literary, legal, and historical genres have largely been forgotten except by specialists" (Chabás and Goldstein 2003: 230). Even previous to the compilation of the *Alphonsine Tables,* dated from 1263 to 1276, Alfonso X had sponsored various astronomical studies which were carried out by his two Jewish collaborators, Judah ben Moses ha-Cohen and Isaac ben Sid; and it seems clear that Alfonsine scholars used astronomical sources credited to Spanish-Arabic authors (Bossong 1987: 13). A large part of medieval astronomy consisted of compiling and using tables consulted mainly for astrological purposes. Therefore books of tables listing positions and rising times of heavenly bodies and rising signs were vitally important. Unfortunately because of the erratic movements of the planets, such as retrograde motion, the astronomical tables needed continual corrections to keep them current.

The Alfonsine Tables

King Alfonso X was the driving force that resulted in the publication of the *Alphonsine Tables*, which were considered, "the most important astronomical charts of the Middle Ages". Yet modern scholarship has determined that Alfonso X was not in fact the author of these tables; although the wise king was greatly involved in their production, his main function was as patron of his large team of scholars, not as the creator. The *Alfonsine Tables* were a series of astronomical records designed to chart the chronology of the sun, moon, stars, and planets, accompanied by explanatory introduction chapters; they were needed to update the earlier *Toledo Tables* that had become outdated. The books of tables of positions and rising times of heavenly bodies and signs were accessed consistently, particularly in devising astrological horoscopes and for medical diagnoses. Astronomical tables lay behind all horoscopes cast to determine the most propitius times for everything from bloodletting through marriage, births to the foundation of universities and churches, even the best time for prayer. Even St. Peter's in Rome had a foundation horoscope prepared. Accurate observations of the heavens were always difficult especially with no clocks or chronometers. The "Toledo School of Translators" set the start of their revised astronomical charts at January 1, 1252, the date of Alfonso's coronation. The entire arrangement of the astronomical material was based on the Arabic-Islamic *zijes*. The *Tables* also contained a star catalog, which was basically Ptolemy's star chart from the *Almagest*, using the translation of Gerard. Although Alfonsine scholars relied on Arabic precedents and consulted Arabic sources, the *Alfonsine Tables* can be considered the first European attempt to develop original research in astronomy.

The original *Alfonsine Tables* were not written in Latin but in Castilian; previous to Alfonso's innovation, the texts of all scholarly and religious manuscripts were expressed in the exclusive domain of the traditional language, which of course was Latin in medieval Christian Europe. Until the second half of the thirteenth century composing a manuscript on legal, philosophical, or scientific matters in a Romance vernacular rather than the official Latin would have been inconceivable (Bossong 1987: 13).[47] This convention began to change around 1250, but not in a consistent manner; in Spain, the transition from Latin to the local vernacular began to take place at a very early date.

In just a few years after his accession to the throne, Alfonso X directed that scientific thought and scholarly treatises were to be written in Spanish/Castilian. Because of the language consideration, the Castilian astronomical corpus was not widely circulated, and the new assemblage of astronomical data inspired no major influence outside Spain at that time. However a Latin version of the *Alfonsine Tables* was put together at Paris in the 1320s and copies of this later translation spread throughout Europe. Shortly after the Alfonsine charts were compiled in Paris, these revised Latin Tables, still credited to Alphonso X, were rapidly replacing the earlier *Toledan Tables* (Chabás and Goldstein 2003: 5).[48] In their book, Chabás and Goldstein analyze and compare the relationship between the Castilian version and the later Parisian version. The *Alfonsine Tables* then became the most influential handbook of practical astronomy until revised again in the late sixteenth century (Bossong 1987: 14).[49]

Most references to the *Alfonsine Tables* in astronomical literature pertain to this later Latin version. The first verification that the Parisian *Alfonsine Tables* arrived back in Spain is found in Madrid, BN MS 4238 (Chabás and Goldstein 2003: 292).[50] No official census of the surviving manuscripts has been undertaken, but it is thought that at least several hundred copies of the *Alfonsine Tables* were produced and updated versions were published for hundreds of years. This set of astronomical tables awaits a modern editor. During the later years of his reign, Alphonso realized that writing in the vernacular greatly restricted circulation of his works and returned to the older practice of publishing scientific works in Latin, that way the writings from his team of scholars would more likely reach the attention of European centers (Bossong 1987: 16).

The Alfonsine *Zij* does not include a description of the corpus of constellations; therefore constellation illustrations would not be expected or found in the *Alfonsine Tables*. The *Alfonsine Tables* was the first European book to mention the *Andromeda Nebula*, not included in Ptolemy. Although the *Tables* are concerned with the constellations, particularly the signs of the zodiac, it seems quite possible

---

[47]The Romance languages were confined to the inferior status of oral vernaculars until the thirteenth century; they were appropriate mainly for literary purposes, especially for poetry.

[48]The *Toledan Tables* had been assembled in al-Andalus by S_ 'id al-Andalus and his circle, using the Muslim calendar, and were translated into Latin in the twelfth century. The authors also traced the probable source for the appearance of the *Alfonsine Tables* in Paris in the early 14th century, pp. 245–47.

[49]"Besides that, we have to wait more than a century before we find an astronomical treatise in English: Geoffrey Chaucer's famous *Treatise on the Astrolabe* was composed in 1391. He justifies his use of English since he is addressing his little son who still does not know enough Latin" (Bossong 1987: 14).

[50]"In contrast to the vast number of copies in libraries elsewhere in Europe, only a few manuscripts containing the Parisian version of the Alfonsine Tables have been preserved in Spanish collections" (Chabás and Goldstein 2003: 292).

that their function was more related to astrology and magic than science. One of the chief motives that inspired the king to request the preparation of an acceptable star catalog was the current fascination with astrology. Alfonso X was greatly concerned with astrology and also had a strong interest in talismanic magic, which was associated with the preparation of astrological images (Samsó and Comes 1988: 72). Surviving law codes show mixed feelings toward astrology and magic during that era. In one of the legal codes promoted by Alfonso X, the *Partidas*, mandates that those who employ magic for mischievous purposes are to suffer the death penalty, but those who use magical forces with good intentions and achieve beneficial results are to be highly praised (Samsó and Comes 1988: 72).[51]

Alfonso X and his extensive royal patronage were important factors in the history of astronomy, and the medieval king too has been immortalized by modern astronomers as a lunar crater bears his name. The first printed edition of the *Alfonsine Tables* appeared in 1483, and a second edition in 1492. The *Alfonsine Tables* continued as the most popular astronomical charts, and were widely disseminated throughout Europe; hundreds of manuscripts are extant. A facsimile of the *Alfonsine Tables* called *Tablas de las constelaciones de Alfonso X el Sabio* with full-page miniatures was published in 2006.[52]

Although numerous copies of Alfonso's works survive, the disappearance of Spanish Arabic manuscripts can be attributed to the Reconquista. In order to eliminate copies of the Qu'ran from the conquered Muslim areas after they were defeated, the Church ordered the total destruction of all Arabic manuscripts. The only Arabic books to survive were those few that had been moved to safety. Only one illustrated manuscript survives from Islamic Spain; the lone Arabic book is a romance, therefore historians cannot generalize about the development of book illustration in the Iberian Peninsula (Bloom 2001: 179).

Astronomical Illustrations in European Royal Courts and the Tudor Court of King Henry VII

Although not an impressive scholar in the mode of Frederick II or Alphonso X, King Henry VII (1457–1509) was renowned as a patron of the arts culminating in his commission of the Lady Chapel, now known as Henry VII Chapel, within Westminster Abbey where he and his wife, Elizabeth of York (1466–1503) are buried. Henry VII continued the widely held tradition of astronomical and astrological interests and associations in European royal courts. Through the late Middle Ages and Renaissance, astrology and divination was traditionally associated with

[51]The manuscripts concerned with talismanic magic and the properties of stones and mineral are called *lapidarios* and may contain illustrations of the constellations.

[52]The facsimile is an edition of Berlin, Staatsbibliothek Preuß. Kkulturbesitz, Kupferstichkabinett 78 D 12 Valencia: Patrimonio Ediciones.

kings and courts of the nobility for the most part. Before the availability of inexpensive printed books, only aristocrats had the means to commission astrological manuscripts and employ personal astrologers. Having forehand knowledge of future intrigues, attacks or natural disasters would be enormously desirable for ruling monarchs but the Church's stance against astrology, that it denied the omnipotence of God and was an affront to God's gift of free will to humanity, discouraged the practice until the fourteenth century (Carey 1987: 41).

King Charles V (1338–80) of France and his brother Jean duc de Berry (1340–1416), as well as King Wenceslaus II (1271–1305) of Bohemia and of course Holy Roman Emperor Rudolph II (1552–1612)[53] were each well known for the brilliance and sophistication of their courts. What these regal personages had in common includes the possession of a huge number of beautifully illuminated manuscripts and in particular, books on astronomy, astrology, and esoteric practices. It appears that the possession of elegantly illuminated manuscripts focused on the occult sciences became a court fad. When Charles V died in 1380, his vast library held more than 1200 books, housed in the Louvre which he had expanded during his reign. Over 10 % of his manuscripts were related to astrology, geomancy, chiromancy, and necromancy, with an additional 70 volumes that were concerned with astronomy, including essential books needed to study and practice judicial astrology (Carey 1987: 46).[54]

In the late medieval and early Renaissance libraries of Italian royal courts, the collections of astrological and occult manuscripts seem to be even more extensive than those in France or England. The Dukes of Urbino and Ferrara, the Gonzaga, Visconti, Sforza, and Medici families all displayed a great interest in the practice of astrology, the consultation of tarot or tarochi cards, and the study of related occult arts (Carey 1987: 47). Court astrologers continued their practices for several centuries, but the Church finally condemned their activities when the first Papal Bull to ban court astrologers was issued by Pope Sixtus V in 1585.

The first English royal figure to look favorably upon astrology seems to have been King Edward III (1312–1377). He ruled through difficult times in European history, the Hundred Year War, the Black Death, and the Papal schism and its move to France. Edward III owned an illustrated copy of *Secreta Secretorum* in Latin, and astrology was practiced at his court, but his personal opinion on the subject is not known. His natal horoscope was recorded and still exists. The fifteenth-century court of England's first Tudor king, Henry VII, was admired for its international tastes and global interests; humanist scholars were invited to his royal court to tutor his four children and to write England's history. Henry 'took great delight' in having books read to him, letting someone else accomplish the task of actually

[53]Rudolph II was noted for attracting famous scholars and astronomers to his court, including John Dee, Giordano Bruno, Tycho Brahe, and Johannes Kepler, to name just a few.

[54]For information on the fourteenth -century library of Charles V see Léopold Delisle (1907) *Recherches sur la librarie le Charles V*, Paris. The nineteenth-century historian and bibliographer described the royal library in great detail.

deciphering the letters (Carley 2004: 41).[55] Although considered parsimonious until his later years, Henry VII enjoyed collecting manuscripts, especially astronomical and astrological texts, and importing expensive art works and decorative furniture from wealthy cities of Europe. Henry VII has been distinguished as the first English king known to have established a formal court library to preserve the copious manuscripts that he bought, collected and commissioned. In 1492 he appointed an educated librarian Quentin Poulet, a scribe and illuminator from Lille, to maintain his growing collection. At least 19 illuminated manuscripts from Henry VII's assemblage still survive in the Old Royal library. He also purchased numerous printed books including 20 volumes from a Parisian printer, some printed on vellum and painted. Henry VII, like his more famous son, also commissioned sumptuously illustrated manuscripts. In addition to importing volumes and furnishings to his court, Henry VII also summoned a famous astronomer/astrologer from Italy.

Doctor, astronomer, astrologer William Parron, a native of Piacenza, Italy, moved to England around 1487 and for 15 years, Parron and a few other astrologers, provided astrological and medical counseling for court patrons. Eventually as a result of his predictions, Parron succeeded in establishing himself as physician to Henry and produced several books containing general predictions.[56] Henry did not seem particularly devoted to his court astrologers; they continued their duties as physicians, astronomers, and mathematicians, but offered him little advice. The arrival of another astrologer-physician from Italy John Baptista Boerio, assured the appearance of the Italian Renaissance in England. Parron took advantage of the printing offices to sell inexpensive annual predictions that became quite popular in later times (Carey 1987: 55). The career of England's first royal astrologer came to an inglorious end; Parron quietly disappeared from view after he predicted a lengthy life of eighty years for Queen Elizabeth of York (1466-1503). Unfortunately, Elizabeth died a few months later of postpartum infection after giving birth to a daughter who only lived for two weeks; she passed away on her thirty-seventh birthday.

A richly decorated, illuminated astronomical manuscript, British Library, MS Arundel 66, was produced under the patronage of Henry VII in 1490, while Parron was engaged as court astronomer/astrologer, but his contribution to its creation cannot be established with certainty (Scott 1992: 365).[57] Arundel 66 in Latin and

---

[55]Most of the illuminated manuscripts held in the library of King Henry VII were inherited by his son King Henry VIII.

[56]Three decorated copies of astrological works were presented to King Henry VII by court astrologer Parron, see Armstrong (1960). His son Henry VIII also had a court astronomer/astrologer named Nicholas Kratzer who wrote an astronomical treatise *Canones horoptri*, that he dedicated to Henry VIII (Carley 1987: 56). Kratzer also fashioned astronomical instruments and was thought to collaborate with Hans Holbein the Younger who often included different astronomical instruments as props in his paintings.

[57]"In recognition of the technical and artistic demands required to produce and illustrate this astrological book and of the influence of Henry VII, this book was chosen to conclude the *Survey* of later Gothic manuscripts" (Scott 1992: 365). Full digital coverage with images, provenance and bibliography is available for this manuscript: see Digitised Manuscripts at http://www.bl.uk/manuscripts.

French is a compilation of astronomical canons and tables based on the *Alfonsine Tables*, treatises on the heavens, excerpts of Ptolemy, nativity charts of family members, including Henry VIII, and a fully illustrated constellation cycle with 43 miniatures in color and gold. Since the scope of possible interpretations of any horoscope was open to a variety of readings, most horoscopes of specific persons were not interpreted by textual explanations in the manuscript; these were probably given by the astrologer present (Carey 1987: 53) (Figs. 5.10, 5.11 and 5.12).

The constellation miniatures in Arundel 66 were painted by one of the three different artists, who worked on the manuscript; many illuminations were by the least talented artist. Scott relates that the "work is naive in the extreme and unexpected in a book of this quality" (Scott 1992: 366). Many of the 43 constellation miniatures are crudely executed, each in a similar format, framed with a patterned background in red and gold, the heavenly figures and animals are not skillfully drawn, appearing awkward and ill-proportioned. The text into which the illuminations are set differs from the usual mythological descriptions of Hyginus, or the scholia of *Aratea,* or the phrases of Pseudo-Bede, but contains a combination of charts and astrological descriptions. The authors of the texts included are Ptolemy, Guido Bonatti, John Killingworth and Plato of Tivoli and the texts are extract from the *Almagest*, *Liber Astronomiae*, *Liber Arenalis*, and astronomical and geomantic tables and political prophecies.[58]

The constellation picture cycle of Arundel 66 follows the traditional order and layout of most similar astronomical treatises, but has been revised from the astronomical focus and adopts astrological emphasis instead. This royal astrological manuscript was created near the end of the fifteenth century, as the final era of handwritten and hand-painted productions made way for the age of printing. The royal courts were still concerned with commissioning religious manuscripts of course, even printed ones, and they often displayed a matching interest in pursuing astronomical and astrological matters as well.

Summary of Medieval Astronomical Manuscript Traditions

In summary, the transmission of astronomical and astrological knowledge through the thousand years of the Middle Ages is a complicated topic with still much to be discovered; this survey just touches on the highpoints. As we have seen, from Late Antiquity until the twelfth century, there had been only minor advancements in the body of astronomical knowledge held in the Latin West; their goal focused on protecting and passing on ancient knowledge not to experiment, change or advance the 'scientific facts' that were already known. But by the thirteenth century, a new

[58]Arundel 66 was purchased by the British Museum from the Royal Society of London along with 549 additional Arundel manuscripts in 1831. Full digital coverage is available for this manuscript: see Digitised Manuscripts at www.bl.uk/manuscripts.

Fig. 5.10 King Cepheus—Arundel 66 f.34, Courtesy of British Library

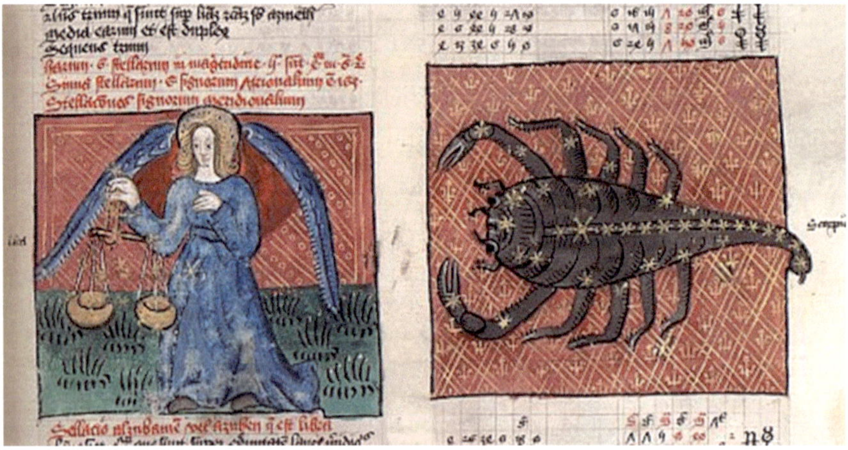

Fig. 5.11 Virgo holding Libra and Scorpio—Arundel 66 f.41, Courtesy of British Library

science of the stars had been instituted, along with an encouraged scientific curiosity regarding the real nature of the heavens. The transmission of Arabic science to Europe was not a singular event deposited at one point in one place. The

Fig. 5.12 Hydra, Crater and Corvus—Arundel 66 f.45, Courtesy of British Library

flow of Islamic knowledge continued over time transferring information that passed in both directions, although the direction was mostly from East to West. The huge output of astronomical manuscripts at that point was more-truthfully inspired by Europe's fascination with the revived and renewed astrological theories and predictions spreading through the royal courts and the learning centers. But the advancements and transitions of Islamic science in astronomy, mathematics, optics, medicine, and scientific instruments have yet to be fully appreciated, even in the Arab world.

At the start of the Middle Ages after the dissolution of the Roman Empire in the West, astrology was rejected as a discipline of 'scientific' study and almost entirely disappeared in the Latin world; but by the end of the twelfth century, nearing the close of the medieval period, astrology had returned to almost total acceptance in the European worldview. With the Hellenistic works lost, the only source for the reappearance of astrology was through Islamic writings, for there were no astrological manuscripts in Latin before the Arabic translation period. After the influx and absorption of Arabic astronomy, a person was not considered learned unless they understood basic astronomical and astrological principles, along with the arcane secrets and practical applications of astrological lore. Even though kings, popes, and cardinals consulted astrologers and commissioned astrological manuscripts for their libraries and celestial murals for their palaces, prophecy was still a very dangerous business in the thirteenth and fourteenth centuries.

An Italian poet, physician, and astrologer, who had worked at the court of Florence and lectured at Bologna on the *Sphaera*, Cecco d'Ascoli was jailed and burned at the stake in 1327.[59] His astrological writings attracted ecclesiastical disapproval and all were condemned; D'Ascoli also created a horoscope for Christ in order to determine actual dates for his birth and death; that was astrology too blasphemous for the Church. D'Ascoli was convicted of heresy by the Dominican Inquisitor in Bologna and then publically burnt to death in Florence, the first university scholar to be sentenced by the Inquisition (Tester 1987: 193).

Current studies emphasis the impact of Islamic authors and their search for and study of the astronomy of India, Persia, and Babylonia, since more and more of these writings have been translated, analyzed, and better understood. The tremendous impact of astronomical advancements formulated by Arabic scholars cannot be exaggerated; its art historical significance requires further study.

The establishment of new universities in the West—Bologna in 1088, Paris by 1150, and Oxford by 1167—increased the demand for instructive books at a time when existing library facilities were deplorable. The thirteenth century saw a dramatic rise in the social and intellectual importance of the expanding universities. Gradually academic libraries increased in size and advanced the body of accumulated knowledge; collecting a personal library became increasingly common among the elite. Lavishly decorated illuminated manuscripts alive with vivid pigments, adorned with golden highlights that are still breathtaking today were created in the wealthy monastic centers and royal courts of Europe. Astronomical and astrological manuscripts were still among the most popular themes, after religious writings and illustrations of course. A great number of these collected astronomical texts were owned not by scientists or those interested in pretty pictures, but many belonged to physicians, medical practitioners, and astrologers (North 2008: 251). By bringing together all of these components, although far from comprehensive, one can appreciate the complicated and varied paths followed in the transmission of astronomical knowledge during the medieval period, from its narrow range of sources in the early centuries to a period of increased intellectual activity, and then an explosion of collected knowledge in the later centuries.

After the great synthesis of the thirteenth century, the fourteenth century tends to be described as a period when intellectual confidence begins to decline due to the huge social problems created by the devastation of the Black Death, the political disasters of the Hundred Years War, and the lengthy religious crisis caused by the papal schism in Avignon. And yet medieval scribes and artists continued to create innumerable illuminated manuscripts whose breath-taking beauty has never been equaled.[60] After the thirteenth century and the introduction of strict astronomical accuracy, constellation imagery changed its focus from its original source which was mythological figures drawn from *Aratea* manuscripts but toward an altered objective. The relative placement of the stars in the illuminations became so

[59]See *Cecco d'Ascoli* by Lynn Thorndike (1964) an article extracted from Thorndike's *The History of Magic and Experimental Science* V4.

[60]See for example Scott (1996) *Later Gothic Manuscripts 1390–1490*; De Hamel (1997) A History of Illuminated Manuscripts; Stones (2013), *Gothic Manuscripts 1260–1420: A Survey of Manuscripts Illuminated in France.*

accurate that the constellation could be plotted on a celestial globe with little effort (Duits 2011: 13).[61]

During the fifteenth century, the demand for illuminated codices in Europe had expanded enormously. The basic activities of university teaching, as established in the thirteenth and fourteenth centuries, were maintained, refined, and elaborated. Commentaries on Aristotle were produced and hotly debated, while Humanist scholars brought great changes in ideas and attitudes including a desire to revive the literary and cultural legacy, as well as the moral philosophy of classical antiquity. Their interest in the literature of ancient Greece and Rome accelerated production of astronomical and poetic texts. Even before Ottoman leaders took possession of Constantinople in 1453, Byzantine intellectuals began arriving in Western Europe bringing their valuable scientific manuscripts. The availability of even more Greek works, which were soon translated into Latin, expanded yet again the range of classical thought available for study; new directions for research and experimentation were taken. In spite of the revival of the study of Greek during the Renaissance, most teachings and writings continued to be presented in Latin.

This segment has presented a broad view of the variety of astronomical manuscript traditions prominent in the Middle Ages and the enormous influence of Arabic writings both their translations and original texts and also their illustrations. The poem of Aratus and the illustrated translations of the *Aratea* were still read and enjoyed, even inserted in the astronomical anthologies—both manuscripts and printed editions, but the descriptions and mythological tales of the constellations had no real place in the growing body of technical astronomical data nor in the popularity of astrological lore. The following chapter focuses on a more narrow view and examines in greater detail the production of *Aratea* manuscripts along with their cultural circumstances in order to investigate their proper role and actual function throughout this vast astronomical tradition.

References

Alexander, J.J.G., Marrow, J. and Sandler, L.F. (2005) *The Splendor of the Word*, London: Harvey Miller Publishers.

Blume, D., M. Haffner, and W. Metzger. (2016) Sternbilder des Mittelalters: Der gemalte Himmel zwischen Wissenschaft und Phantasie. Part Two 1200–1500. Berlin: Akademie Verlag.

[61]The preface to the 2011 article by Duits explains that it is directed toward assessing the importance of Saxl's studies and to demonstrate his development from a near follower of Aby Warburg to an independent theorist, applying his own methods and approach and his reply to one of the main issues that Warburg raised, which is how to assimilate the former view of the Renaissance as the revival of classical Antiquity with the new understanding of the Renaissance as the birth of the modern rational world introduced by Burckhardt in the nineteenth century. In this article he discusses the huge changes in astronomical and astrological imagery in the fifteenth century.

Bosch, L. M.F. (2000) *Art, Liturgy and Legend in Renaissance Toledo*. University Park, PA: Pennsylvania University Press.
Bossong, G. (1987) "Science in the Vernacular Languages: The Case of Alfonso X El Sabio", in *De Astronomia Alphonsi Regis*, Proceedings of the Symposium on Alfonsine Astronomy held at Berkeley (August 1985), Barcelona.
Burnett, C.F.S. (1997) *The Introduction of Arabic Learning into England*, Toronto: University of Toronto Press.
Burnett, C.F.S. (2003) "The Transmission of Arabic Astronomy via Antioch and Pisa", in the *Enterprise of Science in Islam: New Perspectives*, J. Hogndijk and A. Sabra, Eds. USA: The MIT Press, pp. 23–51.
Carey, H. (1987) "Astrology at the English Court in the Later Middle Ages", in *Astrology, Science and Society*, P. Curry ed., Woodbridge, Suffolk: The Boydell Press.
Carley, J. P. (2004) *The Books of King Henry VIII and His Wives*, London: British Library.
Carmody, F. J. (1952) Ed. *De Motibus Celorum: Critical Edition of the Latin Translation of Michael Scot*, Berkeley: University of California Press.
Carmody, F.J. (1956) *Arabic Astronomical & Astrological Sciences in Latin Translation*, Berkeley: University of California Press.
Chabás and Goldstein, B. (2003) *Alfonsine Tables of Toledo*, Dordrecht; Boston: Kluwer Academic Publishers.
Clark, V. (1979) *The Illustrated Abridged Astrological Treatises of Albumasar*, dissertation University of Pittsburgh.
Cochrane, L (1994) *Adelard of Bath: The First English Scientist*, London: British Museum Press.
Duits, R. (2011) "Reading the Stars of the Renaissance, Fritz Saxl and Astrology", *Journal of Art Historiography*, University of Birmingham, United Kingdom.
Eastwood, B. S. (1984) Review of Hermann of Carinthia, Burnett, C., *Speculum*, Vol. 59, No.4, pp. 911–13.
Flint, V. (1990) "The Transmission of Astrology in the Early Middle Ages", *Viator, Medieval and Renaissance Studies*, 21, pp. 1–27.
Gambel, B. R. (1992) "Jews, Christians, and Muslims in Medieval Iberia: *Convivencia* through the Eyes of Sephardic Jews" in *Convivencia: Jews, Muslims, and Christians in Medieval Spain*, V. Mann, T. Glick and J. Dodds, eds. New York: George Braziller, Inc.
Goldstein, B. R.(2000) "Theory and Observation in Medieval Astronomy", *The Scientific Enterprises in Antiquity and the Middle Ages*, University of Chicago Press.
Goldstein, B. R. (1986) "The Making of Astronomy in Early Islam", *Nuncius: Journal of the History of Science* 1, pp. 79–92.
Goldstein, B. R. (1991) "The Blasphemy of Alfonso X: History or Myth?" in *Revolution and Continuity: Essays in the History and Philosophy of Early Modern Science*, P. Barker and R. Ariew, eds. Washington, DC: Catholic University of America Press, pp. 143–153
Gousset, M-T. and J.-P. Verdet (1989) *Liber Astrologiae, Georgius Zothoros Zaparus Fendulus*, Paris: Editions Herscher.
Haskins, C. H. (1922) "Science at the Court of the Emperor Frederick II", *American Historical Review* XXVII, July, Washington D.C.
Haskins, C. H. (1927) *Studies in Mediaeval Science*, Cambridge: Harvard University Press.
Herman of Carinthia, *De essentiis*, ed. and trans. C. S. F. Burnett. (1982) (Studien und Texte zur Geistesgeschichte des Mittelalters, 15.) Leiden and Cologne: E. J. Brill.
Kunitzsch, P. (1974) *Der Almagest: die* Syntaxis Mathematica *des Claudius Ptolemaus in Arabisch-Latinischer Uberlieferung*. Wiesbaden: Otto Harrassowitz.
Kunitzsch, P. (1989) "Star Catalogues and Star Tables in Mediaeval Astronomy", *The Arabs and the Stars*, UK: Variorum reprints.
Kunitzsch, P. (1993) "Arabische Astronomie im 8. Bis 10. Jahrhundert", *Science in Western and Eastern Civilization in Carolingian Times*, P. Butzer and D. Lohrmann, eds. Basel: Birkhäuser Verlag, pp. 205–220.
Kunitzsch, P. (2004) *Stars and Numbers: Astronomy and Mathematics in the Medieval Arab and Western Worlds*, Burlington, VT: Ashgate.

Lemay, R. (1962) *Abu Ma'shar and Latin Aristotelianism in the Twelfth Century: a Recovery of Aristotle's Natural Philosophy through Arabic Astrology*, Beirut: American University of Beirut.

Lemay, R. (1987) "The True Place of Astrology in Medieval Science and Philosophy: Towards a Definition" in *Astrology, Science and Society*, P. Curry ed., Woodbridge, Suffolk: The Boydell Press.

McCluskey, S. C. (1998) *Astronomies and Cultures in Early Medieval Europe*, Cambridge: Cambridge University Press.

Menocal, M. R. (2002) *The Ornament of the World: How Muslims, Jews and Christians Created a Culture of Tolerance in Medieval Spain*, Boston: Little, Brown and Company.

Montgomery, S. L. (2000) *Science in Translation: Movements of Knowledge through Cultures and Time*, Chicago: University of Chicago Press.

Neugebauer, O. (1957) reprint in 1968, *The Exact Sciences in Antiquity.*, New York: Harper.

North, J. (2008) *Cosmos: an Illustrated History of Astronomy and Cosmology*, Chicago: University of Chicago Press.

Quinlan-McGrath, M. (2001) The Foundation Horoscope for St. Peter's Basilica, Rome, 1506, Choosing a Time, Changing the Storia, *Isis: A Journal of the History of Science*, 92, pp. 716–741.

Samsó, J. and Comes, M (1988) "Al Sufi and Alfonso X" in *Archives Internationales d'Histoire des Sciences* 38: pp. 67–76. (Reprinted in Samsó, 1994, Chapter XVII)

Scott, K. (1996) *Later Gothic Manuscripts 1390–1490*: *A Survey of Manuscripts Illuminated in the British Isles*, 6 – 2 vol., London, New York: McMillan.

Michael Scotus. *Liber introductorius*, (14th century copy). Munich: Bayer. Staatsbib.MS. 10268.

Stones, A. (2013) *Gothic Manuscripts 1260 – 1420: A Survey of Manuscripts Illuminated in France"*. London: Harvey Miller Publishers.

Tester, S. J. (1987) *A History of Western Astrology*, New York: Ballantine Books.

Thorndike, L. (1949) *Sphere of Sacrobosco and Its Commentators*, Chicago: University of Chicago Press.

Thorndike, L. (1965) *Michael Scot*, London: Nelson

Chapter 6
Astronomical Symbolism in the Carolingian Period

Abstract Illuminated manuscripts produced during the Carolingian revival played a crucial historic role in the transmission of not only astronomy, but also in the survival of ancient classics of literature, philosophy, history, and science and appropriating them for new uses. This chapter focuses on some general characteristics of the surviving *Aratea* manuscripts and the changes in the textual component through time. The overarching purpose is to highlight the presence of astronomical interest in the production of numerous pertinent manuscripts and the continued usage of astronomical imagery to legitimate the power and prestige of the monarchy under the rule of the Carolingian kings in the ninth century. This section details particular instances of astronomical iconography, citing some of the recent studies on its many aspects. It also describes the historic cultural circumstances and the idea of rebirth and renewal of educational and cultural life initiated by Charlemagne and its influence on *Aratea* manuscript production. To learn, *stadium discendi,* became a driving force during his extensive forty-six-year reign. This era also saw the first appearance of royal portraiture in religious manuscripts and was a time of great advances in artwork, looking back to the Classical period but adding their distinctive touches and also in creating textual uniformity for easier reading and retention.

This section will discuss the historical background and importance of the earliest significant time period of Aratean manuscript production singled out for specific consideration. The particular historical time periods, already discussed in general for their transmission of astronomical knowledge, are the ninth, twelfth and fifteenth centuries, chosen because these eras stand out as exceptional for their progress in 'science', more accurately referred to in earlier ages as 'natural philosophy'. These three centuries are recognized for their advancements in astronomy and their prominent position in transmitting astronomical knowledge. The two particular manuscripts created in each of these centuries were chosen for a variety of reasons. In the case of the ninth-century Carolingian examples, the two selected subjects are considered among the most important *Aratea*, if not the most important, for their texts survive relatively untainted by later additions or

M. Dolan, *Astronomical Knowledge Transmission Through Illustrated Aratea Manuscripts*, Historical & Cultural Astronomy,
DOI 10.1007/978-3-319-56784-6_6

'improvements'. Both Aratean manuscripts retain the characteristics of their antique prototypes and served as exemplars, influencing much of the Aratean art that followed. The twelfth century historic background and the progress of astronomical knowledge were discussed in Chaps. 4 and 5; the fifteenth century background and *Aratea* production will be discussed in Chap. 7.

The Corpus of Surviving *Aratea* Manuscripts

A select few of the sixty extant *Aratea* manuscripts have received a great deal of attention from art historians and text scholars, but until recently the majority have gone unnoticed in art historical literature. The entire Aratean phenomena had not been studied as a corpus of illustrated manuscripts, for its poetic content, for its educational value, for its reception, nor for its rich cycle of illuminations. The popularity and longevity of the poetry of Aratus' *Phaenomena* and the Aratean authors, copied over and over from the ninth to the fifteenth centuries, is not unique. Other classical authors have also received worthy consideration from medieval book production centers, and were consistently copied by scribes and illustrated by artists; the works of Homer, Hesiod, Virgil, Ovid and Pliny circulated in great numbers and many were fully illustrated. First-hand knowledge of the *Iliad* and the *Odyssey* declined in the Middle Ages along with the ability to read or write ancient Greek; readers had to rely on an abridgement in Latin of Homer's *Iliad*. Because of misunderstandings, mistakes and adaptations by scribes, the writings by classical authors in circulation were not always reliable copies of the original versions. On occasion medieval authors were encouraged to write "moralized" versions of the classics, especially of Ovid's *Metamorphoses*; these revisions presented an ethical interpretation of each pagan episode that was more consistent with Christian morality. The poem of Aratus has remained mostly intact from its origin but the text in Aratean manuscripts underwent extensive changes.

When examining the sixty extant *Aratea* manuscripts, it was easy to see that other than the ninth century manuscripts, most versions of the *Aratea* do not stand alone in a single manuscript; they were included in anthologies and bound along with various additional astronomical treatises. Within these textual compilations, a few celestial maps and diagrams are usually included, but the *Aratea*n astronomical poems are almost always totally illustrated with the full constellation cycle. The quality of each manuscript belonging to this large corpus varies greatly in its size, format and artistic appeal; some of the illuminations are imaginatively painted by expert artists, but many more are repetitious, sketchy or poorly executed by less skilled craftsmen. In general, most *Aratea* manuscripts include miniatures that were illustrated with modest effort, adding lightly-colored line drawings without elaborate borders or decorative trim. Typically the *Aratea* manuscripts were not illuminated by exceptional artists; their paints were not created from costly, semi-precious gems such as lapis lazuli; their covers were not adorned with gemstones, pearls or cameos. For the most part, the higher the quality of a manuscript and its cover, the better were its chances for protection, security and survival. Protected from sunlight and moisture in monastic and palatial libraries, the paintings in luxury manuscripts often survive in pristine condition. In contrast, many Aratean texts were included with inexpensive schoolbook texts whose survival rate is much poorer. Since very few *Aratea* were luxurious books decorated with costly gold or silver accents, and they obviously endured much usage, the large survival numbers seem remarkable.

The illuminated manuscript was considered an artistic whole in which the binding and calligraphy were as important as the paintings. To study or admire the artwork in isolation from the text would have been thought very odd to the original owners. Scripts too are extremely important, not only for the words they express, but because the writing style is often distinctive with specific abbreviations or letter formations; as a result, these precise characteristics can be comparatively dated and sometimes even traced to a specific monastery or production center. The meticulous scripts used by scribes of the ninth century were greatly admired: "Carolingian calligraphy was of such elegance that it was copied by the humanists of the Italian Renaissance and became the basis for the earliest Italian printing" (Dodwell 1993: 40).

An enormous number of decorated Bibles, Books of Hours, chant books and liturgical codices survive from the Middle Ages. In spite of the prevailing supremacy of Christianity and the great variety of religious texts produced during the medieval period, this astronomical poem, with its cast of pagan gods and non-Christian characters borrowed from ancient mythic tales, has survived periods of Christian zeal, when other Greek literature such as the third epic story, the "*Margites*", credited to Homer, and pagan religions, such as the Cult of Mithras, have been practically erased from history. The *Aratea* manuscripts continued to be reproduced over and over through the centuries; this fact suggests that a very great number of *Aratea* manuscripts were circulating throughout the Middle Ages. The following shows an itemization of extant *Aratea* and *Aratus Latinus* manuscripts by centuries: ninth century—thirteen; tenth century—seven; eleventh century—seven; twelfth century—four; thirteenth century—one; fourteenth century—four; and fifteen century—twenty-seven. Contemporary facsimiles have been published for at least three outstanding *Aratea*: Leiden Voss.lat.Q.79 (ninth century); Madrid BN Codex de Metz 3307(ninth century); and London BL Cotton MS Tiberius B.V. (tenth or eleventh century). The surviving *Aratea* manuscripts in general, fall into two groups: the English component and the European component which extended to Italy where production blossomed in the fifteenth century.

Like most surviving classical literature, the earliest *Aratea* manuscripts date to the ninth century; credit Charlemagne and his successors for prioritizing the preservation and reproduction of ancient poetry, mythology, literature and natural philosophy. The Carolingian copyists must have produced numerous Aratean manuscripts, since a surprising number, thirteen, are still extant; but two of these are not illustrated.[1] The most renowned Carolingian book production centers were the Palace Schools of Charlemagne, his son Louis the Pius, and his grandson Charles the Bald, which were located at Soissons, Rheims, Tours, Metz, Lorsch, and St. Gall; but to pinpoint a particular manuscript to its place of origin with any certainty is challenging as codices, artists, and scribes were known to move from one scriptorium to another, sometimes taking unfinished manuscripts to be completed elsewhere.

Originating in the Carolingian palaces and monastic scriptoria, numerous *Aratea* manuscripts circulated among other monastic libraries and centers throughout Europe where early copies were created. A large number of Cicero's *Aratea* were produced in England; while Germanicus' *Aratea* circulated widely and was frequently copied in France and Germany/Austria. A branch of the Germanicus version has been traced to Monte Cassino Monastery in the twelfth century, which then spread to royal courts other learning centers in Italy. The Iberian Peninsula does not seem to have played a role in the transmission of *Aratea* manuscripts; the five *Aratea* codices currently held in Spanish libraries were all made elsewhere.

Cicero, Germanicus, and Avienus Texts

As mentioned, the examination of the textual tradition of the *Aratea* manuscripts is not a focal point of this study; the additions of scholia and the text of other authors through time becomes quite complex and has been studied by classical text scholars.[2] When comparing the numbers of medieval manuscripts by the three Arateans, the translation of Germanicus is "copiously represented" and outnumbers the other two authors by far (Le Boeuffle 2003: xxxiv).[3] A brief explanation of the variety of *Aratea* texts associated with the cycle of astronomical illustrations is required to understand the tradition; for a more detailed examination of the various textual presentations, consult Blume et al., *Sternbilder des Mittelalters* (2012, 2016), the scholarly publications and the text editions. A further explanation of the Aratus and Aratean poetic texts in conjunction with their images is included in Appendices B and C.

Through time, texts found in manuscripts of the three Latin translations of the *Phaenomena* are often accompanied by and intermingled with the works of other authors, such as Hyginus, and also with different types of scholia. When scholars speak of scholia, they are usually referring to excerpts from commentaries found in the margins of early medieval writings. Throughout the Classical and medieval

[1]The thirteen *Aratea* manuscripts from the ninth-century are listed in Appendix A. The two that do not include pictures *Aratea* contain the text of Avienus and are held in Vienna, Codex Vindobonensis Palatinus 107 and in Paris, BNF Lat. 7886.

[2]See Kidd and Martin for analysis of Aratus' *Phaenomena* in Greek, and Gain, LeBoeuffle and Soubiran for the *Aratea* in Latin. Buescu, V. (1941) *Cicéron. Les Aratea.*

[3]Le Boeuffle, "*La texte des Aratea de Germanicus est copieusement représenté*".

periods, large collections of scholia had become associated with religious, literary, scientific and especially legalistic texts. On occasion these side commentaries were copied along with the original writings and at times even circulated as independent works, which could be consulted in addition to the texts they explained. Since both the original poem by Aratus and the Latin translations had a minimal amount of mythological material, scholia was often added in order to provide more specific information about the myths associated with the constellations (Cameron 2004: 19). By the third century at the latest, substantial scholia on Germanicus' translation had already been produced; the *Scholia Basiliensia,* which was added a bit later, included both myths and star catalogs for each constellation (McGurk 1981: 318) (Fig. 6.1).

As mentioned previously, a Latin prose version, derived from the Greek poem of Aratus, was produced by an anonymous author at some point in the early Middle Ages. This rendition of the *Phaenomena* begins with a series of texts presenting the biography of Aratus and a small amount of astronomical data; then sections of the poem are interspersed with scholia on the myths, descriptions, and number of stars. The earliest prose version, which is known as the *Aratus Latinus*, dates from the first quarter of the ninth century (McGurk 1981: 318).[4] The *Aratus Latinus* itself was probably revised in the early ninth century by a writer who had access to the third-century Germanicus scholia; but in these texts the translator merely substituted Latin words for the Greek. A cycle of illustrations was added to this ninth century version, which is called the *revised Aratus Latinus* or *Recensio Interpolata*. The *Aratus Latinus* and the *revised Aratus Latinus* were divested of the sophistication and striking verses of classical poetry; thus the descriptions of the constellation figures became simplistic and formulaic.

The Carolingian copyists were aware of all of these Aratean texts: the three Latin verse translations of Cicero, Germanicus and Avienus, the Germanicus scholia, and the two prose editions (the *Aratus Latinus* and the *revised Aratus Latinus*) and also the mythical writings of Hyginus that had become attached to the *Aratea*. Three of these variations were not illustrated in the Carolingian period: the Avienus, the *Aratus Latinus* and the Hyginus; all other versions included the constellation cycle miniatures. Of course the writings of Hyginus were fully illustrated in later centuries (McGurk 1966: xxii).[5] Although McGurk states that no early Hyginus manuscripts include miniatures, research uncovers a listing of eleven Hyginus manuscripts from the ninth century alone; several are cited in text editions as illustrated.

---

[4]McGurk cites Maass 1898, pp. 172–306; and Breysig 1867, pp. 105–232. The descriptions of the stars are mainly a Latin adaptation of the so-called *Catasterisms* of Ps-Eratosthenes, a mythological companion to the *Phaenomena*. See also McGurk, P. (1973) "Germanici Caesaris *Aratea* cum Scholiis.

[5]"The earliest Hyginus manuscripts with pictures date from the eleventh century, possibly the earliest being the Ademar of Chabannes book, Leiden Voss. Lat. Oct. 15 fasc. XIII, made at St Martial near Limoges about 1025." (McGurk).

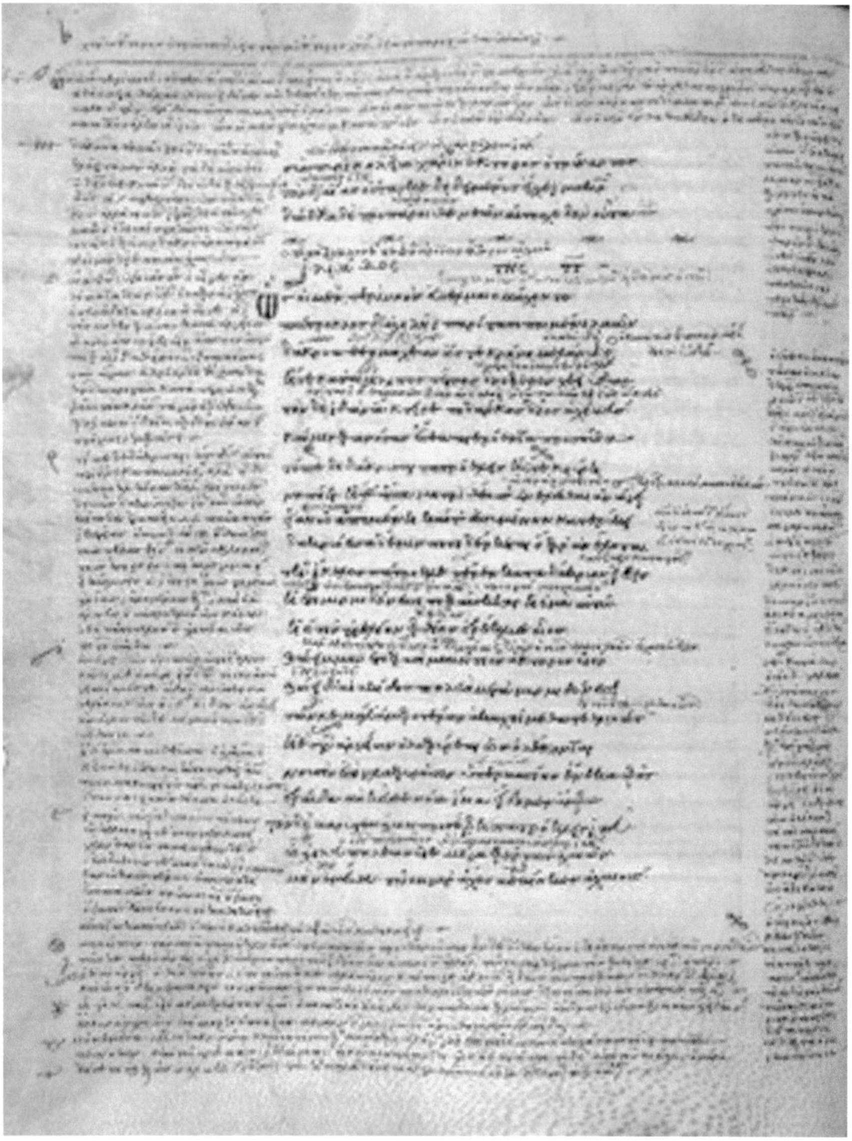

Fig. 6.1 Scholia Overtakes Original Text, Public Domain

Incipits, the opening words of a manuscript, are important for identifying the textual portions of the various astronomical codices. Therefore the common incipits for Germanicus, Cicero, and Hyginus follow, but there are occasionally minor variations because of the complicated text situation. Most Germanicus *Aratea* texts have the incipit, *Ab Iove principium magno deduxit Aratus carminis*. Some Cicero manuscripts have the incipit, *Incipiunt vers ciceronis de signis*, but his texts vary greatly. The incipit for Hyginus' *De astronomia*, Book II, in which the constellations are described, is stated: *Sed quoniam quae nobis de terra positione*.

Although both Cicero and Germanicus are concerned with the same subject matter, through time their translations and treatments of the celestial material became quite different. Germanicus wrote his didactic poem in a concise and accurate rendition especially the relative positions of the constellations and the mythical tales. In comparison, Cicero's incomplete text is more formulaic, repetitious and verbose than that of Germanicus. Dodwell states, "From both the literary and artistic point of view the most important source was the *Aratus* of Cicero" (Dodwell 1954: 61). I would disagree on both counts. Since Cicero's *Aratea* only survives in fragments, it cannot be the most important as the literary source; and its artistic influence was mostly confined to England, as the earliest Ciceronian *Aratea* manuscript left the continent in the tenth century. But the accumulation of large amounts of scholia to the translations of both Cicero and Germanicus has altered the pure poetic quality of their original versions. The poetry included does not flow from start to finish, but is often broken into small sections of six or seven lines, and then inserted amidst the various other prose material. The illuminations are mainly placed before, but occasionally after, the textual descriptions. Because of the complicated circumstances, only the format and placement of the text in respect to the images will be discussed.

Poetic Text and Constellation Images

In general, the pictorial cycle of constellations in all illustrated versions of the *Aratea* follow a similar pattern, each image easily recognized by its standard iconography. The cycle of images in Germanicus' versions becomes canonical in quantity and in appearance, as approximately the same number of constellations are discussed and pictured. In most examples, they follow in the same order, as originally set down by Aratus. Because segments of Cicero's *Aratea* have been lost, his illustration cycles contain fewer images than Germanicus' manuscripts, but later copies add some of the missing constellations. Many of the illustrations in *Aratea* manuscripts are simplified, sketchy images inserted into the body of the text, probably looking back to the papyrus scrolls and Late Antique antecedents. Treatises dealing with astronomy and cosmology have been accompanied by illustrations and diagrams from at least the early medieval period; the influence of illustrations from the scroll can be easily recognized in the ninth-century astronomical and computus texts. When observing the images in the few surviving ancient scrolls, the illustrations were placed precisely within the short lines of the columns of script, as shown in Chap. 2. But in the codex, since the script area changes and the lines of text lengthen, the space left free for the illustration becomes much too broad for a picture inherited from a scroll. A faithful reproduction of the model would not possibly fill the wide spaces created on the page, and empty gaps would result. In order to fill these gaps (horror vacui),[6] it became necessary to change the model; sometimes two figures are placed side by side to solve the problem.

Just as alterations of the Latin poem occur through the addition of scholia, as well as some misreading and mistakes by scribes; so in the transmission of pictorial models, modifications have also occurred. Concerning the pictorial element, the assumption of exact fidelity to the original is problematic, and cannot be taken as an absolute fact. Changes in the text will occur through misreading, carelessness and the scribe's preconceptions; in the same way the transmission of pictorial models can be altered by misunderstandings and by the skill of the artists. Both text and image are most often corrupted through time. "Any new interpretation is a translation into forms specifically intelligible and intended for one's contemporaries," (Pächt 1984: 24). In spite of these problems, some classically-inspired pictures found in the Carolingian *Aratea* manuscripts were copied with scrupulous care, and are reflections of illustrations from a vanished past, generally agreed among art historians to be the fourth and fifth centuries. The classical essence of the illuminations survived through the Middle Ages, but in the twelfth century, the more exotic Byzantine and Islamic influences begin to appear, particularly in manuscripts from the Mediterranean area (Panofsky and Saxl 1933). As mentioned, changes also occur in the placement of the stars in the constellation figures; prior to the twelfth century, the stellar positions are random in number and location, except for a few examples. If the purpose for including stars in the illustration was a practical concern for recognizing the patterns of star groups, you would expect a certain scientific accuracy. But this was not a consideration in the Latin West, until after the assimilation of the more-accurately illustrated Arabic manuscripts. Blume et al. list eight surviving al-Sufi Latinus manuscripts produced between the twelfth and fifteenth centuries (Blume et al. 2016: 57–65). These manuscripts include the more exacting star information and the Islamic-influenced garb and drawing style.

In spite of the complications, variations, and changes in *Aratea* manuscripts, the illustrations clearly fulfill and expand the textual material, included for both aesthetic and informational purposes. These representations influenced not only later book production, but also astronomical art in other media; painting, sculpture, frescoes and even textiles. In the strongly religious environment of the medieval period, the astronomical illuminations in *Aratea* manuscripts contained and transmitted a wealth of classical symbolism. The images of the constellations and cosmological diagrams are replete with classical allusions, personifications of the Sun and Moon, river gods, the earth, the winds, the seasons, the Labors of the Months, and the classical gods and goddesses. The *Aratea* manuscripts most often begin with a map of the entire night sky, another design replete with classical personifications, another astronomical tradition that has been carefully studied for

[6]*Horror vacui* also called kenophobia in Greek is 'fear of the empty', covering the entire space with artistic details was very prominent in medieval visual art, mainly in manuscript illuminations.

the first time by Dekker in her rigorous examination of medieval globes and celestial maps.[7]

Illustrations of the Celestial Sphere

A panorama of the celestial sphere appears at the beginning of most *Aratea* manuscripts but not all; these become more prevalent in the later medieval period. Possibly a map of the heavens also circulated in the *Phaenomena* of Aratus in the Classical era (McGurk 1981: 324), or perhaps the graphic image had been added to the fourth or fifth-century prototypes consulted by the Carolingians. Within the *Aratea* manuscripts, the celestial map has no standard arrangement but appears in various formats; there are at least four different types of celestial illustrations that can be found in Aratean codices.

One type of global map appears to be more scientific, displaying the summer and winter sky in two separate images; the sky is divided into zones. Some celestial maps depict the figural representations of the constellations but with no indication of individual stars. In order to include more constellations, certain celestial maps are drawn divided into two separate hemispheres demarcating the signs of the zodiac on the ecliptic, which divides the northern and southern skies. This style of global map is often included with the *Revised Aratus Latinus* text; these globe illuminations are followed by the *scholia Sangermanensia* (Dekker 2013: 192). This map tradition was based on an early Eudoxan globe; six examples of this type survive. The majority of the constellations appearing on celestial maps are simplified images, depicted from a frontal view rather than the reversed god-like view of most celestial globes. This is another indication that this type of map was taken from a scientific rather than a literary source.

Other maps appear more picturesque and include the whole sky in one panoramic view. The representation of a celestial sphere found in a fifteenth-century Germanicus *Aratea* manuscript with *scholia Strozziana* (BL Add. MS 15819, f.3), made in Florence, depicts the sky divided into zones, but there is no interest in accurate locations of the constellations. A planetary diagram is another option, such as one that appears in Dijon Bibliothék Municipale MS 448, depicting the ancient Pythagorean harmonic proportions between the orbs of the planets, the music of the spheres, not meant to be audible, but is more a mathematical, harmonic or religious theory. In this eleventh-century *Aratea*, the seven planetary gods are pictured with identifying attribute, overseeing their heavenly realms (Figs. 6.2 and 6.3).

A fourth and most intriguing variety is a celestial globe, often included in the *revised Aratus Latinus* manuscripts, which presents a completely different format.

[7]See Dekker (2013) *Illustrating the Phaenomena: Celestial cartography in Antiquity and the Middle Ages.*

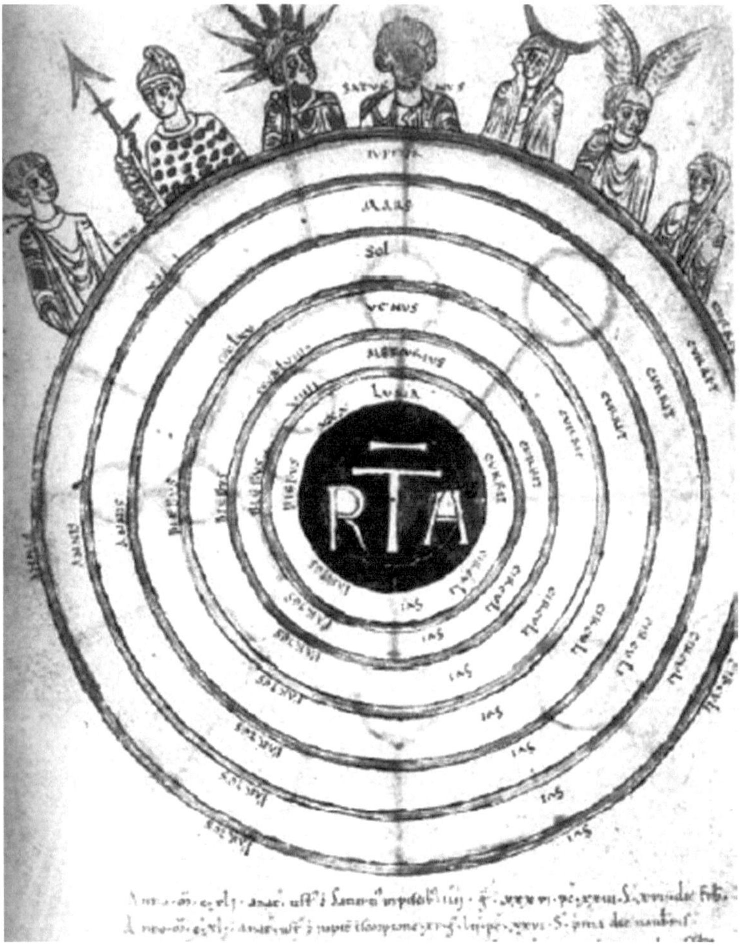

Fig. 6.2 Celestial Diagram with Planetary Gods, Dijon Bible, f.14, Public Domain

This type of celestial sphere is clearly derived from antiquity and shows a *sphaera solida* supported on a stand (McGurk 1981: 324). This large public celestial globe is reminiscent of the Farnese Atlas globe that was discussed in Chap. 2 as another possible source for constellation imagery. Could this possibly be a representation or remembrance of an actual temple-like structure that held either a celestial or terrestrial globe in Roman times?

Most of the human depictions on these celestial maps are classically-derived, almost fully-naked, and depicted from a reverse view; individual stars are rarely included on these Aratean celestial maps. This brief description of the variety of celestial diagrams and maps demonstrates the multiple options available to the ninth-century artists; the tradition of including a view of the entire heavenly sphere continues in later *Aratea* manuscripts, particularly in fifteenth century copies.

Fig. 6.3 Celestial Sphere, f.6 Revised Aratus Latinus, Lombard, 14th or 15th Century, Gottweig Stifsbibliothek, Cod.146, Courtesy of Gottweig Stifsbibliothek

Historical Background and Significance of the Ninth-Century Carolingian Revival

Although illustrated *Aratea* manuscripts survive from each century between the ninth and the fifteenth, addressing each individual manuscript and every pertinent time period in an extensive manner would be a lengthy and repetitious process.

Each *Aratea* manuscript becomes a product characteristic of its unique time, society, and culture; each copy also reflects the available levels of technology, expertise and material resources. When examining the accompanying illustrations, each miniature conveys the current techniques and popular styles of expression in addition to the obvious competence and talent of the artists. Therefore in keeping with current trends in scholarship, this study of *Aratea* manuscripts takes particular notice of the societal, cultural and religious contexts within which these manuscripts were commissioned or copied. In what ways did *Aratea* manuscript production reflect the needs or interests of the communities that created them? Each of the three centuries singled out for special consideration were periods of significant astronomical advancements, each era in its own way inspired a significant renaissance.

During the era of the Carolingian dynasty, the first 'renaissance' of Western Europe, scholars worked to restore classical knowledge, reclaiming many ancient texts. The Carolingian monastic and court scriptoria produced huge numbers of literary, scientific and astronomical manuscripts, including those especially important for this study, the *Aratea* of Cicero, Germanicus and Avienus. Two of the most illustrious *Aratea* manuscripts were produced in the ninth century and decorated with outstanding illuminations; they will be examined in connection with their cultural conditions. The thirteen extant *Aratea* manuscripts dating to the ninth century show a great variety in their texts and painted images. Their considerable number demonstrates the enormous interest from all areas of the Carolingian Empire in finding, copying and preserving the available Late Antique versions of Aratean manuscripts.

Beginning in the late eighth century, the Latin West experienced a striking acceleration of scholarly activity, thanks to the sponsorship and patronage of Charlemagne who came to the throne as king in 768 and as holy Roman emperor in 800. Pope Leo III (r. 798–816) crowned Charlemagne at St Peter's in Rome and provided his imprimatur to the *renovatio* beginning in northern territories. A few areas of Western Europe still held residues of ancient Roman culture, but the intellectual and educational resources had become almost nonexistent from neglect, disinterest and barbarian invasions. Under the guidance of Charlemagne and his descendants, the best-known scholars, theologians, artists and architects gathered in the Carolingian courts, and monastic centers in order to fulfill their grand designs; the *renovatio* was underway throughout the Frankish lands. Charlemagne's main court and his splendid palatine complex were established during the 790 s in Aachen (Aix-la-Chapelle). "In the age of Charlemagne until the time of Charles the Bald, the artistic composition of manuscripts reached its first peak in the medieval period" (von Euw 1993: 251).

Fortunately, the courts of Charlemagne and his descendants ensured widely-based intellectual interests, and desired to enrich their libraries with secular as well as religious manuscripts. Both the size and the artistic quality of a manuscript reflect the status of the codex and add to the prestige of its owner. Around 780, Charlemagne himself may have stimulated the creation of writing centers when he asked to receive copies of any rare or unusual books that might be stored within monastic libraries throughout his realm (McKitterick 1989: 10). Classical texts found housed in civic or monastic libraries were a significant link in the rediscovery of Late Antiquity. In any historic period, the process to choose which texts should be copied was highly selective, which ancient authors would be transferred from crumbling papyrus scrolls onto the more durable parchment codices. The Carolingians were very selective concerning which Classical or Late antique texts they decided to copy and illustrate.

Of course their principle concern was directed toward preserving and reproducing religious manuscripts and the writings of Church fathers. Beyond the various religious works, they choose from among the secular texts that were considered important at that time, but their resources were limited to what was actually available in the limited libraries of ancient Gaul. There was some borrowing from neighboring countries, but the real choice of which manuscripts to copy was restricted by the taste and interests of the monastic leaders. “Their selections have affected our present intellectual horizons,” (Pächt 1984: 16). Fortunately, the scholars associated with Charlemagne, even though they were Christians and clergymen, held a deep attachment to the literature of Rome’s classical history (Dodwell 1993: 47). The Carolingians were intrigued with the antique representations of the constellations for both their astronomical and mythological features. The sciences of astronomy and cosmology were fundamental topics in the early medieval intellectual worldview.

It is our good fortune that the *Aratea* of Cicero, Germanicus and Avienus all made the cuts and survived that critical selection process. In fact, Carolingian artists and scribes preserved not only the Latin *Aratea* versions and their illustrations; they are responsible for preserving almost all of the classical works that are extant from antiquity. “The Carolingian revival saved for us so much of the heritage of Rome and proceeded with such impetus and vigor that by the end of the ninth century the classical tradition had been securely re-established” (Reeve 1983: xvii).

Manuscript Production During the Reign of Charlemagne

The previously-held view that before the revitalization that occurred during Charlemagne’s reign, the book culture and manuscript production in the Latin West had virtually perished, has now been revised by historians. In reality, some Roman institutions continued to be functional after the fifth century, and the heritage of the written word continued as a valued medium (Shultz 2004: 138). In his re-analysis of Carolingian cultural history, Schutz depicts the traditional interpretation of the Carolingian Renaissance as a reductionist response based entirely on its renewal of Classical Antiquity without crediting the significant influences of the Merovingian, Frankish, Germanic, Celtic, and Insular cultures (Schutz 2004: 3). The Carolingians were not merely “slavish followers and restorers of Roman ideals, agendas and practices”. Schutz demonstrates that, in denying the complicated development of the cultural milieu of the Carolingians era, historians have simplified the real situation, presenting an interpretation of the ninth-century "Roman Revival" in a much too narrow basis. Although Carolingian art does have its roots in Late Antiquity, it was not immune to influences derived from cultural artworks created in neighboring areas as its style developed through the ninth century. This became the artistic period during which the fundamentals of medieval art and future art forms as well originated.

The manuscripts produced during the Carolingian revival display these myriad cultural traditions and influences in their broad creativity and decorative variations. Just as architecture, artworks, and decorative adornments were manifestations of wealth and status, so too were the lavishly illustrated, gilded and decorated manuscripts. It is impossible to determine exactly how many books were produced from the late eighth through the ninth century, but well over 7000 Carolingian manuscripts, including most of the classical authors, survive from this time period; a staggering total of about 50,000 may have existed at that time (Schutz 2004: 155). The enormous output of luxurious manuscripts by the Carolingians demonstrates their power and material wealth which in turn reveals their expertise, scholarship and organization.

Charlemagne's administrators encouraged the highest standards for the production of books including both text and illustrations. The textual contents were expected to be accurate and the script was required to be clear, precise and legible; therefore the competent scribes developed small, neat and standardized forms of the Carolingian miniscule alphabet, enabling manuscripts to be written more rapidly and read with little difficulty. Standard versions of the various treatises were made from the oldest and best texts to insure that the copies of both secular and sacred works were authentic and reliable. The illuminators achieved a high level of artistry and instituted new methods of decoration in the magnificent illuminated codices created in the scriptoria of the Carolingian Empire. The diversity of their artistic creations displays more than just a reproduction of the surviving classical and Late Antique treatises. The variety of images and innovative formats that they developed are witnesses of their innovations which would influence medieval as well as future Western art styles. They expanded the decorative and narrative possibilities of the decorated and historiated initials and extended the range of artistic styles. The surviving manuscripts demonstrate the diversity of pictorial productivity much more than merely a repetition and renewal of classical and Late Antique codices.

Many types of astronomical diagrams were utilized by Carolingian astronomers, variations of antique models and many entirely original; both types were created and included in their manuscripts. The diagrams were teaching tools, as they are not images of the real positions of planets, but are abstract representations of theoretical elements which present a visual way of thinking and teaching. When Carolingian scholars created diagrams to help explain and portray texts, they on occasion gained new and deeper understanding. Bruce Eastwood published the first in-depth study of how the diagrams included in Carolingian astronomical manuscripts performed as a mechanism for interpretation and for instruction directed to the readers (Eastwood 2007). Eastwood's analysis of the most important Roman authors of astronomical texts—Pliny the Elder, Macrobius, Martianus Capella and

Calcidius[8]—shows how these texts were studied by Carolingian scholars and then how those scholars used Roman astronomy as a basis for their understanding of the celestial phenomena.

Although some of the new Carolingian techniques were derivative, much of their artwork was highly creative and entirely innovative. The *Aratea* manuscripts display this variety of artistry, and the interest of ninth-century leaders and scholars in the course of the stars. Several pictorial variations in the constellation cycle are known from the Carolingian period, and the stellar images constituted by far the largest body of secular imagery (McGurk 1981: 317). Charlemagne's patronage brought together leading Irish, English and Italian scholars; although they were all Christians, these sages held a sincere commitment for the literature—both prose and poetry—produced in Rome's classical past. Many of the initial scholarly accomplishments achieved during the Carolingian renewal were orchestrated by the brilliant monk and scholar, Alcuin, who was especially well-versed in astronomy.

Alcuin, English Monk and Scholar

Alcuin (ca. 730–804), master of the cathedral school at York, was already fifty years old when Charlemagne summoned him from his northern post and appointed him to a principal position in the royal court at Aachen. In England, Alcuin had taught a scholarly program based on the liberal arts, including astronomy, with an emphasis on the specific knowledge needed for the practice of computus. Alcuin introduced English learning into the Frankish schools which included the writings of Bede. Alcuin wrote his own computus, as well as a short treatise on the motions of the Moon emphasizing its daily passage through the signs of the zodiac, called *Ratio de luna* (Kühnel 2003: 104). Since Carolingian manuscripts exhibit the earliest known representations of the zodiac in Christian art, Alcuin may have been a factor in their appearance within a religious context. Kühnel suggests that the revival of pagan zodiacal representations in religious art during this time could quite likely be traced to Alcuin's positive attitude to the ancient concept (Kühnel 2003: 104).[9]

Charlemagne's personal interest in astronomy can be verified by his correspondence with Alcuin in the years 797–800 (Lohrmann 1993: 79–114).[10] Eight

[8]Calcidius was known for his translation of the *Timaeas* as it was the only extensive writing of Plato available to Latin scholars for almost 800 years. He also wrote a commentary on the astronomical knowledge held by Greek scientist.

[9]Kühnel includes one known example of zodiacal art, a sixth-century mosaic in the Catholic monastery of Lady Mary in Bet She'an. (See Chap. 2—Possible Classical Models for Aratea Illustrations).

[10]See D. Lohrmann, "Alcuins Korrespendenz mit Karl dem Grossen über Kalender und Astronomie" in *Science in Western and Eastern Civilization in Carolingian Times*; also Borst, A "Alkuin und die Enzyklopädie von 809" in the same volume.

letters survive that were written by Alcuin and one by Charlemagne, although several more can be surmised from allusions made in the contents. In one of the letters, Alcuin asks Charlemagne to send him Pliny's *Natural History* (McCluskey 1993: 163).[11] This series of letters between Alcuin and Charlemagne concerning various aspects of astronomy reflects an atmosphere of great interest in celestial phenomena and lively discussions at the court (Kühnel 2003: 107). After the death of Alcuin, another scholar named Dungal became the resident authority in matters concerning astronomy and computus; he too corresponded with Charlemagne, especially regarding two solar eclipses that occurred in the year 810.

Computistical Anthology of 809

In 809 under the auspices of Charlemagne, a colloquium gathered together prominent astronomers and computists, probably coordinated by Adalhard of Corbie (751–827), a cousin of Charlemagne and an expert on computus (McKitterick 1994: 23). The colloquium was an extensive intellectual and scientific project focused on refining the calculation of time and compiling the 'official computus'. The reform conference resulted in the production of an illustrated manual, probably completed in 810, which added antique texts on science and mythology to the traditional computistical data (McCluskey 1998: 135–38). Although the compendium was compiled five years after the death of Alcuin, his scholarly influence was surely felt. The anthology included a calendar noting important celestial and liturgical events, and computistical tables with explanatory texts that provided instructions for calculating religious dating, mainly Easter. Although the original compilation has now been lost, the anthology is replicated in four extant but incomplete ninth-century copies. Two related but different versions of the works survive: known as the three-book-computus and the seven-book-computus.

The three-book computus, also called the *Computus of 809*, consists of a compilation of astronomical material found in two manuscripts,[12] both written in the early ninth century (McCluskey 1993: 139–160).[13] Another anthology, an expanded seven-book computus, was created in Salzburg around 818 containing similar astronomical material, perhaps intended as an improvement to the earlier version. The seven-book computus contains additional sections and paraphrases

[11]McCluskey cites Bischoff, *Hofbibliothek*, p. 42.

[12]The two manuscripts are Vienna ÖNB Cod. 387 and Munich BSB Clm 210.

[13]According to McCluskey, Book I of this anthology is devoted to practical computistical matters in 99 chapters; much of its material was taken from Bede which was ultimately derived from Pliny. Book II deals with planetary excerpts from Pliny; and Book III consists of Bede's *De natura rerum* in 51 chapters.

from Macrobius and two very brief excerpts of Martianus Capella on the size of the earth and the Moon (McCluskey 1993: 139).[14] Macrobius explains how to measure the diameter of the Sun which he shows is twice the diameter of the earth. This computus also contains Martianus' explanation of Eratosthenes' determination of the size of the earth.[15]

The earliest of the surviving manuscripts of the seven-book computus is Madrid Nacionale Codex de Metz MS 3307, made at Metz between 820 and 840 by the order of Drogo (823–855) Archbishop of Metz, illegitimate son of Charlemagne. One of the most important scientific codices of the ninth century, Metz 3307 originally consisted of seven books but only four have survived complete on 76 folios; the entire sixth book is missing and parts of the first and seventh. The anthology of astronomical works contains a summary of calculations and astronomical treatises written to establish the reform of the astronomical and liturgical calendar including patterns of 56 cycles of 19 years corresponding to the years 1 to 1063; diagrams of weights and measures and concurrent tables of liturgical festivals; compilation of calculation and astronomy; extracts of the "Astrology" of Arato; and astronomical fragments taken from various authors. The Germanicus text and constellation illustrations are found in the second chapter of Book V, entitled *De ordine ac positione stellar* in signis. The figures exhibit the influence of Late Antique prototypes; the number of stars in each form is listed but they are not included in the images.

An important development in the manuscripts from the 809 colloquium, showing their avid interest in ancient astronomy, was the inclusion of Pliny's planetary astronomy plus their addition of diagrams and illustrations. Only one apsidal diagram is included in the three-book, whereas four diagrams accompany the text in the seven-book computus. The descriptions of the zodiac and the constellations are illustrated, not by the interpretation of their actual locations or brightness, but with their associated mythological characters (Kühnel 2003: 108).[16] The illustrations of the constellations are thought to be faithful reproductions of a late antique original creation.

Nevertheless, the computus compositions were an important resource for transmitting astronomical information as were the writings of Plato. The *Timaeas*

---

[14]A full color facsimile was published in Madrid, 1993 with a commentary volume by Manuel Sanchez Mariana.

[15]Another Carolingian copy of the seven book computus is Città del Vaticano, BAV Vat. Regin.309, the only complete copy: the others are Città del Vaticano, BAV Vat. Lat. 645 (s. IX); Monza Bibl. capit. MS F.9.176 (ca. 869) and Paris BNF nal. 456 (s. IX/X). See Appendix B for further information.

[16]Kühnel states in an analysis of the anthology of 809 (or the Aix in Chappelle encyclopedia) that "the zodiac in the heaven of fixed stars is depicted in detail, then the motions of the planets through the signs of the zodiac, following Aratus and Pliny." Pliny may be a correct reference, but Aratus does not refer to the signs of the zodiac separately as a special grouping, only as they appear among the other constellation, and he emphatically refuses to discuss the planets at all. Pliny never mentions the planets' appearance in the zodiac or the principles of astrology, which in the medieval period, was the main reason for recording in which zodiacal sign a planet is located.

by Plato was actually the only original Greek philosophical work available during the early medieval period, but the important treatise containing cosmological theories had received little attention until the revival of Greek learning during the ninth-century. Interest in Plato's writing was spurred by the work of philosopher and translator John Scottus Eriugena (815–877), the Irish clergyman, philosopher and poet. Although incomplete, Latin translations of the *Timaeas* by Cicero are extant in ninth- and tenth-century Frankish manuscripts (McKitterick 1994: 24). Thus the Carolingian scholars and scribes assured the continuation of the cosmology of Aratus, Pliny and Plato into the eleventh and twelfth centuries when the translations of Arabic scientific texts began in earnest.

Holy Roman Emperor Louis the Pious and Book Production

The vigorous revitalization program undertaken during the reign of Charlemagne was continued by his son Louis the Pious (778–840). Although in historical accounts Louis had not been credited with the administrative and ruling capabilities of his father, according to current scholarship the early years of Louis' reign have been described as the "absolute summit of the Carolingian Empire" (Butzer 1993: 455). The main contemporary source of information concerning Louis' sovereignty can be found in *Vita Hludovici Imperatoris,* written by his anonymous contemporary biographer and court chronicler, simply and enticingly called "the Astronomer". In the emperor's biography, "the Astronomer" writes of Louis' interest in the heavens and even relates the account of Louis' observation of a worrisome comet.[17]

Charlemagne's policy of sponsoring education continued through the reign of Louis the Pious and that of his three contentious sons, who also sponsored 'schools' and maintained scholars, libraries, and scriptoria at their respective royal courts (Stevens 1995: 175). The important monastic learning centers, including those at Corbie, Tours, Ferrières, Fulda and Reichenau, began their expansion and manuscript production at this time. Two of the most outstanding *Aratea* manuscripts were most probably created during the reign of Louis the Pious. Mütherich asserts that a historian may be tempted to consider what impression the comet in the year 837 made on the ruling Carolingians and which may well have justified the dedication of an astronomical picture book to a member of the court (Mütherich 2004: 107).

[17]In the Astronomer's, *Vita Ludovici PL* 104, col. 971–72, Louis asks the meaning of the comet, aware of the dire predictions attached concerning disasters or the death of the ruler. In reply, "the Astronomer" quotes from the Old Testament that one should not fear signs in the heavens. The worrying comet that Louis witnessed turned out to be Halley's Comet, which was depicted on the Bayeux Tapestry two visits later.

The extensive library of Charlemagne is known to have included books on astronomy; he and his successors were thought to collect and reproduce as many Late Antique manuscripts as they could (Katzenstein and Savage-Smith 1988: 9).[18] The fate of Charlemagne's famous library remains unknown, but an impressive number of exceptional Late Antique manuscripts were still available at the Carolingian court after his death. David Wright states in regard to the Roman Vergil discussed in Chap. 2, "it is a reasonable speculation that it may have been one of the Late Antique manuscripts collected by Charlemagne when he was in Italy, perhaps at the time of his coronation in 800, and taken to his court in Aachen" (Wright 2001: 68).[19] Referring to the royal library Bischoff writes, "It is rather unlikely that it completely ceased to exist after 814 in accordance with the instruction in Charlemagne's will to sell his books" and to donate the money to the poor (Bischoff 1994: 73).

Also a book lover, Louis the Pious maintained a library and a librarian at his residence in the Kingdom of Aquitaine and another at the royal court at Aachen. Manuscripts that can be linked with Louis the Pious with certainty are rare, but at least four illuminated manuscripts, which can be attributed to the central region of the Carolingian Empire, were actually produced at the royal court. All four date from the second quarter of the ninth century; all are unique works and all are copies of antique manuscripts with impressive picture cycles (Mütherich 2004: 104).[20]

Two of these treasured ninth-century manuscripts are Germanicus' *Aratea*, Leiden Voss Q.79 and Cicero's *Aratea*, Harley 647. These two totally different astronomical texts with their classical imagery of the constellations provide a glimpse at the capability of book illuminators among the royal courts of Louis the Pious. Both illuminations show Sagittarius, a hybrid fusion of half-man and half-horse, grasping a bow and arrow in his human hand; his weapon is highlighted by bright stars, and aimed at the dangerous Scorpion, whose tail is raised and pointing. Their treatment in each *Aratea* manuscript is quite different; the body in the Harley version is made up of perfectly written text while the body in the Leiden is naturalistic. Their human heads are both skillfully drawn but with very different results. Just as the astronomical pursuits and illustrated book production continued after the death of Charlemagne and after Louis, a huge number of superior manuscripts were produced throughout the reign of his youngest son Charles the Bald as well (Figs. 6.4 and 6.5).

[18]Katzenstein says, "There is reason to believe that the Virgil manuscripts now in the Vatican Library once belonged to Charlemagne".

[19]Wright continues on Charlemagne's library, "Then when his library was dispersed after his death in 814 it seems to have gone to the royal abbey of Saint-Denis, where it certainly was during the Middle Ages".

[20]Mütherich writes, "The oldest is the Vatican Terence (Vatican lat. 3868), written about 825, and also the Agrimensores codex of the Vatican Library, a collection of illustrated treatises by Roman surveyors.".

Fig. 6.4 Sagittarius f.52v, Germanicus *Aratea,* Leiden Voss Q.79, Ninth Century, University of Leiden Library, Courtesy of University of Leiden Library

Holy Roman Emperor Charles the Bald and Book Production

Thanks to the efforts of his mother Empress Judith of Bavaria (c.797–843), second wife of Louis the Pious, Charles the Bald (823–877), the third son, came to be the best-educated of the Carolingian rulers.[21] Judith requested that Walafrid Strabo (808–849), a poet and scholar who had studied at Fulda from 827 to 829 under

[21]The first wife of Louis the Pious was Ermengard of Hesbaye; they married in 794 and she died in 818. In 820, Louis married Judith for her beauty, intelligence and influential family. "She adorned the robe of her husband Louis so that he might shine like a hero in the eyes of the people". (Dodwell 1954: 30) Louis had six children with his first wife, two with his second.

Fig. 6.5 Sagittarius f.6, Cicero *Aratea*, BL MS Harley 647, Ninth Century British Library, Courtesy of British Library

Hrabanus Maurus, be appointed tutor to the young prince at the royal palace in Aachen. Walafrid was entrusted with the education of his youthful student for nine years.[22] When Charles began his reign in 843, Walafrid and many of the most-successful Carolingian scholars worked under his sovereignty: Odo of Metz, Einhard, Adalhard of Corbie, Dicuil of Iona, John Scottus Eriugena, Lupus of Ferrières and his student Helperic of Auxerre, and finally Remigius of Auxerre, who may be considered the last of the Carolingian scholars (McCluskey 1998: 157).[23] Interested in astronomy like his grandfather and namesake, Charles the Bald

[22]After Charles reached adulthood, Walafrid returned to Richenau where he was named abbot; he continued his scholastic dedication promoting manuscript production and exchange, making their library and scriptorium most famous. After 829 he became chaplain to Empress Judith.

[23]John Scottus Erigena (c.800–877), medieval poet, philosopher and teacher, was the most significant scholar living in the early Middle Ages. He was already mentioned in this study as a promoter of Plato. He wrote a poem *Aulae siderae* for the dedication of Charles' palace church, drawing upon "the traditional bond associating rulers with astronomy" (McClusky 1987: 157). His poem was charged with astronomical metaphors and numerical harmonies containing spiritual significance. He wrote of the solstices and equinoxes and their relationship to the conception and birth of Jesus Christ and St. John the Baptist. He also suggested that the planets Mars and Jupiter,

was considered a great bibliophile and served as the patron of the 'school' of Reims where he commissioned many exceptional deluxe religious manuscripts (Mütherich 2004: 318).

King Charles the Bald appears to be the first European monarch actually depicted in many religious manuscripts, especially those from Tours and St Denis (Kessler 1977: 125). His portraits appear as dedication images which was appropriate since the king was usually either the patron or recipient of the manuscripts. But the royal representations from the Carolingian period (and from most early carvings and coins as well) appear to be idealized portrayals intended to exhibit the majesty and power of the monarch at his physical peak rather than as true portraiture. One of the purposes of Charles 'portraits' was to show his resemblance to his father and grandfather, Louis and Charlemagne (Nelson 1992: 11). Considering his contentious relationship with his combatant brothers and divided loyalties, the glorified image of the royal presence referred to his function as the legitimate supreme head of the empire, fulfilling the biblical models and signifying cosmic associations (Fig. 6.6).

Astronomical Imagery as a Source of Power and Prestige in the Carolingian Court

The Carolingian intention to idealize the Christian kingdom, the *Imperium Christianum*, by reproducing the Heavenly Kingdom on earth, assigned a primary role to education and religious imagery (Schutz 2004: 375). But the architectural and artistic ideals of the Carolingians were always reaching back to Christian Rome, even promoting the cult of relics to reinforce their religious status and redirecting the authority of the Eastern empire. Their concept of depicting majesty or elevated status was borrowed from Roman prototypes, and so the return of the imperial pictorial image reappeared under the heirs of Charlemagne. The Carolingian emperors elevated their aura by including imperial portraits in illuminated religious manuscripts and in minting coins with royal profiles and iconography. The idea of demonstrating elevated status was derived from Roman relief sculpture and mosaics that included much astronomical imagery, which we saw in Chap. 1.[24] Carolingian kings furthered their claim to legitimacy and divine approval by associating themselves with the earlier Christianized Gallo-Roman

(Footnote 23 continued)

along with Mercury and Venus which was promoted by Martianus Cappella, orbit the Sun rather than the earth.

[24]Astronomical images carved on Carolingian ivories are discussed in the following section; also see "The Ivory Throne of Charles the Bald" in Nees, *A Tainted Mantle.* Constellation images appear on a casket that survives at Bayerisches Nationalmuseum, Munich, and also on a comb in the Victoria and Albert Museum. Certainly royal association with astronomical imagery derives from ancient Rome, Mesopotamia and other earlier cultures as well.

Fig. 6.6 Charles the Bald, BnF MS Lat. 1152 fol.3v, Psalter of Charles the Bald, 842–865 Public Domain

areas of the Empire, and also with the royal figures of the Old Testament, the ancestors of Christ.[25] (Schutz 2004: 15) (Fig. 6.7).

Artistic representations by the reigning monarch demonstrate the desire to establish imperial continuity with the biblical and classical past, not only to validate the Carolingian dynasty, but to assure its continuation in the future by claiming both earthly and heavenly authority. These illustrious regal depictions and obvious biblical comparisons included in religious manuscripts contributed significantly to

[25]Although Schutz received negative reviews of this book, *The Carolingians in Central Europe, their History, Arts and Architecture,* this statement is reliable.

Fig. 6.7 Equestrian Statue of Charles the Bald, ninth century, Louvre Museum Public Domain

the prestige and to the evolution of medieval kingship, especially evident when utilized to honor Charles the Bald.

Art historians, such as Krautheimer and Panofsky, recognized and wrote that medieval artists drew primarily on the art of late antiquity rather than reaching back to the earlier classical period. These artists drew upon manuscripts and artworks of the fourth or fifth century and additionally looked to Christian sources over classical pagan models; artworks associated with Roman Emperor Constantine were especially admired and imitated.[26] Historical scholarship more recently has stressed the predominance of Old Testament rulers as models for Carolingian monarchs (Nees 1991: 6). Nees argues that the poetry and artworks created for both Charlemagne and Charles the Bald present an analogous message; they wanted to follow the true path of David rather than be seduced by a dangerous imitation of Roman Emperors.

Implying symbolic relationships with biblical King David was novel, and the attempt to create a new, revived Davidic kingdom on earth, populated with a new chosen people, a renewed but Christianized Roman Empire, centered on a New Jerusalem. Alcuin had originally given Emperor Charlemagne the *nomen* of the admired biblical King David appropriating not only his name but his honor and dignity and implying an allusion to Jesus who was descended from the House of David. The successors of Charlemagne, especially his grandsons Charles the Bald and Louis the German, were also honored with biblical associations; each was revered as a second King David.

Vivian Bible *Illustrations*

The stunning *First Bible of Charles the Bald*, commonly called the *Vivian Bible* (Paris BNF, Lat. MS 1) was produced at the Abbey of Saint Martin at Tours and presented to Charles the Bald early in his reign, perhaps the "most beautiful of all the books produced at Tours" (McKitterick 1981). This magnificent Bible was commissioned by Lay-Abbot Count Vivian (844–851) and dedicated to King Charles II; the completed manuscript was presented to him when visiting the monastery. Following his ancestor's pattern and aiming to enhance his legitimacy with his bellicose older brothers after the Treaty of Verdun was signed in 843,[27] Charles the Bald associated with and modeled himself on biblical King David through texts and illustrations. Within the *Vivian Bible*, he is compared to King David three different times (Kessler 1977: 109).

King David himself appears appropriately as the frontispiece to the Psalms, one of the eight elaborate illustrations in the *Vivian Bible*. The celestial scene depicts a

[26]Constantine was greatly admired in the Middle Ages and almost considered a saint, as he was the first Roman emperor to allow and even promote public displays of Christianity by donating buildings that became churches, such as St John Lateran Basilica, cathedral church of the Bishop of Rome. Constantine's mother Helena and daughter Constantia are both considered saints, even though formal canonization was not officially instituted by the Church until many centuries later.

[27]As the youngest of Louis the Pious' three combatant sons (the older two were already adults when he was born), Charles may have needed heavenly reassurance to maintain control of his share of Charlemagne's Empire, apportioned, after three years of fraternal warfare, by the Treaty of Verdun in 843. The vast kingdom established by Charlemagne was divided into three major kingdoms; Lothair the eldest received the eastern portion along with the Papal States and the title Holy Roman Emperor; Louis the German gained the central portion; and the westernmost territory was allocated to Charles.

bearded King David as an enlarged central figure practically nude except for a flowing cloak, crown and military ankle guards; he stands poised dramatically stroking a lyre which alludes to his reputation as composer of the Psalms. David is accompanied by six musicians and armed soldiers alluding to his poetic, musical and sovereign competences and personifications of the imperial virtues, Prudence, Justice, Fortitude, and Temperance. The celestial scene pictures the seven male figures each supported by wispy clouds afloat in the deep-blue of the sky and

Fig. 6.8 King David with Musicians, Frontispiece to Psalms, Vivian Bible, f.215v, Public Domain

Fig. 6.9 Charles the Bald Presentation Scene, Vivian Bible, f.423, Public Domain

enclosed by a mandorla-shaped form usually reserved for images of Christ in Majesty (Mütherich and Gaehde 1976: 79) (Figs. 6.8 and 6.9).

Painted as the final folio inserted in the *Vivian Bible* (fol. 423) is an impressive 'portrait' of Charles the Bald, who sits regally crowned and enthroned.[28] The presentation illumination portrays Charles enthroned in an architectural setting, as a cosmic king and semi-god, appointed by God himself as his representative. Since the power of the monarch was understood to come directly from God, good Christians were required to be obedient to the authority of their rulers. With a definite political agenda and divine endorsement of his kingship, Charles the Bald merges his reign with the heavens, projecting himself seated amid the clouds rather than on the ground, simulating an honored constellation figure, an intermediary between God and his people. The resemblance of King Charles surrounded by his entourage and King David by his musicians could not have been accidental. This narrative scene actually took place at Tours in 846 (Mütherich and Gaehde 1976: 26).

The tonsured monks of St. Martin's of Tours encircle the king, gesturing as they sing his praises; the abbot presents the completed manuscript to him as he extends his right hand to accept their gift. On either side of his throne, two armed escorts stand guard reinforcing his military supremacy. Reaching down from on high, the Right Hand of God, *manus dei,* (an accepted image representing God in both Christian and Jewish paintings at that time and even earlier), extends rays of golden light over Charles' crowned head. This representation of divine approval reaffirms that the young king was divinely chosen and protected. The hand of God reinforces his sacred right to rule, creates associations with the heavens and alludes to the astronomical symbolism evident among Roman emperors. A third illustration in the *Vivian Bible* also includes astronomical symbolism, to which I now turn.

Vivian Bible, *St. Jerome's Prologue, Initial D, and the Knot of the Heavens*

The fascinating interplay of script, decoration and images occurs as a unique characteristic of medieval illuminated manuscripts. From this interrelationship, a distinctive art form evolved—the decorated initial. Letters that begin a new paragraph or a new chapter matured from a discrete slightly-oversized and colored letter (usually alternating red and blue) into an exquisite artwork in itself. "Both practical and aesthetic factors were significant in the actual origin of the initial, but the aesthetic soon triumphed and the initial thus came to occupy an autonomous sphere midway between script and picture." (Pächt 1984: 45). An initial at the start of a

[28]Dutton and Kessler have published a study of the eleven poems and eight miniatures in the Vivian Bible; they state that the manuscript was begun in the scriptorium at St. Martin's of Tours with an original plan of four illuminations and accompanying tituli. When Charles decided to visit the monastery in 845, the canons added four additional miniatures and elaborate verses in order to present the Bible as a compliment to honor Charles and at the same time attain their own ambitions. The authors believe the monastic community hoped for a *quid pro quo* from the king, the right to elect their own officials in the future.

particularly noteworthy section (such as the birth of Christ in the New Testament) usually receive more flourishes and space on the text page; in luxurious manuscripts, an entire folio might be devoted to a single historiated initial or to the first word. An initial that merely furnishes a space for an image, or provides a frame for the figure within the body of the exaggerated letter is called an inhabited letter. A decorated initial that provides a display area for more distinctive figural or narrative representations is referred to as a historiated initial. The elaborately decorated initials found in Hiberno-Saxon manuscripts overflow with intricately interlaced knots, entwined spirals and embedded figures; the most famous Insular examples are of course the Book of Kells in Trinity Library, Dublin and the Lindisfarne Gospels at the British Library in London. In fact the earliest appearance of an historiated initial can be found in Insular art from the early eighth century. Perhaps influenced by these expressive models, enlarged decorated initials became very popular in certain late-Carolingian learning centers; the resulting artistic productions can be admired in many French manuscripts (Pächt 1984: 77).

The impressive royal portrait of Holy Roman Emperor Charles the Bald in the presentation scene and the image of King David in the *Vivian Bible* just discussed have both received much attention and analysis by art historians but another less prominent illustration appears in that Bible which has attracted little attention. The astronomical implications are confounding. This cosmic historiated initial has been enlarged, covering almost half its folio in the Carolingian Bible. The size and decoration of the initial offers a further clue to both its importance and location. Some viewers might call this decorated letter an inhabited initial since at first glance it holds tiny, unadorned zodiacal constellation figures, not an identifiable narrative scene from the Bible or from a mythological or historical event.

Decorated or historiated initials in religious manuscripts generally illustrated a narrative scene relating to the text within which it appears, although the sides and base of the letter often morph into hybrids, grotesques or mythological animals. Surprisingly the decorated initial "D" at the opening of St. Jerome's *Preface to the Pentateuch* in the *Vivian Bible* has not been enlarged to display an appropriate Biblical illustration but includes an astronomical/astrological zodiacal image. Furthermore, the opening initial holds hitherto unrecognized astronomical information.[29] This decorated initial "D" is especially significant since, of all the Turonian Bibles illustrated in Köehler and Mütherich volumes[30]; this is the only one to include the signs of the zodiac within the enlarged "D" of the Desiderius initial. Repeating the question asked concerning the zodiacal mosaics in Jewish synagogues in Chap. 2, what is an image of the zodiacal constellations with the pagan god of the Sun Apollo and the goddess of the Moon Luna doing in a Christian Bible dedicated to Charles the Bald? (Fig. 6.10).

---

[29]P. McGurk, in "Carolingian Astrological Manuscripts", draws attention to the zodiacal initial, but does not comment on the special placement of Pisces. p. 317.

[30]See W. Köehler and F. Mütherich *Die karolingischen Miniaturen.*

Fig. 6.10 Desiderius Initial "D" *Vivian Bible* f.8, Public Domain

The decorated initial "D" appropriates almost half the page; ten signs of the zodiac are pictured within the borders of the letter and easily recognized by their traditional iconography. But actually eleven signs are embedded into the "D" since Virgo and Libra have been conflated into a single figure of a woman holding the scales. The symbolic illustrations of the Sun and Moon are drawn standing in their fiery chariots, positioned in the open center of the historiated "D". The honored Apollo and Luna iconic images have each been encircled to indicate their heavenly orbits. Note that the two fish representing the constellation of Pisces have been displaced from their normal sequence among the signs of the zodiac to a position of honor in the center of the initial. This decorated initial has been mentioned by J.J.G. Alexander, who commented on the classical appearance of the folio and the gold-on-purple panels of the folio, typical of the Carolingian illuminators. Alexander states; "the imaginary space apparently provided in the center of the 'D' for the Sun and Moon is negated by the flanking Fish" (Alexander 1978: 52). Alexander suggests that because the historiated initial appears at the beginning of the Bible, it may refer to the Creation of the Sun, Moon and Stars in Genesis 1: 16, which is possible. But this celestial illustration could have an entirely different meaning; I suggest that the image encapsulates essential astronomical/astrological information.

The relevance of the special placement of the two fish representing the constellation Pisces has been overlooked by historians. This particular zodiacal sign, I suggest, was not randomly chosen merely to fill an empty space, as Alexander says, "*horror vacui*". Note that the Sun is situated precisely on the cord that attaches the two fish (the cord indicates a string of stars between the fish). This association of the Sun with that specific star in the constellation of Pisces alludes to a particularly eventful occurrence in the heavens.

Ancient and medieval astronomers were well aware that a significant shift takes place in the heavens approximately every 2100 years, as this phenomenon was discovered by Hipparchus in the second century BCE.[31] As the Sun slowly transits its path through the millions of stars, it rises each day against the background of stars of a particular zodiacal constellation, which changes as it moves along the ecliptic through the twelve zodiacal constellations. The equinoxes travel westward along the ecliptic relative to the fixed stars, a movement that is opposed to the motion of the Sun along the ecliptic. The Sun moves at a specific rate that passes through the entire plane of the ecliptic; this gradual progression through the signs takes about 25,000 years to return to its calculated starting point among the fixed stars. Although this epochal change from one sign of the zodiac to the next occurs very slowly, it was recognized by early Hellenistic astronomers. This gradual anti-clockwise movement of the Sun through the signs of the zodiac is referred to as the Precession of the Equinoxes, the ultimate time-keeping event that announces a new astrological age. The arrival of the Sun rising within the stars of a new zodiacal sign signals the start of a new age.

During the first century BCE, there was an earlier transformation in the heavens as the Sun completed its journey rising amid the Sign of the Ram. The solar rising occurred within the constellation of Pisces for the first time, having recessed

---

[31]According to John North, "There is no evidence, however, to support the idea that anyone before Hipparchus both recognized the slow drift in star positions and also described it with reference to a precise astronomical coordinate system"; although some look to a Babylonian discovery. (North 2008: 14).

backward through the thirty or so degrees of Aries (according to the tropical zodiac) into the next zodiacal sign. The initial verification of this slow progression that the Sun had definitely moved into the new astronomical age was confirmed when the Sun, at the spring equinox, rose in alignment with the first star of Pisces, *Alpha Pisces,* or what Hellenistic astronomers referred to as *Nodus Caeli*, the Knot of the Heavens (DeCallatay 1996: 1–13).[32] De Callatay, "The Knot of the Heavens", *Journal of the Warburg and Courtauld Institutes.* This particular star in the constellation of Pisces is the nearest to the stars of the constellation of Aries.

In his article, DeCallatay points out the fact that in the *Phaenomena*, Aratus mentions the constellation of Pisces twice in the poem, the second time he mentions 'the knot of the heavens' the verse appears on line 365, a noteworthy number.[33] Even though a minor star in brightness, Alpha Pisces, which represents the knot, is praised by Aratus in the poem for its importance, bringing more attention to its stature and emphasizing its special significance as the astronomical marker announcing the new Age of Pisces, just as we are now, over two thousand years later, moving slowly into the Age of Aquarius, an era special mainly to astrologers. Since Hipparchus was born in Turkey in 190 BCE and Aratus died in Macedonia in 240 BCE, how would he have attained this knowledge? Aratus based his poem on the writings of Eudoxus which are now completely lost, he could have been a possible source.

Exalting the sign of Pisces in the initial "D" of the *Vivian Bible* could possibly indicate the Sun sign of a particular person, but the birthday of Charles the Bald, the honoree of the Bible, was June 13. But since Apollo sits directly on the knot of the cord, I propose that refers to a crucial moment in astronomical time. The transitional shift from the Age of Aries into the Age of Pisces was a period of particular note for Hellenistic astronomers who were aware of this celestial event.[34] The Age of Pisces was especially meaningful to Christian astronomers because this eventful solar shift in the heavens heralded the birth of Christ. This 'coincidence' was not lost on Christian scholars, as the symbolism of a fish was associated with the ministry of Christ and is still seen as an iconic symbol of Christianity.

This particular emblem designating the new astrological age may have been designed and encoded in various scrolls or codices by Hellenistic or Roman illustrators. The classical or Late Antique features in the letter 'D' could have been copied and/or adapted by the medieval illuminator of the *Vivian Bible*, totally unconscious of its special significance. Or perhaps "the Astronomer", or another knowledgeable astronomer, was still at the royal court of Charles the Bald, and consciously designed this initial for the *Vivian Bible*; just as he may have designed the significant planetary diagram included in the Leiden *Aratea*, which will be

[32]For more information on the significance of the *Nodus Caeli* in the *Phaenomena* of Aratus, see G.

[33]Douglas Kidd translation.

[34]According to Ulansey, this precessional event in the heavens was secret knowledge, valuable to members of the cult of Mithras, and hidden in the astronomical iconography of the Tauroctony scene. See Chap. 2.

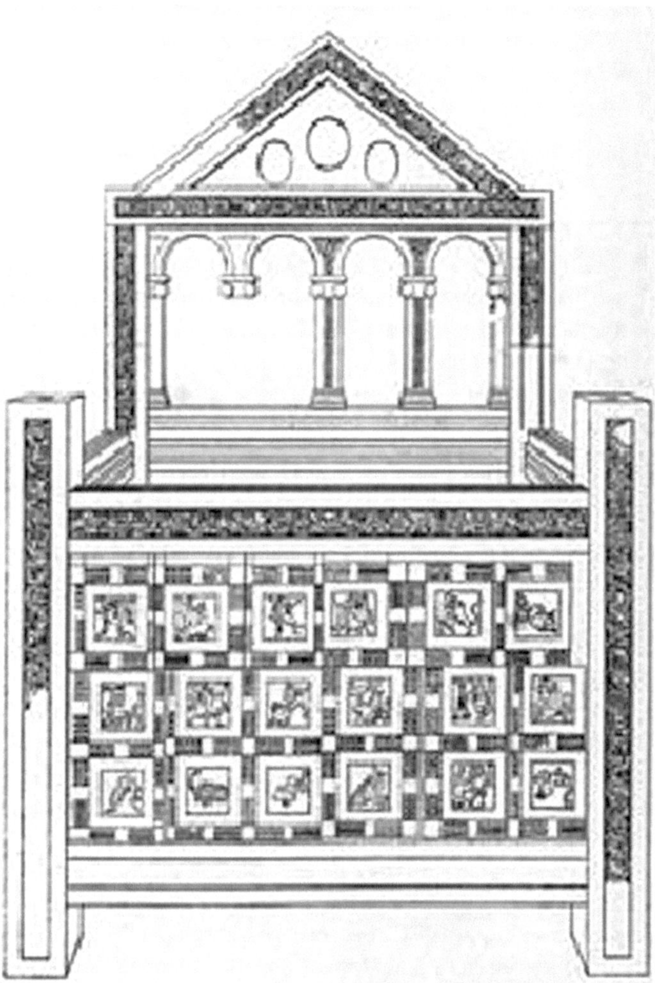

Fig. 6.11 Drawing of Ivory Throne of Charles the Bald, ninth century, *Cathedra Petri*, St Peter's Basilica, Vatican, Public Domain

discussed in Chap. 7.[35] This special treatment of Pisces in Christian art will be explored further in Appendix C.

Astronomical symbolism appears again on the series of carved ivories created for the throne of Charles the Bald. Some scholars suggest that these cosmological ivories on the throne indicate a Carolingian revival of royal apotheosis, which was

[35]As a final note on this extraordinary Bible, the famed library at St Martin at Tours where it was housed was tragically destroyed in a raid by Viking marauders; only one manuscript survived—that was the *Vivian Bible*.

artistically pictured as the heavenly reward for early Roman emperors as shown in Chap. 1. This artifact too has associations with the *Aratea* (Fig. 6.11).

Cathedra Petri a*t St. Peter's Basilica, Rome*

A famous relic held in Saint Peter's Basilica, the *Cathedra Petri* has been honored as a treasured antique throne associated with Saint Peter, the first bishop of Rome, since at least the thirteenth century. The honored seat was considered an emblematic symbol of the power of Saint Peter and his papal successors, not over those at the heavenly gates but over the Christian world, from which the popes ruled the faithful and proclaimed the religious doctrines of the Church.

Gian Lorenzo Bernini, the most impressive Baroque sculptor and one of the many architects of Saint Peter's, was commissioned by the pope to create a monumental sculpture which would hold, honor and protect this precious relic. Bernini worked on the *Cathedra Petri* artwork from 1657 until 1666, when it was installed with great honor. The magnificent Baroque monument encasing the throne of Saint Peter is positioned high in the apse of the Vatican basilica hovering above the main altar, which itself is covered by Bernini's spectacular *Baldacchino*. The *Cathedra Petri* includes an ornate chair in the center of the sculpture to indicate what it holds inside; golden putti hold the papal crown atop the throne. Once it became encased inside the gilt bronze sculpture, the venerated chair was hidden from the eyes of Christian worshippers and also from the research of inquisitive art historians.

In 1968 the *Cathedra* was removed from its monumental hiding place and at last made available for scholarly study; the décor of the chair and the iconography and decorative style of the attached ivory panels inset on all four sides were immediately recognized by specialists; Saint Peter's seat was not carved by Roman artists but by Carolingians (Nees 1991: 149–50). As further evidence, a male image, carved on the ivory insert at the top of the pediment of the wooden throne, resembles other surviving depictions of Charles the Bald with his distinctive long mustache.[36] Although this proposed identification cannot be positively proven, it has been generally accepted by most historians that the chair was made originally for Charles in the ninth century, not for Peter in the first century.

Toward the end of his reign, "furthering his divine aspirations", Charles the Bald was crowned Holy Roman Emperor by Pope John VIII in Saint Peter's on Christmas Day, in the year 875, paralleling his famous grandfather's coronation exactly seventy-five years later. Historians propose that the royal throne may have

[36]Nees writes, "The portrait features are closer to those of Charles the Bald than to any other monarch of the period and much closer than portraits of Ottonian rulers generally portrayed either clean-shaven or with full beard." Also the most likely time period for the carving of the vine-scroll patterns with figures on the chair is the late Carolingian schools of Metz or the court school of Charles the Bald or both.

been transported to Rome by Charles and his entourage on the occasion of his imperial coronation, but there is no documentation or explicit evidence that the esteemed throne had been used for his coronation (Nees 1991: 151). Most probably the Carolingian throne was created as a special gift for the current as well as future popes.

In addition to the royal ivory portrait carved on the pediment at the back of the chair are carved cosmological images, personifications of the Sun, Moon, and Earth, and depictions of some constellations appear on its sides but not the full constellation cycle. These stellar representations appear to be accurate copies of their possible Carolingian models; although they did add depictions of small putti. In his study of the carved ivories Nees states, "Weitzmann showed that the constellation figures of the pediment gable are based upon illustrated Aratus manuscripts". He adds that for other models the artists used, "perhaps a different astronomical manuscript for the images of the constellation Engonasin, first identified with the hero Hercules in the early ninth-century manuscript in Cologne" (Nees 1991: 153–4).[37] Four rows of ivory panels decorate the bottom front of the throne; the upper two rows depict the Twelve Labors of Hercules. Janet Nelson writes that the ivory plaques were almost certainly commissioned by Charles; the Hercules scenes represent a theme with special meaning for a man beset with continual struggles (Nelson 1992: 235). The bottom row plaques carry images of fantastical beasts, which include a scorpion but the beasts are mythological and hybrids not astronomical. The ivory plaques were once inlaid with gold or jewels. As demonstrated, celestial associations and astronomical manuscripts were quite common during the Carolingian dynasty and that use of astronomical symbolism as propaganda continued into the Ottonian period as well.

Astronomical Robe of Emperor Henry II

A final brief example of the royal attribution of astronomical iconography is the adoption of the antique custom of stellar ceremonial robes. These decorated mantles had a long historic tradition of presenting majestic rulers to their subjects enrobed with celestial symbols, inviting comparisons with the wise leaders of the Old Testament, such as Solomon or Aaron, or with the philosophic expertise of Plato. In

[37]The other possible model is Cologne, Dombibliothek, Cod. 83 II (ca. 805), which holds a close likeness to Hercules with the Golden Apples. The study by Nees is mainly concerned with the illustrations of the Labors of Hercules which are attached to the front of the throne. In addition to the Hercules scenes, there is a carving of Scorpio and also some mythological animals, definitely not zodiacal. He compares the carving of Hercules and Achelous with the illustration of Eridanus from the Cicero *Aratea*, Harley 647. There is some likeness, but that example is seen in many astronomical manuscripts. See also Weitzmann, K. (1975) "The Iconography of the Carolingian Ivories of the Throne" in Maccarrone et al., *Nuove ricerche sulla cattedra lignea di S.Pietro.*

Fig. 6.12 Sternmantle of Holy Roman Emperor Henry II, early 11th century, Bamberg Cathedral Treasury, Public Domain

Late Antiquity, billowing capes filled with stars were also pictured on images of celestial and mythical gods, particularly Mithras.

The earliest surviving astronomical robe dates to the reign of Ottonian Emperor Henry II (r.1002–1024) now in the Bamberg Cathedral Treasury. The ceremonial Star Mantle was made originally of deep purple silk damask and embroidered in appliqué and gold thread; until recently, scholars assumed the mantle was made for Henry's coronation but current investigations are dating it later than that event and question whether Henry ever wore the robe. An essay by David Ganz published in 2014 relates the results of his comprehensive study of the Star Mantle and assessment of the damages brought by an extensive restoration that took place in the fifteenth century.[38] The astral robe bears the label, *descriptio tocius orbis*, providing an image of the entire celestial sphere with many of the constellations pictured encircled with gold and marked with identifying labels. An image of Christ surrounded by the symbols of the four evangelists appears at the top center with the representations of the Sun and Moon flanking him; constellations of the starry heavens are scattered in no significant pattern (although the images have been moved significantly); a gold inscription decorates the edges of the regal cloak (Fig. 6.12).

The *Sternemantle of Henry II,* by appropriating artistic celestial images, exploited both religious and astronomical symbolism as an expression of Henry's or later monarchs' cosmic power. Interestingly the celestial images displayed on the cape were closely associated with an *Aratea* manuscript. A dissertation written on

[38]The detailed study by Ganz (*Pictorial Textiles and their Performance: the Star Mantle of Henry II*) has been downloaded on Academia.edu. The treatise explains how the major restoration preserved the mantel and yet destroyed it at the same time.

this Star Mantle has determined that the constellation figures embroidered on the royal robe were derived from a ninth-century illustrated astronomical compendium that contained texts of both Hyginus' *De astronomia* and the *revised Aratus Latinus* not Cicero's *Aratea* (O'Connor 1980: 63–66).[39] Text on the edge of the mantle reads: *Pax Ismaheli qui hoc ordinavit*, indicating that the person who ordered it was either Jewish or from the East. "The Emperor's mantle was a result of that new movement which spread to the north from Spain and southern Italy, the two countries where Greeks and Arabs mingled freely with people of the western world".[40]

These surviving examples: the numerous copies of the *Aratea*; the appropriation of astronomical symbolism; and the Carolingian throne from the ninth century; demonstrate the significance of astronomical affiliation, imagery and knowledge that links the earthly reign of the Carolingian rulers and Ottonian kings with the celestial realm. The contribution of the Carolingian scholars to the transmission of astronomical knowledge and astronomical art is immeasurable. This segment has presented a broad view of the prominent various roles that Carolingian astronomical manuscripts and artworks played in reinforcing the power and right to rule in the early Middle Ages. Chapter 7 focuses on a more narrow view and examines in greater detail the production and impact of *Aratea* manuscripts along with their cultural circumstances in order to determine their proper role in this vast astronomical tradition.

References

Alexander, J. J. G. (1978) *The Decorated Letter*. London: Thames and Hudson.

Bischoff, B. (1994) *Manuscripts and libraries in the age of Charlemagne*, Gorman, M. and trans. Cambridge: Cambridge University Press.

Blume, D., M. Haffner, and W. Metzger. (2012) *Sternbilder des Mittelalters: Der gemalte Himmel zwischen Wissenschaft und Phantasie*. Part One: 800–1200. Berlin: Akademie Verlag.

Blume, D., M. Haffner, and W. Metzger. (2016) *Sternbilder des Mittelalters: Der gemalte Himmel zwischen Wissenschaft und Phantasie*. Part Two 1200–1500. Berlin: Akademie Verlag.

Butzer, P. L. (1993) "Mathematics in the West and East" in *Science in Western and Eastern Civilization in Carolingian Times*, P. Butzer and D. Lormann eds. Basel: Birkhäuser Verlag.

Cameron, A. (2004) *Greek Mythography in the Roman World*. Oxford: Oxford University Press.

De Callatay, G. (1996) "The knot of the Heavens", *Journal of the Warburg and Courtauld Institutes,*, Vol. 59, pp. 1–13.

Dekker, E. (2013) *Illustrating the phaenomena: Celestial cartography in Antiquity and the Middle Ages*, UK. OUP Oxford.

[39]The 'Star Mantle of Henry II' is the topic of a doctoral dissertation by E. O'Connor, (1980) Columbia Univ. O'Connor found close relationships with the astronomical illustrations of the mantle with those in two ninth-century "star catalogs", St. Gall Stiftsbibliothek MS 902 and its copy, St. Gall MS 250, both have the text of *revised Aratus Latinus*, pp. 63–66 and 108. The Star Mantle is also discussed in great detail in McCluskey (1998) 141–144.

[40]Saxl, "Belief in Stars in the Twelfth Century" in *Lectures*.

Dodwell, C. R. (1993) *The Pictorial Arts of the West 800–1200*. New Haven: Yale University Press.
Dodwell, C. R. (1954) *The Canterbury School of Illumination, 1066–1200*, Cambridge: Cambridge Univ. Press.
Eastwood, B. S. (2007) *Ordering the Heavens: Roman Astronomy and Cosmology in the Carolingian Renaissance*. Leiden: Brill.
Eastwood, B. S. (2002) *The Revival of Planetary Astronomy in Carolingian and post-Carolingian Europe*. UK: Ashgate Publishing Company.
Gans, D. (2014) *Pictorial Textiles and their Performance: the Star Mantle of Henry II*, downloaded by Ganz at www.academia.edu/14531241.
Katzenstein, R. and Savage-Smith, E. (1988) *The Leiden Aratea, Ancient Constellations in a Medieval Manuscript*, the J. Paul Getty Museum, CA.
Kessler, H. (1977) *The Illustrated Bible from Tours, Studies in Manuscript Illumination* 7, Princeton: Princeton University Press.
Kidd, D. (1997) *Aratus*: *Phaenomena*. Cambridge: Cambridge University Press.
Köehler, W. and F. Mütherich (1971) *Die karolingischen Miniaturen* vol. 4: *Die Hofschule Kaiser Lothars*. Berlin: Cassirer.
Kühnel, B. (2003) *The End of Time in the Order of Things: Science and Eschatology in Early Medieval Art*. Regensburg: Schnell & Steiner.
Le Boeuffle, A. (2003) *La texte des Aratea de Germanicus est copieusement représenté*: Paris: Les Belles Lettres.
Lohrmann, D. (1993) "Alcuins Korrespendenz mit Karl dem Grossen über Kalender und Astronomie" in Science in Western and Eastern Civilization in Carolingian Times, Basel; Boston: Birkhäuser Verlag.
Martin, J. (1998) *Aratos*. Phénomènes, Paris: Les Belles Lettres.
McClusky, S. C. (1998) *Astronomies and Cultures in Early Medieval Europe*. Cambridge University Press.
McCluskey, S. C. (1993) "Astronomies in the Latin West from the Fifth to the Ninth Centuries" in *Science in Western and Eastern Civilization in Carolingian Times*, P. Butzer and D. Lormann eds. Basel: Birkhäuser Verlag, pp. 139–160.
McGurk, P. (1966) Catalogue of astrological and mythological illuminated manuscripts of the Latin Middle Ages. London: Warburg Institute University of London.
McGurk, P. (1981) "Carolingian Astrological Manuscripts", in *Charles the Bald: Court and Kingdom*, eds. M.T. Gibson and J.L. Nelson. *BAR International Series 101* Oxford, pp. 317–332.
McKitterick R. (1994) ed. "Script and Book Production" in *Carolingian Culture: Emulation and Innovation*. Cambridge: Cambridge University Press, pp. 221–247.
McKitterick, R. (1989) *The Carolingians and the Written Word*, Cambridge: University of Cambridge Press.
McKitterick, R. (1981) "The Palace School of Charles the Bald", in *Charles the Bald: Court and Kingdom*: papers based on a colloquium held in London in April 1979, eds. M.T. Gibson and J. L. Nelson, BAR International Series 101, Oxford.
Mütherich, F. (2004) "Book Illumination in the Court of Louis the Pious" in *Studies in Carolingian Manuscript Illumination*, The Pindar Press (reprinted in 2004).
Mütherich, F. and J. E. Gaehde (1976) *Carolingian Painting*, New York: George Braziller.
Nees, L. (1991) *A Tainted Mantle: Hercules and the Classical Tradition at the Carolingian Court*. Philadelphia: University of PA Press.
Nelson, J. (1992) *Charles the Bald*. London: Longman Group.
North, J. (2008) *Cosmos: an Illustrated History of Astronomy and Cosmology*, Chicago: University of Chicago Press.
O'Connor, E. (1980) *The Star Mantle of Henry II*, PhD. Dissertation, Columbia.
Pächt O. (1984) *Book Illumination in the Middle Ages*. London: Harvey Miller Publishers, Oxford University Press.
Panofsky, E. (1955) *Meaning in the Visual Arts*. University of Chicago Press.

Panofsky, E. and Saxl, F. (1933) "Classical Mythology in Mediaeval Art", *Metropolitan Museum Studies*.

Saxl, F. (1957) "The Belief in Stars in the Twelfth Century" in *Lectures* 2 vols. The Warburg Institute.

Schutz, H. (2004) The Carolingians in Central Europe, their History, Arts and Architecture. Leiden: Brill

Stevens, W. M. (1987) *Cycles of Time and Scientific Learning in Medieval Europe*. Aldershot, UK: Variorum Collected Studies Series.

Stevens, W. M. (1995) *Cycles of Time and Scientific Learning in Medieval Europe*, Aldershot, UK: Variorum.

von Euw, A. (1989) "Der Liedener Aratus, Antike Sternbilder in einer karolingischen Handschrift", exhibition catalog. Munich: Bayerische Staatsbibliothek.

von Euw, A. (1993) "Die künstlerische Gestaltung der astronomischen und komputistischen Handschriften des Westens", in *Science in Western and Eastern Civilization in Carolingian Times*, P. Butzer and D. Lohrmann, eds. Basel: Birkhäuser Verlag, pp. 251–272.

Wright, D. (2001). *The Roman vergil and the origins of medieval book designs*. Toronto: University of Toronto Press.

Chapter 7
Specific *Aratea* Manuscripts

Abstract The seventh chapter looks at some of the cultural and historical factors that influenced the production (or lack of production) of *Aratea* manuscripts. It also examines in more detail six representative *Aratea* manuscripts placed within their historical context. The six selected Latin *Aratea* manuscripts from three different centuries will be examined more closely in this section; the overview will look at the history, style, iconography, and probable places of production of the manuscripts. Because the survival numbers are low from the twelfth century, the manuscripts have been chosen for availability and access in viewing the illustrations. The two fifteenth-century examples were chosen principally because they are part of a group of four deluxe manuscripts created at the same time for the same royal court, which allows for effective and informative comparisons. As shown, these three time periods are the most fruitful and rewarding centuries when tracing the production of Aratean manuscripts and the transmission of astronomical knowledge in the Middle Ages. The illustrations of these particular manuscripts are discussed more extensively in order to compare the artistry and changes through different centuries and even into the period of early printing.

Ninth Century *Aratea* Manucripts

Numerous *Aratea* manuscripts were produced during the Carolingian period for a total of 17 survive. A new interest in the study of the stars was noted with the appearance of the revised version of Aratus's Greek poem, the Aratus Latinus. The revised text was not complete; nor was it an accurate translation; but the renewed interest in classical astronomical writings can be a clue to additional study of the *Aratea* and other similar material under Charlemagne (Eastwood 1997: 238). The appropriation of stellar iconography and symbolism was utilized by Carolingian emperors, as they were known to consciously look back to traditions and values of Roman emperors; they and their courts demonstrated their antiquarian interests in art, architecture, and in creating lavish illuminated manuscripts. *Aratea* texts transmitted and displayed the poetry, mythology, and astronomical capabilities of the Classical world. Some *Aratea* texts were less ambitious and sketchily illustrated, merely included in astronomical anthologies

M. Dolan, *Astronomical Knowledge Transmission Through Illustrated Aratea Manuscripts*, Historical & Cultural Astronomy,
DOI 10.1007/978-3-319-56784-6_7

designed for learning basic cosmological information, especially true after the rise of interest in celestial matters upon the writing of the Computus of 809. In addition, a bright comet (Halley's) appeared overhead in 837; historical accounts say it was one of its closest approaches.[1] "*One truism stands out in the history of comets—mankind has always been transfixed by their appearances.*" (Yeomans (1991: 1) Mütherich states that it is tempting to consider the impression that the comet of 837 made on the Carolingians; this may have justified the dedication of an astronomical picture book to a member of the court (Mütherich 2004a, b: 107).

Four luxurious manuscripts produced in the Court of Louis the Pious still survive; two of these are *Aratea* manuscripts.[2] These two ninth-century *Aratea* selected for specific attention are exemplary manuscripts, both already discussed several times. They are commonly referred to as "Leiden Germanicus" (Leiden MS Voss. Lat. Q.79) and the "Harley Cicero" (British Library Harley MS 647). Both manuscripts were created in the same geographical area, between the eastern and western portions of the Holy Roman Empire, lands later ruled by Lothair (795–855), the son of Louis the Pious, called Lothringia, now Belgium, Netherlands, and Germany. Both were produced in scriptoria where Late Antique codices were readily available as exemplars, as well as the excellent scribes and artists required to create distinctive manuscripts for the elite, educated patrons, who appreciated classical texts (Bischoff 1994: 87). The miniatures found in both manuscripts, though entirely different from each other in design and execution, are outstanding for their artistic quality and their ability to capture the essence of classical art. The illuminations of the constellation figures exhibit traditional portrayals of the heroic figures of myth often seen in Greek carvings and vases, and seem more in keeping with a society that "venerated Jupiter than with one which worshipped Christ" (Dodwell 1971: 23).[3]

Although these are the two most widely known and well-published *Aratea* manuscripts and have been studied for more than a century, it is essential to include them in this study, as they are among the earliest created during the medieval period and have been discussed most extensively in the scholarly literature. They remain true to the imaginative *Aratea* text of Cicero and Germanicus and most copies that followed throughout the Middle Ages, although not exact copies, looked back to these two exceptional originals for models and inspiration. The examination and comparison of these two superior originals is useful in order to gain a fuller understanding of the basis of this astronomical manuscript tradition, including the direction and conclusions of some of the more recent studies by astronomical and art historical scholars. Both of these *Arateas* were copied over and over throughout the Middle Ages. Their influence can be detected in the miniatures of many later

---

[1]See Kronk (1999) *Cometography,* Vol. 1, Cambridge University Press, p. 14.

[2]The other two luxury manuscripts are the Vatican Terence (Biblioteca Apostolica Vaticana MS Vat lat. 3838 and the Agrimensores (Biblioteca Apostolica Vaticana MS Pal lat. 1564). See Mütherich, Book Illumination at the Court of Louis the Pious.

[3]Dodwell adds, "… and they give surprisingly authentic representations of the heroic gods and nubile goddesses of Antiquity". As an example, the author selected a miniature of Andromeda, who was not depicted as a celestial heroine, but an unfortunate princess.

manuscripts of the *Aratea* that are not direct copies, and in complementary astronomical writings as well. Designed at courts where the goals of the Carolingians were defined and implemented, the Leiden and Harley *Arateas* both convey many of their cultural and political aspirations.[4]

Leiden Ms Voss. Lat. Q. 79

History of the Manuscript

The earliest surviving manuscript of the translation of Aratus' *Phaenomena* by Germanicus, Leiden MS Voss.Lat.Q.79 (hereafter the Leiden) has been dated to c.816 and is considered "one of the finest Carolingian reproductions of an antique codex" (Mütherich and Gaehde 1976: 12). Fortunately, the Leiden has been carefully preserved and survives nearly complete in excellent condition. The almost square manuscript appears rather small in size and thickness when compared to the enormous religious and encyclopedic volumes that were being produced in various Carolingian scriptoria and learning centers. The Leiden contains 99 leaves, each measuring 22.5 × 20 cm, and 36 full-page illustrations. In order to determine where a particular manuscript was produced, specialists compare its type of script and artistic decoration; they examine the colophon for contemporary evidence,[5] and also search for subtle references to other religious or historical texts within the manuscript. Since the Leiden possesses none of these telling clues, nor any specific characteristics to indicate its exact place of production, it has been difficult for historians to determine where it was created with confident accuracy (Mütherich and Gaehde 1976: 69). The Leiden had formerly been assigned to a scriptorium at Reims, but "it is the much more exclusively antiquarian element which distinguishes the court manuscripts from those of Reims, where the classicizing forms have been touched by the expressive style characteristic of the school" (Mütherich 2004: 109).[6] Therefore, Reims can be excluded, and the Leiden has been allocated to the Lothringian domain of the Empire, now the province of Lorraine.

The Leiden does possess a personal dedication: "*Vale fidens in domino Christi vestitus amore,*" indicating that it was addressed to a noble patron, but unfortunately the addressee is anonymous. The dedication, in addition to the care lavished on the text and illustrations, assure that it was created for a person of highest standing with an interest

[4]The Leiden and Harley are also listed in Appendix B.

[5]A colophon is an inscription that often but not always appears at the end of a manuscript to record information about its production: names, occasions, patrons or dates.

[6]In her study of Carolingian manuscript illumination in Reims (2004), Mütherich defines the various periods of production in that famous center of spectacular Gospel Books and illusionistic painting. "It was considered the home of copies of classical cycles of illustrations, whose faithful reproduction of Late Antique miniatures is a splendid page in the history of Carolingian manuscript illumination, particular, those of the two most famous of these, the Leiden *Aratea* and the Bern *Physiologus*.".

in classical learning. Art historians have suggested that most probably the poetic manuscript was copied for the royal court of Louis the Pious (r. 814-840) (Mütherich and Gaehde 1976: 69). Others have proposed that Empress Judith of Bavaria (c.798-843), the second wife of Louis the Pious, may have been the patron or the recipient. Judith, the mother of Charles the Bald, was also admired as a great bibliophile (Schutz 2004: 95).[7] Scholars have suggested that the extensively decorated *Utrecht Psalter*, created between 816 and 835, may have been made for Empress Judithas well (Schutz 2004: 268). The Leiden is one of the four outstanding nonreligious manuscripts extant from the Carolingian period that were created at the court of Louis the Pius, who, like his father, had a great respect for antiquity.

Creator of the Leiden Aratea

> It is important to recognize that there is not and never was a total separation between classical and medieval culture, the one pagan and the other Christian, for between the two stands the critically important period known variously as late antiquity, the Patristic Age or the Early Christian world (Nees 1991: 5).

The design and illumination of the Leiden looks back not to the early period of Imperial Rome when Aratus' astronomical poem was first translated by Cicero and Germanicus but to three centuries later. Because the Leiden holds so true to its Late Antique prototype, and has no colophon, nor includes any additional writings, the book provides no conclusive evidence, but only hints at the identity of the patron, compiler or producer. Some historians now point to the anonymous and mysterious court chronicler of Louis the Pious, who wrote a biography of Louis, in which he reveals a provocative interest in the stars and planets, both his own and the Emperor's (Mütherich 2004a, b: 107).[8] This same anonymous person also composed a treatise pertaining to the stars and wonders of the heavens, and also reported the appearance of the great comet in 837. To this day, this scholarly court biographer and scientist still remains unidentified; in academic literature, he is simply called 'the Astronomer'.

---

[7]Louis and Judith were married in February, 819, after the death of his first wife. Louis was greatly enamored of Judith as she was quite beautiful and gracious, admired for her knowledge and intelligence. "Walahfrid Strabo, tutor to Charles the Bald, pays respect to Judith in his poem, *De imagine Tetrici*, comparing her to famous women from the Old Testament, as a person of extraordinary learning and wisdom." When Louis lost favor, Judith was sent to a convent, but later they were both restored to the monarchy.

[8]Anon., *Vita Hludovice Imperatoris*, 63 (Ausgewähle Quellen zur deutschen Geschichte des Mittelalters. Freiherr vom Stein-Gedächtnisausgabe, 5: Darmstadt, 1968), 378. Einhard is another candidate, he left the court in 830; other possible compilers are Gerward, his successor, Grimald, or Walahfrid Strabo.

Surprisingly, this unknown scientist was first uncovered and written about in the seventeenth century by Johannes Kepler (1571–1630), who was himself a historian of science, as well as court mathematician, astronomer and astrologer. No conclusive evidence has been discovered to confirm that 'the Astronomer' had anything to do with the Leiden, but it is tempting to suggest that he may have been involved with the original creation of the manuscript for several reasons. The Astronomer's' interests coincide quite closely with those expressed in the Leiden, and he had the technical knowledge to correlate Germanicus' *Aratea* text, could add its scholia and had the ability to coordinate text and illustrations; most significantly, he acted as court astronomer at the determined time of production (Mütherich 2004a, b: 117). Another strong indicator that at least one of the persons responsible for the creation of the Leiden was proficient in medieval and Classical astronomy is the inclusion of the remarkable planetary diagram (associated with Pliny the Elder) found at its end. This nameless astronomer seems to have been a competent observer and could have made the observation of the planetary positions used in the creation of the astronomical diagram on folio 93v of the Leiden (Dekker 2010: 31). (This will be discussed in more detail further on.) Therefore most circumstantial evidence points to 'the Astronomer' as the ninth-century scholar, scientist, and antiquarian to whom we owe this masterpiece.

Provenance

The partially established provenance of the Leiden manuscript also demonstrates the value and respect with which it was safeguarded. After the close of the Carolingian dynasty, the book had almost certainly remained in northern France, at the monastery of Saint-Bertin in Saint-Omer until the early eleventh century. This has been verified since the manuscript was copied there at least twice during the time of Abbe Odbert (986–1007). After its stay in St. Bertin, the location of the manuscript remains unknown until 1573 when it was bought from a Dutch painter by Jacob Susius/Suys (1520–1596) of Ghent; the notation was inscribed on its flyleaf. Some years later the Leiden *Aratea* appeared in the library of a well-known scholar, editor and bibliophile, Hugo Grotius (1583–1645), who published an annotated, scholarly printed edition at Leiden in1600.[9] His edition included all of the miniatures from the Leiden, reproduced as black and white engravings; these were created by Jacques de Gheyn II (1565–1629), one of the most-talented engravers of his time Grotius.' *Syntagma Arateorum* is considered the first printed facsimile edition of an illuminated manuscript.

[9]Although Hugo Grotius was only seventeen when he published the printed *Aratea*, he had already edited and published a scholarly version of Martianus Capella's treatise O*n the Marriage of Philology and Mercury* at sixteen. He was precocious, an intellectual prodigy and entered University of Leiden when only eleven, known mostly for his writings on political philosophy and international law.

In 1648, the books of Grotius, including the Leiden, became part of the royal library of Queen Christina (1626–1689) the intelligent, free-spirited queen of Sweden. Following her abdication of Sweden's throne in 1654, her collection of codices was transferred to Antwerp where her librarian Isaac Vossius (1618–1689), the Dutch scholar and manuscript collector, conducted an inventory of her holdings. Queen Christina paid her librarian's salary in manuscripts, which included the very precious Leiden *Aratea*. After Vossius' death, his heirs sold his library to Leiden University in 1690, where the famous Carolingian *Aratea* is still housed.

Leiden Text

One of the greatest achievements in book production during the Carolingian 're-naissance' was the implementation of a uniform script for codices that had been perfected at the monastery at Corbie (Brown and Lovett 1999: 67). The introduction of Caroline minuscule script, a flowing cursive style, enabled manuscripts to be written and read more rapidly by using the more legible script. But this Carolingian innovation was not adopted for the Leiden textual material, instead the poetic verses of the *Aratea* were written in a narrow and condensed type of rustic capital, *capitalis rustica*, which was more expensive to produce and reserved for only the most luxurious volumes. Following the model of its Late Antique illustrations, this type of antique text too looks back to earlier codex traditions.[10]

The text of the Leiden is principally that of Germanicus, but his verses were supplemented in some sections by portions of Avienus' more detailed descriptions.[11] Also on the folios are textual additions between the originals in the hands of several later owners. In the late thirteenth or early fourteenth century, someone carefully recopied the *Aratea* text in a Gothic book hand in the upper and lower margins of the book.[12] The original lines of text are written in red; the first letter of each line is enlarged and moved a little to the left. The lines recopied by a later reader follow the same layout, but are written in black ink both above and below Germanicus' original text. The paleography of the Leiden text has been examined and explained by Bischoff (1994). The text of the manuscript seems to be subordinate to the care directed toward the Late Antique miniatures (Fig. 7.1).

[10]The rustic capital had been developed in ancient Rome, based on lettering used for inscriptions that were carved in stone; the distinctive shape of each letter derives from the process of carving in stone with a chisel. Another famous example where rustic capitals were purposely used to give a manuscript an antique appearance is the *Utrecht Psalter*, copied from an Early Christian exemplar. The numerous later versions of the *Utrecht Psalter* did not copy its labor-intensive rustic capital script.

[11]For a thorough explanation of the *Aratea* text of Germanicus see Lott (1983) (The textual tradition of the *Aratea* of Germanicus Caesar: missing links in the « μ » branch), *Revue d'histoire des textes*, Vol. 11, Numéro 1981, pp. 147–158 and see also Appendix B.

[12]The additional text continues on all pages until f. 81, except for f. 74v.

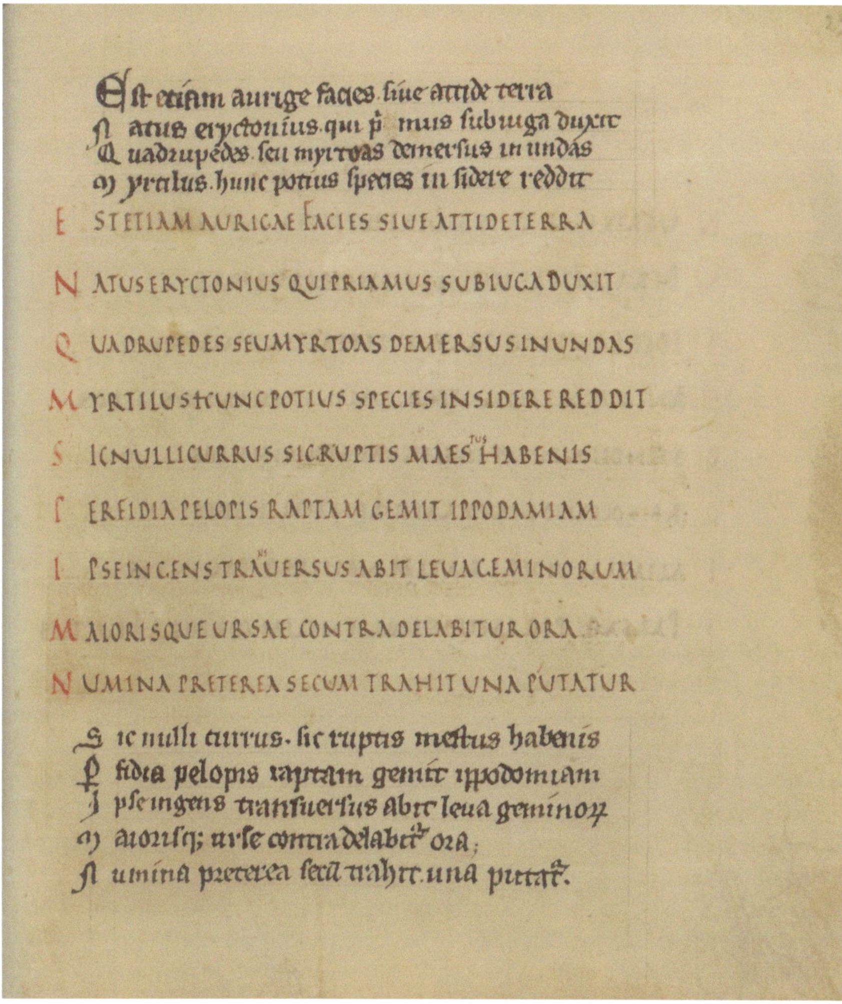

ESTETIAM AURIGAE FACIES SIUE ATTIDE TERRA
NATUS ERYCTONIUS QUI PRIAMUS SUBIUGA DUXIT
QUADRUPEDES SEU MYRTOAS DEMERSUS IN UNDAS
MYRTILUS HUNC POTIUS SPECIES IN SIDERE REDDIT
SIC NULLI CURRUS SIC RUPTIS MAES TUS HABENIS
PERFIDIA PELOPIS RAPTAM GEMIT IPPODAMIAM
IPSE INCENS TRAUERSUS ABIT LEUA GEMINORUM
MAIORISQUE URSAE CONTRA DELABITUR ORA
NUMINA PRETEREA SECUM TRAHIT UNA PUTATUR

Fig. 7.1 Text page f. 23, Leiden Universiteitsbibliotheek MS Voss. lat. Q.79, Courtesy of Leiden University Library

Style, Placement and Iconography

The paintings of the Leiden have been reproduced with such extraordinary accuracy in its antique format, iconography and subject matter, as well as technique and style of painting, that it is difficult to identify any specifically Carolingian features. Because the copyist's Late Antique models of constellation illustrations have been reproduced so faithfully and they exhibit no identifiable idiosyncrasies of script, the manuscript cannot be identified with certainty to any of the major Carolingian scriptoria (Mütherich and Gaehde 1976: 69).[13] Art historian Mechthild Haffner in her dissertation has studied the images of the constellation cycles that appear in Germanicus manuscripts and demonstrates that the artist responsible for the illustrations of the Leiden drew not only on the Late Antique *Aratea* exemplar, but also on other astral cycles. She thinks that the illustrator consulted many sources to create his own constellation pictures, carefully imitating the antique style (Haffner 1997).

When Carolingian artists searched for earlier sources to copy, of course books were the most realistic resource. Parallels can be found in both secular and Christian miniatures of the fourth century and the Leiden reflects those earlier codices. The red frames encompassing each illumination in the Leiden are similar to the framed images in both the Roman Virgil and the Vatican Virgil, dating to the fifth century; these were discussed in Chap. 2. Consequently, the achievements of fourth-century artists and scribes were kept alive and perpetuated as long as parchment codices were produced (Weitzmann 1971: 125). Although the sources of the Leiden illustrations are still debated, the constellation miniatures definitely reflect their fourth or fifth-century exemplars (Verkerk 1980: 261).[14] In the Leiden, the Carolingian illuminator has assimilated to a great extent the antique illusionist style of his exemplars, faithful to their format, technique and manner of painting. The reproduction achieved such accuracy; it could almost be considered a facsimile. The emphasis of this codex seems to be its pictorial content, as the illuminations are much larger than other ninth-century examples, such as Madrid, BN Codex de Metz 3307 (ca. 840), a copy of the Carolingian compendium compiled by astronomers who gathered at Aachen in 809 at the court of Charlemagne to work on the calendar and the computus. In these manuscripts, the small constellation drawings are merely inserted 'papyrus style' between the lines of text with no background or framing. The constellation images in this manuscript are represented in their traditional anthropomorphic or zoomorphic form but have an entirely different appearance than the Leiden.[15] Even though the accompanying text includes the number of stars, none are indicated in these illustrations. The Metz artist avoided the heavier, opaque colors of the Leiden and applied washes of softer colors and loose brushstrokes for shading. These heavenly figures also echo antique archetypes and were copied many times throughout the medieval period.

The concept of separating the text and image and then enlarging each picture to the size of a whole folio was revolutionary for that time period (Weitzmann 1971: 97). In most astronomical texts, the pictures are added merely to illustrate the text. In this manuscript however, the text appears to be ancillary to the artwork, an aid

[13]In 1898 Georg Thiele argued that the Leiden's exemplar originated in the fourth century on stylistic and iconographic evidence.

[14]Verkerk argues that because the text includes sections of Avienus' poem, its model must date to the fourth or fifth century. He adds, "Thiele thought that the finishing touches were only added to the revision of the text in the ninth century by the Carolingian scribe.".

[15]See K. Lippincott, British Museum website for reproductions of the miniatures from Madrid BN MS 3307.

for a clearer understanding of the full-page pictures. But the poetic text and lavish picture cycle serve as close companions, each complimenting, clarifying, and completing the other.

The Leiden survives with 36 full-page illustrations of the constellations (some have more than one on the same folio), as well as depictions of the four seasons and five planets, each group on a single folio; the last two pages are blank. The constellation picture cycle is almost complete except for five lost miniatures: Virgo, Libra (which is often omitted or conflated with Virgo), Centaurus with Lupus, and Corona Australis; also absent are Jupiter and the image of the Sun and Moon together; these were already missing in 1600 at the time Grotius' prepared his printed edition. But fortunately the missing miniatures were preserved in a late tenth-century copy of the Leiden, Boulogne-sur-Mer MS 188, now held at their Bibliothèque municipal. (The total number of illustrations of the Leiden equates with Aratus' original poem that includes 46 constellation descriptions.)

An image of the all-powerful god Jupiter astride an imperial eagle is commonly pictured at the start of Germanicus' *Aratea* manuscripts, as the god was mentioned briefly in the opening verse of Germanicus and honored by a hymn to Zeus in Aratus' poem. *"Ab Ioue principium magno deduxit Aratus carminis*". (LeBoeuffle edition) The missing Sol and Luna illustration appears on folio 32v in the late tenth-century copy of the Leiden now at Boulogne-Sur-Mer. The same textual and pictorial format is used throughout the Leiden; each constellation image follows its poetic description. Every miniature is enclosed in a simple red border, framing each individual constellation as though an independent painting. The background of the constellations exhibits a solid tone, as deep blue as a clear evening sky, decorated with round or squared stars accented with gold. The artist shows no interest in accurate placement of the stars according to their actual position in the sky in any of the Leiden miniatures; they are simply golden squares sprinkled about the figures. In order to identify the constellation in the night sky from the picture their star locations needed to be placed accurately. Therefore they cannot be considered true star charts.

A mind-bending study by Dekker entitled "The Provenance of the Stars in the Leiden *Aratea* Picture Book" asks pertinent questions, "Are the stellar configurations in the Leiden *Aratea* unreliable, or astronomically significant? Were these stellar patterns part of the pictures of the exemplar or exemplars on which the illustrator of the Leiden *Aratea* drew?" (Dekker 2010: 5). She tried to determine whether the fourth-century prototypes had based their constellation description of verbal or scientific sources and if they had access to or referred to the star charts of Ptolemy by comparing each star in each different manuscript type. Through her study of the many variations of the text of Germanicus' translation, including the scholia and other inserts, she came to a conclusion.

> Since only the descriptive part of the Ptolemaic catalogue was used, the stellar configurations lack the precision usually associated with Ptolemaic astronomy. Copying processes may have led to further imprecision and uncertainty regarding the locations of the stars with the result that of the 875 stars marked in the Leiden *Aratea* picture book only 518 can now be linked to specific star catalogues (Dekker 2010: 32).

In addition the drawing of some constellations as seen from the Earth, and others as seen on a globe, would certainly confuse a viewer. Considering all these discrepancies, a person consulting the Leiden text and images as an aid for learning the positions of the stars would find it useless (Dekker 2010: 32). Like many other historians, she too argues in the article that the "roots of the manuscript were first and foremost pictorial" (Dekker 2010: 3).

The astronomy of the Leiden maintains that the Earth is stationary and the heavens revolve about a fixed axis; the course of the sun through the signs of the zodiac is explained. The five planets are not described as gods but merely as heavenly bodies having independent motion without discussing their orbital periods; the moon's orbit, since it is very complicated, is not even mentioned. Each of the planets appears on the same folio as small heads with no particular emphasis or iconography. Since he was not an astronomer, Germanicus did not direct his translation toward achieving scientific precision. Ptolemy had not yet been born (Figs. 7.2 and 7.3).

In spite of the fact that the Leiden was produced in the very Christian Carolingian court, the artistic depiction of the celestial figures, animals and objects conforms to the customary classical designs of the constellations without a shred of Christian imagery added by the artists serving under Louis the Pious, most probably in a monastic scriptorium. The one very slight concession to religious symbolism could be easily missed; tiny crosses were placed atop the small hats of Gemini, their caps were in some early artworks meant to represent the remains of the egg from which they hatched (resulting from the rape of Leda by Jupiter disguised as swan).[16] Castor and Pollux stand focusing on each other, holding attributes, one a club and the other a lyre; the brothers appear with shoulder capes, caps and sandals and nothing in-between. This is a singular depiction of Gemini; the crosses were rarely copied. The illustrations of the warriors Castor and Pollux through the Middle Ages have become intermingled with images of another pair of mythological twins Amphion, a singer and musician, and Zethus, a hunter, sons of Zeus and Antiope in Greek mythology. Their depiction in the Leiden has substituted the iconography of Castor, usually a bow and arrow, with the golden lyre given to Amphion by Zeus; his brother Pollux holds a heavy club rather than a warrior's spear; this switch in iconography is discussed further in Appendix C (Figs. 7.4, 7.5, 7.6 and 7.7).

Following the prototypes of earlier illuminators, Carolingian artists portrayed human bodies and animal figures in a realistic, three-dimensional manner including foreshortening and modeling with white highlights, with heroes often in classical undress. In comparison, the royal family of Ethiopia immortalized in the sky, King Cepheus, Queen Cassiopeia, and their ill-fated daughter Andromeda are depicted fully clothed, but seem to be stiffly frozen within their frames; all three usually have their limbs extended rigidly; fixed forever in the night sky. ("*Behind the Bear Cynosura Cepheus himself is like a man stretching out both his arms*" Kidd, 181).[17] As explained by the

---

[16]The iconography of Gemini and the other constellations is discussed in more detail in Appendix B.

[17]The quotes in italics are from the *Phaenomena* of Aratus, Kidd'.

Fig. 7.2 Five Planets fol.80v, Leiden Universiteitsbibliotheek MS Voss. lat. Q.79, Courtesy of Leiden University Library

accompanying text in the *Phaenomena*, the miniature represents Cepheus, the unhappy Ethiopian king of Greek myth who stands clad in Phrygian garb, a short tunic, with red leggings, laced sandals, and a wind-blown cape of royal purple. His Phrygian hat covers most of his gray hair; he wears a short beard. His wife Queen Cassiopeia sits on her throne in a pink and white gown, feet on a cushion, but because of her vanity, she is destined to hang upside down for half of the year, holding a bright star in each outstretched hand, but manages to cling to her seat. Their daughter Andromeda, her arms chained to the rocks as a sacrificial victim, appears mostly naked, her minimal wrap covers only her legs; her feet are bare as she awaits by the sea the arrival of the fearsome sea monster. ("*The two feet of Andromeda will be pointers to her suitor Perseus*" Kidd, 248.)

Fig. 7.3 Gemini fol. 16v, Leiden Universiteitsbibliotheek MS Voss. lat. Q.79, Courtesy of Leiden University Library

Helmeted Perseus, grasping the snaky head of Medusa in the left hand and a long spear in the right, flies down to rescue poor Andromeda; his figure fills the entire space and his head twists almost unnaturally to the left. Clasped at his neck, his fluttering red cape conceals only his shoulder leaving the rest of his body naked; including his feet but for the wispy wings at his ankles, presented to him by Mercury and protected by his magical spear given to him by Minerva. His magical cap was given to him by Atlas which rendered him invisible. (*His right hand is stretched out towards his bride's mother's chair-seat, and as if on some immediate pursuit he takes long strides*, Kidd 250–51.)

Fig. 7.4 Cepheus fol. 28v, Leiden Universiteitsbibliotheek MS Voss. lat. Q.79, Courtesy of Leiden University Library

The gods honored Cepheus, as they had his wife, Cassiopeia and his daughter Andromeda by placing him among the stars after their deaths.[18] As in most Leiden illustrations, the stars of the constellation are indicated by small golden dots; the blue ground of the sky is enclosed by a simple red frame. Weitzmann states:

[18]The royal family has been recognized by Aratus with a full description on verses 179–264 (Kidd); for a discussion of the iconography, myth and the variations of their illuminations see Appendix C.

Fig. 7.5 Cassiopeia fol. 30v, Leiden Universiteitsbibliotheek MS Voss. lat. Q.79, Courtesy of Leiden University Library

The semi-nude Andromeda in a relaxed pose derives from a very good classical model which seems to have been copied so faithfully that the style of the model still shines through. Not only is a frame added, but this frame cuts through the rocks, thus giving the illusion that the rocky landscape continues beyond the limitations of the picture frame. The Leiden miniature gives the impression of a spatial landscape as if seen through a window, the blue ground adds to this illusion. The addition of a landscape setting was one of the effective means of raising miniature painting to a higher artistic level, Weitzmann (1971: 97–98).

Fig. 7.6 Andromeda fol. 32v, Leiden Universiteitsbibliotheek MS Voss. lat. Q.79, Courtesy of Leiden University Library

As mentioned in Chap. 2, there are two different conventions for representing the figures of the constellations, both perspectives are found in the miniatures of the Leiden. The older convention of drawing human figures from the back occurs often, while some figures face forward. Of the 39 extant constellation miniatures, eleven are clearly derived from a celestial globe, with a God-like orientation; the majority are oriented as if viewed in the sky from Earth. Ophiuchus, Perseus, and Orion show us their back side only the Gemini twins show frontal nudity, while Aquarius, Cepheus, and Eridanus are drawn as if they are fixed on the crystalline sphere

Fig. 7.7 Perseus fol. 40v Leiden Universiteitsbibliotheek MS Voss. lat. Q.79, Courtesy of Leiden University Library

looking down on the observer. Numerous characteristics of the illuminations present evidence of an earlier prototype probably several; the Late Antique prototype may have even been a cluster of separate images on one folio.

The constellations of Ophiuchus and Hercules too stand out from the royal blue ground retaining their usual iconic postures; both are pictured as heroic super-figures in their prime in the Leiden. Ophiuchus, also called Effaces or Serpentarius in antiquity, stands completely naked viewed from the back, while Hercules faces forward wearing a shoulder-baring short pinkish tunic, the iconic lion skin over his left arm, a weapon in his right. Hercules, also known as Herakles, the Kneeler, or Engonasin, bends his left knee as if starting to kneel. (*No one is able to say definitely what it is or on what task the man is intent, but they just call him the man on his knees, Kidd 63–63.*)

Ophiuchus is part of a large, complex assembly of stars which is actually two different but intertwined constellations; a male figure and a threatening snake. Ophiuchus stands totally naked, huge and upright atop Scorpio, grasping the writhing body of the lengthy serpent with both hands, which is most often depicted as if wrapped around his waist, but not in the Leiden.

Scorpio has been painted a pale green and dotted with stars; its claws form a circle and its tail is raised as if to strike. (*Both hands struggle to hold the Serpent that writhes round Effaces' waist. He constantly, with a good firm stance, tramples with both his feet the great monster, Kidd 82-4.*) (Figs. 7.8 and 7.9).

Toward the end of the text, the remaining miniatures include a personification of the five planets, one of the four seasons, and the intriguing planetary diagram that includes classical illustrations from the Calendar of 354, an apsidal diagram, and an accurate positioning of the planets as observed in April of 816.

The Planetary Configuration

At the end of the Leiden *Aratea* on folio 93v, a significant planetary diagram appears, completely out of context with Germanicus' astronomical poem. This diagram is a ninth-century original creation, although it may have have been modeled on a similar type of diagram from Late Antiquity. The cosmic illustration comes from an entirely different tradition than the type of celestial map that often accompanies *Aratea* manuscripts described in Chap. 6. This particular diagram has no direct relationship with the astronomical poem, since it depicts the spheres of the planets, and neither the planets nor the planetary orbits were elaborated on by Aratus or by Germanicus. The diagram was definitely not inserted later, for it is written with the "self-same ink" as the rest of the Leiden (Verkerk 1980: 260). The diagram is compelling in its design which encapsulates remarkable information (Fig. 7.10).

The first element that draws your attention is the placement of the planets. Most medieval diagrams of the planetary system show a centrally placed Earth with the planets properly arranged in perfectly circular orbits enclosing it. But this diagram shows Mercury and Venus in orbit around the Sun, rather than round the Earth. This configuration documents essential information for the history of astronomy since it proves that the partial heliocentric theory of the ancient Greeks was still circulating many centuries before Copernicus' revolutionary work. Among Late Antique and early medieval writers, Martianus Capella was the only one to pass on this theory of partial heliocentric orbits of Mercury and Venus from ancient Greek astronomers. The planetary arrangement provides striking evidence, as scholars in the Middle

Fig. 7.8 Hercules f.6v, Leiden Universiteitsbibliotheek MS Voss. lat. Q.79, Courtesy of Leiden University Library

Ages were not previously thought to be aware of the heliocentric theories of Heraclitus until it was published by Copernicus in the 16th century.[19]

[19]Martianus Capella and Macrobius both quoted the theories of the ancient Greek astronomer, Heraclitus, that because Mercury and Venus are never far from the Sun, that indicated they orbit the Sun, not the Earth. The theory was little noted and this arrangement does not appear in any other medieval diagrams that I have seen, except for the close of copy of MS Leiden, Boulogne-sur-Mer MS 188; the other copy of MS Leiden, Bern 88, does not include the diagram; possibly it was lost.

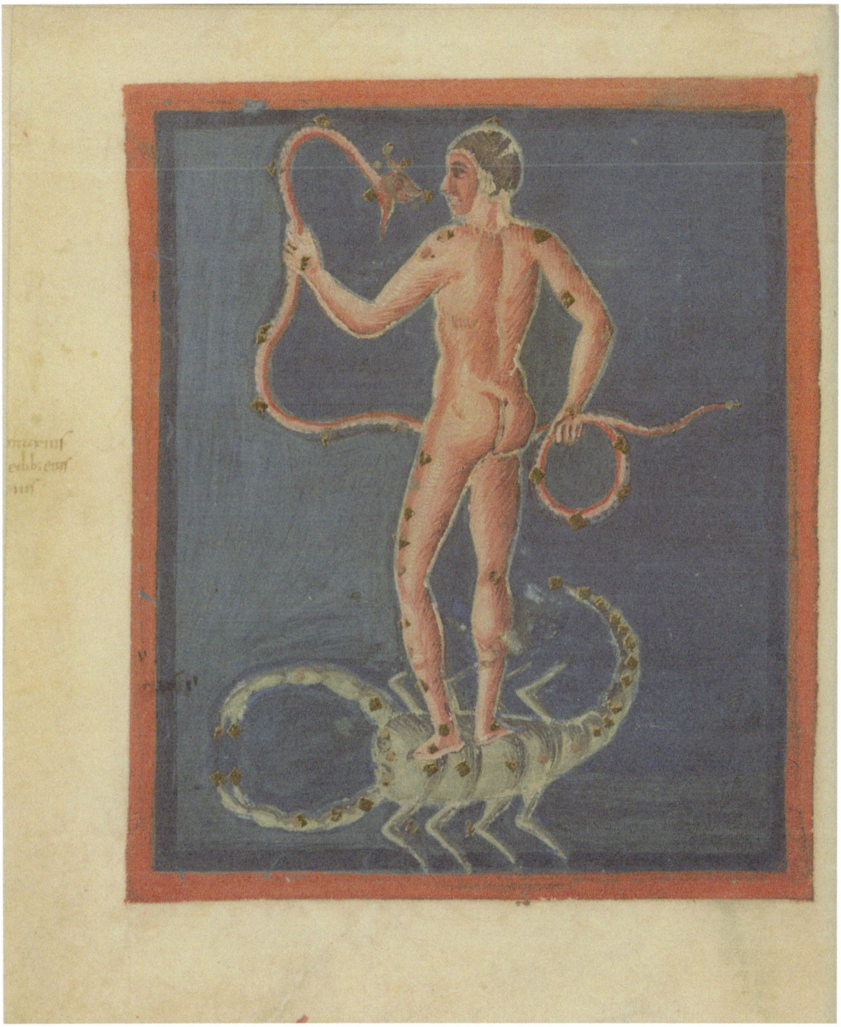

Fig. 7.9 Ophiuchus on Scorpio f.10v, Leiden Universiteitsbibliotheek MS Voss. lat. Q.79, Courtesy of Leiden University Library

The second surprise of the Leiden diagram is that the planets were not randomly placed, but each is purposely aligned to a particular sign of the zodiac. When analyzing the positions of the planets pictured on the diagram, it becomes obvious

Fig. 7.10 Planetary Diagram fol. 93v, Leiden Universiteitsbibliotheek MS Voss. lat. Q.79, Courtesy of Leiden University Library

that the configuration actually represents a map of the sky as it was oriented on one particular night. Historian Bruce Eastwood had recognized the significance of this diagram, and had determined and published the date as March 28, 579.[20] The date

[20]Eastwood (1989) *Aratea, Kommentar zum Aratus des Germanicus, MS Voss. Lat. Q.79* IV 1-40, (1983) "Leiden Planetary Configuration", reprinted in 2002. Eastwood was not the first to publish the heliocentric features of this configuration, he credits Marie-Thérèse d'Alverny, (1975) "Survivances du 'système d'Heraclide' au Moyen Age" *Avant, avec, après Copernic*, Paris: pp. 39–50, for noting the 'peculiarities of its heliocentrism'. She attempts to place it in a tradition

has since been re-examined by R. and M. Mostert and re-fixed at March18, 816 (Mostert and Mostert 1990: 248).[21] My computer-based investigation of the planetary placements had previously established that later date as well. The celestial diagram actually indicates the location of the five known planets on that particular date. In order to verify that the diagram had not been copied from a Roman prototype, I checked the positions of the planets back to 100 BCE; there is no other time when the planets would be even closely aligned to those in the Leiden configuration.

The third surprise found in the cosmic diagram is its technical sophistication. Written in small print around the circular orbit of each planet are Latin quotations derived from Pliny the Elder. These inscriptions state each planet's point of nearest approach (apogee) and furthest point from (perigee) the Earth. The writings also state the exultation of each planet, indicating where it was believed to have the greatest astrological influence. Whoever designed the diagram of the planets was very familiar with astronomical and astrological theories. This type of diagram showing the position of the planets is normally called a planetarium. A planetarium illustrates the motions of the Sun, Moon and planets, or sometimes the planets alone, in their daily or yearly round through the zodiacal signs. Since that is not exactly what is being shown here, Eastwood suggests giving this diagram a more neutral and improved label—a 'planetary configuration'. Many horoscopes survive from antiquity, but this is the earliest known example of a diagram showing the configuration of planets for one particular day. Why did someone in the early ninth-century take such great care to record the position of each planet on this astronomical date and include the diagram in the Leiden manuscript? Who would have had access, permission and the knowledge other than the creator or close contact of the codex? What was the significance of that April date?

The Mostert article offers several suggestions; the most likely is that it was the date that the manuscript was completed or presented to its noble patron. There were no unusual celestial events, such as comets or eclipses at that time, nor any outstanding historical events, births or marriages. I propose a different meaning for that specific date, March 18, was purposely chosen because in the Middle Ages, March 18 was thought to be the date of the creation of the world, a date promoted by Bede in 725. Why did they specify March 18? The only learned sources available to western scholars before the twelfth century were those mentioned many times: Pliny, Macrobius, Martianus Capella, Isidore, Bede, and the Latin commentators on

(Footnote 20 continued)

of heliocentric ideology. Thiele was the first scholar to publish information on the diagram in 1898, in *Antike Himmelsbilder*, p.77 and 90, but he does not mention the heliocentrism. Eastwood corrects the date to March 18, 816 in "The Astronomies of Pliny" (1993) p.173.

[21]They determined that the probability that all the planets were chosen at random, and that they happen to align at this particular date, is the staggering one day in seven million million years.

Plato's *Timaeus*. Their astronomical texts contained systems of thought that were deeply rooted in medieval cosmology, of these only Bede focused on a discussion of time and the calendar. According to his elementary astronomy, logic and reliance on *Genesis*, the stars were created on the fourth day.

> And God said: Let there be lights made in the firmament of the heavens, to divide the day and the night; and let them be for signs and for seasons and for days and years [19]. And the evening and morning were the fourth day. Genesis 1.14,

The Holy Bible, translated from the Latin Vulgate, Douai-Rheims. 1609.

According to Bede, the vernal equinox, March 21, was traditionally considered the start of the celestial year, the beginning of time, but the world was already in operation by then, count back to the fourth day, and March 18 would be the actual date of creation.

> Some have claimed that the first day of the world was the 8th kalends of April [25 March] and others that it was the 12th [21 March]. In both cases they adduce the same argument, namely, the equinox. For it seems reasonable that because God in the beginning divided light and darkness into two equal parts, we should believe that the beginning of the world took place specifically at the point of the equinox. They conduct their inquiry well; /291/but they do not think through their statement consistently. They would act much more knowledgeably were they to assign the equinox not the first day in which light was created, but the fourth, in which the luminaries were made. *De tempore ratione,* trans. Faith Wallis.

Why the year 816 was chosen remains very much open for discussion; possibly that was year the Leiden *Aratea* was produced or when the planetary diagram was created.

This unique diagram holds still more surprises; the tiny illustrations of the planets and signs of the zodiac have been shown to be copied from a fourth-century prototype. The iconography of the planets and signs has been discussed repeatedly in art history literature. In addition to the planets in their spheres, the twelve signs of the zodiac are pictured also in the outermost band of the fixed stars. The Labors of the Months are placed between the signs, each in a small golden medallion. These personifications and those of Terra and the 'seven planets' show a great similarity to early Christian and Roman renderings. A study by Thiele published in 1898 has demonstrated that these tiny figures are based on the famous *Calendar of 354*, discussed in Chap. 2. The lost Carolingian copy of the lost *Calendar of 354* was probably made during those same years that saw the production of the Leiden, the Harley and other *Aratea* manuscripts. The connection of the Leiden diagram with the *Calendar of 354* supplies more evidence of the revival of classical interests and the availability of Late Antique prototypes among the Carolingian manuscript production centers.[22]

[22]This relationship is discussed by Stern, *Le Calendrier de 354*; M. Salzman, *On Roman Times,* and von Euw in "Die Bilder des Leidener Aratus".

The zodiacal signs of the diagram begin with Aquarius in January at the uppermost point (or mid-heaven), but the cycle does not correspond with the Labors of the Month since they proceed in the opposite direction. The diverging directions of the zodiacal and Labor cycles and the stylistic differences among the illustrations in the Leiden point to a variety of iconographic sources. Salzman has also published an extensive study comparing the illustrations in the copies of the *Calendar of 354* with those in the planisphere from the Leiden with its two copies (Salzman 1990: 263-68). She discusses the literature concerning these illuminations beginning with the early study by Thiele, and then states her reservations concerning the positive relationship between the Leiden images and the *Calendar of 354*. Salzman affirms the figures in Leiden diagram "should not be considered a completely trustworthy reproduction of the illustrations of the months in the original Calendar unless corroborated by other manuscript or archaeological evidence" (Salzman 1990: 268).

An exact copy of the planetary configuration in the Leiden was created for its late tenth-century copy, Boulogne-Sur-Mer BM MS188, but four smaller diagrams have been added, one in each corner. The printed Latin copy edited by Grotius, *Syntagma Arateorum*, and its copy published by Cellarius for *Harmonia macrocosmica* have both included the planetary configuration. A slightly different celestial diagram appears in the other eleventh-century copy of the Leiden, Bern, Burgerbibliothek MS 88, folio10v.[23]

This manuscript provides important evidence relating to the content and appearance of a lavishly illustrated fourth-century book. Therefore it is not difficult to conclude that the Leiden is a faithful copy of one or probably several Late Antique Germanicus manuscripts. There are four close copies of the Leiden: two manuscripts and two printed versions. The two illustrated manuscripts were copied at the monastery of Saint-Bertin in Saint-Omer and are discussed in Appendix B. Concerning the two printed editions, the first one, as mentioned previously, was produced in 1600 by Hugo Grotius for his *Syntagma Arateorum;* the other is a copy of Grotius' version published in 1661 by Andreas Cellarius, a Dutch-German cartographer, who included his copy in his star atlas *Harmonia Macrocosmica.* The Leiden *Aratea* of Germanicus has proven to be the most investigated and the most published of all the numerous *Aratea* manuscripts made during the ninth century. Its artwork and text has received much more notice than the innovative Harley Cicero that will be discussed next.

[23]More detailed information on the Leiden and its copies can be found in Appendix B. Literature on these detail: Meyer Schapiro, "The Carolingian Copy of the *Calendar of 354*"; Eastwood (1983) 34; Nordenfalk (1938) 23–30. (Complete citations are listed fully in the bibliography.).

Second Ninth-Century *Aratea* Manuscript

A room without books is like a body without a soul. Marcus Tullius Cicero.

Marcus Tullius Cicero is more famously known for his impressive output of philosophical literature than for his astronomical poetry; those writings were much more popular in his own time. His usage of philosophical terms furthered the development of Latin vocabulary considerably within that discipline, and contributed important advancements in philosophical studies. But during the Middle Ages, his astronomical poem, the translation of the *Phaenomena* by Aratus written c. 80 BCE when only in his teens, circulated widely, for the heavenly constellations were made visual with the addition of startlingly beautiful illuminations by Carolingian artists. Similar to the *Aratea* of Germanicus, the earliest illustrations of Cicero's *Aratea* were the finest and never quite equaled.

British Library Harley MS 647

London British Library Harley MS 647 (hereafter the Harley) contains the oldest and largest surviving portion of Cicero's translation of the *Phaenomena* of Aratus, dated between 820 and 850. Although the Harley manuscript is larger in its dimensions than the Leiden, it is smaller in volume; the extant manuscript is only fragmentary and the text incomplete; as a result this manuscript has many fewer pages and illustrations, only 21 folios (two-sided) holding 24 celestial miniatures, compared to the Leiden which includes 99 folios and 35 miniatures.[24] Folio one was added slightly later (Köehler and Mütherich 1971: 18). The pictorial and textual appearance of the Harley's constellation images is altogether different than the format of the constellations in the Leiden. The variety of illustrations found in these two manuscripts, and the variations seen in other ninth-century astronomical anthologies, reflect the richness of Late Antique exemplars available to Carolingian artists. (The attraction of the Harley paintings was probably too enticing for a Carolingian prince or princess to resist cutting the images from the manuscript for their own pleasure; perhaps that also occurred with many *Aratea* manuscripts through the centuries, but much less likely when confined within a carefully supervised monastic library.) Whether modeled from scrolls, celestial globes, mosaics, calendars or artworks, the plentiful models for astronomical illustrations provide an indication of how imagery of the constellations and signs of the zodiac can "hint at a complex ancient tradition in addition to that of the texts" (McKitterick 2000: 21). Few of the later *Aratea* manuscripts match the caliber and artistry of those produced during the Carolingian era.

[24]The complete list and content of miniatures contained in the Harley can be found in Appendix B.

History of the Manuscript

The Harley is basically an abbreviated picture book with two different sets of text included; the design of both the text and images exhibits a purely classical character; both were most probably derived from fourth-century sources. Like the Leiden, the Harley was produced in Lothringia (now Lorraine), probably at Reims, between 820 and 850 for another or possibly the same royal or religious patron. The Harley is thought to have been at the monastery in Fleury until the early tenth century; later the codex was transported from France to England, possibly as early as the second half of the tenth century (Saxl 1957a, b: 105).[25] By that time, Anglo-Saxon interest in nature and natural science had gained support among monastic learning centers. The wide-spread interest in the writings of Bede had inspired greater interest in raising the level of education in both religious and secular areas.

Exactly how the Harley arrived in England remains a matter of speculation. Both religious and secular manuscripts were often transferred to other production centers specifically for the purpose of making copies. Through such exchanges many Carolingian books were carried to English centers such as Canterbury and Ramsey from French monasteries, particularly from Fleury and St. Bertin. A large number of manuscripts reached English libraries through monastic educational reforms associated with Flanders and Cluny in particular. England had close contacts with Europe in the ninth and tenth centuries, not only between monastic centers, but also among the royal courts. For example, the daughter of Charles the Bald, named Judith after her grandmother, was married to Aethelwulf, King of Wessex in 856, which may have been another potential connection for the transfer of the famous picture book.

Provenance

The provenance of the Harley has been fairly well established by art historians. After being initially produced in northern France during the Carolingian era, the Harley was most probably transferred to St. Augustine Benedictine Monastery, Canterbury, in the late tenth century, as the Carolingian codex was copied there several times. (Ker 1964) The first two folios of the manuscript were added there, as well as several other sections. The Harley remained at St. Augustine at least until the fifteenth century for it was included in the library catalog at that time (Barker-Benfield 2008). After that time, the Harley became the possession of Francis Babington, a college head who died in 1569; the change in ownership of the manuscript was most probably a result of the Dissolution of the Monasteries, which

[25]As proof that the Harley was in England at that time, Saxl reports that the first page of the Harley was apparently lost, and was replaced by an English hand in the second half of the tenth century.

took place from 1536–41 when many monastic libraries filled with irreplaceable illuminated manuscripts were destroyed. The Cicero *Aratea* became part of the Harley Collection, formed by Robert Harley (d.1724) and Edward Harley (d. 1741), the first and second earl of Oxford, prominent book collectors and patrons of the arts. The library was inherited by Edward's widow, and then by their daughter Margaret Cavendish Holles (d.1785), who sold the Harley collection to the nation for a minimal amount. The enormous Harley collection included more than 2200 illuminated manuscripts created throughout the Middle Ages and early Renaissance; the collection was initially housed at the British Museum which had recently opened in 1759. The 7660 Harley manuscripts, 14,000 legal documents and over 500 rolls became one of the foundation collections of the British Library.

Harley Text

Even though the Harley contains the largest segments of Cicero's Latin translation of the *Phaenomena*, the poem still lacks significant portions. Because Cicero's text was fragmentary even in the ninth century, and contained few details about the stars or their myths, the surviving poem was augmented with scholia by the Carolingian producer, mainly derived from a manuscript of Hyginus' *De astronomia*. The additional mythological and astronomical texts contain portions of the legendary tales along with more specific information about the numbers and positions of the stars in each constellation. Sections of the astronomical text were obviously taken from the seven-book edition of the Carolingian Compilation of 810, called by Charlemagne (McGurk 1966: xvi–xix). In addition to the lines of text from Cicero and Hyginus, the Harley also contains excerpts from Pliny the Elder (Reeve 1983: 23).[26] As a result, the Harley has two separate sets of explanatory text associated with each constellation image; the scholia are carefully written within the contours of the constellation figures. One large constellation figure appears on each folio and the appropriate verse of Cicero's *Aratea* is written at the bottom of the page. The scholia appearing within the body of each constellation is written in the old-fashioned rustic capitals or majuscule in red and sepia ink; the verse is centered at the bottom is written in Caroline minuscule in the typical verse format; the first letter of each line is enlarged and set apart, written in dark brown ink, the remainder of each line is written with red ink.[27] Below the verse, the name of the constellation appears in capitals (Figs. 7.11 and 7.12).

Even the lines of Hyginus scholia gently curve to bring a more natural appearance to the human and animal figures, seen in the depiction of Cygnus the Swan, which is

[26]The excerpts from *Natural History*, Chap. 18 appear on ff.16r and 16v; and more excerpts from Pliny's Chap. 2 *Natural History*, Chap. 18 on 18v-20r.

[27]See British Library website prepared by Kristen Lippincott for more details, a black and white image of each page, including the text on each folio and listing the missing constellations—kristenlippincott.com. The BL Catalogue of Illuminated Manuscripts displays each folio in color.

Fig. 7.11 Cygnus f. 5v, BL MS Harley 647, Courtesy of the British Library

especially skillful with its long slender neck and wings tipped in black and the first sign of the zodiac Aries which balances on its two hind legs, both are naturalistic with accurate details. Reaching back to the Classical Era where the style was first created by Greek scribes, this artistic technique of manipulating the tightly written verses to form the body of the figure they are describing is known as *technopaignion* (art game). Its origins can be traced to early Greek patterned poetry called *calligrams* that represented physical objects (Grabar and Nordenfalk 1957: 93).[28] Nordenfalk notes that Constantine's court poet, Publilius Optatianus Porfyrius, spoke of ornate editions of his *carmina figurata* which had been painted with contour lines, and he then wrote the text *picto limite dicta notans*, fitting the written words within outlines indicated by the brush (Nordenfalk 1998: 10). The extremities of the constellation miniatures are naturalistic.

[28]The work pictures were invented by Porfyrius for the benefit of Constantine the Great. The poems were composed so that the verses outlined a shape appropriate to the content.

Fig. 7.12 Aries f. 2v, BL MS Harley 647, Courtesy of the British Library

"The painted parts of the images look back to the illusionistic style of Pompeii through unknown intermediaries (Dodwell 1993: 48)."

The scholia derived from Hyginus, comprising the myths and descriptions of each star group are written in rustic capitals, composing the body of each illumination. The Harley shows various corrections on almost every line of the text, which implies comparisons with another manuscript, and it has been suggested that these were made by Lupus of Ferrières, although this assignment has been questioned.[29] However Mütherich claims it was not Lupus who wrote the corrections and additions in the Harley, as they were written in a "northern French hand of the second quarter of the ninth century". (Mütherich 2004a, b: 107).[30] The text was

[29]This connection was discussed in Chap. 2. Lupus of Ferrières wrote to Prüm in 847 asking to borrow a copy of their 'Tullius in Arato' to complete his own exemplar. The closeness of Prüm to Aachen suggests a relation between the manuscript to which Lupus refers and the Harley Cicero. Cicero's *Aratea* was mentioned in letters of Carolingian humanists as "Tullius in Arato" (Mütherich and Gaehde 1976:12). But that does not eliminate Lupus, as he lived in northern France at that time.

[30]Mütherich, "Book Illumination in the Court of Louis the Pious".

corrected and repaired by a scribe who signed with his name Geruvigus; he was not the scribe of the entire manuscript.[31]

The Harley contains sections of Anglo-Saxon script, later additions to the signs of Scorpio, Sagittarius, Aquarius and Pisces. The manuscript also includes accompanying astronomical texts, some based on Isidore of Seville, and excerpts from Pliny's *Natural History*, Macrobius' *Commentary on the Somnium Scorpiois* and Martianus Capella's *De nuptiis Philologiae et Mercurii*, books six and eight. On folios 1r and 2r, two prayers have been added.

The exact *stemma* of the many surviving *Aratea* manuscripts of Cicero, along with the precise text sources, has been worked out by Soubiran (1973). The Harley was copied at least five times, for that many still survive, confirming the influence of the Harley exemplar; the five copies were certainly created in English scriptoria or royal courts.[32] The British Library holds a total of three copies of Cicero's *Aratea* in addition to Harley 647, including another Harley manuscript MS 2506., along with Cotton MS Tiberius CI and Cotton MS Tiberius B V. The *technopaignion*device of the ancient Greeks became the prototype for later copies and was repeated several times throughout the Middle Age. The textual material of these relatively brief *Aratea* manuscripts has been studied thoroughly by paleographers.

Style, Placement, and Iconography

Like the Leiden and many other Carolingian illuminations, the miniatures of the Harley are also illusionistic, looking back to their Late Antique prototype. According to Saxl, the illustrator of this manuscript "comes nearer to classicism" than does the artist of the Leiden Germanicus (Saxl 1957a, b: 101).[33] He also wrote that the illuminations are closer in spirit to the Pompeian frescoes than anything else made in the West in medieval times (Panofsky and Saxl 1933: 236–7). The Carolingian artists and scribes accepted classical artwork and copied the images as they saw them without changing or rejecting their specific characteristics or iconography from five hundred years earlier.

Because the initial sections of Cicero's *Aratea* were already lost in the ninth century, the first constellation mentioned in Cicero manuscripts is Aries, which is

[31]He wrote the following: "*Ista proprio sudore nomina uno quoque propria. Ego indignus sacerdos er monachus nomine Geruvigus repperi ac acripsi. Pax legentibus*" (f .21v).

[32]The five copies of the Harley are: Cotton Tib. BV; Cotton C 1; Harley 2506; Bodley 614 and Digby 83.

[33]Saxl in "Illuminated Science Manuscripts" goes on to say "this fruit of late antique civilization is a little too fanciful; for if someone was interested in the text he can hardly have enjoyed reading it in this form. The English scribes and draughtsman were therefore sensible enough to drop this mannerism fairly soon; they copied the figures and wrote the commentary in the blank spaces". This is also discussed in Saxl "Belief in Stars", *Lectures,* p.88.

about one-third through the original poem of Aratus. Consequently, the constellation images also begin with the sign of Aries in the Harley rather than Draco, Ursa Major and Minor, which are the usual starting point for constellation cycles. Unlike the Leiden images, the Harley miniatures have neither frames to enclose them, nor bold background colors to embellish them; as a result, the figures seem unanchored, as if free-floating in space. Each image fills about three-quarters of the folio and has a recognizable human or animal form, but only the heads, hands and feet are drawn in accurate detail. The extremities of each constellation are delicately modeled with softly muted colors, sometimes accented in black. Hyginus' text within the figures signifies the exact number of stars found in each constellation; but only rarely does the number match those stars pictured. The markers of the stars are brightly painted with small red circles, but not accurately positioned. Similar to the Leiden, the stars are just randomly placed, but the stars are almost always placed outside the figure in the Harley, so that they do not interfere with the interior text; only Aries and Cygnus have one star placed within the text (Figs. 7.13, 7.14 and 7.15).

Fig. 7.13 Five Planets f. 13v BL MS Harley 647, British Library

Fig. 7.14 Pleiades f. 42v Leiden Universiteitsbibliotheek MS Voss. lat. Q.79

The Harley, just as in the Leiden, includes an illustration of the five known planets and both are treated with minimal representations, showing all five planets merely as small heads in a composite grouping. The planetary gods are not entirely ignored, but their importance has been reduced to a small-scale arrangement. Each planet is encased by diamond-shaped text. Saturn appears older with greyish hair and beard; Jupiter is surrounded with solar rays; Mars wears his iconic helmet; Venus can be identified by her earrings and ribboned hair; lastly Mercury has tiny wings that seem to sprout from his forehead. One of the male heads of the Leiden seems to be tonsured, or depicting Saturn as partially bald; one is encircled with rays of light, which usually signifies Jupiter, the king of the gods. The illustration of

Fig. 7.15 Perseus f. 4 BL MS Harley 647, British Library

the Pleiades in the Leiden has been treated in a like manner, depicting seven small female heads arranged in a tight grouping, but actually only six stars can be seen. Each feminine head wears similar hair treatments, a sophisticated Late Antique style, but with different shading. The sister in the center of the miniature has her head covered with a veil, perhaps indicating that this star is not currently visible. Four of the sisters wear drop earrings, and each displays comparable facial expressions; wispy clouds support each head. Perseus looks away from the frightful head the head of Medusa dripping with streams of blood; his text-filled body lacks clothes, but tiny wings sprout from both his head and ankles.

MS Harley, f.19, pictures a planetary diagram depicting the orbits of the 'seven planets' but is very simplistic, not as detailed as the Leiden planetary configuration, and f. 21 is blank. The last image in the Harley, f.21v, pictures a celestial diagram of the northern and southern hemispheres in red and brown ink, but is indistinct. The celestial map displays all the constellation images with no indication of star placement; this folio is signed with the name of Geruvigus, who does not seem to be the scribe of the entire manuscript. The Harley Cicero and the Leiden

Germanicus both exist as a testament to the talent and creativity of Carolingian scribes and artists and the devotion to Classical antiquity of the patrons.

The painting produced on both sides of the Channel in the last decade of the eleventh century was Anglo-Norman, rather than English and Norman, though small pockets of an almost pure Anglo-Saxon style did survive in England. Moreover England showed the greater stamina, going forward during the first half of the twelfth century to produce some of the greatest manuscript painting in Europe (Dodwell 1993: 196). The Harley has been examined by Dodwell in relation to the role of astronomical manuscripts in the development of Romanesque art in England after the Conquest.[34] He cites the illustrative cycle of the *Aratea* manuscripts as the most popular source of mythological figures from which medieval artists drew inspiration. Credit is given to Cicero's translation as the primary channel from "both the literacy and artistic point of view." This seems to be true in England, since the influential Harley and its copies were reproduced there numerous times; but actually throughout continental Europe, many more manuscripts of Germanicus' translation, with its derivative textual forms, survive today. But the reverse is true in England where six manuscripts of Cicero's *Aratea* are extant, compared to just three of Germanicus, all later arrivals.

During the late tenth and early eleventh centuries, when Ottonian art in Germany was fostered by emperors and powerful princes of the Church, the patron of the most important center of painting in France was not a royal benefactor but Abbe Odbert, head of a small Flemish community, who was himself one of the chief artists illustrating manuscripts in his scriptorium. Odbert of Saint-Bertin, the Benedictine monastery near the north-east coast of France, was in office between 990 and 1007 and we know from inscriptions and paleographic evidence had a number of manuscripts made for his abbey. Text scholars have identified script imitation in certain manuscripts while Odbert served as abbot; some manuscripts, produced by his hand or under his supervision, have archaizing qualities that are so precise they had been inaccurately dated to the ninth century rather than the early eleventh (Wilmart 1924: 169–88).[35]

Odbert's style owed a great debt to Carolingian art, but only gained significance when synthesized by artistic trends from England. When Carolingian influences alone were in effect, the monks merely produced reproductions of the past as is clear from Odbert's illustrations for two copies of the *Aratea* of Germanicus, the finer of which is at Boulogne; the other, whose illustrations are drabber in color and harsher in line, is held at Bern (Bergerbibliothek MS 88). Odbert's model still survives to prove how faithful his copies were; and, as the Carolingian pictures themselves were accurate copies of late Antique illustrations. "It is a little ironic to

[34]*The Canterbury School of Illumination*, 1066-1200, (1954).

[35]Mentioned by J. Crick from 'Script and the Sense of the Past in Anglo-Saxon Traces, ed. J. Roberts and L. Webster, *Medieval and Renaissance Texts and Studies* 405 (Tempe, AZ, 2011) pp. 1–29. Odbert himself was a scribe and artist who decorated six major manuscripts and eight lesser texts; the most important was the hefty St. Bertin Psalter (Boulogne, Bib. Mun., MS. 23) dated 999. Its inscription named Odbert as the illuminator. The Psalter contains three full-page illustrations and 180 decorated initials, 33 of them with colored pen outline drawings of the *Life of Christ*. (Oxford Index).

think of the Benedictine abbot, at some time near the apocalyptic year 1000, painting the virile classical gods and voluptuous goddesses of the astronomical illustrations" (Dodwell 1971: 197).

After the collapse of the Carolingian empire, its Frankish territories fell apart and deteriorated into a patchwork of small and hostile provinces that owed grudging recognition to a series of unfortunate kings. France, as we know it, did not yet exist; in the twelfth century '*regnum Franciae*' usually meant no more than the Ile de France (Dodwell 1993: 191). Another Harley Collection manuscript of Cicero's *Aratea* (Harley 2506), dating to the turn of the eleventh century, is also held at the British Library. The Anglo-Saxon artist who illustrated this copy of Harley 647 visited the Loire. This unidentified artist was influenced by the Carolingian version of the earlier fourth-century style for he continued with the original classical depiction of Perseus, the Centaur, Eridanus and the other constellation figures but ignored the word-picture style, bringing his own talents to create a work of great sensitivity (Dodwell 1993: 116).

The diversity of types and styles of constellation images and textual material seen in the many ninth-century *Aratea* manuscripts indicate that there was a variety of source material for astronomical imagery available from the various astronomical traditions. Unfortunately, after the earlier exemplars were no longer needed, they must have been disposed of by the scriptoria for none survive. Astronomical texts seem to have invited the colorful variety of illustrations and diagrams in Late Antiquity and possibly even earlier, but no proof survives of exactly when the illustrations were actually ascribed to the verses of the *Aratea* of Cicero and Germanicus.

Twelfth-Century *Aratea* Manuscripts

The second 'renaissance' of the medieval period occurred in the twelfth century, a time that was crucial for the advancement of astronomical knowledge in the West as was presented in Chap. 4. During this century, Latin scholars flocked to former Arabic centers of learning to translate the writings and interpret the ancient knowledge of the Greek and Roman scientists held in the libraries of the Muslim-controlled ancient cities. The most important repositories of these scientific and astronomical treatises in the West were in the reclaimed areas of Spain and Sicily, territories formerly dominated by Arab societies.

The large quantity of astronomical manuscripts, mostly translations, produced in the twelfth century, reflects the renewed interest in classical treatises and in more advances scientific and astronomical knowledge. As was demonstrated, this interest was spurred not only by the discovery of the enormous quantity and variety of ancient texts held in Arabic libraries, but also by the progress and applications achieved by Arabic astronomers in their scientific studies. The illustrated *Aratea* manuscript tradition continued through this time of growing astronomical interest and expansion of available scientific writings. Although copies of Arabic astronomical manuscripts, especially those of al-Sufi, were fastidious in the placement, number and magnitude of the stars; but many treatises and the surviving *Aratea* illustrations still lacked any hint of accuracy when indicating individual stars. In contrast to the European lack of scientific accuracy, the artists creating the constellations in Arabic manuscripts oriented the stars within the figures so precisely that the constellation could be easily plotted on a celestial globe.

> Before Saxl, Georg Thiele had already concluded that in this tradition, the accurate rendering of the position of the stars in the constellation was made subordinate to the mythological narratives attached to the constellation images. Saxl showed that this development was progressive. By the twelfth century, manuscripts were made such as Codex 12,600 in the Österreichische Nationalbibliothek. The fanciful illustrator of this book drew the constellations according to his whim. For lack of space, he even reduced quite a number of them to half-figures. It is evident that these illustrations have lost any relation to celestial topography (Duits 2011: 12).

The *Aratea* manuscripts are not conspicuous by their large numbers but by the fact that so few are extant; only four *Aratea* manuscripts survive from the twelfth century and three are from the early part of the century. That minimal interest may be an accident of survival; or more probably, the influx of 'new' astronomical writings deflected interest from the well-known and elementary works to the more enticing and challenging astronomical and astrological material recently recovered, especially the astrological and scientific writings which had entered the Anglo-Saxon world.[36] The rather simplistic *Aratea* manuscripts were already available in abundant numbers and descriptions of the constellations seemed rather lackluster in comparison; consequently new production of *Aratea* manuscripts took a back seat to the 'lost' heritage of the Classical world and the arcane science of astrology recovered from Muslim authors that readers were anxious to discover. Interest in copying the writings of Adelard of Bath and the new and stimulating manuscripts of Michael Scot with his inclusion of more exotic illustrations and astrological lore was much stronger than the interest in the tired and repetitious *Aratea* manuscripts. In continental Europe, only one *Aratea* manuscript survives from the twelfth century, a Germanicus now housed in Madrid, it is doubtful that many were even produced.

By the tenth century, classical representations of ancient Greco-Roman world had been imported to England, by way of constellations paintings in *Aratea* manuscripts. Most copies of Cicero's poem were derived from Harley 647 which arrived in the Canterbury in the tenth century and the extended version which includes additions to the *Aratea* text and additional writings by other authors, Cotton Tiberius B V. Eleven manuscripts of Cicero's *Aratea* are known by editors, but seven of them can almost be eliminated as only fragmentary (Reynolds 1983:

---

[36]Saxl ("Illuminated Science Manuscripts") 110 "The illuminated astronomical manuscripts of the tenth, eleventh and twelfth centuries, do not record any scientific discoveries by English monks. Their work was based entirely on the classical heritage. English manuscripts differ from the Continental ones of the same period in one essential point. The English manuscripts except for two exceptions are based on one classical original".

18). Many types of miniatures were copied in English monasteries until twelfth century with various alterations, innovations, and creative variations. In England, the twelfth century was noted as the greatest period for book production; their decorated codices were large in size, and high in quality with accurate, uncluttered scripts, although not all were illuminated (Kauffmann 1975: 16). Three of the four surviving illustrated *Aratea* manuscripts produced in the twelfth century were created in England: BL Cotton Tiberius C I (dated 1122); and two manuscripts now in the Bodleian Library: MS 614 (first half 12th C.) and Digby MS 83 (mid 12th C.), both of these are close copies of Cotton Tiberius B V. All three English manuscripts are part of the first family of Cicero's *Aratea* and are fully illustrated. The fourth *Aratea* dated to the twelfth century is a fully illustrated Germanicus manuscript made in Italy at the monastery of Montecassino, but is now housed in Madrid, Biblioteca Nacionale MS 19.

First Twelfth-Century *ARATEA* Manuscript Cotton Tiberius C I

Cotton Tiberius C I was chosen as the first manuscript for particular discussion, as it retains the *technopaignion* technique, forming the constellation figures with astronomical text, which was inspired by its exemplar, the ninth-century MS Harley of Cicero. This is the only English representative with that distinctive type of illustrations that survives from the twelfth century; in fact there is just one other *Aratea* of Cicero, (Göttweig, Stiftsbibliothek Codex 7 Bl. 20) produced during the renewal of astronomical interest in the fifteenth century that has adopted this antique convention.[37] The peculiar library title of the Cicero manuscript—Cotton Tiberius—was retained from the designation established by the previous owner Sir Robert Cotton who catalogued his books and manuscripts according to the book-cases in his estate. Each bookcase held the bust of one of the Twelve Caesars and two Imperial women from classical times on top of the many book shelves, fourteen in total, and that name was attached to the books, every manuscript in his vast library was identified by bust, shelf number and placement order starting from left to right in Roman numerals. The British Library maintained the idiosyncratic system of labeling devised by Cotton; thus the identifying label of the Cicero *Aratea* manuscript remains Cotton Tiberius C I, but the manuscripts are no longer shelved in that particular manner. The Cotton collection became a substantial foundation for the large manuscript assembly now held by the British Library; Cotton's collection was not transferred in its totality since some manuscripts were destroyed or badly damaged while in storage at Asburnham House in Westminster during a 1731 fire. The damaged codices were restored in the nineteenth century in the British Museum and made available for study. The British Library now holds

[37]Cotton Tiberius C. I. and its illuminations are fully listed in Appendix B.

1400 medieval manuscripts and 1500 charters, rolls, and seals dating from the fourth to the sixteenth centuries.

History of the Manuscript

Cotton Tiberius MS C I (hereafter Tiberius C I) is a close copy of Harley 647; the scientific codex probably originated at Peterborough, England in 1122. Almost nothing is known of its initial whereabouts until the late seventeenth century when Tiberius C I came into the library of the great book collector, Sir Robert Cotton, who had another copy of the *Aratea* of Cicero in his library as well, Tiberius MS B V. His collection passed to his heirs, son Sir Thomas Cotton and grandson Sir John Cotton, who both added to the collection. Sir John Cotton arranged to donate their manuscript library to the nation upon his death.

Sir Robert Cotton sometimes bound together unrelated manuscripts. This manuscript consists of three different parts; the first part 1a was written in Peterborough around 1122, which is indicated by the annals in Harley 3667 which refer to events that occurred at Peterborough at that time, such as death dates of the abbots, and it includes a pontifical (services conducted by a bishop). The second part 1b was also written in Peterborough in the 1120 s as it appears similar in script and decoration (Ker 1938: 132). The second section contains a copy of the Latin translation of Cicero. While the last section was produced in Germany, most probably it was created earlier in the middle of the eleventh century (Ker 1938: 263). It is thought that Sir Robert Cotton combined the unrelated parts and had them bound together into one manuscript.[38]

Text

The twelfth century saw the last flowering of Cicero's *Aratea* in England, as the popularity of the long manuscript tradition modeled on MS Harley seems to have died out by the end of that century. Only one illustrated *Aratea* manuscript containing the Cicero text is extant from the thirteenth century but survives incomplete; this is Göttweig, Stiftbibliothek MS 146. No *Aratea* of Cicero is extant from the fourteenth century. In the fifteenth century there was a slight flurry of renewed interest, but of the 27 extant Aratean manuscripts of this century only five include any part of Cicero's text. From these five, four merely have a few fragments of

[38]More information on the entire manuscript can be found online at https://www.le.ac.uk/english/em1060to1220/mss/EM.BL.Tibe.C.i.htm.

Ciceronian poetry supplementing Germanicus' *Aratea*. But surprisingly the fifth quattrocento Cicero manuscript, Göttweig Stiftsbibliothek MS 7 Bl.20, was produced at Koster Göttweig and remains in the same location; this manuscript was modeled on Cicero's Harley 647. The Göttweig manuscript revived the format of the Carolingian exemplar in its illuminations, imitating the presentation of the scholia of Hyginus written inside the bodies of the constellations.

Tiberius C I contains an anthology of various astronomical works; in addition to the large fragments of Cicero's *Aratea* with scholia are writings by Macrobius, Bede, Abbo of Fleury, Martianus Capella and others. The manuscript also contains a few Christian liturgical texts and a variety of religious writings, and surprisingly includes the formula for excommunication from Pope Leo IX. The first few pages comprise a miscellany with ecclesiastical computus from around the year 1107, a discussion of the twelve signs of the zodiac, and several diagrams and charts with a *Sphere of Apuleius or Pythagoras*. The *Sphere of Apuleius* diagram, similar to the *Vita Mors* diagrams during the Middle Ages, was also called the *Sphere of Life and Death* and attributed to Apuleius, Pythagoras, Plato, or Hippocrates and was consulted to foretell the duration and outcome of an illness. The diagram was consulted by doctors to determine the prognosis of an ailing patient; convoluted calculations including the number of letters in the name of the patient or his mother, the day and month of the start of the illness, and other information, then adding and dividing the sums to produce auspicious numbers that predict recovery; resultant negative numbers on the chart predict the patient's death even narrowing down the days or weeks. The Sphere of Apuleius pictured was included in the twelfth-century *Liber Floridus;* the Sphere consists of a large circular diagram with text inside, divided into sections, and a small circle in the center. The chart appears to have been well used through the centuries (Fig. 7.16).

In Tiberius C I, the words of Cicero do not appear until the top of folio 20 v "*Ex opera Ciceronis de Astronomia*" which is written in red ink, most of the manuscript's text is written in black. The lines derived from Cicero's *Aratea* are found at the bottom of the page, but the text has no longer been arranged with one line for each poetic verse; it has been modified according to Cicero's scholia and is written continuously as prose. The first line reads, "*e quibus hunc subter possis cognoscere fultum*", which corresponds to line 229 in Soubiran. The first letter of each descriptive paragraph has been enlarged and decorated. Similar to MS Harley 647, Tiberius C I has two different sets of writings for each constellation figure, the text of Hyginus was arranged to form part of the illustration, sometimes filling in the body and sometimes forming an architectural setting. The lines of Hyginus scholia that fill the figures in Tiberius C I appear straighter, stiffer and clumsy compared with the graceful, curving lines of text seen in the Harley and the figures are outlined in dark ink, not done in the Harley. On some illuminations, the text reaches down into the extremities making it impossible to read. The rustic capitals used in the text of the original were not retained; lower case letters were chosen instead.

Fig. 7.16 Sphere of Apuleius, *Liber Floridus*, Rijksuniversiteit te Ghent Bibliotheek, MS 92, Twelfth Century, Courtesy of Ghent University Library

Style, Placement and Iconography

The astronomical drawings in Tiberius C I, like the textual arrangements, follow the Harley as a close but inferior copy. When compared to its exemplar, the general format of Tiberius C I follows its design, but there are differences in the execution. The constellations are drawn with definite, heavy outlines by a firm hand. The text within the images appears uneven and inconsistent, too widely spaced in some figures or extensive and overdone (almost choking the figures) in others; some text overwhelms the whole folio. In comparison, the text written within the Harley constellations is more subtle and evenly balanced. The extremities of the Tiberius C I constellation figures are simplified and color was dispensed with almost entirely except for the red stars; the faces show no modeling or shading, nor the exacting illusionistic manner of its exemplar. When matched with its twelfth-century imitator, the ninth-century Harley stands out in its superiority.

The 26 constellation miniatures can be found on folios 21–32v followed by an illumination of the five planets as individual heads and one of Apollo and Luna in their chariots. As in most versions of Cicero's *Aratea*, the first miniature to appear in Tiberius C I is the constellation of Aries, which occurs before the start of its descriptive text. This depiction of Aries appears out of the ordinary, for the Ram does not stand upright, but lies with its feet toward the gutter of the manuscript; Pisces too takes this unusual vertical form, drawn with the bottom of the fish aimed toward the gutter. The eagle, Aquila, pictured on folio 26v with Sagitta over its head, is one of two famed birds from Greek mythology that were honored by being immortalized in the celestial realm. Cygnus, the beautiful swan that spreads its wings across the heavens, has the Milky Way as its background; sometimes called the Northern Cross, it also appears in all constellation cycles. Although several alternate myths exist, these star formations represent two birds hunted by Hercules and also characterize the avian forms assumed by Jupiter (interpretations of Leda and the Swan were popular artistic themes). The myth of Jupiter abducting Ganymede, the beautiful son of the king of Troy, to be the cupbearer to the gods is repeated in the inscription within the constellation miniature (a theme also depicted in paintings and Roman mosaics). The constellation of Canis Major, the Greater Dog, sits at the feet of mighty hunter Orion; Sirius, the brightest star in the northern hemisphere, marks the dog's head (Figs. 7.17, 7.18 and 7.19).

The remaining constellations are drawn in the usual upright positions. Tiberius C I, like most Cicero manuscripts, is missing the colorful characters, such as Cepheus and his ill-fated family, Pegasus, Boötes, Hercules, Auriga, and Ophiuchus.[39] The illuminations mostly appear at the top of the folio with Cicero scholia at the bottom. The total number of constellations in Tiberius C I is only about one-half the number usually found in Germanicus' *Aratea*.

The full-page illustration of Sol and Luna (folio 33v) depicted after the cycle of constellations was designed in an unusual manner, unique to this particular manuscript, not found in the Harley or within any *Aratea* manuscripts. This non-traditional portrayal of the Sun and Moon has no text and is reminiscent of an ancient chariot race with Luna in the lead. A full-figured Helios stands within his chariot holding a whip and the reins of four horses, all pulling to the right. His cape blows in the wind while strong rays of light project from his head. Luna stands in her cart pulled by the usual pair of oxen; her cloak too billows over her head as she holds two long flaming torches. The depictions of the Sun and Moon have classical references and do not seem uncharacteristic in themselves; but the fact that they are sketched very close together and not set apart by a framing devise is rarely seen. In addition, personifications of the four winds and the four seasons, shown as small heads, are aligned along the upper margins of the picture. These iconographic features indicate that this Sun and Moon illustration was borrowed from a different manuscript tradition, perhaps a Ps-Bede. The individual stars were indicated with the figures as in MS Harley, mostly as round red dots distributed along the outlines of the bodies, in order not to interfere with the text, but occasionally placed amidst

---

[39]The constellations included in Tiberius C.I. are listed in Appendix B.

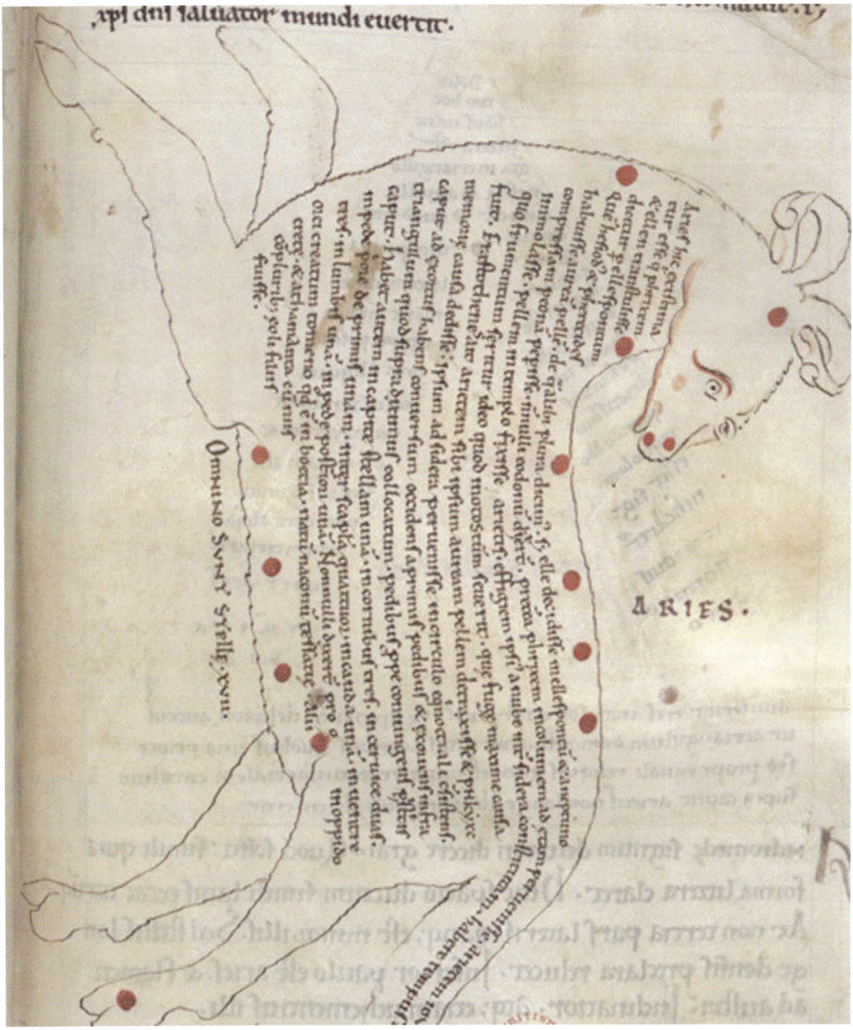

Fig. 7.17 Aries f. 21 BL Cotton Tiberius MS C I, Courtesy of the British Library

the inner text. The numbers of stars is always stated in the description, but does not always agree with the number drawn on the figures.

The second twelfth-century manuscript for closer attention was also produced in England and carries the text of Cicero' *Aratea* with scholia, but its design and illustrations are quite different. The two manuscripts were created about the same time both in English monasteries, but obviously the scribes or patrons involved had a choice in the arrangement of the folios and the type of constellation illuminations they could create in their codices.

Fig. 7.18 Aquila, f. 26v BL Cotton Tiberius MS C I, Courtesy of the British Library

Second Twelfth-Century *Aratea* Manuscript

By the middle of the twelfth century, there had been a definite revival of Platonic cosmology through the recent recovery of his original works, chiefly in conjunction with the 'school of Chartres'. These Platonic influences could be clearly seen in the first appearances of Arabic astronomy in the Latin West (Haskins 1927: 82).[40] By the end of the twelfth century, the translations of Ptolemy's *Almagest*, the tables of al-Khwarizmi and al-Zarkali, and treatises of Abu Ma'shar and al-Sufi, along with the quantity of astrological works from early Hellenistic and Latin sources, had conveyed the vast body of ancient Greek and medieval Arab astronomy to the Latin educational centers. This newly acquired knowledge from pure historic sources,

[40]This Platonic influence can be demonstrated by the *Questiones* of Adelard of Bath and in *De essentiis* of Hermann of Carinthia.

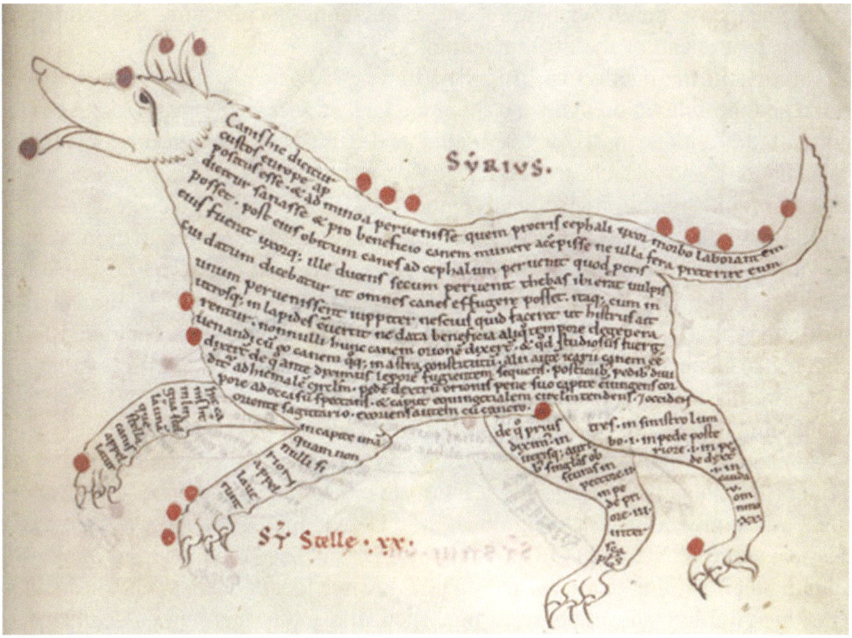

Fig. 7.19 Canis Major f. 28 BL Cotton Tiberius MS C I, Courtesy of the British Library

along with the availability of more accurate celestial observations and mathematics by Arab astronomers, brought great advances as well as controversies that occupied scholars for centuries. The intellectual climate generated a great leap in available educational sources, which at the beginning of the twelfth century was confined to small local monastic communities, cathedral schools, and peripatetic scholars; by the close of the century, the earliest universities at Paris, Oxford, and Bologna had been founded and begun to flourish.

The sharp contrast in the complexity of astronomical manuscripts produced at the beginning of the twelfth century with those created at the century's end confirm the intellectual advancement of this productive period. Instead of simply recopying the 'miserable' writings of Hyginus and the watered-down astronomy of Macrobius, Martianus and Ps-Bede, scholars now had access to the precise mathematical astronomy of Ptolemy and treatises of other Greek and Hellenistic astronomers as well as the writings, advancements and commentaries from generations of Islamic scholars. In addition, the interest in literature and scientific knowledge, which had been confined almost entirely to the royal courts and monasteries, began to spread through broader layers of society; in addition new trade routes opened to exciting areas of the world, such as India and other Asian ports. Great changes were occurring in cultural society too, as the more realistic Gothic style in artistic renderings took precedence over the stylized, exaggerated embodiments of the Romanesque era, dominating architecture, sculpture and painting by the end of the twelfth century.

Throughout the intellectual history of Europe, advances toward improved education and the cultivation of art and literature had, of necessity, always looked to the glory of the antique past for inspiration and guidance. The illustrations of the constellations and planets too had always been modeled on those copied from antique originals. This transitional century saw that change, as the simple basic astronomy found in the manuscripts of Pliny, Hyginus and *Aratea* traditions, gave way to the more sophisticated, theoretical astronomy long-practiced in the Arabic scientific world. But in spite of these advances and new techniques, the classical illustrations that had been preserved by Carolingian illuminators continued to be copied and retained in *Aratea* manuscripts. Although only four that were produced in the twelfth century survive today, the antique *Aratea* manuscripts were retained as a continuing part of this tradition of looking back to a Golden Age.

Compared to the Latin West, the Arab conception of the constellations and the signs of the zodiac was much more innovative, since they incorporated the traditional astronomy of their native groups, from Babylonia, Egypt, India as well as the classical sources. Naturally the Arabic artists Orientalized their celestial images by adapting their familiar garments and their distinctive iconography to classical styles. After the Arabic manuscripts were translated into Latin, this huge assemblage of astronomical data was dispersed among Western learning centers. Until that point, the names of the 48 Ptolemaic constellations were generally universal but little notice was paid to the individual stars. The long-established Arabic star names now became known and adopted in the Latin world (Kunitzsch 1989: 69–74). After exposure to copies of Islamic illuminated manuscripts, especially al-Sufi, Abu Ma'shar, and Fendulus, twelfth-century scribes and artists in Europe continued to create the same general format and appearance in their astronomical manuscripts, but Islamic influences can be observed in the attire and iconography of figural images. In the thirteenth and fourteenth centuries, contacts continued with the Islamic culture in trade and social interchanges, and through consultations among Arabic, Jewish and Christian scholars, as they absorbed the advanced theories and tried to solve problems in astronomical theory.

Bodleian Library MS 614

The first choice from among the remaining three twelfth-century *Aratea* manuscripts was Madrid BN, MS 19 created in Italy at the scriptorium of Monte Cassino. This *Arate*a carries the Germanicus text and comes from a completely different tradition than the Cicero Tiberius C I and the English family of manuscripts based on Harley 647. But unfortunately little information has been published on MS 19, and the codex survives in a deteriorated condition. No one is permitted to handle it, and the microfilm is unreadable. The next choice was a manuscript now at Oxford Bodleian Library MS 614 (hereafter Bodl 614) created in England in the mid-twelfth century, whose text is very similar to the other English *Aratea* at Oxford, Digby MS 83, made in the last quarter of the twelfth century. Both Bodleian *Aratea* manuscripts were modeled on British Library Cotton Tiberius B V, an early eleventh-century *Aratea* of Cicero; Bodl 614 was produced after Tiberius C I.[41] The second twelfth-century manuscript carries the text of Cicero, but its design and treatment of its illustrations appears quite different. The two Bodleian manuscripts were produced about the same time in English monastic centers, but obviously the scribes or patrons involved had individual choice in the arrangement of the folios and the type of constellation images they could include in their codices.

History of the Manuscript

The two twelfth-century Bodleian manuscripts are the last of the Ciceronian *Aratea* constellation cycles to descend from Harley 647, (except for the lone fifteenth century copy) and the last of the handmade books to contain the popular astronomical text of Cicero, before the large-scale penetration into the Latin West of Arabic and Greek science. Traces of these influences can be found in these twelfth-century Aratean codices, as they already contain the Arabic names of the stations of the moon, and also Hebraic and Arabic names for the planets. The exact year and the scriptorium where Bodl 614 was produced have not yet been definitely determined but that manuscript survives in excellent condition where its Cotton model was burned around its edges. Saxl has pointed out that these two Bodleian manuscripts are the first medieval Western manuscripts to include references to astrological theory (Saxl 1957a, b: 88). These manuscripts have textual additions, including more exotic information not available in their model Tiberius B.V. Saxl wrote concerning Bodl 614:

> It contains a number of Arabic words, and there is evidence to show that the compiler intended to supplement his traditional book of stars with another one explaining how to foretell a person's future from the stars. The author must therefore have been familiar with Spanish or Catalan texts dependent in turn on Arabic sources (Saxl 1957a, b: 88).

Text

This astronomical compendium consists of 76 folios, and opens with a list of the months on folio 1; Bodl 614 continues with the calendar pages that show a few illustrations of the Labors of the Month, followed by computus charts. The Cicero *Aratea* fragments along with *Scholia Strozziana* and *Sangermanensia* begin on folio 10. Like its prototype Cotton Tiberius B V, this codex holds excerpts from

[41]Bodl 614 and its complete miniatures are listed in Appendix B.

Hyginus, Pliny, Isidore, Bede and Abbo of Fleury, and it also contains an illustrated copy of the *Marvels of the East*.[42] The text is no longer completely Cicero's, but consists of a new compilation of several astronomical sources mainly from Hyginus and even some Germanicus scholia. Saxl contends, "Neither an astronomer nor a modern philologist would give high praise to the author of this treatise" (Saxl 1957a, b: 108).[43]

Astronomical texts of the twelfth century often depended on the traditional transmissions from early sources, but Bodl 614 opens to a greater variety of writings recently made available as they were translated from Arabic libraries. The text of the *Aratea* manuscripts continues to degenerate from the pure poetry of the original Latin versions of Cicero and Germanicus to just small bits of poetry inserted between large amounts of scholia. Like Tiberius C 1, the textual portions of Bodl. 614 have entirely lost the poetic form with just a few fragments of verse included. But the alternating rhythm between the astronomical text and the constellation illustrations continues its symbiotic relationship, maintaining close associations. The opening words that begin the astronomical writings of Bodl 614 are: "*Sol du igne natura fre..*", written before the illumination of Helios.

Style, Placement, and Iconography

The illustrations of both Bodleian manuscripts follow the tradition of the Anglo-Saxon codex, Tiberius B.V, but definitely not a direct copy. The arrangement of the Bodl 614 illustrations is different than the usual Ciceronian format. The *Aratea* section opens with an illustration of the Sun and Moon on the same folio but each in its own compartment. The standing figures are pictured in their chariots; Helios stands naked and holding a staff with a flag with his right hand and the reins of the quadriga with his left. A fully clothed and veiled Luna holds the lunar orb instead of a torch, and a crescent moon is pictured behind her head. The drawings are stiff and ill-proportioned, lacking artistry; the large flaming crown of Apollo overwhelms his body. The following miniatures are executed with more artistic skill; the garments of the constellation figures in particular are reproduced quite accurately. The constellation images begin with Aries, as usual, but instead of following the order of the original Greek poem, as MS Harley does, Bodl. 614 continues listing the 12 signs and completes the zodiacal constellations. The sign of Scorpio was created in an unusual manner as the Scorpion holds the Scales

[42]*Marvels of the East* was a popular medieval codex which was more a book of monsters and monstrous races, describing odd creatures that they thought lived in unknown regions of the world, such as cyclops, blemyai, and sciopods, which reach back to Pliny the Elder. The tale only appears in a few medieval manuscripts. The *Marvels of the East* can be found in Cotton Tiberius B V and in its copy Bodl 614. These wonders are fully illustrated in both manuscripts and were produced for entertainment.

[43]Saxl, "Illuminated Science Manuscripts" in *Lectures* 108.

of Libra, Scorpio is painted with a human face and hands, and with four legs on each side; his signature raised and threatening tail is missing. The untraditional grouping of the twelve zodiacal constellation points to the recent emphasis on astrological texts as the topic comes to the forefront of astronomical interests.

The zodiacal signs are followed by a composite picture of the seven planets with the Sun encircled in the center; the other planets are aligned above and below within smaller circles. The circles of the planets are interlocked and labeled, but could be identified by their typical appearances, although the wings of Mercury could be growing out of his eyebrows. The illuminations then retreat to the standard beginning point with Draco, Ursa Major and Ursa Minor and continue through the Ptolemaic cycle, depicting 27 constellations in all. Folio 19v depicts a star group not usually illustrated in the constellation series, the Hyades. This open cluster of stars is visible to the naked eye and found near the Pleiades, but appears dimmer and blurry; both star clusters are included among the stars of Taurus. The Hyades are depicted in the same manner as the Pleiades, seven individual female heads, but they are veiled, perhaps indicating that they are less bright. The total number of illuminations, although not complete, is more comprehensive than the 24 images found in Harley 647.[44] The constellation pictures that are missing were substituted from a Hyginus manuscript. At the end of the constellation illustrations in Bodl 614, "the writer says explicitly that these images are not to be drawn indiscriminately, as they indicate certain positions of the stars in the sky and should therefore be carefully copied" (Saxl 1957a, b: 108).[45]

Romanesque book illumination and Romanesque sculpture are rich in distorted forms, distinctive interlacing and exaggerated postures. The bodies of the constellation figures in Bodl 614 are drawn in normal proportions, not elongated or twisted; but the artist has covered the conventionally nude, or partially nude, bodies of Orion, Hercules, Boötes, and Andromeda et al. with weighty layers of concealing clothing. To avoid both giving pleasure to the eye and the imagination, and providing an occasion for temptation, the nakedness commonly seen in Aratean illuminations has been concealed in the English scriptoria. Although the male heroes stand in threatening poses, they do not appear capable of mighty deeds. The constellation figures in this manuscript show close parallels with many artistic conventions of the twelfth century rather than a close reproduction of earlier Aratean illustrations.

> This decomposition of the classical type was not the result of any increasing respect for the scientific and true position of the stars (which were still placed as arbitrarily as ever), but was due to a purely stylistic and intellectual evolution (Panofsky and Saxl 1933: 238).

The drawings of the constellations in Bodl.614 are mostly framed with colored ground, and all but a few of the constellation illustrations are similar to Digby 83. They differ greatly from the images in Tiberius C 1, as they are more fully drawn,

---

[44]The constellations illuminated in Bodl. 614 are listed in Appendix B.

[45]Saxl, "Illuminated Science Manuscripts" in *Lectures*.

and the text has been omitted from the bodies of the figures. The artist of Bodl 614 often combined three, four, or more constellations and crowded them together on one folio.

The skillfully executed drawings are of classical derivation, but, rather than half-naked, have taken on a new guise, heavily clothed (perhaps reflecting English rather than Mediterranean weather) and portrayed in a typical Romanesque manner. The draperies of the figures are drawn in fully developed clinging curvilinear and V-folds, as seen in the illumination of Cassiopeia. Interestingly, the stars have been placed in more determined positions, a change from the usual outlining of the figure. Even though Cassiopeia is framed, the base of her throne, feet and head extend beyond the borders. The iconographic motifs and designs retain some of their classical illusions, thus they are readily recognized, but the newly added attire is indisputably attributed to twelfth-century preferences, and characteristic of Romanesque illuminations. Orion in Bodl 614 has been brought down to Earth and placed in the doorway of an architectural setting as if ready to set off for a hunt in the woods, most uncharacteristic for one who dwells in the sky. Neither his faithful dog Canis Major nor the Hare are shown at his feet, and his garments completely hide the powerful physique of the mighty hunter. A few stars appear upon his clothing, but his iconographic three-star belt has not been accentuated, but partially hidden by his sword.

The portrayal of brave Hercules in Bodl 614 seems far from herculean, more dainty than godlike. He has lost his identifying attributes, the tawny lion skin over his arm; he stands wielding only a rough club. Like Orion, his hefty, naked torso is not on display for God-like admiration, but instead stands, not kneels, fully covered; the short ruffled skirt of Hercules' garment is unbefitting the powerful hero of antiquity. The threatening serpent that guards the Golden Apples has been replaced with a small, winged dragon. Hercules attacks the mild-mannered, contorted dragon with his club, scarcely different than a Romanesque St Michael slaying a dragon. Both Bodl 614 and Tiberius C 1 were modeled ultimately on Harley 647 and both include the same constellation cycle, but the creative spirit and artistic ability of each artist resulted in the production of two entirely dissimilar series of miniatures both of lesser quality than the original.

During the medieval period, artists restored, reinvented, and recopied images that had been preserved by Carolingian illuminators from Late Antique sources. These conservative classical illustrations were continually recycled in the various *Aratea* manuscripts. The thirteenth and fourteenth centuries, replete with wars, plagues and famine, also exhibited little interest in creating new versions of the *Aratea,* as only five survive, all five contain Germanicus' text. Another copy of Cicero's *Aratea* was not produced until the fifteenth century (Stiftsbibliothek Codex Göttweig 7 Bl.2). The historical situation changed dramatically in the fifteenth century since as many as 26 *Aratea* manuscripts survive. What factors motivated this remarkable surge in the production of *Aratea* manuscripts during the quattrocento, and who were the patrons commissioning, collecting and reading these astronomical compilations?

Fifteenth-Century *Aratea* Manuscripts

The third and last period of 'renaissance' in the Middle Ages under study refers to the Italian Renaissance of the fifteenth century, the most significant period of resurgence and rediscovery of classical texts. Artists in the quattrocento transformed their artworks from the parameters of Byzantine and Gothic styles to the more classical Greek and Roman and were first to incorporate scientific perspective. Fortunately, a total of 26 *Aratea* illuminated manuscripts survive from this time, which doubles the total from the ninth century, the second most survivals. The third pair of manuscripts discussed in the present study date from this pivotal time in the history of astronomy, science and art. During this period, the ongoing revival of ancient knowledge and manuscripts was strongly stimulated by the development of humanism and an intense interest in the classics which spread throughout the major cities of Italy from its beginnings in Florence. The huge leap in the number of manuscripts extant from the fifteenth century indicates that this was another cultural period with a strong interest in astronomical and astrological writings; a time newly recovered from the devastation of the Hundreds Year War and the Black Death. This era became especially rich in new and renewed versions of the classic literature and in recreating imagery of the glories of the historic past and the reappearance of classical gods and goddesses.

During the fifteenth century, the illuminations on the pages of religious manuscripts were being magically transformed by early Renaissance artists who had mastered the depiction of light, illusion, texture, modeling and three-dimensional space; but in spite of these huge advances, the illustrations in most *Aratea* manuscripts maintained their traditional antique appearance. Secular codices of highest quality and sophistication appeared in connection with new developments in careful observation of nature, depictions of epic tales, journeys, battles and historic events were authentic and naturalistic. Decorated borders and historiated initials in manuscript painting developed in richness and variation; but these innovations too had a minimal impact on the illuminations of the *Aratea* manuscripts. The two-fifteenth-century examples of the *Aratea* demonstrate how the manuscripts retained their medieval incarnations, acquiring some Eastern or Islamic influences, but continually looking back to the classical past. The *Aratea* were no longer enjoyed, if they ever were, for their poetic verses or 'scientific' material, but for their myths, pictures and historical appeal. The long pictorial tradition of Aratean manuscripts was instrumental in sustaining the antique mythological images that were treasured in the Classical, Medieval, and Renaissance eras.

Surprisingly, the illustrations found in the quattrocento *Aratea* manuscripts are fundamentally conservative and show no noticeable influence or traces of the late Gothic, nor of the great Renaissance artists painting elaborate frescoes, carving magnificent statues and decorating churches and palaces in Italy at this time. The advances in artistic techniques such as scientific perspective, harmonic proportions of figures, backgrounds with naturalistic far-off settings, and creative use of chiaroscuro, do not appear in the miniatures of *Aratea* manuscripts.

But he long-established history of the Renaissance, centered upon the notion of the revival of classical antiquity and the reinstatement of the ancient images of the Greek and Roman heroes, gods and goddesses, no longer stands as the simplistic commonly accepted explanation. Applying the more recent concept of deconstructionism sets forth a more realistic picture of the Renaissance instead, as the "birth of the modern national world presented by Burkhardt during the nineteenth century" (Duits 2011: 13). The present view of the Italian Renaissance brings to light a wider spectrum of stimuli and more varied goals. When studying the variety of creative representations of astronomical images that became available at that time, it is clear that an entirely different set of influences are determining the type of illustrations that appear in the many different astronomical manuscripts currently circulating.

> Saxl's theory can be criticised on many points. For example, research since Saxl has made it clear that there never was one single set of antique prototypes from which all the constellation images were derived; the Ptolemaic tradition of images on celestial globes, from which the Arabic illustrations were derived, was always different from the type of illustrations with mythological content that was reproduced in the *Aratea* manuscripts (Duits 2011: 15).

The artistry of manuscript traditions of Michael Scot, Sufi Latinus, Abu Ma'shar and Fendulus and many other influences changed the constellation images into new characterizations barely recognizable when compared to the classical depictions. But the *Aratea* manuscripts remained the exception to these new and inventive images appearing in astronomical and astrological anthologies, such as images of the "children of the planets", scenes depicting the types of occupations related to the physical influences attributed to each planet, including the Sun and Moon.[46] Other innovative astrological illustrations include the depiction of personified decans and the paranatellonta discussed in Chap. 4.

Although gaps in our knowledge of classical constellation images still exist and their transmission through the centuries also is still being studied, some of the earlier theories can no longer be substantiated. Because of the quantity of information now readily available from new studies, and particularly thanks to the huge variety of constellation illuminations that are readily accessible from the scholarly work in the *Sternbilder des Mittelalters* volumes, more extensive resources are accessible. The accepted theory that the Renaissance saw primarily a revival of classical antiquity has been shown untrue in the creation of astronomical manuscripts in general. Enormous variations in the depiction of constellation and planetary iconography can be seen in reference works and online illustrations. Incentives from Arabic and Indian manuscripts and from astrological compilations inspired various choices for illustrators to model and blend past images, in addition

[46]The most beautifully decorated manuscript devoted to the presentation of the 'children of the planets' is now knows as the *De Sphaera D'Este* created about 1470. It had no original title; the author and scribe are unknown but the illuminator was Christoforo de Predis, who mainly worked at the court of the Sforza family. All scenes illustrating the 'children' and their appropriate activities provide helpful insight into everyday life in fifteenth- century Italy. The illustrations from the *De Sphaera* are readily available online.

to their own ideas and artistic creations. Many astronomical illustrations and celestial ceiling frescos were richly painted using the latest techniques of *trompe-l'oeil* and sophisticated procedures. Some constellation images were transformed to an unrecognizable degree.

Even though newly designed choices and astrologically based images decorated many astronomical anthologies, none were introduced into *Aratea* manuscripts beyond a few stylistic differences. The fifteenth century *Aratea* manuscripts were being produced mainly in Italian Renaissance cities, particularly Naples and Florence, where talented artists were creating spectacular new paintings that displayed elegant jewels, stylish clothing and rich brocades. Through all this outstanding progress in contemporary artworks at the height of the Renaissance, *Aratea* patrons and illuminators consciously looked back to antiquity and chose to retain the classical representations, antique clothing styles and traditional iconographic constellation figures, still freely floating in air and displaying their naked backsides.

Morgan Library Ms M. 389

During the second half of the fifteenth century, the royal court at Naples became a hotbed of *Aratea* activity as four similar astronomical anthologies were produced there at the court of King Ferdinand I (1423–1494), commissioned by his chief administrators and humanist scholars. The four illuminated *Aratea* manuscripts from the fifteenth century survive intact and have been honored holdings of the most prestigious libraries. Three of the Naples codices are now held at the Biblioteca Apostolica Vatican Library; the fourth of this group belongs to the Morgan Library in New York.

Although all four manuscripts were commissioned at about the same time under the same royal patronage, their decor and appearance differs dramatically. All four anthologies contain the *Aratea* of Germanicus utilizing the same general format, which alternates sections of verse with sections of prose scholia, a conflation of the astronomy of Hyginus and Germanicus. Two of these royal manuscripts Morgan M 389 and Vat. Barb. 76 were selected for more detailed observation for several reasons. Although the text of these *Aratea* appears similar, the illustrations show that the two artists drew inspiration from different concepts; one is richly colored with a bold palette, the other more neutral and subdued in its artistic treatment. The attire and iconography of the constellation figures display a variety of classicism, Oriental influences and a slight degree of nakedness. Therefore they make a worthwhile pair for closer comparison. Morgan Library MS M 389 (hereafter M 389) is the first of these *Aratea* manuscripts to be discussed.

History of the Manuscript

An essay in *Sternbilder des Mittelalters* relates the fascinating history of these manuscripts produced in Naples and of the many copies they inspired (Blume et al. 2016: 111–121). Between 1465 and 1467 a twelfth century Germanicus manuscript was discovered in Sicily which was transferred directly to Naples where King Ferdinand d'Aragon (or Ferrante) currently reigned after a contentious takeover by his father Alfonso d'Aragon and of the resultant reuniting of Sicily and Naples. Evidence survives that demonstrates the centuries old astronomical manuscript was copied there almost immediately in either 1467 or 1468; it was copied at least three more times by humanist scholars and scribes at the court of Naples. A copy of this Germanicus manuscript was then transported to Florence where it was reproduced at least three times, for the Medici court, for Francesco Sasetti, a banker for the Medicis, and for Frederico da Montefeltro. Unfortunately the original "newly discovered" manuscript is now lost, but text scholars have determined that it was based on a Germanicus manuscript now in Madrid MS 19, which itself had an earlier exemplar. Thus the twelfth century manuscript was regarded as an extraordinary find reaching back to antiquity; it was reproduced multiple times accounting for many of the twenty-six Germanicus *Aratea* surviving from the fifteenth century (Blume et al. 2016: 111–121). The explosion of illuminated astronomical manuscripts during the fifteenth century Italian Renaissance was also inspired in part by the current rise of humanism.

M 389 was copied from this early prototype and illuminated in Naples and dated 1469 by the scribe Joannes M. Velox, who also signed the colophon that appears on folio 117. The name of the artist was not included, but he has been identified as the Master of Isabella di Chiaromonte, who was active in Italy from 1455 to1469. Its colophon also noted that M.389 was created for Antonella Petrucci (d.), the prime minister and secretary to King Ferdinand I, who had also appointed him a baron. (Petrucci was later accused of treason to which he confessed under torture; he and his two sons were executed in 1487.) In that same year, this *Aratea* codex came into the possession of Francesco Catalano of Messina, as noted by an inscription on the back flyleaf. The next owner was Andrea Matteo III, Duke of Acquaviva and Prince of Atri, who added his coat of arms to the title page, painting over the original coat of arms. Ultimately, in 1910 the Germanicus *Arate*a manuscript was purchased by J. Pierpont Morgan from a well-known French bookseller, Léon Gruel (1841–1923), and there it still remains.

Text

The astronomical manuscript M 389 has 117 leaves with the text written in one column, and includes not only the *Aratea* of Germanicus with scholia but also consists of fragments of the usual texts by Pliny and Hyginus. The opening page begins with the incipit: "*Aratus quidem fuit. Athinodori patris*". The frontispiece is written in large capital letters; some sections of the *Aratea* are written in verse format; the first letter of each line is enlarged and set a little apart in the usual manner, but the initial letters are written in alternating pastel shades. The lines of Germanicus' verses are interspersed with sections of the scholia of Hyginus which provide the stellar positions and numbers. The first line of each constellation, after the appropriate miniature, is written in widely—spaced capitals, in alternating colors of pink and pale blue; the remainder of the descriptive text is written in brown ink.

Style, Placement and Iconography

The earliest manifestation of the Renaissance in book production can be seen in the humanistic book which took on its typical appearance in Florence, under Cosimo the Elder (Salmi 1954: 48). In manuscripts considered to be humanist, the first identifying feature is an ornate title page bearing the imprint showing the style of the particular workshop in the marginal decorations and in the enlarged initial letters. The opening folio, rarely any other page, displays an elaborate frame composed of white interlace and twisted vines, imitating the decorated capital letters of the early Middle Ages and Romanesque era, thought to have originated in classical antiquity. Inserted among the white vines and interlace appear coats-of-arms, winged putti, tiny animals, and occasionally medallions. These humanistic features were usually painted in delicate and subdued colors; many fifteenth-century Italian manuscripts include this typical opening format. Some title pages grew extremely ornate with elaborate architectural structures and imagined author portraits, which achieved the desired association with antiquity.

The opening folio of M.389 follows the proscribed humanist format; the page is surrounded with an ornate border of white interlace and vines, with two putti holding a laurel wreath that encloses the coat of arms of Andrea Matteo III in the bottom border. The 'A' of Aratus is decorated with pastel shades and twisted vines, although by now there is very little of the astronomical poetry of Aratus or of Germanicus remaining. Even though the colors are muted in a similar manner as the rest of the manuscript, the humanist frontispiece stands out harshly and does not match the character or the more or less antique style of the entire manuscript. The ornate frontispiece is followed by a celestial map and 39 colored drawings of the constellations that appear at the top, bottom or middle of the page, almost always before the description. The circular celestial sphere with five zones on folio 3v introduces the theme of the *Aratea* and displays the entire night sky with the constellations pictured in their relative positions. The depictions of the constellations in the celestial map are much less classical than the individual portrayals that follow (Fig. 7.20).

The fact that all figures are facing forward, only Boötes is partially turned, reveals that the constellation figures were modeled on medieval exemplars, not Late

Fig. 7.20 Celestial Map f. 3v Pierpont Morgan Library MS 389

Classical, and probably several differing prototypes in addition to the Madrid MS 19. In the M.389 celestial map, only three of the thirteen human figures are entirely naked, (four in the Vatican 76 map), more are partially naked. Most of the figures retain their traditional postures, but have been stripped of their classical garb. The nudity of the pagan gods and heroes enshrined in the sky symbolizes their perfection; like Adam and Eve in Paradise, they have little need for manmade garments. They have shed the heavy clothing of the twelfth century and stand fixed in their identifying postures. A nude and child-like Hercules kneels; Perseus appears naked but for his cape, Medusa's snakeless head his only identifier. The Gemini twins have lost their stance as warriors, and are depicted as two naked children in the celestial map, not seen before the fifteenth century. The first miniature after the sky map reveals Jupiter, partially covered with a purple cloak, as he rests comfortably astride his emblematic eagle.

The illustrations throughout the manuscript seem to be derived mainly from a Carolingian prototype from Sicily (Harrsen and Boyce 1953: 50). Some constellation images, particularly the animals continued without change, but others like Eridanus were replaced with new nonclassical types; he is shown with horns, lying flat on the water; his vase with flowing water is below him.[47] Some innovations derived from the East, such as Cepheus, and some were freely invented, such as the Baby Gemini. "Beginning in the second half of the quattrocento, imitation of the antique gradually reintroduced the classical types" (Panofsky and Saxl 1933: 263). But many different, more exotic and astrological, images appear as well.

The 42 delicately tinted illustrations in M.389, some, for example Ophiuchus, have two or three pictured together, are drawn in 'papyrus' style, and are merely inserted between the sections of writing, with no backgrounds or framing device.[48] The images of M.389, like the frontispiece and the celestial map, are a softly shaded, tinted wash, and are nicely color-coordinated with the letters of the text. The figures pictured on the sphere do not always agree with the same individual constellation miniature. The constellation cycle begins, as do all Germanicus *Aratea*, with Draco intertwined between Ursa Major and Minor.

The constellation image of Gemini depicts the twins as young men within the text, rather than as children shown in the sky map, mostly naked, but partially draped with long mantles. Still the iconography of Amphion is represented instead of Pollux; his cloak covers his left hand that is holding the lyre, a classical iconographic symbol which appears in the Leiden *Aratea* and is commonly seen in Germanicus' versions. The Twins no longer retain the cap and small cross on their heads, as in the Leiden, instead of crosses, small stars adorn their bare heads. The sign of Cancer is drawn to the right of the Twins' feet, looking more like a lobster than the Crab. The illustration of King Cepheus has taken on the appearance of a bearded Arab warrior, rather than the usual medieval monarch dressed in royal attire, only his stiffly extended arms remind one of the antique depictions. The more classical version of Cepheus is found in the Leiden *Aratea*; where the foreign king stands attired in Persian garb, a short tunic trimmed in gold with loose leggings, wearing a Phrygian cap.

In M 389, the miniatures depicting Hercules and Ophiuchus on Scorpio are drawn naked from the godlike position in a three-quarters view. With his green club raised to strike, Hercules threatens the serpent, coiled around the tree of golden apples which he guards; his iconic lion skin takes the form of a heart-shaped shield. Stars outline the bare legs of Hercules. Located near the celestial equator, Ophiuchus or Serpentarius stands atop the menacing Scorpius, encircled round his waist by a muted green snake. He grips Serpens, one of the Ptolemaic constellations, tightly away from his face, as the lengthy serpent and Ophiuchus stare at each other in confrontation. Scorpio, the third constellation in this illustration, lies ready to strike under his feet. The stars on the constellations consist of small red dots

[47]Blume et al., explain the odd introduction of horns to Eridanus in *Sternbilder des Mittelalters* (2016) page 118. This addition of horns may have occurred already in the eleventh century in a manuscript produced in Montecassino which transferred that motif to Germanicus copies in Northern Europe.

[48]The complete list of illustrations can be found in Appendix B.

scattered mostly along the edges of the figures. The entire cycle of *Aratea* images from M.389 can be viewed conveniently online at the Morgan Library Corsair website (Figs. 7.21 and 7.22).

Fig. 7.21 Hercules, f. 14v Pierpont Morgan Library MS 389

Fig. 7.22 Ophiuchus on Scorpio, f. 16v Pierpont Morgan Library MS 389

Biblioteca Apostolica Vaticana Codex—Vat. Barb. Lat. 76

The Vatican astronomical compendium Vat. Barb 76, (hereafter Barb76) was another of those produced in the second half of the fifteenth century, one of the four similar *Aratea* codices created at the royal court of King Ferdinand I at Naples. Barb.76 consists of 100 folios with Germanicus' *Aratea* scholia written in prose format. Similar to M 389, this compendium includes the works of other authors and is illustrated with a vividly colored celestial map and 42 miniatures, again, some illuminations consist of two or more constellations. Another of the four Naples manuscripts, Vat. Barb 77 contains exact same textual and pictorial contents of the Morgan Library *Aratea* and both are now available on the Vatican Library website.

History of the Manuscript

Barb 76 was commissioned by or for Giovanni d'Aragon (1456–1485, son of Ferdinand I d'Aragon (or Ferrante) king of Naples, for his own library, and dated between 1470 and 1475. Giovanni was appointed notary of the pope in 1471; he was raised to cardinal in 1477, and died in 1485 when only twenty nine (Blume et al., 2016: 113). This manuscript along with another from the group of four *Aratea*, Barb.77, was acquired by Cardinal Maffeo Barberini (1568 –1644), later Pope Urban VIII. In 1901 the magnificent Barberini library, rich in illuminated manuscripts, incunabula and rare editions, was acquired for the Vatican Library by Pope Leo XIII (1810–1903), this large collection now constitutes its core section. Following the lead of Pope Nicholas V (1397–1455), Leo XIII enriched the Vatican's book collection and also reformed and adapted the library to better meet the needs of the times, including conservation of the older manuscripts and allowing access for scholarly research.

Text

The text of this manuscript depends on a lost prototype with Germanicus text, a branch of the O family (Gain 1976: 2).The format of the text layout follows the pattern most often seen in fifteenth-century *Aratea* manuscripts and in the incunabula as well. The text begins on folios 1 and 2 with the title, *Genus Arati*, the standard 'Life of Aratus' appears. The incipit is '*Aratus quidem fuit Athinodori patris filius*'. Folios 2v to 66 contain mainly the *Scholia Strozziana* along with extracts of Germanicus' *Aratea*. The incipit for this section reads '*Coelum circulis quinque distinguitur…*' The text of Germanicus' poem begins on folio 6 with the usual *Ab Iove principium magno deduxit Aratus carminis*. Like M 389, the text of the poem, written in verse format, has the larger first initial of each line as a capital letter and slightly larger. Folios 68v to 71 are also extracts from Germanicus*Aratea*; folios 71v to 77 hold more exerpts of *Scholia Strozziana*. The incipit for this section is *solem* per se *ipsum constat moueri*. Since the manuscript was made for a member of the royal family of Naples, the parchment of Barb 76 is pure white of the highest quality; the scribe wrote the text with dark ink in one column only, in a neat humanistic script. Opening initials of each prose section are enlarged and illuminated in gold and various bright colors. This astronomical compendium also contains some passages of the *Natural History* of Pliny the Elder on folios 77v to 86, and extracts from *De astronomia* of Hyginus on folios 86–100. The poetic fragments are divided into sections and alternate with the scholia, the illustrations appear before the textual descriptions.

Style, Placement, and Iconography

Like the similar fifteenth-century codex M.389, this manuscript opens with a humanistic frontispiece; the Neapolitan decoration indicates the influence of the Northern Italian schools; the ornate scroll work made of white interlaced branches has been offset with bright reds and greens. Numerous putti hang playfully from the rectangular garland within the interlace border along with a few animal figures. Inserted in the decorated border are a bust of a woman in a medallion on the left and top portions, and the bust of a man on right. The large ‘A’ of Aratus, decorated with white twisted vines, has been repeated as well. The coat of arms of Ferdinand I of Aragon (and of Giovanni) are surmounted by a royal crown and painted in the bottom border; the arms of Cardinal Barberini were added later. The constellation illuminations in Barb76 are in part the work of Matteo Felice, one of the artists in the employ of King Ferdinand at that time.

The first miniature of the Barb 76 *Aratea*, folio 3, consists of the five-ringed celestial map; similar to M 389, the celestial globes in both are divided into five concentric circles; the twelve signs of the zodiac appear mostly within the third and fourth ring which represents the ecliptic. Only a few details of the iconography vary between the two views of the heavens. The most noticeable difference relates to the color tones, as the Morgan artist chose a softer range of colors which are applied with a light wash, while the illuminations done by the Vatican artist have been painted with a brighter palette of vivid reds, oranges and blues. The background of the M 389 map is not colored, merely showing the natural beige hue of the parchment, compared to the intense royal blue sky setting of the figures in Barb 76. The celestial spheres hold a mixed blend of classical, but mostly iconography contemporary to the time.

The text that describes the heavenly circles and earthly zones—torrid, temperate and frigid—appears on the next folio. The drawing of Jupiter relaxing on his iconic large eagle with spread wings, holding a flaming torch and his royal scepter, appears on folio 6; his cape forms a halo around his head and long hair. The first six

Fig. 7.23 Celestial Map f. 3, Apostolica Vaticana Codex, Vat. Barb. Lat. 76

lines of Germanicus' poem are written below this image. The following text provides the usual descriptions of the constellations, including the positions of the stars and their total number.[49] The individual constellation images in Barb 76 are presented in a different manner from M 389; they are set apart from the text by a brown, square or rectangular framing device in the same manner as the Leiden *Aratea*. The painting technique appears different as well; while M 389 has simplified figures drawn with pale pastels, Barb76 has been painted with a richly colored palette in a more bold and dramatic manner. The images in M 389 have no background and appear quite plain and conservative, but the Barb 76 artist has added interest with more details. The background space within the border of the Barb 76 constellations is completely filled with mottled white shading that invokes cloud cover, which then dissolves into the deep blue of the sky (Figs. 7.23, 7.24 and 7.25).

The placement and order of the constellation figures is almost exactly the same as M 389. The sign of Libra is missing in both, which is not unusual, as the Scales were often held by Virgo or considered the 'Claws' of Scorpio. The prose format and bolder script treatment of Barb.76 both compliment the constellation images.

[49]The complete listing of constellation illustrations can be found in Appendix B.

Fig. 7.24 Aquarius and Capricorn f.14v, Apostolica Vaticana Codex, Vat. Barb. Lat. 76

The illustration of Aquarius in Barb 76 stands facing the zodiac sign that precedes him, Capricorn; but his head turns the other way, looking over his shoulder. Aquarius appears dark-skinned and has adopted Arabic-styled attire, short tunic, with his long head-covering wrapped around his neck and wearing sandals. He pours water from an urn, which is shown as a double series of closely spaced stars. Some influences from medieval Islamic illustrations can be detected, but since the Arab astronomical traditions were mostly derived from Greek and Roman sources, the differences are minor. The sign of Capricorn appears in the same frame, almost as large as Aquarius; the stylized treatment of his tail looks back to details acquired in the Romanesque period. The individual stars in the Barb 76 constellations are bright red, much larger and more plentiful than those in M389. In both manuscripts, stars are generously pictured on all the constellations with no indication of size or magnitude, nor are the stars accurately placed.

Miniatures of the Sun and Moon, as usual, are positioned near the end of the *Aratea*. Helios, or Apollo, carries the light while directing his quadriga drawn by four horses. Rays of light outline his head; his long flowing hair can almost be mistaken as feminine. With cloak flying in the wind, the framed personification of Sol in Barb 76, guides his quadriga pulled by four horses, while holding his flaming torch aloft. His long hair reaches to his shoulders; wispy clouds surround him. Excerpts from Pliny are written between the images of two luminaries. More often, the Sun and Moon are painted on the same or adjoining folios, not separated by pages of text. Following the full constellation picture cycle, the last section of the Barb 76 anthology comprises the writings of Pliny's *De natura* and Hyginus' *De astronomia*; these texts are complemented by only three miniatures.

The third and fourth *Aratea* manuscripts produced at the court of Naples were copied from their twelfth century exemplar in quite similar formats, and Cologny, Cod. Bodmer 7 which is a fifth copy of this twelfth-century Germanicus

Fig. 7.25 Sol in Quadriga, f. 73 V, Apostolica Vaticana Codex, Vat. Barb. Lat. 76

manuscript. The textual content of other Germanicus *Arateas*, Apostolica Vaticana Codex, Vat. Barb. Lat. 77 and Apostolica Vaticana Codex, Vat. Lat. 1358 were composed in a like manner, only the treatment of the illuminations varied in their design and choice of color palette. These manuscripts also open with the ringed celestial sphere filled with stellar figures and contain the full constellation cycle. The miniatures of each codex, vary slight as to whether they are famed or not and the selection of their colors, but the figures or animals are drawn in the same manner. Virgo stands wearing softly flowing gown and endowed with large feathery wings. Andromeda is chained to the rocks in courtly robes, Eridanus lies naked with curved horns, and Pollux holds his lyre. The outstanding *Arateas* of Germanicus created in Naples and their exemplar were copied many times in Italian cities inspiring many of the fifteenth-century *Aratea* manuscripts. *Sternbilder des Mittelaters* provides more information and illustrations of a few of these manuscripts; Appendix B supplies fundamental information concerning the other surviving *Aratea* manuscripts.

Summary

The three historical periods investigated here are each considered renaissances, stimulated by diverse forces and driven by different incentives. For the purposes of this study, it became necessary to focus on a particular viewpoint, concentrating mainly on the production of 'scientific' manuscripts, on the transmission of astronomical knowledge, and on how the images related to the text. During the ninth century, the classical poem of Aratus through its Latin incarnations was revived by the Carolingians in a variety of formats that were for the most part inspired by Late Antique prototypes. The two earliest examples of this inquiry, the Leiden A*ratea* of Germanicus and Harley Cicero are emblematic of the effort and expertise expended in creating codices during the Carolingian period. These manuscripts contain only the original poem of Aratus translated into Latin with a bit of Hyginus prose added to Germanicus' poetry and were decorated with lively miniatures of the constellations. Through the centuries, most *Aratea* illuminations included the sky map, images of Sol and Luna, the planets and the Pleiades. The illustrated *Aratea* manuscripts must have been enjoyed and treasured in the ninth century, as thirteen examples are still extant more than a thousand years later.

During the twelfth century, a full corpus of ancient Greek and Hellenistic astronomical data, especially the *Almagest* of Ptolemy, became available to the Latin West through the collaboration and erudition of Christians, Jews and Arabic scholars. The modest number of *Aratea* manuscripts extant from that time demonstrates that astronomical knowledge was available then from numerous new, more exciting sources. The astronomical theories of Ptolemy, Plato, Aristotle, Abu Ma'shar, and al-Sufi were copied in large quantities, while fresh versions of the *Aratea* slowed. It is possible that many libraries already possessed a version of the *Aratea* in an astronomical anthology, since numerous manuscripts had been produced and bound together in the previous centuries. During the twelfth century, the *Aratea* tradition survived longer and seems to have been stronger in England than on the continent. The two twelfth-century examples demonstrate that those responsible for creating new Aratean manuscripts had a desire to enlarge Cicero's version to include a more complete cycle of constellations texts and illustrations.

If the surviving numbers are representative of reality, the thirteenth and fourteenth centuries too saw a dwindling interest in the *Aratea* tradition. The manuscripts of Sacrobosco came to the forefront in the thirteenth century and became the primary text for learning basic astronomy. Although the more sophisticated mathematical astronomy and predictive astrological techniques discovered in the newly translated works of Ptolemy acquired a prominent place in manuscript production, along with the Arabic translations of Abu Ma'shar in particular, the popular astronomy found in the *Aratea* manuscripts did not disappear, but was still copied, illustrated and read. The opening image of the celestial sphere and the constellation images beginning with Ursa Major and Minor in Germanicus *Arateas,* and with Aries the Ram in the Ciceronian *Arateas* could still be helpful in learning the rising and settings of the constellations. The text, however, evolved (or declined) gradually through the centuries from the pure classical poetry to pure prose which has become a combination of the *Aratea* and other astronomical writers, with the majority moving away from Germanicus and toward other less poetic text.

During the fifteenth century, another renewal of astronomical scholarship developed; in its wake, the reappearance and renewed appreciation for the gods of antiquity followed. Joscelyn Godwin, writing about the pagan revival, sums up the situation:

> The pagan divinities are a hardy breed. After being subverted by Homer, atomized by Lucretius, and toppled from their pedestals by the Christians, one would have thought them finished (Godwin 2002: 2).

Actually the reverse occurred; new incarnations of the pagan deities decorated the villas and palaces of the wealthy nobles and cardinals. Ancient myths and exploits of the gods became popular themes on the walls and ceilings, in decorative tapestries, and as statuary for fountains and impressive garden decor. In some instances these classical images have taken on entirely new iconography in their reincarnation. The writings of distinguished historians and iconographers such as Abby Warburg, Erwin Panofsky, E.H. Gombrich, and Jean Seznec have examined and theorized on the reappearance of the pagan cosmos in the European imagination.[50] Their systematic studies of symbolism brought to light the connections between the Neoplatonic theories and underlying esoteric messages in fifteenth and sixteenth century artwork.

Constellation images covered the pages of the astronomical anthologies in medieval and Renaissance libraries and cosmological images covered the ceilings of fifteenth-century salons. After two centuries of relative neglect, the vast quantity of astronomical and astrological symbolism lying dormant was brought to light; the Sun, Moon and stars of the *Aratea* were not eclipsed by events and disinterest in the late Middle Ages and early Renaissance. The fifteenth century witnessed an explosion of beautifully illustrated *Aratea* manuscripts, not updated, recycled or improved, but still looking back with nostalgia to its antique roots. Some newly invented iconography did adorn the constellations and planets in astrologically based manuscripts, derived particularly from Abu Ma'shar and Michael Scot. These are varied and plentiful, and would require another complete study to discuss

[50]See Warburg, *The Renewal of Pagan Antiquity*; Gombrich, *Symbolic Images: Studies in the Art of the* Renaissance. But many of their conclusions have been challenged by more recent interpretations of Renaissance art.

accurately. But in *Aratea* manuscripts, the same basic designs and formats from earlier medieval books continued to be copied with a few new influences, and some independent inventiveness by the artists, but without major overhaul or innovation. Why? Obviously the patrons, scribes and artists recognized the value in protecting this particular body of information, both text and image in their classical form, and transmitted that format basically intact into the early modern printed era, when interest in the *Aratea* assemblage persisted and has inspired publications and studies that still expand at present, rather than decline.

Toward the end of the fifteenth century, the age of the spectacular illuminated manuscripts was coming to a close, for the printing press was revolutionizing book production. The handwritten parchment painted with brilliant pigments and decorated with glistening embellishments of gold and silver was replaced with cheaper paper pages and simplistic black and white woodcut images, although deluxe parchment books in large and expensive format were still produced occasionally for wealthy booklovers. From the start of printing, the *Phaenomena* of Aratus and the *Aratea* of Germanicus, Cicero and Avienus were among the first classical poems selected to be published; the watered-down *Aratea* texts of the later Middle Ages were edited and restored as much as possible to their original versions. The printed editions were decorated, not as expected with newly designed Renaissance-inspired stellar images, but with the familiar heavenly figures, nude and semi-nude, passed down from Antiquity. The constellation illuminations from the Leiden *Aratea*and Hyginus manuscripts in particular were accurately copied for the printed editions. The printed books lost in visual appeal but gained in readership because of the much cheaper cost and wide availability, in addition to appealing to a broader, literate readership. The fact that the Latin *Aratea* of Germanicus, Cicero and Avienus continued to be read, studied, and analyzed for two millennia stands as testimony to their valued position as a vital contribution to our classical heritage. The final chapter of this work examines several of the cultural and historical factors that influenced the production (or lack of production) of *Aratea* manuscripts. The numerous functions of the *Aratea* manuscripts in the transmission of astronomical knowledge will be examined.

References

Aratus, Kidd, D. (1997) ed. *Aratus*: *Phaenomena:* Cambridge: Cambridge University Press.

Barker-Benfield, B. C. (2008) *St Augustine's Abbey, Canterbury/Vol. 3, Other sources, appendices and indexes*, London: The British Library.

Bischoff, B. (1994) *Manuscripts and Libraries in the Age of Charlemagne* Gorman, M. ed. and trans. Cambridge: Cambridge University Press.

Blume, D., Haffner M., and Metzger W. (2012) Sternbilder des Mittelalters: Der gemalte Himmel zwischen Wissenschaft und Phantasie. Part One: 800–1200. Berlin: Akademie Verlag.

Blume, D., Haffner M., and Metzger, W. (2016) Sternbilder des Mittelalters: Der gemalte Himmel zwischen Wissenschaft und Phantasie. Part Two 1200–1500. Berlin: Akademie Verlag.

Brown, M. P., and Lovett, P. (1999) *The Historical Source Book for Scribes*, London: British Library.

Dekker, E. (2010) "The Provenance of the Stars in the Leiden *Aratea* Picture Book", in the *Journal of the Warburg and Courtauld Institutes*, pp. 77–90.

Dodwell, C. R. (1954) *The Canterbury School of illumination, 1066-1200*, Cambridge: Cambridge University Press.

Dodwell, C. R. (1971) *Painting in Europe 800-1200*, Hamondsworth: Pelican History.

Dodwell, C. R. (1993) *The Pictorial Arts of the West 800–1200*, New Haven: Yale University Press.

Duits, R. (2011) "Reading the Stars of the Renaissance: Fritz Saxl and Astrology". *Journal of Historiography*.

Eastwood, B. S. (1983) "Origins and Contents of the Leiden Planetary Configuration (MS Voss. Q.79): An Artistic Astronomical Schema of the Early Middle Ages". *Viator*, 14, pp. 1–40.

Eastwood, B. S. (1993) "The Astronomies of Pliny, Martianus Capella and Isidore of Seville in the Carolingian World." *Science in Western and Eastern Carolingian Times*. Eds. P. L. Butzer and D. Lohrmann, pp. 161–180.

Eastwood, B. S. (1997) "Astronomy in Christian Latin Europe, c. 500-c. 1150", *Journal for History of Astronomy* the 28, Cambridge.

Gain, D. B. (1976) *The Aratus ascribed to Germanicus Caesar*, London.

Germanicus. Les *Phénomènes d'Aratos*, trans. André Le Boeuffle, (2003) Paris; Les Belles Lettres.

Gombrich, E. H. (1979) *Ideals and Idols: Essays on values in history and art*, Oxford: Phaidon Press Limited.

Godwin, J. (2002) *The Pagan Dream of the Renaissance*, Grand Rapids, MI: Phanes Press.

Grabar, A., and Nordenfalk, C. (1957) Early Medieval Painting from the Fourth to the Eleventh Century. New York, Skira (Rizzoli Publishing).

Haskins, C. H. (1927) *Studies in Mediaeval Science*, Cambridge: Harvard University Press.

Haffner, M. (1997) Ein Antiker Sternbilderzyklus und seine Traderung in Handschriften vom frühen Mittelater biz zum Humanismus, Hildesheim, dissertation.

Harrsen, M. P. and G. K. Boyce. (1953) *Italian Manuscripts in the P. Morgan Library*, New York: Morgan Library.

Kauffmann, C. M. (1975) *Romanesque Manuscripts: 1066-1190*, London: Harvey Miller.

Ker, Neil R. (1964) *Medieval Libraries of Great Britain, a List of Surviving Books*, London.

Köehler, W. and Mütherich, F. (1971) Die karolingischen Miniaturen, 4. Die Hofschule Kaiser Lothars. Einzelhandschriften aus Lotharingien. Berlin: Cassirer.

Kunitzsch, P. (1989) "The Astronomer Abu 'L-Husayn Al-S_f_ and His Book on the Constellations", *The Arabs and the Stars*. UK: Variorum Collected Studies Series 307, Northampton, pp 69–74.

Le Boeuffle, A. (2003) *La texte des Aratea de Germanicus est copieusement représenté*: Paris: Les Belles Lettres.

McGurk, P. (1966) *Catalogue of Astrological and Mythological Illuminated Manuscripts of the Latin Middle Ages*, London: Warburg Institute, University of London.

McKitterick, R. (2000) "Books and sciences before print", in *Books and the sciences in history*, M. Frasca-Spada and N. Jardine eds. Cambridge: Cambridge University Press, pp. 13–34.

Mostert R., and Mostert M. (1990) "Using astronomy as an aid to dating manuscripts: The example of the Leiden planetarium", *Quaerendo,* 20: pp. 248–261.

Mütherich, F. (2004a) "Book illumination in the court of Louis the Pious" in *Studies in Carolingian Manuscript Illumination*, The Pindar Press, reprinted in 2004

Mütherich, F. (2004b) "Die bilder des leidener aratus" in *Studies in Carolingian Manuscript Illumination*, The Pindar Press.

Mütherich, F. and J. E. Gaehde. (1976) *Carolingian Painting*. New York: George Braziller.

Nees, L. (1991) *A Tainted Mantle: Hercules and the Classical Tradition at the Carolingian Court*, Philadelphia: University of PA Press.

Nordenfalk, C. (1938) *Der Kalender vom Jahre 354 und die lateinische Buckmalerie des IV Jahrhunderts*. Göteborg: Elanders boktryckeri aktiebolag.

Nordenfalk, C. (1988) *Early Medieval Book Illumination*. Geneva, Skira; New York, Rizzoli.
Panofsky, E. and Saxl, F. (1933) *Classical Mythology in Mediaeval Art*, New York: New York Museum of Art.
Reeve, M. (1983) "*Aratea*", *Texts and Transmission*, ed. Reynolds, L. D., Oxford, pp. 18–24.
Salmi, M. and E. M. Borgese (1954) *Italian Miniatures*, New York: Abrams.
Salzman, M. R. (1990) *On Roman Time: The Codex-Calendar of 354 and the Rhythms of Urban Life in Late Antiquity*, CA: University of California Press.
Saxl, F. (1957a) "Illustrated Science Manuscripts in England", *Lectures*, 2 vols. The Warburg Institute.
Saxl, F. (1957b) "The Belief in Stars in the Twelfth Century", *Lectures*, 2 vols. The Warburg Institute.
Schutz, H. (2004) *The Carolingians in Central Europe, their History, Arts and Architecture*, Leiden: Brill.
Seznec, J. (1953) *The Survival of the Pagan Gods*, trans. B. F. Sessions, Princeton: Princeton University Press.
Verkerk, C. L. (1980) "*Aratea*: A Review of the Literature Concerning MS. Voss. Lat. Q. 79 in Leiden University Library", *Journal of Medieval History* 6, pp. 245–287.
von Euw, A. (1993) "Die bilder des Leidener Aratus" in *Science in Western and Eastern Civilization in Carolingian Times*, P. Butzer and D. Lohrmann, eds. Basel: Birkhäuser Verlag.
Weitzmann, K. (1971) *Studies in Classical and Byzantine Manuscript Illumination*, Chicago: University of Chicago Press.
Yeomans, D. K. (1991) *Comets: a Chronological History of Observation, Science, Myth, and Folklore*, New York: Wiley Science Editions.

Chapter 8
Analysis and Conclusions

Abstract This final section of the monograph presents and analyzes the findings of the preliminary Aratean manuscript study and summarizes the various results and conclusions that can be drawn from the vast amount of evidential and scholarly material currently available. The numerous functions of these illustrated astronomical manuscripts are defined, compared, and evaluated for the purpose of exploring the influence of its literary impact and establishing the true significance of this astronomical manuscript tradition through the medieval period and its associated values that assured its millennial-long survival. The great variety of surviving evidence provides an understanding of where Aratus and the Aratean poets belong in the scientific traditions of the medieval Latin West. The considerable variety of functions performed by the constellation illustrations is established by looking at the diverse aspects of the stellar images along with the Latin text that accompanied the pictures. The primary role of the entire oral and literary tradition in the transmission of astronomical knowledge can be demonstrated, yet certainly remains open to argument and supplementation.

This survey has gathered together 60 surviving manuscripts that contain varying portions of Aratean text, all codices were produced between the ninth and fifteenth centuries, nearly all of them illustrated with full constellation cycle. The study has provided basic information on the majority of the *Aratea* manuscripts, as well as including other popular medieval astronomical manuscript traditions. Where accessibility allowed, a more in-depth presentation on details and particular characteristics of the manuscripts was also included. This inquiry looked not only at the manuscripts and illustrations themselves, but at many of the cultural climates and historic periods that inspired their production and garnered information on many related topics. These numerous and varied factors and influences cannot be excluded when endeavoring to gain a more all-embracing conception of the function and astounding survival of so much of this complex Aratean tradition. Although far from comprehensive, this marks an initial step toward bringing a bit

M. Dolan, *Astronomical Knowledge Transmission Through Illustrated Aratea Manuscripts*, Historical & Cultural Astronomy,
DOI 10.1007/978-3-319-56784-6_8

more organization and clarification to this complex and exceptionally long-lasting artistic tradition; much still remains to be determined. Each of the 60 extant illustrated manuscripts could be studied and placed in proper perspective. Rather than focusing on one particular aspect of this manuscript tradition, the work provides an overview of the various functions performed by the *Aratea* in the societies that created them, and the reception of the *Aratea* manuscripts through the constantly evolving medieval period.

The medieval period can be viewed as the Age of Faith; proponents channeled their intense spiritual beliefs and most aspects of the physical world toward devotion to the sacred. The major focus of medieval art reflected that religious zeal, but the material world encompassed human ambitions, desires and doubts, and many questioned nature and philosophical or scientific principles. Nonreligious artworks and interests are often neglected or marginalized in favor of the more spectacular or impressive creations designed for liturgical usage. Arts and sciences, spiritual and secular activities often intermingled; astronomical calculations were essential for determining the date of Easter; monks experimented with flora and fauna and included their results in their gardens, farms and illustrated manuscripts; kings were involved in pious prayer and collecting votive offerings and secular manuscripts, as well as promoting wars to expand their territories As the Middle Ages gave way to a more enlightened age, art and literature began to withdraw from their religious fixation and turn toward the natural world and desires of the human condition. But throughout these major changes in societal interest, the illustrated *Aratea* manuscripts were continually copied and circulated in monastic, cathedral and royal libraries.

Classical antiquity had passed on a large collection of scientific knowledge and associated art, although a huge percentage of ancient Greek knowledge has been lost. Scientific treatises and illustrations, including diagramatic forms and terrestrial and celestial maps, renewed during the Carolingian period and then reconstituted under Holy Roman Emperor Frederick II, seemed a largely separate antiquarian production in many cases. Diagrams relating the circulation of the planets, phases of the moon, the relationship of the elements and humors, tidal patterns, and other natural phenomena remained consistent over long periods (Kessler 2004: 131). Many of the diagrams found in encyclopedic and scientific manuscripts can be credited to Pliny the Elder; these condensed devices were adapted by Isidore of Seville and Pseudo-Bede to elucidate their writings. Diagrams found in the Liber Floridus, a twelfth-century encyclopedia written by Lambert, the canon of St. Omer, presently Northern France, were borrowed from these three main sources (Figs. 8.1 and 8.2) (See Lambert of St. Omer, Derolez and I. Strubbe, ed. (1968)).

This topic of illustrated Aratean manuscripts encompasses at least four inter-related disciplines: history of poetry and literature, history of mythology, history of astronomy, and history of art. *Aratea* manuscripts by their nature institute a meeting place of these distinct branches, all encased in one vast and complex topic. A most essential fact relevant to this phenomenon remains that the original

Fig. 8.1 Courses of the Planets, *Liber Floridus*, Rijksuniversiteit te Ghent Bibliotheek, MS 92, Twelfth Century, Courtesy of Ghent University Library

Fig. 8.2 Phases of the Moon, *Liber Floridus*, Rijksuniversiteit te Ghent Bibliotheek, MS 92, Twelfth Century, Courtesy of Ghent University Library

poem of Aratus and its successive incarnations have survived through centuries of modification and transition, which speaks to its important position in each of these disciplines.

The current study has tried to limit the scope of this huge topic by concentrating mainly on one specific aspect of the *Aratea* manuscripts, its role in the transmission of astronomical knowledge, and by concentrating on three specific time periods, since examining the entire medieval period would be repetitious, even overwhelming. But the long and complicated transmission of astronomical knowledge from the Greco-Roman World to the Latin West cannot be oversimplified or restricted to just a few time periods or to a single manuscript tradition. And so for context and comparison, a brief overview of the main channels for conveying astronomical information within the illuminated manuscript tradition was included. Understanding other sources of astronomical information that circulated during the medieval period can assist in determining what specific role the stellar material passed on within *Aratea* manuscripts actually played in the history of astronomy. In that way, the impact of *Aratea* manuscripts can be evaluated in content and importance by comparison with other conduits of astronomical information.

When undertaking this survey of *Aratea* manuscripts from a holistic viewpoint, numerous factors were considered. This culling of information from historical and scholarly literature and from the manuscripts themselves focused on the wider view, as a more narrow approach could not provide a proper perspective or understanding. Each surviving *Aratea* manuscript belongs to its own particular circumstance of production, created within the cultural, political, and educational milieu that encouraged its manufacture. Even though each codex stands alone, noticeable patterns can be detected and certain conclusions can be drawn. Basing this discussion on art historical and historiographic methods, it was essential to establish the function of the *Aratea* within various disciplines. Much of the material included in this study had not been assembled in one account, or presented in a systematic and comprehensive form. This interdisciplinary inquiry has drawn together the separate aspects of *Aratea* manuscripts and looked at the different roles they played and asked pertinent questions in order to gain a fuller understanding of their place in the history of astronomy and of art.

Summary of the Functions of *Aratea* Manuscripts

Function of the Aratea *in History of Literature and Poetry*

For countless years, humans have observed the nighttime sky in wonder, fascinated by and questioning the nature and continual movement of that unbelievable sight. Many ascribed meaning and purpose to the glistening phenomena overhead; faces were seen in the moon, stellar patterns were established, nearby stars were connected and imagined to form specific pictures of familiar fabled creatures or legendary persons, such as sheep, bull, crab, and scorpion, even mythical gods. Transitory heavenly phenomena were assumed to bring special influences that instigated beneficial or malevolent results to earthly events and human lives.

The semi-religious guide to the constellations composed by Aratus in the third century BCE has endured a lengthy and intriguing life. What factors allowed this astronomical poem to survive and thrive until the present day, when so many ancient works have perished? Although little read today, didactic poetry was enormously popular in the ancient Greco-Roman world. "Obviously, the purpose of didactic poetry is to instruct the reader, usually in a systematic manner, with a single instructor's voice advising or explaining to those he is addressing" (Toohey 1996: 2). Through the easily remembered verses of his poem, Aratus taught his readers the relative positions of the constellations along with the order of their risings and settings and some basic astronomical information. The instructive quality of his original Greek poem was maintained in the initial Latin translations, but the astronomical content cannot be separated from the poetic skill and literary value of the Latin authors. All of these features: the subject matter, the artistry and the sense of history were appreciated by learned readers in the Classical world. During the Middle Ages, famous authors from the Classical world were highly respected for their literary excellence. During the ninth century, much of their writings had been conserved and has been treasured since that time. Fortunately the Greek poem of Aratus and the Latin poems of Germanicus and Avienus survive intact, along with considerable fragments of Cicero, as "an enormous amount of Latin literature was lost even in ancient times" (Reeve 1983: xiv).

Concern for classical literature and its survival was an important factor in the ninth century, inspiring the large number of *Aratea* created then and into the following centuries. During the twelfth century, many essential developments took place in the field of literature. An impressive quantity of Latin poetry was composed to accompany sacred music with religious verses, uniting music and literature. Interest in literature continued to expand as Latin authors edited and rewrote favorite classical epics, such as the *Iliad* and the *Aeneid*, updating them and adding recognizable medieval traits to the characters in the epic poetry (Colish 1997: 182). The adventures of popular heroes of history, Alexander the Great, Charlemagne and King Arthur were reproduced in both Latin and the vernacular. (The first vernacular language to be immortalized in literature and science was Castilian.) The love poetry of Ovid was revived by medieval readers and courtly romance literature entertained audiences in the later Middle Ages. Only four illustrated *Aratea* manuscripts are extant from the twelfth century, but perhaps many earlier manuscripts were already available, and perhaps many have just not survived.

But the pure poetic content and literary excellence of the original Latin *Aratea* translations changed and declined through the long centuries of the medieval period. Already in the Late Antique period, the poems of Cicero and Germanicus had been 'corrupted' and 'improved' with the addition of scholia. The Aratean poems preserved by Carolingian scribes are the versions truest to the original Roman authors (until the age of printing when scholarly editors looked back to the oldest sources). The renderings of the Latin *Aratea* that are carried through the Middle Ages seem quite tainted next to the pure epic poetry of Aratus. What passed through the centuries as the poetry of Cicero and Germanicus was more an assortment of borrowed and re-assembled astronomical writings interspersed with scant portions of poetry. Usually not more than six or seven lines of verse appear together on any manuscript folio without being interrupted by segments of prose and illustrations. By the fifteenth century, a life of Aratus had been added to most copies along with the Revised Aratus Latinus, bits of Germanicus, the various scholia and text of Hyginus, which leads unsuspecting readers to believe the writings actually were the work of Aratus.

In contrast the Latin poem of Avienus persisted without scholia, was less impaired, and endured in a more intact state, but his manuscripts were not illustrated. Therefore his version of the *Aratea* was much less popular. Since only three manuscripts containing Avienus' astronomical poem survive from the medieval period, his writings were not a part of the illustrated Aratean tradition; and so they have not been included in most discussions. This raises the question; did the fact that the *Aratea* of Avienus had no illustrations help keep the poetic content intact? Did the addition of pictures somehow detract from the poetic value and the literary content of the *Aratea* of Germanicus and Cicero, or were separate factors influencing the deterioration of the poetry?

From the tenth to the fifteenth centuries, the Latin poems of Cicero and Germanicus became more and more diluted. The *Aratus Latinus* and the *revised Aratus Latinus* had eliminated the verse format completely; they were written entirely in prose. The brilliance, wit, and erudition of the classical poetry were no longer the core feature of Aratean manuscripts. The literary value had deteriorated and the clever allusions to literature of the past poetry of Homer and Hesiod had been entirely lost. The unity and beauty of the original astronomical vision, a starry sky designed by a benevolent force first envisioned by Aratus, and then transmitted by the Latin translators, had degenerated. Although the literary excellence of the original works was not often conveyed intact, the basic textual content (the description of constellations) was still retained, but in a reduced state. The body of astronomical information, the descriptions of the constellations, myths and images continued to be transmitted, but the literary value of epic poetry survived in a very limited state. In spite of the many alterations, the epic poetry of Aratus and the Latin authors are still studied by classical scholars today and mined for the depth of their contemporary material. The mythical component of the *Aratea* manuscripts plays a significant role in their value as well, both in the medieval period and still today.

Function of Aratea *in History of Myth*

What was the significance or relevance of preserving remnants of ancient Greek and Roman mythology through the Middle Ages? Why, in a culture dominated by Christianity, the Christian Bible and its religious dogma, were the tragic, heroic, and lusty myths of ancient pagan gods and goddesses still of interest? In medieval societies where few people were literate or had access to books, how did people acquire their knowledge of ancient mythology? In the Roman world, interest in the past and in Greek and Hellenistic art and culture was essential and understandable.

> A comprehensive knowledge of Greek myth was essential for anyone who wanted to hold his head up in polite society, throughout the Roman world. Not just for reading the classics or identifying allusions in contemporary poets and orators but for understanding the conversation of peers, scenes on wall paintings, silver plate, mosaic floors and the decoration of the sarcophagi in which loved ones were now interred (Cameron 2004: 222).

Religion, myth, and art were closely allied in classical cultures in an attempt to give expression to the forces of nature that dominated daily existence. Educated persons in the Roman world knew classical myths in the same manner that medieval Christians knew the stories of the Old and New Testaments. Even though by the end of the fourth century, Christianity had supplanted paganism and the many secretive Roman cults, such as Isis, Dionysus, and Mithras; nevertheless literate and illiterate people continued to enjoy the stories of classical mythology alongside those of the Bible.[1] Dismissing thousands of years of antiquity and familiar folklore, became impossible.

In the Middle Ages, the ancient classical myths were acted out and alluded to in performing arts; drama, mime, plays, songs, and musical programs (Cameron 2004: 222). In addition, the fabled places of mythical events were still remembered and revered as medieval pilgrims traveled to well-known celebrated sites throughout the Mediterranean world. The feats and archetypal traits of the planetary gods and constellation figures persisted because they had become deeply rooted in human consciousness. "The myths preserve and transmit the paradigms, the exemplary models, for all the responsible activities in which men (and women) engage" (Eliade 1954: x). The symbolic tales of mythology often contained lessons for understanding human nature and timeless truths, in the same way that stories found in the Bible are now considered mostly symbolic, created to project those eternal truths in a fundamental manner.

"The pagan divinities served as vehicles for ideas so profound and so tenacious that it would have been impossible for them to perish" (Seznec 1953: 149). Undoubtedly the illustrated *Aratea* manuscripts were a dominant factor during the medieval period, credited with helping to keep alive many of those classical traditions with their pictorial images of the planetary gods and goddesses, as well as the numerous characters, animals, and articles from the ancient myths. The mythological material was not entirely adapted from the original poem of Aratus for he included few details about the mythology of the constellations. The myths are limited in number, as Aratus supplies only 14 brief myths to accompany the 43 constellations he discusses. While Germanicus, in his translation, doubles the number of myths included. Much of the mythological material contained in

[1]For more information of this survival of ancient mythology see Selznec, *Survival of the Pagan Gods* and also Weitzmann, *Studies in Classical and Byzantine Manuscript Illumination*.

astronomical manuscripts can be traced to Hyginus whose work, as we have seen, was added to many *Aratea* manuscripts as scholia. The works of the two different authors became intermingled in the Middle Ages, probably indistinguishable and unimportant to a medieval scribe or reader.

In the Christian era, the colorfully painted *Aratea* manuscripts were an acceptable means of transmitting mythological stories, as they combined the allure of science, history, and ancient knowledge. Although the Aratean codices generally did not expand the mythic stories beyond the basic 'facts', the tragic story of Cepheus and Cassiopeia, the sacrifice of Andromeda, the feats of Perseus and Orion, and fate of Virgo could incite the reader's more inclusive remembrance of these archetypal tales. The heroic expedition of Hercules in his pursuit of the Golden Apples and the danger-ridden voyages of the Argo in Jason's quest for the Golden Fleece epitomize the "difficulties of the seeker for the road to the self, to the 'center' of his being… The road is arduous, fraught with perils, because it is, in fact, a rite of the passage from the profane to the sacred, from the ephemeral and illusory to reality and eternity, from death to life, from man to the divinity" (Eliade 1954: 18).

In spite of its changes over time, the text of *Aratea* manuscripts fulfilled a vital function in maintaining the existence of the entire corpus of constellation myths from antiquity to the present day. Not only did the body of mythical material transmit the age-old stories, but the astronomical images captured and retained the visual impact and impressed the images in a way not possible through any other media. This is not the case for the entirety of ancient Greek mythology.

> The images that survive from antiquity represent only a fragment of what once existed, the same is true of the myths they illustrate. Out of the hundreds of tragedies dramatizing myths that were produced in the 5th century BC, we have only thirty-one, the rest having been lost when neglected books ceased to be copied, old manuscripts disintegrated and precious libraries were destroyed (Woodford 2003: 255).

The role of the illustrated *Aratea* manuscripts in the transmission of ancient myths therefore is enormous; not so much for the individual tales, as they are basically brief and limited in their details. Its chief accomplishment lies in relaying intact throughout the entire Middle Ages the complete corpus of astronomical mythology associated with the classical constellations from the ancient world. The *Aratea* manuscripts conveyed the full body of mythological material relating to the constellations, and also retained the depictions and idiosyncrasies of heroic and tragic figures, animals, and objects featured in the myths. Modern culture continues much richer for having access to this crucial corpus of astronomical text along with its valuable illuminating images that have retained their Late Antique characterization.

Function of the Aratea *in History of Religion*

Since the poem of Aratus and the translations of Germanicus and Cicero were all written before the advent of Christianity, they of course have no association or references to the tenets of the Christian religion. In the context of religion, the points of interest for this study are how this 'pagan' poem concerned with the celestial semi-deities from antiquity was received and accepted in the medieval Christian world. Church scriptoria, particularly the medieval monastic communities, were primarily responsible for the preservation of classical learning (McCluskey 1998: x) and also for the survival of the *Aratea* manuscript tradition. The majority of the manuscripts that survive from the early medieval period are Biblical and liturgical writings, especially from the Latin Church Fathers: Augustine, Jerome, Ambrose, and Gregory. When manuscript production increased dramatically in the ninth century, the main focus and effort was directed toward religion and exegesis.

> While the Christian Fathers endeavored to prove that the pagan gods were either illusions or malignant demons (therefore transmitting much valuable information about them), the pagan world itself had become so estranged from its divinities that the educated public had to read up on them in encyclopedias, in didactic poems or novels, in special treatises on mythology and in commentaries on the classic poets (Panofsky and Saxl 1933: 45).

The success of Christianity and the degree to which it gradually engulfed more and more of the intellectual drive and emotional commitment of the Latin West could only have been at the expense of pagan literature and values. In order for the Aratean manuscripts to survive through the thousand-year medieval period, it was necessary for them to find safe haven within the walls of monasteries and cathedrals. Although the attitude of the Christian Church to pagan learning may have been complex and ambivalent, the aura of respect for Greco-Roman intellectual accomplishments permitted the preservation of much ancient literature such as the *Aratea*. The astronomical manuscript tradition was carried forward and new Latin 'editions' of Germanicus and Cicero were reproduced almost every century. Pagan poetry and myths about the ancient gods, goddesses, heroes, and heroic animals who became catastrophized continued to be enjoyed along with the Bible stories. Art historian Bernard Bischoff stated that knowledge of the classical authors became a necessary prerequisite of Christian culture.

Therefore the reception of *Aratea* manuscripts in Christian Europe during the medieval period was one of honor and respect. The fact that *Aratea* manuscripts were continually copied and revitalized, most often in monastic settings, during the earlier centuries, conveys evidence that the classical heritage was valued in spite of, or perhaps because of, its historic subject matter. It stands as a credit to the poetry of Aratus, Cicero, Germanicus, and Avienus that the pagan aspects of their poems were in most cases not edited nor removed, but respected and preserved. Unlike other classical writings and even the writings of Copernicus, Bruno, and Galileo, the *Aratea*, although 'watered down' with scholia, were not significantly 'purged', moralized, Christianized, or placed on the Index. The 'echoes of medieval mysticism' never swayed the moral or immoral stories attached to the constellations. Cepheus was willing to sacrifice his daughter to the sea monster, the vain Queen Cassiopeia insulted the daughters of the sea god by bragging of her beauty, and Cetus the monstrous dragon-whale personified the evils of the world. Even though all manuscripts were produced under the dominance of Christianity, Christian dogma never penetrates the Aratean tradition.

Another 'Aratean' misconception that should be addressed is the fact that astronomical information in *Aratea* manuscripts came to be perceived by modern historians as relevant to the Christian calendar, feast days, and computus. The function of *Aratea* manuscripts in the history of religion has often been misunderstood and its astronomical content has been misinterpreted. Large amounts of misinformation can be found in the art historical literature concerning this mistaken idea. Even well-respected scholars are unclear on the content and meaning of the poem, but in their defense, there have been no comprehensive sources until recently to provide this basic information of the *Aratea*.

In the *Illuminated Manuscript* Backhouse, in her description of Cicero's *Aratea*, BL Harley MS 647, states, "This became the fundamental textbook of medieval astronomy, a science of everyday importance, because correct observance of the Church's feast-days depended upon an accurate understanding of the movements of the heavenly bodies" (Backhouse 1986: 14). This complete misunderstanding, both of the *Aratea* tradition and of the method of determining moveable feast days, mainly Easter became the norm; the *Aratea* manuscripts were not considered textbooks for medieval astronomy, but were included in astronomical anthologies. With the rise of universities in the thirteenth century, Sacrobosco remained the required text for the study of astronomy.

Determining the date of Easter and keeping the calendar current were continual problems during the Middle Ages. But the information found in *Aratea* manuscripts concerns the positions of the constellations and the order in which they rise and set. This data was helpful for orienting your position on Earth or on sea at night, or determining the time of night, but was of no use for establishing the date of Easter or other Church feast days. The date of Easter was determined by a complex mathematical formula based on the 18 ½-year cycle of the moon, not by direct observations of the night sky. Complex computus diagrams were required in order to calculate the date for Easter each year, and these are sometimes found in the same astronomical collections as the *Aratea*, which may account for the confusion. The *Aratea* text does not discuss the cycles of the Sun, Moon, or planets and especially not Christian church calendars.

In another example, D. Kahn in *Canterbury Cathedral and Its Romanesque Sculpture*, looking for prototypes for sculpture in manuscripts, states, "This sea-monster (Cetus) can be found in its proper astronomical context in the *Phaenomena* by Aratus, the fundamental text for the calculation of the liturgical year… Its presence in the library may explain the popularity of this type of grotesque in Canterbury manuscripts and sculpture" (Kahn 1991: 60). The manuscript to which she refers is not the *Phaenomena* by Aratus, but the *Aratea* of Cicero, and the calculation of the church's liturgical year was never a concern of the *Aratea* manuscripts.

As the strongly religious environment of the Middle Ages gave way to the more relaxed attitudes of the fifteenth century, the astronomical illuminations in *Aratea* manuscripts continued to be reproduced, in even larger numbers. They maintained and transmitted an increasing wealth of classical symbolism. The *Aratea* manuscripts were no longer languishing in monastic libraries, but were produced anew and commissioned by cardinals and nobility for their private libraries. But still the naked images of the pagan gods, goddesses, heroes, and heroines created no conflict for Christian leaders.

At the start of the fifteenth century, almost no images of the classical gods had been represented in frescoes or sculptural form for a thousand years, although they were available in many illustrated astronomical manuscripts. By the end of the quattrocento, images of the classical deities, latent in *Aratea* manuscripts, began to decorate the walls, ceilings, and gardens of numerous palatial homes. "A significant change in the reception and status of the pagan gods and a fundamental transformation in the history of art is what happened to idols in the fifteenth and sixteenth centuries" (Camille 1985: 338). Along with the decor of the classical deities and demi-gods, Renaissance palaces of dukes and cardinals also displayed religious art and sacred icons of Christianity. The combining of old and new religious artistry seemed to create no struggle for the conflicting images or for the patrons. The *Aratea* manuscript tradition, having no direct connection with Christianity, passed through the Middle Ages 'watered down', but unlike so much of non-Christian historical literature, survived unscathed by religious fervor.

Function of the **Aratea** *in History of Astronomy*

Today, the popularity of astronomy has dwindled and has become a topic of little concern or relevance for the majority of people, except during the rare appearance of a comet, a 'blue moon', 'super moon', the occurrence of a solar or lunar eclipse, or a rocket into space. In the postmodern era, astronomy exists strictly in a rarified atmosphere, the domain of astrophysicists, celestial mechanics, and theoretical mathematicians. But this was not the prevailing attitude in the world before the onset of light and industrial pollution; in the not so distant past, the view of the stars gliding across night sky and the annual movements of the celestial bodies were an intimate part of the rhythms of everyday life.

The attempt to assess the function, influence, and role of the *Aratea* manuscripts within the discipline of astronomy, required an examination of the various aspects of astronomy separately in order to gain a better understanding of that role. Thus this inquiry looked at contemporary astronomical texts and at similar manuscript traditions. Before the twelfth century, the Latin West had inherited only meager extracts of Greek science and was little concerned with new observations and computations. For the most part, Western Europe made a painfully slow start in producing scientific and astronomical advances.

Through his studies of the history of medieval astronomy, Stephen McClusky has created a helpful model for defining astronomical knowledge in the early Middle Ages. He has determined that there were at least four distinct astronomical traditions that 'flourished and interacted' (McCluskey 1993: x–xi). The first and oldest type was traditional folk astronomy, the main feature of which was an observational solar calendar, dividing the year into quarters by determining the exact times of the solstices and the equinoxes. This type includes the use of the heliacal rising and setting of prominent stars or constellations to denote significant markers through the solar year. This technique is found in oral traditions and in the earliest literature of the Babylonians, Greeks, the Bedouins, and undoubtedly goes back to even more remote times. Evidence for these practices extends to celebrations at prehistoric stone monuments to medieval Christian festivals, for many saints' feast days are celebrated on particular celestial markers.[2] As is well known, the four key dates of the astronomical year were taken over for honoring the conceptions and births of Christ (March 25, December 25) and St John the Baptist (June 24 and September 24).

The second medieval astronomical tradition defined by McClusky was computus, the arithmetical astronomy used to calculate the date of Easter. This discipline required continual calculations and updating to be sure all areas of the medieval world celebrated Easter on the proper day. A third kind of astronomy was centered on the observations of the stars to determine the time of night, a necessary procedure in medieval monasteries to call or awake the monks for ritual prayer (McClusky 1990: 9–22). The fourth tradition involves geometrical astronomy, concerning the continuous motions of the heavens within a geometrical model of the universe. The *Aratea* manuscripts are not a direct component for supplying needed dates or data to any of these four astronomical classifications. When considering the first type, *Aratea* manuscripts provide no specific data for determining calendric dates, merely general locations of the constellations as they relate to each other, and their risings and settings.

As for the computus and calculating the date of Easter, obviously *Aratea* manuscripts have no association with this Christian aspect of astronomy. The material in Aratean manuscripts may have been helpful in the third type, determining the time of night by the position of the stars. But this requires years of careful observation of the heavens since the area of the sky visible each night changes throughout the year and that sort of specific information is not included in the *Aratea*. The fourth astronomical tradition is definitely not a factor for the type of information in Aratean manuscripts has no mathematical aspects. Nor do they contain any information gained from the technical astronomy of other ancient

---

[2]The establishment of equinoctial and solstice markers existed and has been uncovered in almost every advanced culture. Recent excavations begun in 1996 by Klaus Schmidt in Göbekli Tepe, Turkey have revealed astronomically-aligned stone circles, decorated with animal figures, that date back at least 12,000 years.

Greek writers. In what way therefore does the information found in *Aratea* manuscripts become significant in the history of astronomy?

The history of astronomy was influenced and advanced enormously by Islamic scientists who maintained and corrected Ptolemy's writings. For example, Ptolemy determined the obliquity of the ecliptic to be 23°51′20″ which was determined as too large by Muslim astronomers. New measurements took place in Baghdad during the first half of the ninth century; they concluded that the inclination was closer to 23°33′ (Saliba 2007: 81). This finding and many others were established by competent Islamic astronomers; similar updates and observational corrections continued into the medieval period. Later Islamic astronomers benefited from the passing of time to correct the inaccurate astronomical parameters that were encased in the Greek tradition. They named their astronomical studies and findings "the science of the configuration of the world" (Saliba 2007: 175). During the eleventh and twelfth centuries, many Arabic astronomical writings and Arabic versions of Greek scientific and philosophical works were translated into Latin and assumed into the medieval culture of the Latin West. These texts played an important role in the "renaissance" of the twelfth century, but the rediscovery of Ptolemy's great works and other dynamic astronomical advancements played no part in the astronomical text of the *Aratea* tradition.

During the twelfth and thirteenth centuries, newly invented labels were assumed to indicate the various sciences that studied the heavens. The general 'science of the stars' (*scientia stellarum*) described the knowledge of the entire heavens. The 'science of the movements' (*scientia motum*) indicated the classification of astronomy; the third label was a 'science of the judgements' (*scientia idudiciorum* or judicial astrology) to specify astrology. The origin of these new terms can be definitely traced to the translations from Arabic; they also indicate a revival of a scientific curiosity of the motions and influences of the stars (Lemay 1962: 64).

In the course of this research, many direct questions were addressed. In particular, what kind of astronomical information can be derived from the *Phaenomena* and *Aratea* manuscripts, and how was that data received, used and modified through the Middle Ages? Unfortunately the complicated and extensive tradition of *Aratea* manuscripts does not allow a simple answer, as the astronomical content of the original poem of Aratus differs enormously from the information in later *Aratea* manuscripts that were copied and adapted throughout the medieval period. The three Latin translators generally followed the lead of Aratus in the data they included in their versions. To analyze and compare the exact text included in each Latin version would be an enormous undertaking, especially since the surviving manuscripts are far from uniform that will be left for text specialists and paleologists. Scholarly editions of each work are available for that particular research. For a basic understanding of the direct astronomical information in the original Greek *Phaenomena* of Aratus, the following will suffice:

1. The *Phaenomena* states the names and positions of 46 constellations, starting at the north celestial pole and circumpolar constellations, and then moves about the visible night sky defining each of the 46 constellations and explaining how they relate to each other in space.
2. The later section of the poem becomes slightly more technical, relating the design of the heavens with the polar axis and four celestial circles, defining the Tropics of Cancer and Capricorn, the ecliptic and the Milky Way. Kidd (463–545)
3. The poem differentiates the constellations in the northern from the southern celestial hemispheres.
4. It names the 12 signs of the zodiac, beginning with the Crab through the Twins, and explains that the Sun 'goes round this circle'. Kidd (546–555)

The areas that the original Greek poem and Germanicus' version do not touch upon are important as well.

1. The particular stars that make up the constellations are not a concern; for the most part, they are not mentioned individually, since very few stars were even named before the Arabic translation movement.
2. The original poems do not include the number, magnitude, or position of the stars within each constellation, although a bit of this was added through the later scholia.
3. Only minimal mention of the planets is provided, with no specific information about their order in the heavens, their movements, or relative sizes, not even their familiar names or the traditional godlike attributes assigned to them. This is true for both the Greek and the Latin versions.
4. No information is delivered on transitory celestial events, such as comets, solar or lunar eclipses, meteorites, or the changing hours of daylight and darkness. Interest focuses on the perfection of the celestial realm, but the *Weather Signs* of Aratus do offer beneficial data concerning terrestrial phenomena.

In addition to the large amount of direct information that can be found in the *Phaenomena* and *Aratea*, a variety of indirect data can be deduced through calculations and computer reckoning.

Indirect Information Determined by Modern Astronomers

A surprising discovery was brought to light by astronomer Sergey Zhitomirsky in his investigation of the basis of the original celestial observations appropriated by Aratus when composing the *Phaenomena* (Zhitomirsky 1999: 483). His astronomical studies have extracted additional and most unexpected data from the poem.

By scientific analysis of the text of the poem, astronomers have determined exactly how Aratus derived his view of the constellations. You would expect that Aratus would have just looked up at the night sky and described what he saw, but that process is not so easy. One can never get a complete celestial view at one time, only by viewing for the entire yearly cycle, since the positions of the constellations turn through the hours of the night and the months of the year. Even astrologers did not create natal horoscopes directly from their own observations of the heavens; they always looked up the exact positions, the celestial latitude and longitude, in a star atlas or ephemeris.

By examining specific details that Aratus provided in his descriptions, astronomers are able to determine at what latitude, and at what time period, the original observations were made. Since the fixed stars move in a very slow orbit in relation to the Sun, the rising and setting of the constellation positions are like a cosmic clock that can be 'read' by accurate calculations. Using the information that Aratus supplies, such as the constellations that are never seen to rise or set, astronomers were able to determine the approximate date and the latitude at which his descriptions would be accurate. Their results were quite unexpected and actually astounding. In order to view the constellations exactly as Aratus describes them, the appropriate latitude would be limited to 38° or 39° North. That latitude and other technical information in the poem have pointed to the conclusion that the observations recorded by Aratus, which were actually borrowed from the *Phaenomena* of Eudoxus, were documented to have been observed not in Greece, but on the island of Crete.

Investigators assumed that the time period for the appropriate positions of the constellations, that is, when they aligned with the celestial coordinates that are described in the poem, would place them at locations appropriate for the fourth century BCE. This is the time when the positions were fixed and 'published' by Eudoxus. But to the amazement of all (all of us who are interested in this discovery), astronomers have determined that the only possible time when these positions were accurate would move the original observation date to a much earlier time, approximately 2600 BCE (plus or minus 800 years). This is not just one opinion; three separate astronomers have confirmed this ancient date for Aratus' constellation positions. Analysis of the astronomical data from the poem places the time of the celestial reading at least 1500, maybe 2000 or more years before Aratus or Eudoxus lived.

The startling results of this investigation have raised another whole series of fascinating questions. How was this specialized astronomical information held and transmitted over such an enormous time span? How could an astronomer in ancient Greece obtain and record the results of observations of one to two millennia previous? It has been suggested that Eudoxus may have seen an older celestial globe that may have somehow survived. But it is doubtful that celestial globes even existed before the time of Eudoxus; Cicero says that Eudoxus invented them.[3] Did

[3]See Chap. 2 or Dekker's study of celestial globes, *Illustrating the Phaenomena*.

the formation of pictures have a role in transmitting this specific information? Could a written or carved record of some sort have survived from this early time? But who could have written or read it? (Linear B was the ancient script of Crete, based on Linear A, the oldest surviving record of the Greek dialect known as Mycenaean. The script was used from approximately 1500 BCE and 1200 BCE on the island of Crete. [Ancient Scripts.com] This information could be 1000 years earlier.) Another source of transmission may be oral tradition. But that constitutes odd and lengthy information to pass on orally (but then Homer's works were quite long). The means of transmission of this specialized astronomical information are most perplexing, but the evidential data survives.

Function of the Aratea *Manuscripts in Education*

An often-repeated statement found in history and art history books is that the illustrated *Aratea* manuscripts were a fundamental work of reference for learning astronomy in the Middle Ages. One of the goals of this investigation was to determine if this statement was correct. Were the *Aratea* codices the basic source of astronomical information during that time? Were there even any accomplished astronomers in the Latin West during the early to mid-medieval period? It appears that there were few astronomers doing any type of active studies during the medieval period; those students or teachers concerned with celestial phenomena were primarily looking back to the surviving knowledge of the Greeks and Romans, which was minimal until the twelfth century translation movement. The goal during the medieval period was not directed toward expanding astronomical knowledge or making new discoveries, but to study the received knowledge that had been transmitted from antiquity, and preserve it from errors.

The spread of learning inherited from Greek science had been a slow process in two directions; an eastward extension into Asia where it would be absorbed into the Islamic culture, and a weakened westward branch that was transmitted in the Latin West. These two traditions collided in the turbulent twelfth century and the process of dissemination and absorption occurred in the succeeding centuries. After that period of transition and the assimilation of the 'long lost' material uncovered in the writings and theories of Ptolemy, the course of astronomy advanced considerably. At last many medieval scientists "were prepared to challenge or modify significant elements in the tradition that they received" (Goldstein 2000: 213).

During the Carolingian period, Alcuin, Adalhard of Corbie, and 'the Astronomer' stand out for making new observations and commenting on celestial movements and recent cosmological activities. But until the twelfth century, there is very little record of actual astronomical activity, such as observations or calculations. The type of astronomical information that was available was rather simplistic, chiefly the allegorical material in verse from Macrobius and Martianus Capella, the fables of Hyginus in prose, and the *Aratea* manuscripts. "It was of no great importance that the details of the poem correspond exactly with scientific facts. The lines of the poem were well turned and pleasing to the ear, containing graphic and accurate information about the stars, so that they served as an easy guide to the night skies" (Saxl 1957: 100).

Through most of the Middle Ages, the *Aratea* texts appeared in compilations with various other astronomical authors, which speaks to its didactic function. The fact that the *Aratea* were almost always bound in volumes with the same popular astronomical writings by Pliny, Macrobius, Martianus Capella, Hyginus, and Pseudo-Bede, demonstrates that they were read and classified by their astronomical information. They do not seem to be bound with classical literature, religious writings, or other poetic works.

Astronomical subject matter in *Aratea* manuscripts was limited mainly to the constellations, other celestial phenomena such as the waxing and waning of lunar phases, hours of daylight, zodiacal positions, planetary brightness, or periodic transits were not addressed. Cosmological diagrams or rotae are often found in these anthologies, but they are associated with other authors, particularly Pliny or Bede, not with the *Aratea*. Studying a medieval encyclopedic volume, such as the Li*ber Floridus* (discussed in Chap. 3) would supply much more astronomical information than an *Aratea* manuscript. Like the Sacrobosco writings, this massive work included helpful diagrams to explain eclipses, the changing hours of daylight, the zonal divisions of earth, courses of the planets, in addition to the constellation images. Although lacking explanatory diagrams, the *Aratea* usually did contain a celestial map dotted with constellation images as a visual guide to the heavens. The inclusion of a celestial map, a panorama of the complete sky, indicates that medieval readers were interested in seeing the appropriate placements of the constellations in their accurate regions; the majority of the 48 constellations are usually pictured. Yet it is surprising that the Milky Way, although described briefly in the body of the poem, is almost never indicated in medieval celestial maps, nor is the starry path ever pictured in *Aratea* codices.

> If ever on a clear night, when all the brilliant (470)
> stars are displayed to men by celestial Night,
> and at new moon none in its course is dimmed,
> but all shine sharply in the darkness - if ever at such a time
> a wondering has come into your mind (474)
> when you observed the sky split all the way round by a broad circle,
> or someone else standing beside you has pointed out to you
> that star-emblazoned wheel (men call it the Milk), no other circle
> that rings the sky is like it in colour, (Aratus, Kidd translation)

The chart of the entire celestial sphere enabled readers to learn the whole sky of which they could only see a limited section, depending on the terrain, weather, the time of night and the season of the year. Therefore a person who consulted an *Aratea* manuscript could learn the names and relative positions of the constellations from the text, and they could learn the general appearance of the stellar figures from the illustrations and from the northern and southern hemispheres of the sky map. But there is no possible way that the text or pictures could teach you any practical astronomy or help you locate the constellations, if you did not already know where they were, and could identify which stars constituted the group.

Lacking previous astral knowledge, identifying the actual designs of the constellations would be impossible for a medieval reader to learn simply by consulting the text or miniatures in the *Aratea,* because before the Muslim influence, stars were never accurately placed. In addition, there was no indication of stellar brightness, proximity, or color, and rarely was any attention paid to exact numbers of stars, an addition from Hyginus scholia. The constellation pictures, in rare instances, relate to the shape of the constellations, but generally the stars have to be pointed out, as Aratus says, by "someone else, standing beside you," explaining their relative distances, brightness and order. Most of the constellations are illusionary requiring a great stretch of the imagination, since the actual star patterns of the constellations rarely look like the pictures; any number of patterns could be formed or imagined from the same group of stars. Other societies have formed pictures as well, completely different than the classical Greek formations. Others have also attributed different qualities to the planets as well; for example, the ancient Maya consider Venus to be a masculine god of war, the exact opposite of the feminine goddess of beauty and love (Milbrath 1999:93-195). The innumerable stars shining on a clear, dark night form a brilliant carpet mosaic of random orientations; they do not gather into naturalistic shapes except in human imagination.

In order for the *Aratea* manuscripts to relay any elucidative knowledge, it would have been necessary to already know the pattern of each of the star formations. I suspect that most people did learn this basic information about nature at a young age, at least the circumpolar stars and the seasonal markers, such as Taurus, Orion and Virgo, and other sky lore that was necessary for farming and travel. The format of the *Aratea* would relay the order in which the constellations appeared, and the poetic verses would help sky watchers to memorize which constellations would arise next. The pictures and stories in the manuscripts would further reinforce the celestial images and myths.

The analysis of the didactic function of the astronomy in the *Aratea* manuscripts could be compared to the argument about the function of art and stained glass windows in medieval churches and cathedrals. Their function in a mostly illiterate world was carefully evaluated in the well-known article, "Was Art Really the 'Book of the Illiterate'?"

> Words are not perfect and never can be, but they will always remain our most precise, if ever defective, mode of communication. By comparison, pictures can be interpreted as artifacts and be very useful also as 'sources' and as stimulants of new insights, but their correct interpretation can be corroborated only by reference to other sources, by allusion to what one already knows, and by the use of words to conduct an intelligible and fruitful discussion of the meaning of individual pictures. Pictures cannot 'speak' clearly, only words can. That is the long and short of it (Dugan 1989: 251).

Pictures were meaningless unless the viewer already recognized the stories in the paintings or stained glass windows, otherwise the pictures were worthless, had no meaning if one was unfamiliar with that image. In the same way, the form of the constellations had to be known and understood before reading the *Aratea* manuscripts; otherwise the supplied text would be meaningless without accurate guides to tweak out the design of the constellation figures from amongst the hundreds or thousands of nearby stars. Although you could learn the names of the constellations from the poem, you could not actually find them in the sky without additional information; it would be almost fruitless to even try. If you already knew the formations, you could gain helpful information from Aratus or the *Aratea*. The poem does provide guidance markers and techniques to learn certain aspects.

> It takes four signs of the zodiac together for the ocean to receive Boötes' setting. When he is sated with daylight, he occupies more than half of the passing night in the loosing of his oxen, in the season when he begins setting as the sun goes down Kidd (585).

The astronomy lesson would be rather round-about, but sections of the poem could be memorized to learn the relative positions. The poem also says to look for the paranatellonta, the constellations that rise with the zodiacal signs, "if they are darkened by clouds or obscured by a mountain when they rise" (Kidd: 565). If you cannot see the zodiacal sign, look for the constellations that stand nearby. Perhaps the textual changes made throughout the medieval period were to eliminate the more obscure passages and to insert more practical information about the formation of the constellations. Therefore, although the astronomical information in the *Aratea* could not teach basic astronomy; there were many helpful facts that could be learned.

The *Aratea* might possibly have been included in the typical liberal art education before the twelfth century, but astronomical education changed greatly after that time, for the available astronomical texts grew more detailed and challenging. The body of astronomical knowledge had expanded enormously. By the end of the thirteenth century, the astronomy found in the textbooks of Sacrobosco achieved a secure home in the university curriculum, and so it remained for the next four centuries (McCluskey 1998: 196). Sacrobosco's *De sphaera* included a simplified discussion of the structure of the heavens and provided a basic account of spherical geometry to support the mathematical astronomy of Ptolemy and his Islamic commentators. Although the *De sphaera* of Sacrobosco had moved to the forefront as the most influential text for astronomical studies from the thirteenth to the seventeenth century, a wide range of other astronomical texts by astronomers in Greek, Latin and Arabic were also in circulation for consultation by students of astronomy.

During the thirteenth century, the astronomical writings of Robert Grosseteste (ca. 1175–1253) transmitted his original research and discoveries in meteorology, light, color and optics. Roger Bacon (ca. 1214–1294), a Franciscan scholar who promoted the scientific method, included new discoveries in astronomy, astrology and alchemy in his writings. Bacon promoted experimental research in the physical sciences rather than relying strictly on ancient authorities. Latin translations of Islamic astronomical and astrological discoveries and commentaries on Greek science were also widely available.

Consulting *Aratea* manuscripts as sources for attaining astronomical knowledge, if that was ever a true purpose, certainly diminished by the thirteenth and fourteenth centuries. In the company of the scientific writings of these great intellects and learned scholars, the elementary astronomical information found in the *Aratea* manuscripts was of little consequence for the academic world. Scholastic commentaries and guides to the study of arts occasionally mention Aratus and Hyginus, but not their specific texts (McCluskey 1998: 197).

The three periods of renewal discussed in this study were times of actual rediscovery and recovery of earlier writings. These three renaissances were each followed by periods in which scholars were able to exploit the work of their predecessors. Each of the renaissances provided the material needed for the next period of advancement and made important contributions to the body of held knowledge. It is not surprising to find that the greatest writings and most original thought date from soon after one of the cultural revivals. Roger Bacon, Albertus Magnus, and Thomas Aquinas were spurred to their great achievements in scholasticism and scientific method by the rediscovery of the writings of Aristotle and Ptolemy. Copernicus, Tycho Brahe, Kepler, and Galileo realized their greatest discoveries after the renewal of ancient writings of Neo-Platonists and the Greek Byzantine and Islamic scientists during the Italian Renaissance.

This study has shown that the astronomy in the *Aratea* did have educative value, but in a limited way. Some didactic qualities were included, but that does not seem to account for the lasting appeal of this manuscript tradition. The function of the illustrations on the other hand is a fundamental characteristic, since the surviving *Aratea* are rarely without the full cycle of colorful constellation miniatures.

Function of the Aratea *Manuscripts in History of Art*

This study has identified and examined the body of constellation illustrations that appear in *Aratea* manuscripts, dating from the medieval period.[4] Were the illustrations included to explain the text, or was the text merely subsidiary to the pictures? Visual representations of the pagan gods and mythological characters from antiquity were still abundantly available in the early Middle Ages within Roman ruins, temples, and public buildings. The astronomical poem offered an ideal neutral theme for a medieval picture book and was copied over and over. The *Aratea* were as rich in visual themes as they were in celestial data. The poetic descriptions enlightened and entertained viewers, and enabled readers to memorize the pictures and recreate them in their minds, and later project them on a star-filled sky. The stories in the heavens above have since become a large part of our cultural heritage. Without the formation of star configurations and then linking them to

[4]See Appendices B and C.

eternal, entertaining myths that personify human qualities and aspirations, the stars would have remained mute.

For the most part, these illustrations owe their appearance and iconography to the classical heritage; their prototypes in all probability were established in the last centuries of the Roman Empire. "Clearly there must have been illustrated manuals or pattern-books of some sort" (Cameron 2004: 220) for artists to utilize for copying images; many mythological scenes appeared in Roman frescoes, mosaics, statuary, medals and other artistic media. These images appear in an easily recognized standardized form; Apollo/Helios always stands in his quadriga; Luna stands in her cart pulled by two oxen; Hercules kneels with a club in his right hand, a lion pelt over his left arm. The constellation images on the Farnese Globe too correspond exactly with the descriptions of Aratus, which indicates that these consistent pictures were familiar in the Hellenistic and Greek worlds.

As seen in Chap. 2, the *Aratea* were not unique in that they included images, since other classical writings, for instance, Homer's *Iliad,* Virgil's *Aeneid*, and Terence's plays were all abundantly illustrated as well; manuscripts of these famous authors also survive from the early Middle Ages. Illustrations served primarily as functional and didactic companions to the texts and complemented the textual adaptation of the classical sources.

Making comprehensive statements about the entire corpus of Aratean miniatures is not possible for obvious reasons. Although always recognizable by their basic iconography, the constellation miniatures vary considerable in style, details and the artistic ability of the illustrator. Innovative changes occurred in thirteenth-century astronomical manuscript traditions such as Michael Scot and Abu Ma'shur, but the *Aratea* manuscripts remained true to their antique heritage. This illustrative tradition lasted for over 700 years, and manuscripts were produced throughout Europe. Therefore it is more suitable to allow each individual manuscript to stand on its own merits, considering each example as representative of its historic perspective and cultural milieu. But in a broad sense, some conclusions may still be drawn, since all of the manuscripts belonging to the Aratean poetic tradition do share certain characteristics.

Except for the ninth and tenth-century manuscripts of Cicero's *Aratea*, most codices carry the complete cycle of 42–48 constellations, in addition to miniatures of the Sun and Moon together and the five planets on a single folio. The *Aratea* commonly begin with one or more celestial maps, and they more or less follow the same order when describing the constellations, which begin with Draco and the Ursae and end with Procyon. The constellation pictures most frequently appear before the text descriptions, simply inserted between sections of text. Although the miniatures of the Leiden *Aratea* are often full-page, framed, and set apart with colored backgrounds, the majority of illustrations are small and unframed, as if free-floating in space, with no background color or setting. In some manuscripts, the text runs around the images, as they bump into each other. The most common type of painting in *Aratea* manuscripts is a conservative line-drawing and pastel color-wash technique. When reading the paintings compared to the text they illustrate, artists do not seem to utilize the information in the text; for the most part, they do not follow the visual clues stated in the poem.

Aratean Art in the Ninth Century

As described in Chap. 6, the Carolingian Renaissance was a key chapter in the history of scholarship and science. "Classical conceptions, literary, philosophical, scientific and artistic, had survived throughout the centuries, particularly after they had been deliberately revived under Charlemagne and his followers" (Panofsky 1955: 40). The artistic productions and manuscripts that they created matched their advances in other fields and genres. During the ninth century when the process of classical borrowing begins, the intellectual and cultural sphere fostered the collection and translation of classical scientific texts. The miniatures in ninth-century *Aratea* manuscripts captured the essence of classical art, since the accomplished artists were able to clothe classical subjects in classical form. Even though the 13 *Aratea* manuscripts that survive from the ninth century all seem to draw from classical sources, they are far from imitative. This fact points to the existence of a variety of models from which the artists could draw astronomical inspiration, but it also speaks to the abilities and creativity of Carolingian artists. Similarities among some of the ninth-century miniatures point to a common ancient source, perhaps a Late Antique prototype. Although it is tempting to single out the ninth-century illustrations as the most creative with the highest quality, it is unfair to judge the manuscripts from later centuries by those high standards.

Later artists restored, reinvented, and recopied images that had been preserved by Carolingian illuminators from antique sources. These conservative classical illustrations were continually recycled in *Aratea* manuscripts. Artists who produced the tenth and eleventh centuries copies of the Carolingian manuscripts succeeded in recapturing the antique models with surprising fidelity. The quality and faithfulness to earlier models can be observed in the two copies of the Leiden *Aratea*, Boulogne-sur-Mer MS 188 and Bern MS 88. Nevertheless as time goes on, some iconographic changes are introduced into the illuminations. Observations of the sky which could have been used to verify or make the pictures more accurate are not employed.

During the eleventh century, monasteries still functioned as the leading centers of religious, intellectual and artistic life. Bishops and monastic leaders moved to the forefront as patrons of art and culture. The Ottonian court in imperial Germany, as heirs to the Carolingians, desired to recreate that admirable era and sanctioned a huge amount of artistic and cultural production. Like the Carolingians, Ottonian artists looked to Late Antique models and admired their imperial influences but applied a distinct northern touch. The Ottonian revival coincided with an era of growth and reform in the Church; northern monasteries created much of the admired Ottonian artworks, including their outstanding illuminated manuscripts and lavish liturgical objects. Political imagery was sometimes integrated with religious settings, but the political structure of that Empire was established in a different manner than their Carolingian model. The emperor could not legislate or dictate throughout the entire empire, as other powerful realms and dukedoms were ruled independently (Colish 1997: 162). Although seven *Aratea* manuscripts are extant from the era of the Ottonian Empire, it seems that none of these were produced at the royal court scriptoria.

Aratean Art in England

For the most part, production of *Aratea* manuscripts in England was confined to religious centers and to Cicero's translation. There both the text and the picture cycle were expanded to include more of the constellations missing from Cicero's surviving fragments. As stylistic changes developed in English monastic scriptoria, Romanesque characteristics in particular, these preferences were reflected in Aratean miniatures as well. Mythical creatures from the *Aratea*, like the sea monster, centaur, snakes, and dragons, became perfect prototypes for the fantastical contorted beasts populating Romanesque manuscripts, sculpture and decorative motifs. Dodwell has shown how images from astronomical manuscripts were borrowed and then adapted as sources of Romanesque decoration, particularly at Canterbury in the tenth century (Dodwell 1954: 60–62) When investigating manuscripts of Cicero's *Aratea* that survive in England, it is obvious that scribes and artists were correcting details and adding to the small number of constellations, only 24 were painted in their ninth-century prototype, Harley 647. English artists sometimes used their own imaginations, creating images of the constellations that were far removed from their classical prototypes. Costumes worn by Boötes and Aquarius, for example, are drawn from other sources, quite different from those in classic-inspired miniatures. Architectural and landscape settings were sometimes created for the constellation figures, which is definitely out of character for personalities situated in the night sky. The production of illustrated *Aratea* manuscripts in English scriptoria appears to terminate in the thirteenth century.

Aratean Art in the Twelfth Century

Contact with Arabic astronomical art revealed that there were some similarities with western art, since both traditions derived in a great part from Hellenistic sources, but they displayed many differences as well. The fact that Arabic astronomical codices produced from the tenth to the twelfth century were illustrated with an almost identical cycle of constellation figures as the *Aratea* manuscripts points to their use of Late Antique exemplars as well. A prime example is al-Sufi's *Book of the Fixed Stars*, an original scholarly endeavor, designed with constellation illustrations and created in Persia at the end of the tenth century. The original autograph treatise by al-Sufi does not survive, but Bodleian Marsh 144 originally dated to 1009–1010, but now re-dated, is a close copy. Since the work was derived from the writings of Ptolemy and others, the images were rooted in the classical tradition. Even though there was little contact with manuscript traditions of the Latin West, the same zodiac and constellation images appear in Islamic codices. Islamic artists copied illustrations from their own Greek and Hellenistic models, since it is extremely doubtful that Islamic artists were visiting European royal or monastic libraries to view, to copy or borrow their astronomical manuscripts. Although Charlemagne did both receive and send envoys to the court of Hārūn al-Rashīd, an astronomical or *Aratea* manuscript could have possibly changed hands.

Since the original Hellenistic texts, as well as Late Antique copies, had been preserved in Arab libraries, these were most probably illustrated examples that supplied the patterns. The artistic adaptation of classical sources for astronomical art in the Arabic East mirrors precisely the format of astronomical art in the Latin West. The transformation of Late Antique forms and motifs into new, distinctly Arabic manuscript miniatures precisely parallels the adaptation of the classical exemplars in the West. Though the Arabic manuscripts retained the classical images in names, number, and order of the constellations, the clothing, attributes and iconography naturally were derived from the Middle Eastern culture. The drawings in manuscripts of al-Sufi and Abu Ma'shar brought Oriental influences to astronomical art in the West, along with many changes. But these changes were confined to the manuscripts based on these Arabic writings. The constellations did not relay accurate depictions of the stars until al-Sufi's images with symbols graded in size and brightness were translated for the Latin world. Since the *Almagest* held no illustrations, al-Sufi was the first to display the stars as Ptolemy described and classified them, but this scientific accuracy was only included in copies of al-Sufi's *Book of the Fixed Stars*.

Discussing the manuscript illuminations of Michael Scot, Panofsky wrote:

> the genuine images had fallen into oblivion and were replaced by others-partly newly invented, partly derived from oriental sources-which no modern spectator would ever recognize as classical divinities… Wherever a classical image, that is, a fusion of a classical theme with a classical motif, had been copied during the Carolingian period of feverish assimilation, this classical image was abandoned as soon as mediaevel civilization had reached its climax, and was not reinstated until the Italian Quattrocento (Panofsky 1955: 49).

I would disagree as Panofsky was comparing two different manuscript traditions. The classical images of the *Aratea* were not abandoned, but still produced in smaller numbers, retaining the traditional iconography. The newly invented iconography found in the illustrations of Michael Scot and other Arabic derivatives brought new and exotic pictures specific to those traditions. Astrological theories, such as the paranatellonta, are illustrated, and each traditional sign of the zodiac was often overrun with thirty decans, each with its own name and face, a concept of Egyptian origin, especially in the translations of Abu Ma'shar. The seven planetary gods had been assigned 'children', representing attributes of human nature that are governed by the Sun, Moon, and five planets, as they transit the signs of the zodiac.[5] These were an additional series of images designed for newly introduced concepts based on astrology; they did not eliminate the antique motifs, but existed alongside them. Michael Scot's treatises were not classic-based literature, but were written by him in the thirteenth century. They became quite popular and were copied many times. His work incorporates astrological theory chiefly from Arabic sources that he translated; his images were possibly derived from Arabic manuscripts as well, but were updated with a more elegant, court-like appearance.

Aratean Art in the Fifteenth Century

After many centuries of declining interest in *Aratea* manuscript production, it was amazing to find no less than 27 manuscripts surviving from the fifteenth century. The resurgence of interest in the classical past begun in Renaissance Florence included—literature, mythology, the Greco-Roman gods and goddesses, astronomy and art—all components of the Aratean tradition. Apparently the *Aratea* manuscripts, where the classical gods and heroes were still animated and alive, became a vital ingredient in the renewal of pagan antiquity in the intellectual and artistic life of the Renaissance. The mythological images of Hellenistic and Roman astronomy were too deeply ingrained to be changed; pagan imagery was adopted and used in the artistic programs of the fifteenth century. The Renaissance has been described as the nostalgic dream of a lost world of order and beauty. In Seznec's multifaceted study of the survival and metamorphosis of the pagan gods, he states, "How much can be learned of the history of civilization from a study of their metamorphoses and reincarnations" (Seznec 1953: 149). This enthusiasm for the classical past was most felt in Italy; it was there that the revitalization of *Aratea* manuscripts was strongest, but the allure of the past spread to other areas of Europe as well.

Were these astronomical manuscripts considered new symbols of an educated status, picture books for palace libraries, or were they read for their practical or 'scientific' information? Selecting four richly illuminated *Aratea* manuscripts produced at the court of Naples for closer examination demonstrated the intense interest in antique manuscripts and ancient astronomical knowledge. The *Aratea* manuscripts provided the perfect subject matter, learned and lofty, a suitable alternative for royal or private libraries already laden with numerous religious books, Bibles, Psalters, and Books of Hours. Although the topic of another study, perhaps multiple gift-giving accounted for the large increase in *Aratea* production in the quattrocento. A complete investigation of how the *Aratea* functioned and what part they played in this cultural renewal is unfortunately beyond the scope of

[5]A doctoral dissertation has been written on this subject, Gwendolyn Trottein, *The Children of Venus in Late Medieval and Renaissance Iconography,* University of Illinois at Urbana, 1986.

this study. A closer look at how some specific areas were treated in the art can bring a fuller understanding to the tradition.

Treatment of the Planets

In astronomical lore, the constellations were depicted as mythical figures, both larger than-life human personalities and steadfast animals that embodied ancient legends. The main characters of these myths became frozen in time, and their stories were fixed firmly in the heavens and unchangeable. In contrast, the five planets, along with the Sun and Moon, were constantly changing their positions among the fixed stars. When presenting an account of change in the world, Aristotle had written almost nothing about the known planets compared to the Sun and Moon (North 1987: 7). In the Greco-Roman world, the 'seven planets' were treated as divine forces interacting with human life, creating and directing fates and fortunes. The planets were not just human or semi-human characters caught up in moralistic fables like the constellation figures, but were considered gods with superhuman powers. Throughout time, the character traits and both the heroic and deadly deeds attributed to the Greek gods have been sources of metaphor in literature, poetry and daily life. The beauty and allure of Venus, the warlike demeanor of Mars, the good fortune of Jupiter have all been embodied and celebrated in sculpture and artworks. "Where would European poetry be without Venus and Mars, Cupid's dart and Jove's thunderbolt...?" (Gombrich 1979: 12)

In the early Middle Ages, the five planets seemed to have been of little or no significance outside of 'book' astronomy before the Carolingian age. "Only in texts on the liberal arts or in astronomical texts incorporating theoretical elements did the planets put in a more than perfunctory appearance" (Eastwood 1983: 161).[6] After the Arabic-to-Latin translation movement, Abu Ma'shar's theories, that the planets presented the key to the great diversity of change in the world, added new interest to the movement of the planets through the signs of the zodiac. Abu Ma'shar summarized the writings of Aristotle on generation and corruption receptivity in regard to planetary influence on terrestrial events (North 1987: 7).

Countless temples and monuments had been created to honor the pagan gods personified in the sky: Mercury, Venus, Mars, Jupiter, and Saturn in their various manifestations. Ruins of these remarkable temples and remnants of cult statues survived in all areas of the vast Roman Empire. Reminders of the gods of pagan religions could be found in many antique artworks, frescoes, and relief carvings. To combat and eradicate these vestiges of pagan religious cults, the Christian Church needed to reverse the former high regard and reputation of these planetary gods. In the later Middle Ages, these alien images of 'the Other' were represented within the

[6]Eastwood also discusses this issue in his 2002 book, *The Revival of Planetary Astronomy in Carolingian and Post-Carolingian Europe.*

'visual value-system' of the Church in order to control and eventually negate the foreign gods by distortion and effacement. The planetary gods had to be toppled from their pedestals. Michael Camille, in *The Gothic Idol*, expresses the dilemma:

> The Christian Church was never in doubt that idolatry should always be condemned; the perennial debate was how to define and separate 'correct' visual representations from incorrect idols and how to operate within that definition (Camille 1989: xxvii).

The planets were difficult to plot in a predictable course as they are not always visible and their paths fluctuate due to their erratic motions. Usually they progress in a forward direction, but sometimes pause briefly, and then reverse into a retrograde motion in their apparently self-directed heavenly courses. For these reasons, the five planets are only briefly mentioned by Aratus and the Arateans. In view of these complications, the planets were handled in a different manner than the dependable and predictable fixed stars. The illustrations of the five (or seven) planets in *Aratea* manuscripts are treated much differently than the constellations; their depiction appears quite conservative, for obvious reasons.

Just as the planets were short-changed in their textual descriptions, lacking admirable attributes or unusual powers, their illustrations are minimal as well. Depicting images of the constellations in astronomical manuscripts never presented a problem in a Christian-dominated society; Andromeda, Perseus, Virgo, Hercules, and the other heroes appear regularly in illuminated manuscripts, complete with their traditional iconography and characteristics. But in contrast, representing the five planets anthropomorphically, in their easily recognized human guises, could be an affront to Christian morality, and seemingly honor the pagan gods. On other hand, the planets were an important component of cosmology and could not be totally ignored. Statues and representations of the planetary gods were considered pagan idols, expressly forbidden by the Second Commandment. To depict the planets embodied as powerful, superhuman figures could be interpreted as honoring the earlier pagan religions, acknowledging their astrological mastery and control over human destiny and challenging the Christian concept of free will.

This dilemma required careful consideration by medieval illuminators and their patrons. How could you adopt the knowledge, science, philosophy, and mythology of other cultures, but negate and condemn their religious system? Carolingian creativity resolved this difficulty by representing the planets as a group, rather than a full-page miniature for each individually, downplaying their significance and picturing them merely as disembodied heads. This compromise allowed artists to illustrate the five planets in their traditional symbolic representations, but not afford undue attention nor allude to their previous roles as powerful gods, directing the course of human life on earth. The depiction of the five planets in the Leiden *Aratea* is treated in an overall similar manner as the constellations; the same blue background and border frames the planets. But the gods are pictured as tiny heads with little acknowledgement of their former significance; also they have no narrative to elevate them (Image appears in Chap. 7.)

Most *Aratea* manuscripts treat the planets in a dismissive manner, most follow the Carolingian precedence and picture the five planets in a similar way, as a group of small heads; none of the planets appear as full-page or full-body illustrations. In some manuscripts, the illustrators would avoid the whole issue and just represent the planets as stars, as in the *Liber Floridus*. I suggest this treatment of the planets was purposeful, but it is possible that artists were merely continuing the Late Antique prototypes.

Illustrations of the two other 'planets', the Sun and Moon, are included in most manuscripts of Germanicus' *Aratea*. The Sun and Moon are usually painted in action scenes, embodied as human forms with their distinctive iconography, directing their respective chariots across the sky, at times on the same folio and occasionally on adjoining pages. The illuminations of the Sun and Moon most often appear at the beginning of the *Aratea* astronomical text, after the biography of Aratus, when included. Occasionally a miniature of Jupiter riding his eagle is painted at the start of the poem where he was honored in the proem by Aratus. Only one Germanicus manuscript, National Library of Wales MS 735C, includes an image of Jupiter receiving inspiration from his Muse, a quite unusual illustration.[7]

After acquaintance with the illustrated Arabic manuscripts, depiction of the five planets in the Christian West begins to change. *Aratea* codices continue their traditional subdued display of the planetary gods, but in other astronomical manuscripts, images of the planets begin to appear in a variety of new ways. The planets are no longer ignored or downplayed; instead they become the centerpieces in a new cycle of images and begin to circulate in thirteenth-century astronomical anthologies. In later astronomical manuscripts, the personifications of the planets still had few iconographic models; therefore images of each were created from their assigned characteristics in Islamic astrological writings, including a color, an occupation and a day of the week; although these were not often seen.

By the fifteenth century, there are numerous astronomical and astrological illuminated manuscripts produced that freely depict the planets as large scale, full-page celebrated figures. In an astrological manuscript from the library of Francesco Sforza (1400), Jupiter is depicted, within concentric, iridescent circles which include the band of fixed stars.[8] Jupiter appears naked, but for a strategically placed scarf, and holds a scepter in his left hand and arrows in his right with a quiver slung over his shoulder. Smaller medallions at his feet contain the signs which Jupiter rules, Pisces and Sagittarius, who is pictured as wholly human. (A similar illustration of Jupiter appears in a 1512 woodcut of a printed version by Regiomontanus but Jupiter is fully dressed and enthroned.) On Earth, below the realm of the powerful planet, his 'children' are shown taking part in daily-life activities that are ruled by Jupiter, such as trade, agriculture and banking. On another double folio, Mercury (f. 10v–11) is shown holding a sack and a caduceus, symbols of wealth and medical knowledge. Below in the medallions, appear the

[7]This eleventh-century manuscript was probably made in the Limoges region. The text and illustrations can be viewed on online at the Library of Wales website.

[8]The manuscript is titled, *De sphaera*, (1450–60) Modena, Biblioteca Estense, Lat. 209, discussed in *Treasures from Italy's Great Libraries*, Jupiter, (f. 6v).

signs that Mercury rules, Gemini and Virgo. The facing folio represents activities under the influence of Mercury.

Treatment of the Signs of the Zodiac

In *Aratea* manuscripts, illustrations of the 12 signs of the zodiac are not singled out as a separate unit, nor are they treated differently than any other constellation figures. Each sign is addressed in order as its particular area of the sky is reached in Aratus' original systematic arrangement. Aratus discusses the zodiacal constellations and mentions that the planets "circulate all the way through the twelve figures of the zodiac."[9] He emphasized their importance as markers for the transit of the Sun, Moon, and planets, and their significance for calendric purposes.

Although the signs of the zodiac have become closely connected with astrological symbolism and predictions, as mentioned several times, astrology has no association with the *Aratea* tradition, not in the Roman era when astrology was quite important and regularly practiced, nor during most of the Middle Ages when it basically disappeared, not even after the revival and wide-spread adoption of astrological practices in the late twelfth century. An important segment of astronomical art, although beyond the scope of this study, appears in the religious tradition when the motifs of the 12 zodiacal signs are removed from their celestial matrix and joined with the occupations of the months to indicate time and its passage through the years. In artistic representations of the Christian world, the signs of the zodiac are most often linked with the Labors of the Months. This motif had previously existed in the classical world, seen in floor mosaics and in the Calendar of 354, but is another Christian adaption of astronomical images never introduced into the *Aratea*. During the medieval period, the zodiacal signs, along with the Labors, assume a religious connotation, linked with the liturgical year of the Church. The illustrations are purely symbolic of the yearly zodiacal cycle in the heavens and the rotation of agricultural duties and symbolic labors on the earth. In this situation, the signs and occupations signify humanity's place in the rhythmic order and present a connection between the toils of earthly life and the joys that await in the eternal heavens. These patterns of twelve signs and 12 labors were well established in artistic form by the twelfth century, changing gradually to become more experimental in the fourteenth and fifteenth centuries.

The signs and Labors were often carved on Romanesque and Gothic church portals and western facades, such as: Abbaye Sainte-Marie-Madeleine de Vézelay, the Abbey Church of St. Denis, Chartres Cathedral, and Amiens Cathedral, to name just a few. Manuscript illustrations of the signs and Labors became more impressive through time, advancing from tiny roundels to sometimes taking over the whole folio. Among the earliest surviving medieval examples that represent both signs and

---

[9]Line 456, Kidd translation.

Labors is a ninth-century manuscript of Bede's *De rerum natura* made in Regensburg.[10] The most magnificent example is the *Très Riches Heures*, illuminated by the Limbourg Brothers for Jean Duc de Berry in 1413–16.[11]

In spite of the strong presence of this double motif in Christian art, the motif was not conveyed to *Aratea* manuscripts. The signs of the zodiac receive no special treatment; the Labors of the Months were not included originally nor added later. The signs retain their classical iconography, which is discussed further in Appendix C.

The Iconography of Gemini

The depiction of the constellation of Gemini in manuscript illuminations has been especially fruitful as a subject of investigation. The iconography of the twins, Castor and Pollux, also known as the Dioscuri, has at times been intermingled with a completely different set of twins from Greek mythology, Amphion and Zethus.[12] These not so famous brothers were twin sons of Zeus and Antiope associated with the founding myth of Thebes and the construction of its city walls.[13] The musician Amphion played enchanting tunes on his golden lyre given to him by the god Hermes, which magically caused the stones of the walls to rise and set themselves into position. Brother Zethus married the nymph Thebe, for whom the city was renamed. Thus Amphion is celebrated in myth and art for the magical power of his music. The 'real' Gemini twins are Castor and Pollux, who were honored with many temples and worshipped in antiquity as protectors of mariners, particularly during the Roman period.[14] This interchange of identities and iconography can be traced to ancient Greece. In Christianity, Castor and Pollux have become associated with twin brothers who became Early Christian martyrs in the early fourth century, Saints Cosmas and Damian.

---

[10]Munich, Bayerische Staatsbibliothek, Clm.210.

[11]A publication discusses the appearance of the signs of the zodiac in calendars found in Books of Hours and in sculptural form decorating medieval cathedrals, *Time in the Medieval World: Occupations of the Months & Signs of the zodiac in the Index of Christian Art*, Colum Hourihane ed. Princeton University, 2007. The publication provides an extensive list of manuscript illustrations of the zodiacal signs.

[12]I would like to thank Professor Drew Armstrong, University of Pittsburgh, for pointing out the dual identities of Gemini.

[13]Reid, Jane D., *Oxford Guide to Classical Mythology in the Arts, 1300–1990*, Oxford University Press, 1993. Amphion and Zethus were twin brothers abandoned at birth and raised by a shepherd. They have been the subject of numerous paintings, poems, musical scores and even an opera, pp 93–5.

[14]According to the *Oxford Guide*, they are mentioned in the Iliad, Pindar, Euripides, Ovid, and Hyginus. In addition to the *Aratea* manuscripts, they are illustrated in numerous paintings and remembered in poems. pp 369-73.

There is a variety of depictions for the brothers Gemini, as Castor and Pollux; they are often naked with just a cloak, carrying weapons or a shield. Their arms are usually intertwined indicating their love and care for each other. In later manuscripts, they are fully clothed with contemporary military garb. Depictions of their alternate identity show Amphion holding the lyre that enticed the stones and Zethus carrying a curving sickle. In *Aratea* manuscripts, Gemini are commonly represented as two male warriors, sometimes naked, holding long spears or other symbols of warfare; but often they take on the iconography of the lesser-known mythological pair. So far, the earliest medieval depiction of Gemini found in their alias disguise appears in the zodiacal floor mosaic at the Jewish synagogue at Sepphoris, dating to the fifth century CE, discussed in Chap. 2. The Leiden *Aratea* prefers the Amphion and Zethus iconography for Gemini.

Treatment of the Stars

The individual stars of the constellation miniatures are of little concern in the majority of *Aratea* manuscripts. The presence of specific stars is indicated by small squares or circles, but their placement is completely random, far from faithful representations of the real stars in the heavens. None of the constellation illustrations includes coordinate lines and no effort was extended to place the stars accurately. Precise details about singular stars, such as their positions, numbers, brightness, or even their names were not included until the scholia was added in later centuries. There is one exception in the naming of individual stars, the Pleiades, a group of seven stars within the constellation of Taurus, which assume their mythological names. Today the Pleiades are considered a star cluster, but the *Aratea* manuscripts treat the group as a separate constellation. Another small star cluster can also be found in Taurus, called the Hyades; it seems to be included in just one manuscript, Oxford, Bodleian Library, MS 614, where it too is treated as a constellation.

Although the scholia in the *Aratea* manuscripts usually state the correct number of stars in each constellation and even their position, i.e., one on the head, two on the feet, the artists rarely match the number or positions of the stars they depict with the text. For the most part, there is no connection between the illustrations and the real stars in the sky, even when designated correctly in the text; the intent was merely decorative. If the purpose for including stars was a practical concern for recognizing the outlines of star groups, you would expect scientific accuracy. But this was not a consideration in the Latin West, until after the assimilation of illustrated Arabic manuscripts. Before that time, individual stars were merely sprinkled within the constellation figures or along the outlines. If viewers were interested in the actual formations of the stars in the sky, these manuscript illustrations would be of little value.

The care and accuracy of the placement of stars in Arabic constellation images demonstrated that it was possible to depict and learn accurate positions of the individual stars within each constellation. The Arabic miniatures indicated the exact formations of the stars, relative to each other, and in addition gave accurate representations of the relative brightness of each individual star by varying their size. They also included the names of especially bright stars, which were easy to identify, and even marked those stars with noticeable colors. The scientific precision of their star locations would have made the recognition of each constellation much more likely. Some Western manuscripts adopted this more scientific placement of stars, for instance Paris, Bibliothèque de l'Arsenal MS1036, which was a copy of al-Sufi.[15] But many artists that illustrated fifteenth-century *Aratea* codices, unaware or uninterested of scientific precision, continued their random scattering of stars.

Numerous aspects of this illustrative tradition could be a separate research topic, tracing the changes through the extensive length of time. The variety of celestial maps and each of the forty-some constellation figures that decorate *Aratea* manuscripts are worthy of more study. The Labor of Hercules, the tragedy of the foreign king Cepheus, his wife Cassiopeia and daughter Andromeda, or the depictions of Orion could be tracked and investigated, as their iconography varies greatly through the various periods and cultures. In examining astronomical manuscripts, some of the most fascinating constellation images by far are the illustrations of the celestial ship (or half-ship), Argo sometimes called Navis. Unfortunately acquiring and including pictures of so many constellations to show these continually changing figures and objects became impossible. But many of these illustrations are available within the two-part volumes by Dieter Blume et al. *Sternbilder des Mittelalters*. Each intriguing marine image was entirely different, reflecting the current state of shipbuilding during the period. The ships alone contain an abundance of nautical information and would produce a fruitful study.

Mnemonic Function of the Illustrations

Rhyming verses from the *Aratea* were memorized to announce the time of night, the time of the year, and the seasons, as the mythological figures slowly followed each other on their nocturnal journey across the heavens. The configuration of the constellations into distinctive and easily remembered pictorial images is itself a semiotic device to assemble the countless and indistinguishable stars into a more recognizable form. Words and pictures are intimately and collaboratively related as devices for composing thoughts and memories (Carruthers and Ziolkowski 2002: 2). Pictures can be found in the words themselves. "Moreover it is not so apparent where one medium leaves off and the other begins, for many of the pictures are visual puns and pictures of words and many of the words are verbal paintings and drawings" (Carruthers and Ziolkowski 2002: 2). This is especially true in Cicero's

[15]This manuscript was made in Bologna in the third quarter of the thirteenth century.

Harley 647, where the descriptive words of Hyginus actually form the constellation pictures.[16]

Beyond their astronomical significance, the planets and constellations had an additional important function in the past as a memory device. The zodiac has had a long association with mnemotechnic (*ars memoriae*) traced back to the early Roman author Quintilian, who used the constellations of the zodiac, and other specific locales, as a source of memory places to practice the art of memory (Carruthers and Ziolkowski 2002: 229). In her important study of memory in medieval cultures, Mary Carruthers expands upon the groundbreaking publication of Francis Yates (Yates, *The Art of Memory*, 1966). By stressing the great importance of memorization and a trained ability to remember, she shows how that skill was not made redundant by the availability of codices. But at the same time, books were considered the only way to remember and attain knowledge. She examines the use of devices, such as the alphabet, numbers and the signs of the zodiac applied as memory aides by medieval scholars, such as Thomas Bradwardine, Ramon Lull, and Roger Bacon (Carruthers 1990: 95, 110), and later of course by Giordano Bruno with his 'memory theater". Thomas Aquinas too was renowned for his prodigious memory and the ability to dictate to three or four secretaries at the same time and never lose his place in each different essay or lecture.

The signs of the zodiac and the constellations were utilized as 'sets of images' assigned as 'memory places' where the needed information could be 'stored' and then later 'retrieved' when visited through the mind. "Ordered lists of this sort, I propose, were deliberately memorized in order to serve as potential mnemonic heuristics, 'seats' into which one could place the variety of diverse material one would acquire in one's education and reading" (Carruthers 1990: 110). In the early fourteenth century, the constellations of the zodiac served as an elementary exercise in creating mnemonic images in artificial memory systems (Carruthers and Ziolkowski 2002: 16). Bradwardine (ca. 1290–1349) used mental pictures borrowed from the traditional representations of the signs of the zodiac, as well as bestiaries, for mnemo-technical purposes (Carruthers 2002: 126). Hugh of St Victor (d. 1141) also used mental pictures of "a map and several different cosmological diagrams, including the zodiac, the Six Days, and the angelic circles" as part of his devises for mnemonic purposes.

These particular examples of the art of memory reveal an interesting ancillary function of the constellations and signs of the zodiac, but these distinctive uses were not as well known, and were limited to a certain few medieval scholars, clerics and orators, who applied these memory aides for formal training, exercising and expanding their memories. When considering a wider variety of the population, the more useful purpose for retaining the order and positions of the constellation names

[16]Only two other manuscripts follow this convention, Cotton Tiberius C. 1 (12th C.) and Göttweig, Stiftsbibliothek Codex 7 Bl. 20 (15th C).

and pictures was for calendric and geographic purposes, to learn the specific time of day or night and your exact location and direction on land or sea.

Astronomical Decorative Programs

Cosmic themes and constellation cycles were very frequently appropriated as ornamental schemes for decorating buildings both inside and outside, since architecture is midway between the macrocosm of the universe and the microcosm of the world of humans. Domes and apses by their very design allude to the hemispherical vault of the heavens and were often adorned with heavenly scenes featuring Christ in Majesty, the Last Judgment, and the choirs of angels. During the fifteenth century in particular, astronomical pictorial programs that had been mainly confined to illuminated manuscripts and cathedral ceilings in the past expanded to decorate the palaces, chapels, and salons of the aristocracy. Without repeating specific examples, it has been sufficiently demonstrated that astronomical iconography was customarily adopted by rulers of church and state, from the time of Augustus through the fifteenth century. The list includes the Carolingian and Ottonian kings, Alfonso X, Frederick II, Ferdinand II, Henry VII, and numerous members of the Italian nobility. Abundant representations can be found in books devoted to the art and architecture of the Renaissance.

Mentioning just one especially interesting example, Saxl examined the ceiling fresco of the Villa Farnesina, Rome, built by the papal banker Agostino Chigi (1466–1520) and painted in 1510–11 by Baldassare Peruzzi (1481–1536). The ceiling panels in the *Sala di Galatea* are arranged with the planets, the signs of the zodiac, and the paranatellonta, which Saxl recognized as encoding the date of Chigi's birth. Saxl calculated the particular date as December 1, 1466, which was later verified when Chigi's baptismal certificate was discovered.[17] The decorative cycle of the *Aratea* manuscripts seems quite conservative when compared to the extravagant display seen in the astronomical programs of Renaissance palaces. The numerous applications of astronomical art all lead ultimately to the *Aratea* manuscripts, the main conduit for depictions of the heavens.

The constellation illustrations therefore fulfilled a wide variety of functions; in a broader sense, they provided entertainment and amusement to the medieval readers and listeners. In addition they provided pictorial evidence of classical literature and mythology from an earlier 'golden age'. The illustrations were a visual connection to an historic past with the illusions inherent in learning about the ancient gods, goddesses, and heroes and their mighty or foolish deeds. Perhaps too the readers felt they were attaining ancient 'scientific' knowledge through the manuscripts. The illustrations simplified and reinforced the learning process when studying basic astronomical principles.

[17]Saxl, "Villa Farnesina" in *Lectures*, 197–98.

Observations and Conclusions

The original naive estimation of illuminated *Aratea* manuscripts was about 15 or 20 survivals that would be the subject of this inquiry; the large number of 60 or so surviving manuscripts has been rather overwhelming, most of which are illustrated with a full constellation cycle. This chosen research focus quickly became a topic of galactic proportions; to investigate all sixty in precise detail would be unmanageable. Therefore this investigation has taken a broader approach, establishing the foundations needed for further investigations. The concentration has been directed on the larger picture of determining the scope and function of this manuscript tradition within the disciplines of art and science. Much thought has been extended in acquiring and consolidating the huge amount of new information recently available on this body of work, and synthesizing it into one coherent picture in order to determine what avenues could be investigated. This initial accumulation of data on numbers, locations, and authors was necessary to establish a broader understanding of the elementary factors of the *Arate*a tradition.

In addition to comparing the text of the *Aratea* and other medieval astronomical writings, it was necessary to investigate the significance of the art work inspired by the various traditions. A rich corpus of constellation images can be observed through the manuscripts of the Aratean tradition and was a large component of this study. "The relation between science and the visual arts, in general, and astronomy and the visual arts, in particular, has never been treated seriously and systematically" (Kühnel 2003: 17). Recognizing this need, the current study has investigated the long life, history, appearance, and functions of this important body of medieval astronomical manuscripts and its accompanying illustrations.

The second chapter brought together, for investigation and comparison, the possible sources for astronomical imagery prevalent in the Late Antique and Early Christian periods. This additional background material pointed out the large variety of astronomical art that was available, and that may have motivated artists when creating the constellation cycles for the numerous ninth-century *Aratea* manuscripts. The sources for the archetypal astronomical images were many; such as frescoes, statues, ceramics, and calendars, in particular the *Calendar of 354*, which was shown to be a model for some of the astronomical illustrations in the Leiden *Aratea*. Other models may have been celestial globes one of which survives, the *Farnese Atlas*, upon its surface 41 constellations are carved. The enormously popular cult of Mithras used astronomical art to symbolize and conceal its secret doctrines. Frescoes and sculptures filled with celestial imagery were associated with their worship centers that were scattered throughout the Roman Empire, many can still be seen. For some yet undiscovered reason, zodiacal mosaics appeared in many early Jewish synagogues. All of these representations were possible sources for *Aratea* illustrations, but the most probable sources were Late Antique manuscripts prototypes of the *Aratea* itself.

Important Components

Some intriguing aspects of this investigated topic are singled out and reiterated:

- the large number and variety of *Aratea* produced in the ninth century, 13 survive
- the retention of the classical forms of the illustrations throughout the medieval period, including partial or full nudity
- the planetary configuration on folio 93v in the Leiden *Aratea*, which holds vital data for the history of astronomy
- the lack of Christian incursion into the text or into the illustrations
- the huge number of Aratea manuscripts produced in the fifteenth century, twenty-seven, the majority were made in Italy

Related Issues not Discussed

Many related topics have merely been touched upon that require further investigation; for instance, the role of Byzantine influences on the *Aratea* manuscripts. The *Aratea* manuscripts had a separate tradition in the East, especially Constantinople. A Latin version of the *Aratea* was translated back into Greek, probably in Constantinople, and the cycle of constellation pictures was attached to these manuscripts, two examples survive in the West, at the Vatican and El Escorial, Madrid. This is an important, yet unstudied, component for the study of the *Aratea*, but left for another investigation. A body of literature consisting of ancient commentaries and elaborations on Aratus' *Phenomena* in Greek and Latin still survives and much of this has been studied by classical textual specialists. This contemporary material has not been included or discussed.

Celestial maps and explanatory diagrams of celestial phenomena such as phases of the moon, and zonal descriptions, the colours, and latitudes also constitute an essential component of astronomical manuscripts. The various types of celestial spheres and the separate images of the northern and southern hemispheres found in Aratean manuscripts had been another topic barely investigated until the all-encompassing research of Dekker was published in 2012, the first in-depth listing and analysis of these important astronomical artifacts. The analysis examines all known survivals of both celestial globes and maps of the heavens found in astronomical manuscripts with particular attention paid to the large number of Islamic globes that survive.[18]

Astrology too is an immense topic in the transmission of astronomical understanding and in the illustrative traditions. Much of the attention, study and mathematics in the field of astronomy had been conducted, not for purely scientific

[18]Dekker also cataloged the National Maritime Museum's large collection of globes in Greenwich, England.

interests, but for astrological purposes. The history and role of astrology was only briefly touched upon in this study, as it had no role in the transmission *Aratea* manuscripts; but its subject matter remained an enormous factor in the history of astronomical art. Medieval illuminators created pictures in order to illustrate the concepts introduced through the beliefs of planetary and zodiacal theories. Some of these 'new' types of images can actually be traced to ancient Egypt, Persia and India. Other astrological concepts that were illustrated were the paranatellonta and decans, mentioned by Aratus, but brought to the Latin West through the writings of Abu Ma'shar. The images of the planets accompanied by their 'children' became another astrological theory abundantly illustrated at the end of the Middle Ages. There have been few recent studies conducted in these areas since the work of Warburg, Saxl and Panofsky in the early twentieth century.[19] There is still much research to be done in analyzing and comparing the numerous astronomical and astrological manuscripts traditions.

Keys to Recognize Aratea *Writers*

Hopefully this study has made clear the difference between a manuscript attributed to Aratus and an Aratean manuscript. To restate the issue, the *Phaenomena* of Aratus is written in Greek and was not in circulation in the Latin West during the Middle Ages except in a poor and partial translation. Any identification or discussion of the poetry of Aratus that appears in a medieval Latin manuscript is derived from the Latin translations by Cicero, Germanicus or Avienus, or the corrupted Latin versions derived from Greek translations, the *Aratus Latinus* or more often the *revised Aratus Latinus*.

One of the goals of this inquiry was to determine if there was a particular model or distinctive version for each of the three Latin authors that could be pointed out and easily recognized to simplify and categorize the various texts and illustrations. Would it be possible to establish guidelines or specific criteria for determining to which astronomical author, treatise, poem, or tradition unidentified writings belong? But as discussed in the earlier chapters, the process of accurately specifying the textual component of an *Aratea* or other astronomical manuscript is quite difficult, since the Latin versions became less and less pure and then intermingled through time. The complicated textual situation is best left to text specialists and paleographers. Most surviving *Aratea* have already been identified and analyzed by text scholars, but the information is widely scattered and must be traced. The easiest of the Latin translators to recognize would be Avienus, considering that his Latin versions remain in verse format with no scholia and no illustrations. But the

[19]Two studies have been published on the illustrations of Abu Ma'shur, a PhD dissertation by Vicky Clark *The Illustrated Abridged Astrological Treatises of Albumasar*, University of Pittsburgh, 1979 and a study by Marie-Thérèse Gousset, and J.-P.Verdet, *Liber Astrologiae, Georgius Zothoros Zaparus Fendulus*, Paris: Editions Herscher, 1989.

circulation and interest in Avienus' *Aratea* was quite limited, and only three of his manuscripts survive.

When analyzing the history and circulation pattern of Cicero's *Aratea* manuscripts, there are basically two main channels; one in England and one in Italy. Before the thirteenth century, the circulation of Cicero's *Aratea* was primarily centered in England where seven of his 13 surviving manuscripts are now found. To summarize the listings in Appendix B, six manuscripts of Cicero belong to the first group, ultimately traced to Harley 647. Four Cicero manuscripts belong to the second family of texts produced in Italy, in which the *Aratea* material is very fragmentary. The remaining three are not part of the established groups, but are also derived from Harley 647, but through an intermediary. The distribution of Cicero's *Aratea* may have been more extensive in the Middle Ages, but we can only base this examination on those that are now extant. A possible way to distinguish Cicero's *Aratea* could be established when examining the pattern of illustrations. Because of the fragmentary survival of Cicero's text, many of his versions begin with the constellation Aries rather than Draco and the Bears. But this arrangement changes in later Cicero manuscripts as scribes and artists filled in the missing sections. Some of Cicero's *Aratea* show Aries as the first miniature and then the other 11 signs of the zodiac were inserted, before continuing with the following constellations. The Oxford manuscript, Bodleian Library MS 614, follows this particular format.

Germanicus' version became the most widespread *Aratea* tradition, since his manuscripts circulated throughout much of Europe. It is also the most complicated with at least 29 extant manuscripts that have been classified by text scholars into six different groups. There are four manuscripts in the first group of Germanicus' *Aratea*, derived from the Leiden *Aratea*, all can be seen in Appendix B. Surprisingly five different variations of Germanicus' *Aratea* survive from the ninth century; again examination of the complex textual issue is best left to experts in that area. Although the poem and scholia vary in Germanicus manuscripts, the illustration cycles seem to follow the same basic pattern. After the initial celestial map, Germanicus' *Aratea* often begin with a miniature of a huge eagle with wings widely spread, Jupiter sits atop his eagle in a commanding position, as Germanicus references the invocation of Zeus by Aratus, even though Germanicus invoked his "father" Augustus instead. The full cycle of 42–48 constellation miniatures follow, inserted among the textual material. The illustrations included mostly retain their original classical appearance and iconography. There are no distinctive traits to characterize them as belonging to Germanicus manuscripts only, since the illustration cycles in Hyginus and Ps-Bede manuscripts are often quite similar.

The chief reception and audience for the *Aratea* manuscripts seems to concentrate around two obvious centers, the royal courts and the monastic communities. A large proportion of the surviving manuscripts can be traced to the European nobility who commissioned luxurious copies for royal libraries or for gift-giving. The second dominant setting for *Aratea* production was monastic centers, in England especially until the twelfth century. Both patronage centers reduced their output or survival became very limited in the thirteenth and fourteenth centuries, the devastation of war and disease dominated that period. Both versions were revitalized in the fifteenth century, particularly at the noble courts of Italy and the monastic communities of Austria. Although these two locales commissioned a large percentage of secular and religious manuscripts in general, it raises questions as to the particular function of the *Aratea* in these environments.

Continuing Life of the Aratea *Tradition*

Surprisingly, the popularity of the poem did not end with the fifteenth century, and demand for illustrated copies of the *Aratea* did not cease when the technology of printing arrived in major European cities. By that time, calendars, almanacs, encyclopedias, and celestial maps became readily and cheaply available; and the type of astronomical information found in the *Aratea* was no longer expedient for memorizing the patterns of the night skies. This change leads one to question whether the didactic quality was truly one of its practical functions. If its primary information was no longer needed to learn about the heavens, why did its popularity rise so substantially?

The epic poetry of Aratus and the *Aratea* were prime prospects for the newly invented presses, and were among the most popular early printed books. The astronomical poems were re-instituted into their original classic forms and again became 'best sellers', as outstanding examples of classical poetry. In Venice, Erhard Ratdolt, who became the finest publisher of scientific material in the incunabula age, produced the first printed editions of numerous astronomical works. Ratdolt published the first printed illustrations of the constellations that accompanied Hyginus' *Poeticon Astronomicon* in 1482. He also printed editions of Abu Ma'shar and Sacrobosco. An edition of the *Aratea* with woodcut illustrations based on the Leiden *Aratea* manuscript images was printed by Hugo Grotius in 1600. Even in the printed illustrations, there was no attempt to place the stars accurately. These illustrated incunabula have become collectors' items now worth thousands of dollars. When examining the collections of major university and city libraries, many possess an early printed edition of Aratus or an *Aratea*. The survey of the printed life of the poem with its illustrations is another aspect of the *Aratean* tradition that is worthy investigation, but beyond the scope of the present study.

The Role of Illustrated Aratea *Manuscripts in the Transmission of Astronomical Knowledge*

The primary focus of this investigation, ultimately, was directed toward determining the role of the *Aratea* manuscripts in the transmission of astronomical knowledge in the Middle Ages. The process of transmission is essentially that of finding, preserving, and renewing the intellectual legacy from the past. This monograph has defined the main streams for transmitting astronomical information, comparing the type of information carried in each different tradition. But as in all aspects of the extensive *Aratea* tradition, there is no simple conclusion to explain its role in the transmittal of knowledge, as it is not easily defined. When evaluating the overall role of the *Aratea* manuscripts in the transmission of astronomical knowledge, you need to define the type of knowledge concerned. The role itself became complex with implications reaching into the many different fields. And of course the role, or roles, of the *Aratea* developed and changed during the long course of transmission. There is no doubt that the information in Aratean manuscripts was valued, for they were retained, revised, copied, and recopied for over two thousand years. As seen above, the tradition remains important among modern scholars for its information reached into many different disciplines. But how were the manuscripts and their illustrations viewed by medieval readers?

After looking at the wide volume of sources, uses, functions, and transitions of the most important science, the following lists a summary of the conclusions that can be drawn from the surviving evidence. The astronomical knowledge available in the early medieval period consisted of popular astronomy that was transmitted mainly through poetry and allegories. The advanced mathematical astronomical writings and heritage from the Greco-Roman world had been mostly lost to the Latin West. In that light, *Aratea* manuscripts, from the ninth century onward, did have a large role in conveying popular astronomy including some elementary information on how the universe is structured. These manuscripts were one of the few sources available for reading about celestial topics and the data and stories were entertaining, but the miniatures cannot be considered true star charts. European production centers created at least 14 new manuscripts, and probably many more that were lost through time, for royal and monastic libraries in the tenth and eleventh centuries.

After the twelfth century, the popular and simplistic astronomy found in the *Aratea* was of little interest and only one tattered manuscript survives from all of continental Europe, in addition to three created in England. The newly translated and transmitted astronomical material from Hellenistic sources brought more sophisticated technical information and the material from Arabic astronomers and its sources conveyed large amounts of technical, practical and astrological material. The *Aratea* no longer fulfilled their previous function of communicating desired astronomical knowledge, if that was ever a primary function. The constellation text and pictures took a back seat to astrological procedures and predictive texts for a few hundred years.

During the fifteenth century, production of *Aratea* manuscripts flourished, but that renewed interest had little association with a quest for astronomical knowledge. The body of astronomical knowledge in the Latin West had expanded immensely and encompassed a much wider spectrum and more advanced calculations and celestial mechanics. Astronomy was on the threshold of huge advances and earth-shaking paradigms to be brought in the next century. The long-held ideas of the nature of the cosmos proposed by Aristotle and Ptolemy, crystalline spheres, perfect circular motion, epicycles, heavenly harmonies, and a stationary earth, were doomed. The strong involvement with the planets, classical constellations, and sky lore through the Middle Ages can be credited to the pursuit of classical antiquity and re-establishment of its remnants. The astronomical subject matter itself was desirable reading material, but not for advancing or even transmitting the more advanced scientific knowledge now available.

Therefore judging from the available evidence, except possibly for the early medieval period, I would argue that the *Aratea* actually had nothing, or very little, to do with learning astronomy or with transmitting astronomical knowledge. With its continual text revisions, any scientific content in the poem became virtually nonexistent. Evidence demonstrates also that, contrary to many opinions, its astronomical information was not vital for agricultural and navigational purposes. It is most likely that every medieval farmer was closely in tune with the cycles of nature and knew when to plant and when to harvest without memorizing the verses of the *Aratea*. The same situation applies to sailors, who were intimately connected with the rhythms of the sky and sea. Though it is not impossible that certain sections of the poem had become part of common folklore, memorizing over 700 lines of poetry, does not seem advantageous or probable for typical seamen. No indications exist that any *Aratea* manuscripts were taken or needed on ocean voyages. Constellation patterns can be learned in other ways than through studying classical Latin poetry. Probably most knowledge of basic astronomy was acquired at a very young age, since you could not find your way at night without an elementary understanding of the sky and its association with directions of earth. It is incontrovertible that the *Aratea* were of absolutely no significance for the reckoning of computus. If *Aratea* manuscripts had no scientific, educative, or even practical functions or qualities, what then accounts for their continuous popularity and lengthy survival?

The more probable reason for its prolonged life cycle was the appealing subject matter of Aratean manuscripts. Just as favorite movies are viewed again and again, musical scores, plays, and operas are enjoyed repeatedly, and well-loved stories and family anecdotes are retold, the lofty themes of the heavenly realms bear repetition, savoring, and protection. The familiar characters and heavenly deities were comforting in their predictability and reliability in an uncertain medieval world, susceptible to unpredictable weather, disease, and famine. The variety of celestial topics had tremendous appeal as morality tales, exotic adventures, and love stories. Young or old, wealthy or poor, lord or abbot, queen or nun, all could relate to the legendary personalities, deeds and misdeeds, and the mythical subject matter. Foremost evidence for their value as books or fables for entertainment remains the cycle of illustrations that consistently adjoins the textual descriptions. It has been demonstrated that the constellation pictures, without accurate star locations, had no 'scientific' value. The fact that Avienus' version had not been illustrated, and therefore was not so appealing supports the entertainment value of the other two authors.

Numerous *Aratea* manuscripts were created throughout the medieval period, not to learn astronomy or classical poetry, as the actual significance of both topics had been fundamentally lost, but to amuse and entertain, and at the same time, reinforce a favorite topic, the nocturnal motions and classical myths, adventures and personalities fixed forever in the heavens. The text and images of *Aratea* manuscripts were sources of pleasure and reflection with unbroken ties to classical past, the Golden Age of antiquity that instilled within the constellations an extraordinary power and spiritual mystique. Although *Aratea* manuscripts are basically mythological picture books, they were utterly successful in preserving and transmitting the entire body of astronomical images intact from antiquity, perhaps much more ancient than we imagine, to the present day, which is a most phenomenal success story.

Aratus created an imaginative vision of the heavens, crammed with poetic symbols and artistic signs that might be used for the benefit of those who wished to learn. Today astronomical imagery and poetry has been relegated to historians and the historical past, no longer a vital part of the body of astronomical knowledge, but thankfully the writings of Aratus and the Latin translations have survived. Hopefully this investigation has filled some of the gaps in the present literature, revealing the fascination and relevance of astronomy through time, and the need for further research especially in the significance of Arabic contributions. This material may open outlooks to help orient other historians in understanding this complicated manuscript tradition. The investigation of knowledge transmission is far from complete as the subject matter is astronomical in all aspects. There are still existing lacunae in this inquiry, but hopefully no black holes.

References

Backhouse, J. (1986) *The Illuminated Manuscript*, London: Phaidon Press Ltd.

Blume, D., Haffner M., and Metzger W. (2012) Sternbilder des Mittelalters: Der gemalte Himmel zwischen Wissenschaft und Phantasie. Part One: 800–1200. Berlin: Akademie Verlag.

Blume, D., Haffner M., and Metzger, W. (2016) Sternbilder des Mittelalters: Der gemalte Himmel zwischen Wissenschaft und Phantasie. Part Two 1200–1500. Berlin: Akademie Verlag.

Cameron, A. (2004) *Greek Mythography in the Roman World*, Oxford University Press.

Camille, M. (1989) *The Gothic Idol: Ideology and Image-making in Medieval Art*, Cambridge University Press.

Camille, M. (1985) "Seeing and reading: some visual implications of medieval literacy and illiteracy." *Art History* 8, no. 1, March, pp. 26–49.

Carruthers, M. (1990) *The Book of Memory: A Study of Memory in Medieval Culture*. Cambridge: UK, Cambridge University Press.

Carruthers, M. and J. Ziolkowski, eds. (2002) *The Medieval Craft of Memory: An Anthology of Texts and Pictures*. Philadelphia: University of Pennsylvania Press.

Colish, M. L. (1997) *Medieval Foundations of the Western Intellectual Tradition 400-1400*, New Haven: Yale University Press.

Dekker, E. (2013) *Illustrating the Phenomena: Celestial Cartography in Antiquity and the Middle Ages*. Oxford: Oxford University Press.

Dodwell C. R. (1954) *The Canterbury School of Illumination, 1066-1200*, Cambridge: Cambridge University Press.

Dugan, L. G. (1989) "Was Art Really the 'Book of the Illiterate'?" *Word & Image*, Vol. 5, No. 3, July–September, pp. 227–251.

Eastwood, B. S. (2002) *The Revival of Planetary Astronomy in Carolingian and Post-Carolingian Europe*. Aldershot: Ashgate.

Eastwood, B. S. (1983) "Origins and Contents of the Leiden Planetary Configuration (MS Voss. Q.79): An Artistic Astronomical Schema of the Early Middle Ages", *Viator*, 14, pp. 1–40.

Eliade, Mircea. *The Myth of the Eternal Return: Or, Cosmos and History*, trans. W.R. Trask, Princeton University Press, 1954.

Goldstein, B. (2000) *Astronomy in the Iberian Peninsula: Abraham Zacut and the Transition from Manuscript to Print.* Transactions of the American Philosophical Society, 90.2. Philadelphia: American Philosophical Society, [with José Chabás].

Gombrich, E. H. (1979) *Ideals and Idols: Essays on values in history and art*, Oxford: Phaidon Press Limited.

Kidd, D. (1997) *Aratus*: *Phaenomena:* Cambridge: Cambridge University Press.

Kessler, H. (2004) *Seeing Medieval Art*, Canada: Broadview Press.

Kühnel, B. (2003) *The End of Time in the Order of Things: Science and Eschatology in Early Medieval Art*. Regensburg: Schnell & Steiner.

Lambert of Saint-Omer. (ca.1120) *Liber Floridus: Codex Autographus Bibliothecae Universitatis Gandavensis*, A. Derolez and I. Strubbe, ed. (1968) Ghent, Bib. Rijksunivsiteit. MS. 92, Story-Scientia.

Lemay, R. (1962) *Ab_ Ma'shar and Latin Aristotelianism in the Twelfth Century*, Beirut.

McClusky, S. C. (1998) *Astronomies and Cultures in Early Medieval Europe*, Cambridge University Press.

McCluskey, S. C. (1993) "Astronomies in the Latin West from the Fifth to the Ninth Centuries" in *Science in Western and Eastern Civilization in Carolingian Times*, P. Butzer and D. Lormann eds. Basel: Birkhäuser Verlag, pp. 139–160.

McCluskey, S. C. (1990) "Gregory of Tours, Monastic Timekeeping and Early Christian Attitudes Toward Astronomy". *Isis*, LXXXI: 9–22.

North, J. (1987) Medieval Concepts of Celestial Influence: A Survey, in *Astrology, Science and Society*, P. Curry ed., Woodbridge, Suffolk: The Boydell Press.

Panofsky, E. (1955) *Meaning in the Visual Arts*, University of Chicago Press.

Panofsky, E. and Saxl, F. (1933) "Classical Mythology in Mediaeval Art", *Metropolitan Museum Studies*.

Reeve, M. (1983) "*Aratea*", *Texts and Transmission*, ed. Reynolds, L. D., Oxford, pp. 18–24.

Saliba, G. (2007) *Islamic Science and the Making of the European Renaissance*, Cambridge, MA: MIT Press.

Saxl, F. (1957) "The Belief in Stars in the Twelfth Century", *Lectures*, 2 vols. The Warburg Institute.

Seznec, J. (1953) *The Survival of the Pagan Gods*, trans. B. F. Sessions, Princeton: Princeton University Press.

Toohey, P. (1996) *Epic Lessons: An Introduction to Ancient Didactic Poetry*, UK: Routledge.

Woodford, S. (2003) *Images of Myths in Classical Antiquity*, UK: Cambridge Univeersity Press.

Yates, F. A. (1966) *The Art of Memory*, Chicago: University of Chicago Press.

Zhitomirsky, S. (1999) "Aratus' *Phainomena*: Dating and Analysing its Primary Sources", in *Astronomical and Astrophysical Transactions,* 17, pp. 483–500.

Appendix A
Listing of Surviving *Aratea* Manuscripts

This appendix consists of a listing of surviving Latin *Aratea* manuscripts of Cicero and Germanicus, giving their current location and the century or date produced, also included is a listing by century of production. The listing makes it possible to determine elementary patterns of their creation and circulation. In addition to handwritten copies of the *Aratea*, numerous incunabula survivals argue for the continued interest in this astronomical poem into the printed era. Among the surviving *Aratea* manuscripts, only four are not illustrated, and two manuscripts have Greek text, but are illustrated. Surprisingly, another medieval Germanicus *Aratea* has come to light in Iceland, but it is fragmentary, making the total 61.

Aratea Manuscripts

Austria

1. Göttweig, Stiftsbibliothek Codex Göttweig 7 Bl.20 (15th C.) Cicero
2. Göttweig, Stiftsbibliothek Cod. 146 (Rot) 190 (second-half 15th C.) *Aratea* (Recensio interpolata) Genus Arati, Cicero, Hyginus, made in Italy
3. Salzburg, Universitätsbibliothek M I 36 (15th C.) Undetermined text
4. Vienna, Österreichische Nationalbibliotek Codex Vindobonensis Palatinus 107 (9th C.) Avienus
5. Vienna, Schottenkloster, Cod. Scotensis (521) 454 (15th C.) Genus Arati, Germanicus (1470-1480) made in Naples

France

6. Boulogne-sur-Mer, Bibliothèque Municipale MS 188 (late 10th C.) Germanicus, made in St Omer, northern France
7. Dijon, Bibliothèque Municipale MS 448 (11th C.) Germanicus
8. Montpellier, Bibliothéque del 'Ecole de Medicine, Cod. H. 452 (late 15th C.) Genus Arati, Cicero, Germanicus, Hyginus, Italian Part 1—France Part 2 (13th C.)
9. Paris, Bibliothèque nationale de France Lat. 5239 (10th C.) Germanicus with scholia

M. Dolan, *Astronomical Knowledge Transmission Through Illustrated Aratea Manuscripts*, Historical & Cultural Astronomy,
DOI 10.1007/978-3-319-56784-6

10. Paris, Bibliothèque nationale de France Lat. 7886 (late 9th) Germanicus with scholia
11. Paris, Bibliothèque nationale de France Lat. 7887 (ca. 10th C.) Germanicus with scholia
12. Paris, Bibliothèque nationale de France Lat. 12957 (9th C.) *revised Aratus Latinus*
13. Paris, Bibliothèque nationale de France Ms. N. Acq. Lat. 1614 (early 9th C.) *revised Aratus Latinus*
14. Strassburg, Bibliothéque Universitaire et Regionale 326. Lat.275 (10th C.) Germanicus

Germany

15. Berlin, Deutsche Staatsbibliothek Cod. Phillipps 1832, Lat.130 (873) Germanicus
16. Berlin, Deutsche Staatsbibliothek Lat. Oct. 149 (15th C.) Germanicus
17. Cologne, Dombibliothek Cod. 83 II (ca. 805) *revised Aratus Latinus*
18. Dresden, Sächsische Landesbibliothek Dc.183 (9th C.) Cicero with scholia (destroyed)
19. Freiberg, Andreas-Möller-Bibliothek, MS Cl. XI. 4°.9 (15th C.) Germanicus, Hyginus, made in Padua
20. Wolfenbüttel, Herzog Augustbibliothek Codex Gudianus 132, (10th C.) Avienus

Holland

21. Leiden, Rijksuniversiteitsbibliotheck MS Voss. lat. Q. 79 (9th C.) Germanicus with Avienus (c.816)//
22. Leiden, Rijksuniversiteitsbibliotheck Voss. Lat. f.121 (11th C.) Cicero with scholia

Iceland

23. Reykjavek, Arni Magnusson Institute for Icelandic Studies, GKS 1812 4t0, (12th - 14th C.) Germanicus with scholia

Italy

24. Florence, Biblioteca Medicea Laurenziana Cod. Strozzianus 46 (14th C.) Germanicus with scholia
25. Florence, Biblioteca Medicea Laurenziana Cod. Laur. Plut.89 Sup.43 (15th C.) Germanicus with scholia, Hyginus (1482-83)
26. Palermo, Biblioteca Communale Cod. Panormitanus 2Q.q.E.11 (15th C.) Germanicus with scholia, Cicero
27. Milan, Biblioteca Ambrosiana Cod. Ambrosiana D.52 inf. (15th C.) Germanicus, Avienus, Cicero
28. Naples, Biblioteca Nationale Cod. XIV D 37 (15th C.) Genus Arati, Germanicus with scholia (1470-1480) made in Florence

29. Città del Vaticano, Biblioteca Apostolica Vaticana Cod. Vat. Barb. Lat. 76 (15th C.) Genus Arati, Germanicus, made in Naples (1470-1472)
30. Città del Vaticano, Biblioteca Apostolica Vaticana Cod. Vat. Barb. Lat. 77 (15th C.) Genus Arati, Germanicus, (1470) made in Florence
31. Città del Vaticano, Biblioteca Apostolica Vaticana Cod. Reg. Lat.1324 (15th C.)*revised Aratus Latinus* with fragments of Cicero and Germanicus (1460–1480) made in Italy
32. Città del Vaticano, Biblioteca Apostolica Vaticana Cod. Vat. Urb.Lat.1358 (15th C.) Genus Arati, Germanicus scholia, Hyginus, made in Florence (1479–1480)
33. Città del Vaticano, Biblioteca Apostolica Vaticana Cod. Vat. Urb.Lat.1653 (15th C.) Germanicus
34. Città del Vaticano, Biblioteca Apostolica Vaticana Cod. Vat. Lat. 3110 (14th C.) Germanicus and Hyginus, made in Florence (1379–1380)//
35. Città del Vaticano, Biblioteca Apostolica Vaticana Cod. Vat. Reg. Lat. 1801 (15th C.) Germanicus
36. Città del Vaticano, Biblioteca Apostolica Vaticana Cod. Vat. Gr.1087 (1320–1330) Greek Aratus, Eratosthenes, Catasterism, illustrated
37. Siena, Biblioteca Comunale Degli Intronati, Ms. L.IV. 25 (second-half 15th C.) *revised Aratus Latinus,* fragments of Cicero, made in Italy

Spain

38. Madrid, Biblioteca Nacional, MS 19 (12th C.) Germanicus with scholia. Made in southern Italy or Sicily
39. Madrid, Biblioteca Nacional, Cod. 4691 (1465) Aratus Greek text
40. Madrid, Biblioteca Nacional, Cod. 8282 (15th C.) Genus Arati, Germanicus (1470–1480) made in Florence//
41. Madrid, Biblioteca Nacional, Cod. de Metz 3307 (ca. 840) Germanicus
42. El Escorial, Scorialensis Σ III 390 (1490) Genus Arati and *Aratea*, Greek text illustrated

Switzerland

43. Basel, Universitätsbibliothek Cod. A.N. IV.18 (9th C.) Germanicus and *Aratus Latinus*, made in Fulda (820–830)
44. Bern, Burgerbibliothek Cod. 88 (beginning 11th) Germanicus, with gloss from ps. Bede, made in northern France, St Bertin
45. Cologny, Bibliotheca Bodmeriana, Cod. 7 (1469) Genus Arati, Germanicus, Hyginus, made in Naples
46. Einsiedeln, Stiftsbibliothek Cod. 338 (10th C.) Germanicus
47. St. Gall, Stiftsbibliothek Cod. 250 (ca. 889) *revised Aratus Latinus*, with scholia Sang, computus, Cicero fragments
48. St. Gall, Stiftsbibliothek Cod. 902 (ca. 837) *revised Aratus Latinus*, with scholia Sang. Computus, Cicero fragments

United Kingdom

London—British Library

49. Harley MS 647 (mid-9th C.) Cicero with scholia, made in Aachen
50. Harley MS 2506 (10th C.) (988–994) Cicero with scholia made in Fleury
51. Cotton MS Tiberius B.V. (991–1016) Cicero with scholia, made in Canterbury?
52. Cotton MS Tiberius C.1. (1122) Cicero with scholia, computus, made in Peterborough
53. Arundel MS 268 (14th C.) Germanicus, Sacrobosco Hermann of Carinthia, made in Italy after 1303
54. Additional MS 15819 (15th C.) Genus Arati, Germanicus with *Scholia Sangermanensia* (1470–1480) made in Florence
55. Egerton MS 1050 (15th C.) Genus Arati, Germanicus with *Scholia Sangermanensia* (1470) made in Rome

Oxford—Bodleian Library

56. MS Bodl. 614 (S.C. 2144) (first-half 12th C.) Cicero with scholia
57. Digby MS 83 (S.C. 1684) (mid 12th C.) Cicero with scholia
58. MS Laud Misc. 644 (second-half 13th C.) fragments of *Aratea* with ps. Bede (1268–1274) made in France (Bayeux)

Cambridge

59. Trinity College, MS R.15.32 (ca. 1000) Cicero with scholia

United States

60. Cambridge MA, Harvard University, Houghton Library MS Typ 43 (ca. 1425) Anthology with *Aratea,* Alfonso X, Michael Scot, made in Austria
61. Cambridge MA, Harvard University, Houghton Library MS Typ 18 (15th C.) Greek text of Aratus
62. New York, Morgan Library MS M. 389 (1469) Genus Arati, Germanicus with scholia, made in Naples

Wales

63. Aberystwyth, National Library of Wales MS 735C (first section-early 11th C: second section-late 11th C.) Germanicus with *scholia,* Hyginus

Aratea Manuscripts by Century

Ninth Century

Basel, A.N. IV 18
Berlin, Codex Phil.1832
Cologne, Dom.83 II
Dresden, Dc. 183

Leiden, Voss. Lat. Q.79
London, BL Harley 647
Madrid, Metz 3307
Paris, BNF nouv. acq. lat.1614
Paris, BNF Lat. 7886
Paris, BNF Lat. 12957
St. Gall 902
St. Gall 250
Vienna, Scotensis-Vindobon. Pal. 107
Tenth Century
Boulogne-sur-Mer 188
Einsiedeln, Codex 338
Paris, BNF Lat. 5239
Paris, BNF Lat. 7887
Strassburg, B. U. et Regionale 326. Lat.275
Wolfenbüttel, Codex Gudianus 132
Eleventh Century
Aberystwyth, National Library of Wales 735 C.
Bern, Codex 88
Cambridge, Trinity MS R.15.32
Dijon, BM 448
Leiden, Voss. Lat. f. 121
London, BL Cotton Tib. B.V.
London, BL Harley 2506
Prague, Universitatesbibliothek MS. IX.C.6
Twelfth Century
London, BL Cotton Tiberius C.1
Oxford, Bodl. 614
Oxford, Digby 83
Madrid, B. N. Codex Matritensis 19
Thirteenth Century
Oxford, Laud Misc. 644
Fourteenth Century
London, Arundel 268
Florence, Codex Stroz. 46
Vatican, Vat. Lat. 3110
Vatican, Vat. Gr.1087
Fifteenth Century 27
Berlin, Lat. Oct.149
Cologny, Bibliotheca Bod. Cod.7
El Escorial, Scorialensis Σ III 390
Florence, Laur. Plut.89
Freiberg, MS Cl. XI. 4°.9
Göttweig Stiftsbibliothek 146 (190)
Göttweig, Stiftsbibliothek 7 Bl 20

Harvard U., Houghton Lib. Typ 18
Harvard U., Houghton Lib. Typ 43
London, BL Add.15819
London, BL Egerton 1050
Madrid, Matritensis 4691
Madrid, Matritensis 8282
Milan, Cod. Amb. D.52
Montpellier, H. 452
Naples, XIV.D.37
New York Morgan Library M.389
Palermo, 2 Qq. E. 11
Salzburg, M I 36
Siena, Ms. L.IV. 25
Vatican, Vat. Barb. Lat. 76
Vatican, Vat. Barb. Lat. 77
Vatican, Vat. Lat. 1324
Vatican, Vat. Urb. Lat. 1358
Vatican, Vat. Urb. Lat. 1653
Vatican, Vat. Reg. Lat. 1801
Vienna, Schottenkloster Scot. (521) 454

Appendix B
A Brief Description of Surviving *Aratea* Manuscripts

This appendix provides an abbreviated descriptive listing of the sixty-one extant Latin *Aratea* manuscripts, including which are illustrated. For many of the manuscripts, there was only brief mention in various text editions, and there is little published information available. The manuscript traditions of Cicero, Germanicus, and Avienus are each listed separately. Almost none of the surviving *Aratea* manuscripts consist of pure versions of the original poem by author, without scholia or insertions of additional material, the least diluted texts are mainly from the ninth century. The majority of the manuscripts listed are anthologies of astronomical texts that include the writings of one or more *Aratea* authors, accompanied by sections of Hyginus and/or fragments of other known and unknown authors that have been added to the prototypes through the centuries. Although this manuscript tradition includes many uncertainties, the surviving manuscripts of all three Latin versions of the *Aratea* point to a beginning in northeastern France during ninth century; and very soon after that time another branch appeared in Italy (Reeve 1983: 23). The variety of texts and illustrations demonstrates the unceasing interest in antique culture and in the *Aratea*.

List of *Aratea* Manuscripts

Greek Manuscripts

Fifty manuscripts of the original Greek *Phaenomena* are currently known, and all are described in J. Martin, *Histoire du texte des Phenomenes d'Aratos*. Most of these fifty Aratus manuscripts have Greek text and are not illustrated. There are two exceptions; one illustrated Greek Aratus manuscript is in the Vatican Library, Vat. Gr.1087 (15th C.) and one in Spain, el Escorial, Scorialensis Σ III 390 (1490). These two Greek versions may be the *Phaenomena* of Aratus, or possibly a Greek translation of Germanicus or Cicero that circulated in Byzantine society. The fact that they are illustrated makes me suspect the latter. There is little information available whether the Latin versions of the *Aratea* circulated in the Byzantine culture. There are also some surviving scraps of papyri containing the Greek text copy of the *Phaenomena* listed in Martin's text history as well. Because the Greek

M. Dolan, *Astronomical Knowledge Transmission Through Illustrated Aratea Manuscripts*, Historical & Cultural Astronomy,
DOI 10.1007/978-3-319-56784-6

texts are not commonly illustrated and are not part of the Latin *Aratea* tradition in the West, they were not included in this study.

Latin Manuscripts

This appendix provides a brief description of each Latin Aratean manuscript, arranged by author, and by related types of text, where known; not all scholars agree on these textual groupings. Only a select few of the illustrated manuscripts have been studied and an abundance of information has been published for these; but the majority have not been studied, but the volumes of *Sternbilder des Mittelalters* by Blume et al present much needed data available for new investigations. The focus of this study was not to examine every manuscript individually, but to compare the contents of the *Aratea* manuscripts with other astronomical anthologies to determine their role and function in the Middle Ages. Now that many *Aratea* manuscripts are available online, the task of consulting a variety of astronomical codices is becoming much more attainable. The chief concern of this examination of the entire corpus of *Aratea* manuscripts was to understand its function or influence in medieval society, its importance in the history of science, and especially its role in the transmission of astronomical knowledge. But in order to gain an understanding of the whole *Aratea* tradition, it is necessary to gather together the individual, very complicated parts of that manuscript tradition.

Already in the ninth century, there were at least five different versions of the *Aratea* circulating among European libraries. Bischoff groups the ninth-century *Aratea* manuscripts into five different text families:

Group 1. Leiden *Aratea* (text family Z; Bischoff, a.a.O., S. 89–91)

Boulogne, BM MS 188
Bern, Burgerbibl. MS 88

Group 2. Illustrated Germanicus manuscripts (text family O; Bischoff, a.a.O, S.89–90)

Basel, Univ. B. MS. A N IV 18
Madrid, BN MS.19

Group 3. *Aratea* of Cicero

London, BL Harley 647

Group 4. Madrid, BN Cod. 3307 (Germanicus scholia)
Group 5. *Recensio Interpolata* or *revised Aratus Latinus*

Paris, BNF, lat.12957
St. Gallen, Stiftsbibl. Cod. 902 (Mütherich 2004: 153)

The Germanicus text became much more convoluted through time than the Cicero text. For a full explanation of the Germanicus *Aratea* texts see Lott, E. S. (1983), "The textual tradition of the *Aratea* of Germanicus Caesar: missing links in the «μ» branch», in the *Revue d'histoire des textes*, Vol. 11, Numéro 1981, pp. 147–158. Lott relates that in 1831 J.-C. Orelli published a major edition of Germanicus' *Aratea*; he was the first to split the *Aratea* of Germanicus into two different families, the "O" and the "Z". The "O" is most often accompanied by scholia and "Z" contains interpolations from a fourth-century translation of Germanicus by Avienus. When comparing both branches, the "Z" is 'smaller and more homogenous', but survives in only four manuscripts. The texts of the "O" family are more numerous and more complicated, since its manuscripts have been divided into two separate families "ν" and "μ". The oldest manuscripts of the Germanicus text are dated to the ninth century. The latest editions of Germanicus are written by le Boeuffle (1979) and Gain (1976). Full citations are available in Chap. 7 references.

Ninth-Century *Aratea* Manuscripts

It was quite surprising to discover that there are at least 13 Carolingian *Aratea* manuscripts extant from the ninth century; all but two are illustrated. None of the 13 manuscripts has a 'pure' text, each has varying degrees of scholia, and most are included in anthologies with additional 'popular' astronomical writings. Ten have Germanicus or Germanicus scholia, and two contain the text of Cicero and one of Avienus. The various textual complications were discussed in more detail in Chap. 6. When observing the illustrations, Carolingian artists copied classical models; but they demonstrated their skill by drawing on their own creativity and ingenuity as well. A basic feature of Carolingian art is 'the interplay between old and new, drawing on Roman and Early Christian precedents' and then transforming them according to contemporary tastes and interests. Although not able to view examples of the illustrations from every ninth-century survival, there must have been many visual traditions of constellation imagery from which the Carolingian artists may have modeled and drawn inspiration when creating new manuscripts and illuminations. This is obvious since the ninth-century miniatures are not based on the same prototype, but have great variety in their execution.

The following is a listing of the ninth-century *Aratea* manuscripts, more details can be found in their respective groupings.

1. Cologne, Dombibliothek Cod. Ms 83 II (ca. 805)
2. Paris, BNF Nouv. Acq. Lat. 1614 (early 9th C.)
3. Basel, Universitätsbibliothek Codex A.N. IV. 18 (early 9th C.)
4. Madrid, Biblioteca Nacionale Codex de Metz 3307 (ca. 820)
5. Leiden, Rijksuniversiteitsbibl. MS Vossianus lat. Q.79 (After 816–840)
6. London, British Library Harley MS 647 (9th C.)
7. Paris, BNF Lat. 12957 (9th C.)
8. St. Gall, Stiftsbibliothek MS. 902 (mid 9th C.)
9. St. Gall, Stiftsbibliothek MS. 250 (ca. 889) copy of MS.902
10. Dresden, Landes. MS. Dc. 183 (late 9th C.).
11. Paris, BNF Lat. 7886 (late 9th C.)
12. Berlin, Deutsche Staatsbibliothek Codex Phillipps 1832, lat.130 (873)
13. Vienna, Österreichische Nationalbibliotek Scotensis-Vindobon. Pal. 107 (9th C.)

Manuscripts with Germanicus Text

Germanicus—Group I

Leiden, Rijksuniversiteitbibl. MS Vossianus lat. Q.79 (816 to 840)
This manuscript is discussed in Chap. 6, material presented there will not be repeated. MS Leiden is one of the almost 'pure' Germanicus poetic texts with no scholia, but it does have additions of Avienus' poem.

f. 1: missing

f. 2: Germanicus text

ff. 3–93v: Germanicus text with additions of Avienus text and full constellation cycle

ff. 94–95: text

ff. 96–97: blank

Miniatures: ff. 3v Draco with Ursae; 6v Hercules; 8v Corona; 10v Serpentarius; 12v Boötes; 16v Gemini; 18v Cancer; 20v Leo; 22v Auriga; 24v Taurus; 26v Cepheus; 28v Cassiopeia; 30v Andromeda; 32v Equus; 34v Aries; 36v Deltoton or Triangula; 38v Pisces; 40v Perseus; 42v Pleiades; 44v Lyra; 46v Cygnus; 48v Aquarius; 50v Capricorn; 52v Sagittarius; 54v Aquila; 56v Delphinus; 58v Orion; 60v Canis with Sirius; 62v Lepus; 64v Argo; 66v Cetus; 68v Eridanus; 70v Pisces Magnus; 72v Ara; 78v Hydra with Corvus and Crater; 78v Anticanis; 80v Five Planets; 82v Four Seasons; 93v planetary diagram. Missing: Jupiter on eagle (1v), Virgo (13v), Centaurus (74v), Helios with chariot (95v) and Luna with chariot (96v).[1]

Literature: McGurk, P. (1966) *Catalogue of Astrological and Mythological Illuminated Manuscripts of the Latin Middle Ages*; Le Boeuffle, A. (2003) *La texte des Aratea de Germanicus est copieusement représenté*: Paris: Les Belles Lettres, Josephe-Henriette Abry, '*La constellation du flueve dans le ciel Greco-Roman'*, (1994) 103–110; Smeyers, *Flemish Miniatures from the 8th to the mid-16th Century*; Maurits, *The Medieval World on Parchment*, (1998) 38; Arblaster-Homer (2002) *Medieval Mastery*: *Book Illumination from Charlemagne to Charles the Bold: 800–1475*, trans. Lee Preedy, 136–37; Smeyers (1998);, Maurits. *Vlaamse miniaturen van de 8ste tot het midden van de 16de eeuw*. *De middeleeuwse wereld op perkament,* Leuven Seznec (1980) 136; Reeve (1983) "*Aratea*", in *Texts and Transmission* 20; Colette Jeudy and Yves-Francois Riou (1989), *Les manuscrits classiques latins des bibliothéques publiques de france,* Editions du centre national de la recherche scientifique, Paris 308–09; Dekker, E. (2010) "The Provenance of the Stars in the Leiden *Aratea* Picture Book", *Journal of the Warburg and Courtauld Institutes*, pp. 77–90; Dodwell, C. R. (1993) *The Pictorial Arts of the West 800–1200*; Mostert R. and Mostert M. (1990) "Using astronomy as an aid to dating manuscripts: the example of the Leiden planetarium", *Quaerendo*, 20: pp. 248–261; Mütherich, F. (2004) "Book Illumination in the Court of Louis the Pious" in *Studies in Carolingian Manuscript Illumination*, Bischoff, B.

[1]Folio numbers of missing illustrations as per von Euw, *Die Builder*, 154–186.

(1994) *Manuscripts and Libraries in the Age of Charlemagne*, Gorman, M. ed. and trans; Kühnel, B. (2003) *The End of Time in the Order of Things: Science and Eschatology in Early Medieval Art*; Blume, Haffner, and Metzger (2012) *Sternbilder des Mittelalters: Der gemalte Himmel zwischen Wissenschaft und Phantasie*. Part One: 800–1200.

Boulogne-sur-Mer, Bibliothèque municipale MS 188 (late 10th C.)

Most probably executed at Abbey of St. Bertin, under Abbot Odbert, at current location since 1799. Reproduces Leiden *Aratea* or possibly a copy of third manuscript, a lost exemplar, source of MS Leiden as well. Caroline minuscule, 33 ff, decorated initials, first letter of each line enlarged, alternates in color. Text and images in excellent condition. Colors are green, orange with some violet and black. Images do not fill whole page, aligned side by side with poem. Thirty-nine illustrations of constellations. Miniatures well-executed, similar to MS Leiden, follow in exact order, but often two or three on page.

Includes five miniatures missing from MS Leiden, Sun and Moon, both on same folio; Helios wears flowing red mantle, stands in chariot with four rearing horses and holds orb of Sun in left hand. Luna appears with two silver oxen pulling cart, carries unusually large torch. Has exact copy of planisphere from MS Leiden; with gold-leaf applied to Labors and planets. Planetary configuration includes four additional diagrams in each corner. Eastwood identifies four smaller discs as Isidorian rotae. (Eastwood 1983: 25) Upper left disc shows five latitudinal zones of earth, (not usual Macrobius map), three habitable and two uninhabitable, arranged around central disc-like flower petals. Upper right disc derived from Isidore's *De natura rerum*, represents the paths of planets. Lower left disc shows hours of sunlight in three circles, at equinoxes, summer, and winter, derived from Martianus Capella. Lower right disc has round T-O map of earth at center of disc, twelve radial segments for winds.

ff. 1–19: Computus tables and calendar

ff. 20v–33: Germanicus *Aratea* text, no scholia, *Ab Iove principium magno deduxit Aratus*

ff. 30v–32v: *'Prognostica'* of Germanicus

ff. 32v–33: '*Prognostica*' of Avienus

Miniatures: f. 20 Celestial map; 20v Jupiter on eagle; 21 Hercules; 21v Serpentarius and Boötes; 22 Virgo and Gemini; 22v Cancer and Leo; 23 Auriga and Taurus; 23v Cepheus and Cassiopeia; 24 Andromeda and Pegasus; 25 Pisces and Perseus; 25v Pleiades and Lyra; 26 Cygnus and Lyra; 26v Aquarius and Sagittarius; 27 Aquila and Delphinus; 27v Orion, Lepus and Canis; 28 Argo and Cetus; 28v Eridanus and Pisces Magnus; 29 Ara, Centaurus and Corona; 29v Hydra, Corvus, Crater and Anticanis with Planets; 30 planetary diagram; 32v Helios in chariot (top), Luna in chariot (bottom).

Literature: Josephe-Henriette Abry, '*La constellation du flueve dans le ciel Greco-Roman'*, (1994) 103–110; *The Utrecht Psalter in Medieval Art*, Exhibition Catalog, Utrecht (1996) 200–201; Smeyers, *Flemish Miniatures from the 8th to the mid-16th Century*; Maurits, *The Medieval World on Parchment*, (1998) 38, *Medieval Mastery*: *Book Illumination from Charlemagne to Charles the Bold: 800–*

1475, trans. Lee Preedy, Greta Arblaster-Homer, Turnhout: Brepols, (2002)136–37; Smeyers, Maurits. *Vlaamse miniaturen van de 8ste tot het midden van de 16de eeuw. De middeleeuwse wereld op perkament,* (1998); Seznec (1980) 136; Reeve "*Aratea*", in *Texts and Transmission* (1983) 20; Colette Jeudy and Yves-Francois Riou, *Les manuscrits classiques latins des bibliothéques publiques de france,* Editions du centre national de la recherche scientifique, Paris (1989) 308–09; Blume, D., Haffner M., and Metzger W. (2012) *Sternbilder des Mittelalters: Der gemalte Himmel zwischen Wissenschaft und Phantasie*. Part One: 800–1200.

Bern, Burgerbibliothek MS 88 (ca. 1000)

Also made at Abbey of St. Bertin, early eleventh century, presented to St. Mary's, Strasbourg, by Werinhar, bishop (1001–28). Listed in Bern Library catalog in 1760. Probably a copy of MS 188. Has Germanicus text, with added scholia in extreme margins. Carefully written and colorfully illuminated, not as well-known as prototype, Leiden *Aratea*. Illustrations imitate Carolingian model. For example, representation of Boötes appears awkward missing the illusionism portrayed in MS Leiden. MS 88 also contains Jupiter on eagle, Helios and Luna missing from MS Leiden.

Miniatures: f.1v Jupiter on eagle; 2v Draco with Ursae; 2v Hercules and Boötes; 3v Gemini, Cancer and Leo; 4v Pleiades and Lyra; 5 Aquarius; 5v Sagittarius; 6v Lepus; 10v Helios and Luna; 11v planetary diagram. Many constellations missing.

Einsiedeln Stiftsbibliothek Codex 338 (10th C.)

Another of three descendants of MS Leiden *Aratea*, Germanicus text, also includes sections of Avienus. Consists of 22 folios; no illustrations or scholia, in abbey library since 14th C.

Germanicus—Group II

LeBoeuffle assigns the following five Germanicus manuscripts to the same family.

Basel, Universitätsbibliothek Codex A.N. IV 18 (9th C.)

Made at Fulda, Germanicus text with scholia, Caroline minuscule script. Has 45 folios with 36 illustrations, also includes other astronomical writings. Thought to be modeled on manuscript dated before 6th C., as discussed in Mütherich, which pictures four illustrations. Illustrations sketchily executed in papyrus style between lines of text, no frame. Aquarius and Capricorn pictured close together. Illustrations of constellation cycle show signs of antiquity, similar to MS Leiden *Aratea*. Stars in constellations carefully marked.

Miniatures: Hercules 14v, Serpentarius 16, Boötes 16v, Virgo 18v, Gemini 20, Auriga 22, Taurus 23, Cepheus 24r, Aries 26v, Perseus 28, Cygnus 29v, Aquarius 31v, Sagittarius 32v,

Orion 34, Lepus 35v, Eridanus 37, Ara 38. Some missing are Jupiter on Eagle, and Pleiades.

Literature: K. Escher, *Die Miniaturen in den Basler Bibliotheken, Museen und Archiven*, Basel (1917) 46; Bischoff, "Katalog I", a.a.O., Nr. 258, S.57; McGurk, *Carolingian Astrological Manuscripts*, (1981) 317–33; Thiele (1898) 138–41; Lehmann, "Die alte Klosterbibliothek Fulda" and "ihre Bedeutung", *Erforschung des Mittelalters*, II, Leipzig (1941) 217; Byvanck, *De platen in de Aratea van Hugo de Groot*, (1949) 213–9; F. Mütherich, "Der Fuldaer Buchmalerei in der Zeit des Hrabanus Maurus", *Studies in Carolingian Manuscript Illumination*, reprint (2004) 153.

Paris, Bibliothèque nationale de France MS Lat. 7886 (late 9th C.)

Produced at Corbie, Germanicus with scholia, 36 folios, no illustrations, large capital letters, no colored initials. Description of Zodiac begins at 19v. Same lacunae as Basel, Univ. Cod. A.N. IV 18. Incipit: *Ab Iove principium magno deduxit Aratus*.

Aberystwyth, National Library of Wales MS 735 C (11th C.)

Contains *Aratea* of Germanicus and *Scholia Basilensia*; illustrated, with two parts, first part probably made in Limoges region early eleventh century, second part, in Insular hand of late eleventh. Descendant of Basel A.N.IV.18 or its exemplar (McGurk 1973: 197). Text and illustrations studied by P. McGurk, published in *National Library of Wales Journal*. In same journal a few years later, B. Eastwood published full description and analysis of single planetary configuration that appears at beginning of manuscript, folio 4v, is accompanied by verses. Diagram fills upper two-thirds of folio; verses concern seven planets, band of fixed stars written on lower third. Signs of the zodiac illustrated in outermost ring. Small face representing each planet placed within its circular sphere. Placement of inner rings shows interest in theories of position of orbits of Mercury and Venus, discussed by Martianus Capella and Macrobius. Illustrated with 24 constellations, some joined into one group, also has miniature of five planets.

Celestial map of entire sky also included. Iconography of constellations closer to Pseudo-Bede catalog than to any other group of Germanicus' illustrations.[2] Has portrait of Aratus with traditional Muse of astronomy, Urania, found in one other *Aratea* manuscript, Madrid MS 19. Miniature of Jupiter on eagle also included, common in Germanicus, but not in Cicero picture cycles (As expected, since that initial text is absent.). Many variations found in representations of constellations, such as placing Cancer between feet of Gemini instead of on right, which McGurk calls 'very curious'. Iconography reveals some antique features; such as, Sagittarius shown as satyr, and representation of aselli or dot formation on back of Cancer. Folios and images online at National Library of Wales website.

Part A:

ff.1–3: historical texts, principal vices, and Priscian

f. 3v: drawing of Southern Hemisphere

[2] McGurk (1973) 198 NLW 735C may have associations with two other manuscripts from the Limoges area, Paris lat. 5543 and Paris lat. 5239. These may have been known to Adémar of Chabannes and used in his copy of Hyginus, Leiden Voss. Oct. lat. 15.

f. 4: drawing of Northern Hemisphere
f. 4v: drawing of zodiacal circle wih planets
ff. 5–11: text of Sallust and Macrobius
ff. 9, 11, 25v, 26, 26v blank
Incipit: *Ab Iove principium magno deducit Aratus.*

Miniatures: f.10v planisphere, 11v Aratus and Muse; 12 Jupiter and eagle; 13 Draco and Ursae; 13v Hercules; 14 Corona; 14v Serpentarius on Scorpio; 15 Boötes; 16 Virgo; 17 Gemini with Cancer; 17v Leo; 18 space for Auriga; 18v Taurus; 19 Cepheus; 19v Cassiopeia and Andromeda; 20 Pegasus; 20v Aries; 21 Hydra, Crater, Corvus and Anticanis; 21v five planets; 25 diagram of planisphere with names of constellations. Part B ff. 17–47: Hyginus, *De astronomia*, not illustrated.

Literature: P. McGurk, (1973) "Germanici Caesaris Aratea cum Scholiis: A New Illustrated Witness from Wales", *The National Library of Wales Journal XVIII* 197–216; B. Eastwood, "Notes on the Planetary Configuration in Aberystwyth N.L. W. Ms. 735 C, f 4v", *National Library of Wales Journal*, (1981)129–140.

Madrid, Biblioteca Nacionale MS 19 (also known as A.16) (12th C.)

Compiled at Montecassino includes Germanicus' *Aratea* and scholia. Contains 199 ff. and is illustrated with 43 miniatures. Based on Beneventan script, Catalan expressions in text, perhaps a Catalan monk at Montecassino. Manuscript in deteriorated condition, has microfilm, but unreadable. Illustrations in this manuscript, or a copy similar to this one, have been recognized as influential on work of Michael Scot in Sicily.

"The almost square format typical of the early codex gave the illustrator the chance to copy from monumental painting more complex compositions with other figures being added to that of the author. One such instance is the group in which a man of letters is inspired by a muse. A twelfth-century Aratus manuscript in Madrid shows the seated poet with the muse Urania standing in front of him, both holding scrolls and pointing at the celestial globe between them. This composition is very similar to the Aratus-Urania group in the Monnus mosaic in Trier which dates from the middle of the third century. The archetype of the Aratus miniature must be assumed to have been close in date to the Filocalus Calendar." (Weitzmann 1971: 116–7)

Miniatures: 55r Jupiter on eagle, 56r Draco with Ursae, 56r Hercules, 57r Serpentarius, Boötes 57r, Virgo 57v, Gemini 58r, Taurus 59 v, Cepheus 59v, Cassiopeia 60r, Andromeda 60r, Aries 61r, Pisces 61v, Perseus 61v, Pleiades 62r, Lyra 62v, Cygnus 62v, Aquarius 63r, Sagittarius 63v, Orion 64v, Lepus 65v, Ara 67v.

London, British Library Arundel MS 268 (13th C.)

Germanicus *Aratea* with scholia, written on paper with 103 folios, in small, compressed manner, difficult to read. *Aratea* in first section with other treatises following; written as prose in two columns with glosses on each page. No illustrations or diagrams, nor any spaces left blank to include them. Later in the manuscript, associated with another treatise, spaces are left blank, some circular for illustrations or diagrams that were never completed. It is one of only two *Aratea* surviving from thirteenth century.

Germanicus—Group III

Scholars have reconstructed another group of Germanicus manuscripts, all related to a lost source called π.

Berlin, Deutsche Staatsbibliothek Codex Phillippicus 1832, lat.130 (late 9th C.)

Made at Metz, fragments of Germanicus with scholia on ff. 86r–90v. Has illustrations of constellations on beginning folios. Astronomical compendium also includes Bede's *De temporum ratione.*

Paris, Bibliothèque nationale de France MS Lat. 5239 (10th C.)

Made at Limoges, Germanicus with scholia. Computus charts, calendars, chronicles and consanquinity. Also contains work of Isidore and parts of Bede's *De natura rerum*, *De temporibus* and *De temporum ratione* with many diagrams. Description of constellation precedes illustrations.

ff. 1–215: astronomical texts, T-0 maps, and cosmological diagrams.

Miniatures: 215v Draco and Ursae; 216 Hercules and Corona; 216v Serpentarius on Scorpio; 217v Gemini and Cancer; 218 Leo and Auriga; 218v Taurus and Cepheus; 219 Cassiopeia; 219v Andromeda and Pegasus; 220 Aries, Triangula and Pisces; 220v Perseus and Lyra; 221 Cygnus and Aquarius; 221v Capricorn and Sagittarius; 222 Aquila and Delphinus; 222v Orion and Sirius; 223 Lepus and Argo; 223v Eridanus and Pisces Magnus; 224 Ara and Centaurus; 224v Hydra, Corvus, Crater and Anticanis.

Strassburg, Bibliothéque Universitaire et Regionale 326 (lat.275) (10th C.)

Germanicus with scholia. No information available at this time.

Germanicus—Group IV

The following is another group of Germanicus manuscripts that are related to an ancestor that migrated to Monte Cassino, where it was copied and created much interest.

Florence, Biblioteca Medicea Laurenziana Codex Strozzianus 46 (late14th C.)

Made in Florence, Germanicus with scholia based on version in Madrid, Biblioteca Nacionale, MS 19, which was made at Monte Cassino. Not sure if it is illustrated, its model is. One of only 4 *Aratea* produced in 14th C. Not listed in McGurk, *Astrological Manuscripts in Italian Libraries.*

Palermo, Biblioteca Communale, Codex Panormitanus 2Q.q.E.11 (15th C.)
Related to Madrid, Biblioteca Nacionale, MS 19 with Germanicus and scholia, but includes fragments of Cicero as well, is illuminated. Not listed in McGurk, *Astrological Manuscripts in Italian Libraries.*

Città del Vaticano, Biblioteca Apostolica Vaticana Codex Vat. Lat. 3110 (15th C.)
Compendium of miscellaneous astronomical treatises made in Florence, fifteenth or possibly late fourteenth century for Coluccio Salutati, canon of Florence. Name of Pope Paul V written in gold in front, name of Cardinal Scipione Borghese at end. Linked to prose descriptions of Hyginus that appear after Germanicus. Line drawings have classical appearance, well-executed with some shading, no borders nor any color added. Miniatures appear after text, often two on page. Text is one-column, written in black, drawings mostly one shade—brown. Stars added in red, usually matching number given in scholia, but placement is random. J. J. Alexander mentions Vat. 3110, "The positions and movements of the figures in the copy of the Hyginus illustrations at NY Public Library almost invariably correspond to similar ones in this manuscript".

ff. 1–42: portions of Hyginus

ff. 42v–78: *Aratea* of Germanicus, also excerpts of Martianus Capella.

Incipit: *In Iove principium magno deduxit Aratus.*

Miniatures: f. 65 Draco with Ursae; 65v Boötes; 66 Corona; 66v Hercules; 67 Lyra; 67v Cepheus; 68 Cassiopeia; 68v Andromeda; 69 Perseus; 69v Auriga without cart; 70 Serpentarius; 70v Sagitta, Aquila and Delphinus; 71 Pegasus; 71v Aries and Triangula; 72 Taurus and Pleiades; 72v Gemini and Cancer; 73 Leo; 73v Virgo, Libra and Scorpio; 74 Sagittarius and Capricorn; 74v Aquarius; 75 Pisces and Canis Major; 75v Eridanus and Lepus; 76 Orion and Canis; 76v Anticanis and Argo; 77 Centaurus; 77v Ara; 78 Hydra, Corvus, Crater and Southern Fish.

Literature: Saxl and Meier (1966) I, *Verzeichnis astrologischer und mythologischer illllustrierter Handschriften des latinischen mittelalters* I. pp. 88–90; LeBoeuffle (1975) XL; Viré (1981)159–276; Eastwood (1982)150; Reeve, *Texts and Transmission*, (1986) 21–22; K. Lippincott, "The Astrological Vault of the Camera di Griselda for Roccabianca", in *Journal of the Warburg and Courtauld Institutes*, 48 (1985) 43–70; Alexander (1978) *The Painted Page*, 120: Alexander (1994) *The Decorated Letter*; M. Buonocore (1996) *Vedere i Classici*, Fratelli Palombi Editori pp. 251–53.

Germanicus—Group V

This group of four manuscripts was all made about the same time in the second half of the fifteenth century at the court of Naples; all are fully illustrated but the artwork in each is treated differently.

New York, Morgan Library MS M. 389 (1469)

Italian, Naples, made for Ferdinand I, king of Naples, (d. 1494) or court member. Astronomical anthology: *Aratea* of Germanicus, Pliny the Elder, Hyginus. 117 folios, 40 delicately tinted drawings by Master of Isabelle di Chiaromonte (active 1450–60s in Naples). Similar to Vat. Barb. 76, Barb.77, and Vat. Urb. 1358, characteristics of illustrations derived from Carolingian prototype found in Sicily. Light but firm text written by Joan Marco Cinico. Includes planisphere with constellations of northern and southern celestial hemispheres. Discussed fully in Chap. 4.

f. 1–2v: Frontispiece and biography of Aratus.

Incipit: *Aratus quidem fuit. Athinodori patris*

f. 2v–5v: *Scholia Strozziana* with Germanicus *Aratea* excerpts.

f. 6–68: Germanicus, *Aratea* with *Scholia Strozziana.*

Incipit: *Ab Iove principium magno deduxit Aratus carminis*

f. 69v–71: Germanicus excerpts.

f. 71v–77: *Scholia Strozziana* with Germanicus *Aratea* excerpts.

f. 77v–116: Pliny, *Natural History,* Hyginus, and other astronomical texts.

f. 117: colophon

Miniatures: f.3v celestial map, f.7 Jupiter on eagle, f.11v Draco with Ursae, f.14v Hercules, f.15v Corona, f.16v Ophiuchus with serpent standing on Scorpio, f.19 Boötes, f.20v Virgo, f.23v Gemini with Cancer, f. 27 Leo, f.28v Auriga, f.31 Taurus, f.33 Cepheus, f.34v Cassiopeia, f.35v Andromeda, f.36v Pegasus, f.38 Aries, f.40 Triangula, f.41 Pisces, f.42v Perseus, f.44 Pleiades, f.46 Lyra, f.47 Cygnus, f.48v Aquarius, f.51 Capricorn, f.53v Sagittarius, f. 55v Aquila with Saggita, f.57 Delphinus, f.59. Orion, f.62 Canis, f.64 Lepus, f.65v Navis, f.67v Cetus, f.68v, Eridanus, f.70v Pisces Magnus, f.72 Ara, f.73v Centaurus, f.75 Hydra with Corvus and Crater, f.77 Helios in chariot, f.83 Luna in cart, f.100 another Centaurus.

Literature: De Ricci, (1962) *A Census of Medieval and Renaissance Manuscripts in the United States and Canada.* NY: H. W. Wilson Co., 1935–40. *Supplement.* W. H. Bond ed. NY: v.2, p.1439; De Marinis (1952) *La biblioteca napoletana dei re d'Aragona*, Milan: p.50; Martin (1956) *Histoire du texte des Phénomènes d'Aratos*, Paris; De Marini (1969) *La biblioteca napoletana dei re d'Aragona*, Supplemento, Verona: p. 38; Harrson and Boyce (1953) *Italian Manuscripts in the Pierpont Morgan Library*, NY: Morgan Library, no.8; *Masterpieces of Medieval Painting*: *Pierpont Morgan Library*, University of Chicago Press (1980) 42; G. Toscano (1995) "Matteo Felice: un miniatore al servizio del re d'Aragona di Napoli", *Bollettino d'Arte*, pp. 93–94; Blume, Haffner, and Metzger (2016) *Sternbilder des Mittelalters: Der gemalte Himmel zwischen Wissenschaft und Phantasie*. Part Two: 1200–1500.

Città del Vaticano, Biblioteca Apostolica Vaticana Codex Vat. Barb. Lat. 76 (second-half 15th C.)

Italian, Naples, a manuscript made for Giovani d'Aragon, son of Ferdinand I, king of Naples. Later owned by Cardinal Barberini. Astronomical anthology: *Aratea* of Germanicus, Pliny the Elder, Hyginus. 100 folios, forty miniatures, illuminated in gold, bold colors, (reds, blues, greens, oranges), compared with softly colored Morgan MS M 389. Discussed more fully in Chap. 7.

f. 1–2v: Biography of Aratus.

f. 2v–5v: *Scholia Strozziana* with Germanicus *Aratea* excerpts.

Incipit: *Coelum circulis quinque distinguitur.*

f. 6–66: Germanicus, *Aratea* with *Scholia Strozziana.*

Incipit: *Ab Iove principium magno deduxit Aratus carminis*

f. 68v–71: Germanicus excerpts.

f. 71v–77: *Scholia Strozziana* with Germanicus *Aratea* excerpts.

f. 77v–85: Pliny, *Natural History* excerpts.

f. 85v–86: *De polis mundi.*

f. 86: Pliny, *Natural History* excerpts.

f. 86–100: Hyginus, *De astronomia* excerpts.

Miniatures: f.3 celestial map, f.6 Jupiter, f.9v Draco with Ursae, f.12 Hercules, f.13v Corona, f.14v Ophiuchus with serpent standing on Scorpio, f.16v Boötes, f.18 Virgo, f.21 Gemini with Cancer, f. 24 Leo, f.25 Auriga, f.27 Taurus, f.29 Cepheus, f.30r Cassiopeia, f.31 Andromeda, f.32 Pegasus, f.33 Aries, f.35 Triangula, f.35v Pisces, f.37 Perseus, f.38v Pleiades, f.39v Lyra, f.41 Cygnus, f.41v Aquarius and Capricorn, f.46 Sagittarius with Sagitta, f. 47v Aquila, f.48v Delphinus, f.49v. Orion, f.52v Canis, f.55 Lepus, f.56 Navis, f.57v Cetus, f.58, Eridanus, f.59v Pisces Magnus, f.60v Ara, f.61v Centaurus, f. 63 Hydra with Corvus and Crater, f.73v Helios, f.77 Luna, f.79 another Centaurus (missing Anti-canis or Procyon). Morgan M.389 has the same series of miniatures.

Literature: Vire, (1981) 'La transmission du De astronomia d'Hygin jusqu'au XIII siecle': RHT 11, pp. 173–276; M. Buonocore, (1996) *Vedere i Classici*, Fratelli Palombi Editori pp. 486–88; Pellegrin, et al (1975) *Manuscrits Classiques Latins de la Bibliothèque Vaticane*, Tome I, Centre National de la Recherche Scientifique pp. 123–25; de Marinis, (1952) *La biblioteca napoletana dei re d'Aragona*, 1 Milan: p.158; Martin, (1956) *Histoire du texte des Phénomènes d'Aratos*, Paris: p. 39; Silverstein, *Medieval Latin Scientific Writings in the Barberini Collection*, Chicago (1957) 26; A. Dell'Era (1979) Una miscellanea astronomica medioevale: gli Scholia Strozziana a Germanic: MAL 23, p. 152; Marucchi (1964) *Stemmi di possessori di manoscritti conservati nella biblioteca Vaticana*, Vatican: pp. 31–32. Manuscript was on view at the 'Miniatures of the Renaissance' exhibit at the Apostolic Vatican Library, (1950) *Quinto centenario della Biblioteca Vatican. Minature del Rinascimento. Catalogo della mostra*, Città del Vaticano and described briefly in the exhibition catalog on p. 25; Blume, Haffner, and Metzger (2016) *Sternbilder des Mittelalters: Der gemalte Himmel zwischen Wissenschaft und Phantasie*. Part Two: 1200–1500.

Città del Vaticano, Biblioteca Apostolica Vaticana Codex Vat. Barb. Lat. 77 (15th C.)

Italian, Naples, for King Ferdinand I or court member. Original owner not identified, arms at bottom of page one, eagle holding baton in beak, motif designed in the sixteenth century. Owned by Cardinal Barberini, second *Aratea* MS in Vatican Barberini Collection. Astronomical anthology with 59 folios: *Aratea* of

Germanicus, Pliny the Elder, *Natural History*, extracts of Hyginus, *De astronomia*. Humanistic frontispiece, bordered with white vines on three sides, title inscribed in gold letters. Humanistic script, textual contents identical to Barb. 76. Thirty-nine miniatures, appear before descriptions. Opens with map of celestial sphere, f. 2v.

f. 1–2: Biography of Aratus, title *'Arati fragmentum cum comento incipit'*.
f. 2–4: *Scholia Strozziana* with Germanicus *Aratea* excerpts.
f. 4v–41: Germanicus, *Aratea* with *Scholia Strozziani*.
f. 41v–43: Germanicus extracts.
f. 43–46v: *Scholia Strozziani* with Germanicus *Aratea* excerpts.
Constellation miniatures on folios 2v–40v, 44v, 47, 52.
f. 47–51v: Pliny, *Natural History* extracts.
f. 51v–52: *De polis mundi*.
f. 52: Pliny, *Natural History* extracts.
f. 52–59v: Hyginus, *De astronomia* extracts.

Miniatures: f.3 celestial map, f.4 Jupiter, f.5v Draco with Ursae, f.7 Hercules, f.7v Corona, f.8v Ophiuchus with serpent standing on Scorpio, f.10v Boötes, f.11 Virgo, f.13 Gemini with Cancer, f. 14 Leo, f.15v Auriga, f.17 Taurus, f.18 Cepheus, f.18v Cassiopeia, f.19 Andromeda, f.20 Pegasus, f.21 Aries, f.22 Triangula, f.22v Pisces, f.23v Perseus, f.24v Pleiades, f.25 Lyra, f.26 Cygnus, f.26v Aquarius and Capricorn, f.29 Sagittarius with Sagitta, f.30 Aquila, f.31 Delphinus, f.31v. Orion, f.33 Canis, f.34v Lepus, f.35v, Navis, f.36v Cetus, f.37, Eridanus, f.38v Pisces Magnus, f.38v Ara, f.39v Centaurus, f. 40v Hydra with Corvus and Crater, f.44v Helios, f.47 Luna (missing Anti-canis or Procyon).

Literature: Saxl and Meier (1966) I, *Verzeichnis astrologischer und mythologischer illllustrierter Handschriften des latinischen mittelalters*,pp. 6–7; M. Buonocore, (1996) *Vedere i Classici* p. 413; E. Pellegrin, et al, (1975) *Manuscrits Classiques Latins de la Bibliothèque Vaticane*, Tome I, Centre National de la Recherche Scientifique pp. 125–27; J. Martin, (1956) *Histoire du texte des Phénomènes d'Aratos*, Paris: p. 39; T. Silverstein, (1957) *Medieval Latin Scientific Writings in the Barberini Collection*, Chicago (1957) 26–27; Viré (1981) "La transmission du '*De astronomia*' d'Hygin jusqu'au XIII siècle", in *Revue d'histoire des textes*, 11 pp. 173–76; Blume, Haffner and Metzger (2016) *Sternbilder des Mittelalters: Der gemalte Himmel zwischen Wissenschaft und Phantasie*. Part Two: 1200–1500.

Città del Vaticano, Biblioteca Apostolica Vaticana Codex Vat. Urb. Lat.1358 (15th C.)

Italian, Naples, fourth of group made for court of Naples. Frontispiece has floral decorations with putti and medallion with insignia of Federico da Montefeltro (1422–82). Also owned by Pope Leo XIII (1878–1903) and Cardinal Giovanni Battista Pitra (1869–1889), their names inscribed in gold. Astronomical anthology with 166 folios: *Aratea* of Germanicus, Pliny the Elder, *Natural History*, extracts of Hyginus, *de Astronomia*. Comparable to Barb. Lat. 76, Barb. Lat. 77, Morgan MS M.389, but more folios and more illustrations. Opens with celestial map, f.3 Constellation cycle appears on folios 3–47v (41 miniatures of *Aratea*) and 123–139

(30 miniatures of Hyginus). Miniatures drawn in mono-color reddish-orange with shading, similar coloring to Barb.77, but bolder.

f. 2–2v: Biography of Aratus.

Aratvs sicionius in sicilia noviter repertus incipit feliciter (in red)

f. 2v–4v: *Scholia Strozziana* with Germanicus *Aratea* excerpts

f. 4v–37: Germanicus *Aratea* with *Scholia Strozziani*

f. 37–39: excerpts of Germanicus *Aratea*

f. 39–41 Germanicus with *Scholia Strozziani*

f. 41–47: Pliny, *Natural History* excerpts

f. 47–47v: *De polis mundi.*

f. 47v: Pliny, *Natural History* excerpts

f. 47v–152: Hyginus, *De astronomia*

f. 152–152v: Martianus Capella, *De nuptiis,* excerpts.

Miniatures: f.3 celestial map, f.4v Jupiter, f.6v Draco with Ursae, f.8 Hercules, f.8v Corona, f.14v Ophiuchus with Serpent standing on Scorpio, f.10v Boötes, f.11 Virgo, f.12v Gemini with Cancer, f. 14 Leo, f.15 Auriga, f.16v Taurus, f.17 Cepheus, f.17v Cassiopeia, f.18 Andromeda, f.18v Pegasus, f.19v Aries, f.20v Triangula, f.21 Pisces, f.21v Perseus, f.22v Pleiades, f.23v Lyra, f.24 Cygnus, f.24v Aquarius and Capricorn, f.26v Sagittarius with Sagitta, f. 27v Aquila and Sagitta, f.28 Delphinus, f.28v. Orion, f.30 Canis, f.31v Lepus, f.32 Navis, f.33 Cetus, f.33v, Eridanus, f.34 Pisces Magnus, f.35 Ara, f.35v Centaurus, f. 36v Hydra with Corvus and Crater, f.40 Helios, f.41 Luna (missing Anti-canis or Procyon).

Literature: Saxl and Meier (1957) I,103–06; M. Buonocore, (1996) *Vedere i Classici*, 414; Viré, (1981) 175–76; *Manuscrits Classiques Latins de la Bibliothèque Vaticane*, Tome II-2, Centre National de la Recherche Scientifique, (1982) 675–77; Blume, Haffner, and Metzger (2016) *Sternbilder des Mittelalters: Der gemalte Himmel zwischen Wissenschaft und Phantasie*. Part Two: 1200–1500.

Germanicus—Group VI (Miscellaneous Remainders)

These remaining manuscripts of Germanicus have not been grouped into families, and they are not related, other than containing the text of Germanicus with scholia. They are arranged by century.

Madrid, Biblioteca Nacionale Codex de Metz 3307 (ca. 840)

Anthology of astronomical works composed around 840 at Metz during the tenure of Drogo (823–855) Archbishop of Metz, illegitimate son of Charlemagne, with 76 folios written in Carolingian script of early 9th C. with the titles in uncials. Compilation was result of gathering of church scholars to attempt reform of calendar and computus. This assembly and Charlemagne's role discussed in Chap. 6. Efforts lead to creation of manual, now lost, but reflected in four incomplete copies from ninth century. This is one of surviving but incomplete copies. Includes Germanicus' *Aratea*. Has one of earliest extant copy of Pliny's text, with apsidal diagram. Origin of this composite, apsidal diagram may date back to Late Antiquity (Kühnel 2003: 80). Contains finest miniatures of the 4 copies, only example with fully painted figures. Constellation cycle derives from palace exemplar (Mütherich 1976: 89). Figures lack marking of stars, which are mentioned in text, clearly reflect Late Antique prototypes. This compendium was working manual kept up to date by later entries, recording solar eclipse in 840, another in 876. Full color facsimile published, Madrid, 1993, commentary volume by Manuel Sanchez Mariana.

ff.1–53: early copy of 'Aix-la-Chapelle encyclopedia', Bede's *De tempore ratione*, computus, astronomical charts and tables.

ff. 54–66: excerpts of *Aratea* of Germanicus with scholia

f.65v: Pliny's text, with apsidal diagram.

f. 66: planetary diagram

ff. 66v–76: astronomical texts

Miniatures: f. 54v Draco and Ursae; 55 Hercules; 55v Serpentarius with Scorpio; 56 Boötes and Virgo; 56v Gemini and Cancer; 57 Leo and Auriga; 57v Taurus, Cepheus and Cassiopeia; 58 Andromeda and Pegasus; 58v Aries, Triangula, Pisces; 59 Perseus, Lyra and Cygnus; 59v Aquarius and Capricorn; 60 Sagittarius, Aquila and Delphinus; 60v Orion and Canis; 61 Lepus, Argo, Cetus and Delphinus; 61v Eridanus, Pisces Magnus and Ara; 62 Centaurus, Hydra, Corvus and Crater; 62v Corvus and Crater again, Anticanis.

Literature: McGurk (1981) 331; Koehler, *Die karolingische Miniaturen*, vol. 3 Berlin (1960) 100–101, 119–127; Mütherich, *Carolingian Painting* (1976) 89; von Euw, "Die kunstlerische Gestaltung (1993) fig. 23; B. Kühnel, *The End of Time in the Order of Things: Science and Eschatology in Early Medieval Art* (2003) 80: Mariana, *Codex de Metz 3307, Tratado de computo Y Astronomia*, Estudio Critico, Testimonio Compania Editorial, Madrid (1993) facsimile, commentary by Sánchez Mariana.

Reykjavik, Arni Magnusson Institute for Icelandic Studies GKS 1812 4t0 (12th–14th C.)

36 folios, divided into 4 unequal parts, in the collection of King Frederick III (11609–1670) in year 1662 and in Royal Library of Copenhagen. Returned to Iceland in 1984: 15 constellation images, f. 3r—4r and 7v; written in Old Icelandic and Old Norwegian; Aratea in Latin; diagram attributed to Macrobius; also contains computus. For more information see Academia.edu, Christian Ethridge: "A Possible Source for a Medieval Icelandic Astronomical Manuscript on the Basis of Pictorial Evidence".

Madrid, Biblioteca Nacionale Codex Matritensis 8282 (15th C.)

Germanicus *Aratea* with commentary, humanistic border on frontispiece with angels, green and gold. Illustrations are unframed, placed between sections of text.

f. 1: title page

ff. 1v–2: biography of Aratus

f. 2v: celestial map, figures all youthful

Miniatures: f. 4 Jupiter on eagle, 6v Draco and Ursae; 8 Hercules; 9 Corona; 9v Ophiuchus with Serpent on Scorpio; 11v Boötes; 12 Virgo; 13 Gemini and Cancer; 15v Leo; 16v Auriga; 18 Taurus; 19 Cepheus; 19v Cassiopeia; 20v Andromeda; 21 Pegasus; 22 Aries; 23 Triangula; 23v Pisces; 24v Perseus; 25v Pleiades in 2 rows; 26 Lyra; 27 Cygnus; 28 Aquarius and Capricorn; 30 Sagittarius and Sagitta; 31 Aquila; 31v Delphinus; 32v Orion; 34 Canis; 35v Lepus; 36v Argo; 37v Cepheus;

38 Eridanus; 39 Southern Fish; 39v Ara; 41v Centaurus; 42 Hydra, Corvus and Crater; 46v Helios in chariot with 4 horses; 49 Luna in chariot with 2 oxen; 55 another Centaurus.

Literature: Blume, Haffner, and Metzger (2016) *Sternbilder des Mittelalters: Der gemalte Himmel zwischen Wissenschaft und Phantasie*. Part Two: 1200–1500.

Göttweig, Stiftbibliothek Codex Göttweig 146 (190) (Second-half 14th C.)

Germanicus with scholia, 31 folios, not listed with any definite groups. First illustration is celestial globe on pedestal standing within small temple-like structure. This arrangement is similar to BNF Lat.12957 which dates to ninth century. Two different styles of miniatures f.7 to17 have traditional figures, colored drawings; f.18 to 29 has constellations similar to Harley 647 with text filling the bodies of the figures. Has some verses of Germanicus; descriptions are before the illustrations. No stars are indicated on constellations. HMML has microfilm of Göttweig 146, some miniatures are online at HMML website

ff.1–6v: astronomical texts

Incipit: *Arati et que videntur aste Sionem.*

ff. 32–33v: astronomical texts

Miniatures: f.7 Celestial globe; 7v Ursa Major; 8 Ursa Minor; 8v Draco with Ursae; 9 Hercules; 9v Corona and Ophiuchus with Serpent on Scorpio; 10 Scorpio and Boötes; 10v Virgo holding Libra; 11 Auriga and Taurus; 11v Cepheus; 12 Cassiopeia; 12v Andromeda; 13 Pegasus and Aries; 13v Pisces and Triangula; 14 Perseus; 14v Pleiades; 15 Lyra and Cygnus; 15v Aquarius and Capricorn; 16 Southern Pisces and Ara; 16v Centaurus; 17 Hydra, Corvus, Crater, and Anticanis; 18 Triangula; 18v Pisces; 19 Perseus; 19v Pleiades; 20 Lyra; 20v Cygnus; 21 Aquarius; 21v Capricorn; 22 Sagittarius; 22v Sagitta; 23 Aquila; 23v Delphinus; 24 Orion; 24v Canis; 25 Lepus and Argo; 25v Cetus; 26 Eridanus; 26v Southern Pisces; 27 Centaurus; 28v Hydra, Corvus and Crater; 29 Anticanis; 29v five planets; 31 Helios at top, Luna at bottom.

London, British Library Egerton MS 1050 (1470)

Made in Rome, the majority of fifteenth-century *Aratea* are made in Italy, has 125 ff. Purchased from Jos. Lilly, Jan, 1844. Lines of poetry alternate with prose commentary throughout. Constellation cycle pictures 41 miniatures. Frontispiece has humanistic border, scrollwork on three sides, gold trim on edges, first letter A decorated with same scroll, red, blue and green color added. Miniatures all line drawings, tinted with limited palate, red, blue and green. No stars indicated. Each constellation description is marked by large initial in light ink.

ff.1–3: title page, biography of Aratus

ff. 3v–50: Sections of Germanicus' *Aratea* with *Scholia Strozziana and Sangermanensis*.

f. 4: Incipit: *Ab Jove principium magno deduxit Aratus*.

ff. 50–64: excerpts from Pliny's *Natural History*, Chap. 18.

ff. 55v–125v: excerpts of Hyginus' *De astronomia,* no illustrations or diagrams.

Miniatures: f. 3v Jupiter on Eagle; 6 Draco and Ursae; 7v Hercules; 8v Corona; 9 Ophiuchus with Serpent on Scorpio; 10v Boötes; 11v Virgo; 13 Gemini and Cancer; 15 Leo; 16 Auriga; 17v Taurus; 18v Cepheus; 19v Cassiopeia; 20 Andromeda; 20v Pegasus; 21v Aries; 22v Triangula; 23 Pisces; 24 Perseus; 25

Pleiades; 26 Lyra; 26v Cygnus; 28v Aries; 30 Sagittarius and Sagitta; 31v Aquila and Sagitta; 32 Delphinus, 33 Orion; 35 Canis; 36v Lepus; 37v Argo; 38v Cetus; 30 Eridanus; 40v Pisces Magnus; 41 Ara; 41v Centaurus; 42v Hydra, Corvus and Crater; 43v Helios; 47 Luna; 55 a second Centaurus.

Literature: Saxl and Meier III (1966) 140; *Catalogue of additions to the Mss. in the BM.*, 1841–1845 (London 1850) 152; Blume, Haffner, and Metzger (2016) *Sternbilder des Mittelalters: Der gemalte Himmel zwischen Wissenschaft und Phantasie*. Part Two: 1200–1500.

Vienna, Schottenkloster Codex Scotensis Vindobonensis 545 (521) (15th C.)

Germanicus *Aratea* with scholia, text has many large foliated initials, alternate red and blue. Has 90 folios and 40 well done illustrations, based on two different models, two or three to a page. Beginning on f.18 illustrations change from fully colored images to Harley 647 style with text making up the body of constellation. Five planets shown as heads in v-shaped sections surrounded with text, similar to Harley 647. HMML has microfilm, can be viewed online at Vivarium.

Florence, Biblioteca Medicea Laurenziana Cod. Laur. Plut. 89 sup. 43 (15th C.)

Humanistic manuscript, probably made in Florence, second-half fifteenth century, 171 ff; produced for Guardaroba Medicea. Medici arms inscribed on f. 3. Fully illustrated with 43 constellations, first miniature is Jupiter on Eagle, Sun and Moon are on separate folios at end. Excerpts of Pliny and Hyginus follow *Aratea,* Hyginus text is fully illustrated also.

ff. 1–2: blank

f. 2v: frontispiece

ff. 3–4: life of Aratus

Incipit: *Incipit liber arati de signis celestibus*

ff.4–6: excerpts from *Scholia Sangermanensia*

Incipit: *Ab Iove principium magno*

ff. 6v–54v: Germanicus' *Aratea* with scholia

ff. 55–169: excerpts of Pliny's *Natural History*, Hyginus with another constellation picture series.

Miniatures: f. 4v Celestial sphere; 6v Jupiter on Eagle; 9 Draco with Ursae; 11 Hercules; 11v Corona; 12v Ophiuchus with Serpent on Scorpio, 14v Boötes; 15 Virgo; 17 Gemini and Cancer; 19 Leo; 20 Auriga; 22 Taurus; 23 Cepheus; 23v Cassiopeia; 24v Andromeda; 25 Pegasus; 26 Aries; 27v Triangula; 28 Pisces; 29 Perseus; 30 Pleiades; 31 Lyra; 32 Cygnus; 32v Aquarius and Capricorn; 35v Sagittarius; 36v Aquila and Sagitta; 37v Delphinus; 37 Orion; 39 Canis; 41 Lepus; 42 Argo; 43 Cetus; 43v Eridanus; 44v Pisces Magnus; 45v Ara; 46v Centaurus; 47v Hydra, Corvus and Crater; 52 Helios; 55 Luna.

Literature: McGurk (1966) *Astrological Manuscripts in Italian Libraries*, 26–29; Leonardi (1960) "Icodici di Marziano Capella", *Aevum*, 34, 45–46; Blume, Haffner, and Metzger (2016) *Sternbilder des Mittelalters: Der gemalte Himmel zwischen Wissenschaft und Phantasie*. Part Two: 1200–1500.

London, British Library Additional MS 15819 (c. 1470)

Germanicus *Aratea* fragments (verses 1–430) with scholia Strozziana, annotations and, possibly, diagrams by Bartolommeo Fonzio, made in Florence, Italy, contains 71 ff, fine quality vellum, very thin and white. Text in one column only. Frontispiece similar to BL Egerton 1050, humanistic border on three sides, scroll work outlined in blue. Has illustrations of 38 constellations, several folios have 2 constellations on same leaf. Prose commentary and poetry alternation continues throughout, similar to Egerton 1050. Constellation miniatures drawn before text descriptions. Also Pliny's *Natural History* XVIII, and Hyginus' *De astronomia*. Purchased from Jechener, Paris May 1846.

ff. 1v–2: Prose text with spaces left for images, but not drawn.

Incipit: *Arathi fragmentvm cvm commento noviter repertvm in sicilia incipit*

ff. 2–4: excerpts *Scholia Sangermanensia*.

f. 3: celestial map with faint color added to constellation figures.

ff. 4v–55v: *Aratea* of Germanicus with Scholia Sangermanensia

Incipit: *Ab Jove principium magno deduxit Aratus,*

ff. 55v–71: excerpts of Pliny's *Natural History* and Hyginus' *De astronomia*

Miniatures: f. 4v Jupiter on Eagle; 6 Draco with Ursae; 8 Hercules; 9 Corona; 9v Ophiuchus with Serpent on Scorpio, 11v Boötes; 12 Virgo; 14 Gemini; 16 Leo; 17v Auriga; 19 Taurus; 20v Cepheus; 21v Cassiopeia; 22 Andromeda; 23 Pegasus; 24 Aries; 25v Triangula; 26 Pisces; 28 Perseus; 29 Lyra; 30 Cygnus; 30v Aquarius and Capricorn; 33v Sagittarius; 35 Aquila; 35v Southern Pisces; 36v Orion; 39 Canis; 41 Lepus; 42 Argo; 43 Cetus; 43v Eridanus; 45 Pisces Magnus; 48 Hydra, Corvus and Crater; f.49 space left for picture but none drawn; 52v Helios; 55v Luna (unfinished); 62 Centaurus.

Literature: Saxl and Meier III (1966) 51–53; *Catalogue of additions to the mss. in the B.M.* 1846–47, London (1864); Haffner (1997) *Sternbilderzyklus*, p.26; Blume, D., Haffner M., and Metzger W. (2016) *Sternbilder des Mittelalters: Der gemalte Himmel zwischen Wissenschaft und Phantasie*. Part Two: 1200–1500.

Città del Vaticano, Biblioteca Apostolica Vaticana Codex Vat. Urb. Lat.1653 (15th C.)

Belonged to library of Dukes of Urbino, noted for outstanding illuminated manuscripts. Acquired for Vatican Library by Pope Alexander VII in 1658. Illustrated Not included in *Vedere i Classici*, no microfilm at St Louis University Library.

Città del Vaticano, Biblioteca Apostolica Vaticana Codex Vat. Reg. Lat. 1801 (15th C.)

Excerpts of *Aratea* of Germanicus, commentary and biography of Aratus. Not illustrated. Astronomical manuscript not included in *Vedere i Classici*, no microfilm at St Louis U. Library.

Naples, Biblioteca Nazionale MS XIV. D. 37 (third-quarter 15th C.)

Made in southern Italy, humanistic cursive, has 55 ff. Begins with life of Aratus, continues with Germanicus' *Aratea* in verse with sections of *Scholia Sangermanensia*. Colored miniatures start on folio 3v; 36 constellations are illustrated, also Jupiter on eagle, Sol in chariot, Luna in cart. Pliny and Hyginus included but not illustrated.

ff. 1–1v: Life of Aratus

ff.1v–3v: excerpts from *Scholia Sangermanensia*

ff. 4–43: *Aratea* of Germanicus (verses 1–430) and scholia

ff. 43–55: excerpts from Hyginus and Pliny's *Natural History*

Miniatures: f.2v celestial map, 3v Jupiter; 6 Draco with Ursae; 7v Hercules; 8v Corona; 9 Ophiuchus with Serpent standing on Scorpio; 10v Boötes; 12v Leo; 13v Auriga;.14v Cepheus; 16v Cassiopeia; 17 Andromeda; 18 Pegasus; 18v Aries; 19v Triangula; 20 Pisces; 21 Perseus; 22 Pleiades; 22v Lyra; 23v Cygnus; 24 Aquarius and Capricorn; 26v Sagittarius with Sagitta; 29v Canis; 31 Lepus; 32 Navis; 32v Cetus; 33 Eridanus; 34v Pisces Magnus; 35 Ara; 35v Centaurus; 36v Hydra with Corvus and Crater; 40v Helios and chariot; 44 Luna and chariot.

Literature: P. McGurk (1966) *Astrological Manuscripts in Italian Libraries* 62–63; Blume, D., Haffner M., and Metzger W. (2016) *Sternbilder des Mittelalters: Der gemalte Himmel zwischen Wissenschaft und Phantasie*. Part Two: 1200–1500.

Berlin, Deutsche Staatsbibliothek Lat. Oct. 149 (15th C.)

No information available on this *Aratea* manuscript other than it has Germanicus text.

Paris, Bibliothèque nationale de France Lat. 7887 (10th C.)

Germanicus *Aratea* fragments, life of Aratus, commentaries, 60 folios, no illustrations, bound with works of Bede.

Manuscripts with Cicero Text

Earliest surviving manuscript with text of Cicero, BL MS. Harley 647, produced in Lotharingia/Lorraine area arrived in England by middle of tenth century. In English centers, a few later manuscripts modeled on Harley 647, but no exact copies. Additions attached to text, and constellation cycle was enlarged to include more images. It appears that only Cicero's version of *Aratea* circulated in medieval England, unless the Germanicus at Aberystwyth arrived there in Middle Ages, which McGurk doubts. Three Germanicus *Aratea* currently in England arrived there much later, with manuscript collectors. Illustrations of constellations in Cicero manuscripts often begin with Sign of Aries, rather than with usual group of Draco and Ursae, as this is where his surviving text begins. By the twelfth century, this changes. Because Harley 647 was missing descriptions of numerous constellations, some copies supplemented the lacunae with other sources, adding constellation myths that did not appear in the exemplar. Along with additional text, missing images were inserted to compliment stories.

After twelfth century, it seems that no more copies of the *Aratea* in Harley mode were made, at least none survive. At that time astronomical interest was attracted by 'new', exotic astrological and Arabic writings. Thirteen extant illustrated manuscripts have fragments of Cicero's poem or Cicero recension. Seven now in England; six housed in other European libraries. Four surviving Cicero manuscripts are in British Library: Harley 647; Harley 2506; Cotton Tiberius B.V; and Cotton Tiberius C.I; two at Oxford: Bodleian 614 and Digby 83; and one at Cambridge, Trinity College, MS R.15.32.

London, British Library Harley MS 647 (9th C.)

Carolingian manuscript, 21 folios, discussed in Chap. 7, information will not be repeated. Like MS Leiden *Aratea*, Harley 647 was modeled on unknown Late Antique prototype, probably from fourth or fifth century. Though incomplete, has one of the most 'pure' texts of Cicero's *Aratea,* much scholia was added to later versions.

ff. 1–16v: excerpts of Cicero's *Arate*a and illustrations.

Incipit: *E quibus hinc subter possis cognoscere fultum.*

ff. 16v–21v: Pliny's *Natural History*, Macrobius, Martianus Capella, and Macrobius.

Miniatures: f. 2v Aries; 3 Triangula; 3v Pisces; 4 Perseus; 4v Pleiades; 5 Lyra; 5v Cygnus; 6 Sagittarius; 6v Sagitta; 7 Aquila with Sagitta; 7v Delphinus; 8 Orion; 8v Canis; 9 Lepus; 9v Argo; 10 Cetus; 10v Eridanus; 11 Pisces; 11v Ara; 12 Centaurus; 12v Hydra with Crater and Corvus; 13 Anticanis; 13v five planets; 20 solar system diagram; 21 celestial map or planisphere of Geruvigus with figures but no stars.

Literature: Buescu, (1941) *Cicéron. Les Aratea*; *Cicéron: Aratea Fragments Poètiques*, Ed. Soubiran, (1972); Barker-Benfield, B. C. (2008) *St Augustine's Abbey, Canterbury/Vol. 3, Other sources, appendices and indexes*; Mütherich, F. and J. E.Gaehde. (1976) *Carolingian Painting*; Soubiran, (1972) *Cicéron: Aratea Fragments Poètiques*; Arblaster-Homer, (2002) *Medieval Mastery*: *Book Illumination from Charlemagne to Charles the Bold: 800–1475*, trans. Lee Preedy, 136–37; McGurk (1966) *Catalogue of Astrological and Mythological Illuminated Manuscripts of the Latin Middle Ages*; Mütherich, (2004) "Book Illumination in the Court of Louis the Pious" in *Studies in Carolingian Manuscript Illumination*, *Sternbilder des Mittelalters: Der gemalte Himmel zwischen Wissenschaft und Phantasie*. Part One: 800–1200.

The following manuscripts are five other representatives of this family of Cicero texts.

London, British Library MS Cotton Tiberius B. V. (995–1016)

Large volume of miscellaneous works, 87 ff. In possession of Battle Abbey by 1100, material added there. In possession of Sir Robert Cotton by 1621.[3] Often called close copy of Harley 647, but really not comparable. Astronomical text is similar, but arrangement and artwork is quite different. Probably copied from Winchester exemplar, at Winchester or possibly Christ Church. Unlike Harley 647, which includes no other texts, Cicero text is small segment, bound with works of Priscianus, Pliny, Bede, Macrobius, Martianus Capella, and short Anglo-Saxon tracts. Presumably composed before 1016, since in list of West-Saxon kings, total number of years of Aethelred's reign not filled in; he died in 1016. That fact and other evidence suggest a date early in eleventh century. Scribe included two sets of descriptions for each constellation, like Harley 647. Verses written in dark brown at

[3]McGurk, P. (1966) Catalogue of astrological and mythological illuminated manuscripts of the Latin Middle Ages, London: Warburg Institute University of London, 106.

bottom of page, first letter set off and larger. First letter of each paragraph of Hyginus scholia is much enlarged, but text has been removed from within figures to above.

Has three different picture cycles: calendar with Labors of Months; *Marvels of the East* with griffins and dog-headed people; and partial constellation cycle (26 only) of *Aratea,* missing first sections of constellations. Calendar has miniatures with Labors of Months at top of each page; medallions with signs of zodiac are placed at bottom. Originally had three maps: world map, celestial map, (now lost) and Macrobian-style zonal map. Has miniature of Sun and Moon that does not appear in Harley 647. Drawings quite large, filling between half and three-quarters of page, not framed; sometimes overlap into text, which is arranged around artwork, with one constellation per page. Colors include dull blue, orange, brown, and some green. Figures have firmness and solidity characteristic of English drawings. Male figures fully dressed with identifying emblems. Garments do not retain antique elements, instead are 'modernized' with proper Romanesque ruffled edges and V-folds. In classical representations, Perseus and Aquarius mostly naked, Orion in backward pose. Orion removed from heavens, stands under pediment supported by four columns. Innovations show constellations were not directly copied from Harley 647, artists or patrons were not interested in reproducing antique art forms. Stars are soft-red dots, still aligned along perimeter of figures, not accurately placed. Five planets shown as heads, encircled, and designed in diamond format, each bound with knotted frame. Has been analyzed under direction of Patrick McGurk, who produced facsimile edition.

f. 1: three *rotae*

ff. 1v–19: Calendar and computus

f. 22: list of West-Saxon kings

ff. 32v–49v: large fragment of Cicero's *Aratea* with scholia.

Incipit: *A quibus hinc subter possis cognoscere fultum.*

ff. 49–87v: excerpts of Priscianus, Pliny, Bede, Macrobius, and Martianus Capella

Miniatures: f. 32v Aries; 33 Triangula; 33v Pisces; 34 Perseus; 34v Pleiades; 35 Lyra; 35v Cygnus; 36 Aquarius; 36v Capricorn; 37 Sagittarius; 37v Sagitta; 38 Aquila; 38v Delphinus; 39 Orion; 39v Sirius; 40 Lepus; 40v Argo; 41 Cetus; 41v Eridanus; 42 Pisces Magnus; 42v Ara; 43 Centaurus; 43v Hydra, Crater and Corvus, 44 Anticanis; 44v five Planets; 47 Helios top, Luna bottom.

Literature: Hooper (1777) *A Catalogue of the Manuscripts in the Cottonian Library,* London: British Museum; Planta (1802) *Catalogue of the Manuscripts in the Cottonian Library deposited in the British Museum,* London: the British Museum; Saxl and Meier III (1966) 119–128; Ker (1964) *Catalogue Anglo-Saxon MSS*, no. 193; R.I. Page, (1966) "Anglo-Saxon Episcopal Lists III", *Nottingham Mediaeval Studies* X 2–24; Temple, (1938) *Anglo-Saxon MSS: 900–1066,* no. 87, 104; J.C. Webster, pp. 53–56; Dodwell (1954), *Canterbury School of Illumination*, 18; Saxl, (1957) "Illustrated Science Manuscripts in England", p. 103; Weitzmann, (1971) *Ancient Book Illumination*, 18; Watson, *Catalogue of Dated and Datable Manuscripts,* p.106–7; Köehler and Mütherich (1971) *Die karolingischen*

Miniaturen, vol. 4, p.101; McGurk et al. (1983) *An Eleventh-Century Anglo-Saxon Illustrated Miscellany: British Library Cotton Tiberius B.V. Part 1, Together with Leaves from BL Cotton Nero D. II, Early English Manuscripts in Facsimile*, XXI; Listing and description of the Cotton manuscripts are available through Catalogue of Archives and Manuscripts; Haffner M., and Metzger W. (2012) *Sternbilder des Mittelalters: Der gemalte Himmel zwischen Wissenschaft und Phantasie*. Part One: 800–1200.

London, British Library Harley MS 2506 (ca. 1000)

Produced with 93 ff., about same time as Tiberius B.V, Not made in England, but at Benedictine monastery of Fleury on river Loire. English monastery of Ramsey was aligned with Fleury in late 10th C. Abbo of Fleury went to England to teach, was school-master in 986–8. Abbo was interested in mathematics and astronomy, wrote four short astronomical treatises for students. At that time, important astronomical and computistical manuscripts were brought from Fleury to England, and actively copied. Corrections in Harley 647 had been wrongly attributed to Abbo. Harley 2506 may have reached England through these monastic relationships.

Miniatures in Harley 2506 painted by English artist, who perhaps worked in France, 'the Ramsey master'. His expert painting is seen in other manuscripts, talented in both tinted line drawings and fully colored illuminations. Some of his work was apparently produced on continent, which implies that he traveled between monasteries of England and Europe. (Backhouse 1997: 24) Saxl too compliments this artist, "one of the best draughts men of the end of the tenth century, who belongs to the so-called Winchester school, applied his gifts to the illustration of Cicero text" (Saxl 1957: 102). Saxl adds, "There is nothing more beautiful in mediaeval illustration of the constellations than the work of this artist" (Saxl 1957: 102).

Verses of Cicero's *Aratea* are written on one side of folio in brown, scholia of Hyginus on other side in red. Text is same as Harley 647, but commentaries, written in margins are different. Has 43 constellations pictured, higher total reflects sections of Hyginus that have been added to fill in 16 constellations before Aries. Line drawings use color wash with delicate shading, drawings are outlined in brown ink, touches of red and blue wash. Ramsey Master takes liberties with drawings, showing creative talent. Miniature of Sagittarius is unusual, since artist has chosen older type for this zodiacal sign. Sagittarius more often depicted with full horse body and human from waist up. Here Sagittarius is winged satyr with only two animal legs; artist humorously positions two stars on his private parts. Face of Aquarius is more serious, his hair becomes wings, both figures not usually shown with wings. Depictions of males are mostly nude with delicate facial features in profile. Stars are large red dots, generously distributed, but randomly placed.

ff. 1–30: Hyginus' *De astronomia*

ff. 30–35v: excerpts of Priscian, Abbo of Fleury, and religious texts.

ff. 36r–48v: fragments of Cicero's *Aratea* with scholia and illustrations

Incipit: *A quibus hinc subter possis cognoscere fultum*.

ff. 49–93v: excerpts of Pliny the Elder, Macrobius, and Martianus Capella.

Miniatures: f.36 Aries; 36v Triangula and Pisces; 37 Perseus; 37v Pleiades; 38 Lyra and Cygnus; 38v Aquarius; 40v Delphinus; 41 Orion and Canis; 41v Lepus; 42 Argo and Cetus; 42v Eridanus; 43 Pisces Magnus; 43v Ara; 44 Centaurus, Hydra, Corvus and Crater; 44v Anticanis; 45 planets.

Literature: Saxl and Meier III (1966) 157–60; C. Niver, "The Psalter in the British Museum, Harley 2904", *Medieval Studies in Memory of A. Kingsley Porter* (1939) 44–9; Saxl and Wittkower, "British Art and the Mediterranean" (1948) 37; Saxl, "Illustrated Science Manuscripts in England" (1957) 102–8; F. Wormald, *English Drawings of the Tenth and Eleventh Centuries*, London (1952) 32–3; Temple, *Anglo-Saxon Manuscripts, 900–1066* (1976) 65; J. Backhouse (1997) 24; D.H. Turner, L.E. Webster, eds. *The Golden Age of Anglo-Saxon Art 966–1066* (1984) 120; L. Grodecki, *Le siècle de l'An Mil.* (1973) 242; Haffner M., and Metzger W. (2012) *Sternbilder des Mittelalters: Der gemalte Himmel zwischen Wissenschaft und Phantasie*. Part One: 800–1200.

Cambridge, Trinity College MS R. 15. 32 (ca. 1000)

Probably created at New Minster, Winchester in late tenth or early eleventh century. Third manuscript of Cicero's *Aratea* produced around the year 1000. Astronomical anthology compiled in England along with Hyginus' *De astronomia* and Abbo of Fleury's treatises on astronomy. Contains 110 ff, written in two columns using minuscule of eleventh century. Illustrated, but not listed in Saxl and Meier.

London, British Library Cotton MS Tiberius C. 1. (ca. 1122)

Originated at Peterborough, England in about 1122, illustrated, 42 ff, discussed in detail in Chap. 7. Cottonian and Harleian libraries were rich a source of *Aratea* manuscripts. Robert and Edward Harley, 1st and 2nd Earls of Oxford, and Sir Robert Cotton, were great collectors. Both libraries held two copies of Cicero's *Aratea*. Sir Robert, the elder Harley (1661–1724), Queen Anne's minister, began to collect books when he retired to his family estate in 1700, until 1711, when son Edward (1689–1741) took over. Harley collected many books from English monasteries.

ff. 2–20: excerpts of Macrobius, Isidore, Pseudo-Bede, calendar tables and ecclesiastical rotae.

ff. 21–34: *Aratea* of Cicero and scholia

ff. 36–42v: excerpts of Pliny's *Natural History*, Martianus Capella, and Macrobius.

Incipit: *E quibus hinc subter possis cognoscere fultum.*

Miniatures: f. 21 Aries; 21v Triangula; 22 Pisces; 22v Perseus; 23 Pleiades; 23v Lyra; 24 Cygnus; 24v Aquarius; 25 Capricorn; 25v Sagittarius; 26 Sagitta; 26v Aquila and Sagitta; 27 Delphinus; 27v Orion; 28 Canis; 28v Lepus; 29 Argo; 29v Cetus; 30 Eridanus; 30v Pisces Magnus; 31 Ara; 31v Centaurus; 32 Hydra with Crater and Corvus; 32v Anticanus; 33 five planets; 33v Helios and Luna in chariots.

Literature: Catalogue of Manuscripts in the Cotton Library, London: (1802); Saxl and Meier Catalogue (1966) 128–134; Kauffmann (1975) *Romanesque Manuscripts: 1066–1190,* no. 37; Watson, *Catalogue of Dated and Datable Manuscripts,* p.107; N.R. Ker (1938) "*Membra Disiecta*" in *The British Museum Quarterly* 12; Saxl (1957) "Illuminated Science Manuscripts in England", *Lectures,* pp. 102–08; H. Bober (1956–57) "Illustrated Medieval Schoolbook of Bede's '*De natura rerum*'," Journal of the Walters Art Gallery, XIX-XX pp. 77, 91; Leonardi (1960) pp. 71–72; Morton (1975) pp. 362–82; Bishop (1971) *English Caroline Minuscule*, Oxford, p. 15; McGurk, (1983) *Eleventh-Century Anglo-Saxon Illustrated Miscellany British Library Cotton Tiberius B.V. Part 1, Together with Leaves from BL Cotton Nero D. II, Early English Manuscripts in Facsimile*, XXI. Copenhagen: p.73; Haffner (1997) p.175. *Nero D. II, Early English Manuscripts in Facsimile*, XXI.

Leiden, Rijksuniversiteitsbibliotheck Voss. Lat. f.121 (11th–12th C.)

Last of this family of Cicero's *Aratea*. Derives from a lost copy of Harley 647, but is incomplete. Illustrated

The following manuscripts belong to a second group of Cicero's *Aratea* texts.

Città del Vaticano, Biblioteca Apostolica Vaticana Codex Vat. Reg. Lat. 1324 (15th C.)

Contains fragments of Cicero's *Aratea* with fragments of Germanicus. Illustrations have no frame, no stars pictured, writing is very small. Microfilm at St Louis University, not listed in *Vedere i Classici*. A small illustration of Boötes with one arm turned into a wing appears in McGurk, *Astrological Manuscripts in Italian Libraries*, p.87.

Miniatures: f.27 Celestial globe in temple-like structure; 27v Ursa Major and Minor; 28 Draco and Ursae; 28v Serpentarius on Scorpio, Boötes, Virgo and Gemini; 29 Cancer, Leo, Auriga and Taurus; 29v Cepheus, Cassiopeia and Andromeda; 30 Pegasus, Aries, Triangula, Pisces and Perseus; 30v Pleiades, Lyra and Cygnus; 31 Aquarius, Capricorn, Sagittarius, Sagitta, Aquila and Delphinus; 31v Orion, Canis, Lepus and Argo; 32 Anticanis, Eridanus, Pisces Magnus and Ara; 32v Centaurus, Hydra, Corvus and Crater; 33v five planets.

Milan, Biblioteca Ambrosiana, Codex Ambrosiana D. 52 inf. (15th C.)

Contains fragments of Cicero's *Aratea* with sections of Germanicus and Avienus. Includes illustrations, not listed in McGurk, *Astrological Manuscripts in Italian Libraries*.

Monpellier, Bibliothéque interuniversitaire, Faculté de medicine, MS H. 452 (15th C.)

Germanicus and Hyginus appear first, then Cicero fragments, paper, 173 ff. with 34 illustrations. Descriptive text is preceded by illustration. Also Hyginus' *De astronomia*.

ff. 1–7: text and unfinished diagram.

Miniatures: f.7v Draco and Ursae; 10 Hercules; 11 Corona; 12 Ophiuchus with Serpent; 13v Boötes; 16 Virgo; 17 Gemini; 19v Leo; 21 Auriga; 22v Taurus; 24 Cepheus; 25 Cassiopeia; 25v Andromeda; 27 Pegasus; 28 Aries and Triangula; 30 Pisces; 31 Perseus; 32v Pleiades; 33 Lyra; 34 Cygnus; 35 Aquarius; 38 Sagittarius; 40v Delphinus; 41 Sagittarius shooting Scorpio; 43v Capricorn; 45v Lepus; 46v Argo; 48v Eridanus; 50 Southern Fish; 52 Centaurus; 53 Hydra, Corvus and Crater; 57 Helios; 59 Luna.

Siena, Biblioteca Comunale Degli Intronati, L.IV. 25 (14th C.)
Italian, late fourteenth century, written in Gothic minuscule, 172 ff. Includes *Scholia Sangermanensia, revised Aratus Latinus* and several fragments of Cicero. Only 27 constellations represented, starting with Hercules. Only manuscript containing *Scholia Sangermanensia* with illustrations to survive between ninth–tenth century (4) and fifteenth century (2). Cited in McGurk, *Astrological Manuscripts in Italian Libraries* p.82; miniature of Boötes with his arm covered by an animal skin shown on p.88.

The following manuscripts are Cicero *Aratea* manuscripts that are not a part of the families established by Soubiran.

Oxford, Bodleian Library MS Bodl. 614 (ca. 1120–40)
Aratea of Cicero, made in England, 51 folios, 20 illustrations, also contains Germanicus scholia.

Astronomical illustrations of both Bodl. 614 and Digby 83 go back to model related to Cotton Tiberius B.V. These two also differ from Harley 647 and Tiberius B.V. tradition of a more 'pure' text. In Bodl.614, text and illustrations have been supplemented with material from Hyginus' *De astronomia*. Bodl. 614 covered in more detail in Chap. 4.

ff. 1–14: calendar, computus tables
ff. 18–22: Signs of Zodiac
ff. 22–42: Cicero's *Aratea* with *Scholia Sangermanensia*
Incipit: *Sciendum itaque est Deum in ipso firmam(ento) XII domicilia.*

Miniatures: f.18 Aries; 18v Taurus; 19 Gemini and Cancer; 19v Leo; 20 Virgo, Libra held by Scorpio; 20v Sagittarius; 21 Capricorn; 21v Aquarius; 22 Pisces; 23 planets with Sun and Moon in medallions; 24 Draco, Ursae, Boötes, Corona; 24v Hercules/Engonasin; 25 Lyra and Cygnus; 25v Cepheus; 26 Cassiopeia; 26v Andromeda and Perseus; 27 Pleiades; 27v Hyades; 28 Serpentarius; 28v Auriga; 29 Sagitta and Aquila; 29v Delphinus; 30 Pegasus, 30v Triangula; 31 Orion; 31v Lepus and Canis; 32 Anticanis and Argo; 32v Eridanus; 33 Centaurus; 33v Cetus and Pisces Australis; 42 Sun; 42v wind diagram; 43 rainbow; 43v man holding astrolabe.

Literature: E. Panofsky and F. Saxl, 'Classical Mythology in Medieval Art', *Metropolitan Museum Studies, IV* (1933) 238, 240–1; M.R. James, *Marvels of the East*, (1929); J.C. Webster, *The Labours of the Months in Antique and Medieval Art*, (1938); Saxl and Meier III, xxvi, xxviii, 313–16; Saxl, 'Illuminated Science Manuscripts', and 'Belief in Stars', *Lectures I*, (1957); E. Panofsky and F. Saxl, "Classical Mythology in Medieval Art" (1933) 238; J. J. G. Alexander, and Elzbieta Temple, *Illuminated Manuscripts in Oxford Libraries, the University Archives, and the Taylor Institution*. Oxford: Clarendon Press (1985); C.M. Kauffmann, *Romanesque Manuscripts, 1066–1190*, London: Harvey Miller (1975) 16; Haffner M., and Metzger W. (2012) *Sternbilder des Mittelalters: Der gemalte Himmel zwischen Wissenschaft und Phantasie*. Part One: 800–1200.

Oxford, Bodleian Library Digby MS 83 (last quarter 12th C.)

Produced near Winchester, another copy of Cotton MS. Tiberius B.V. Astronomical compilation has fragments from Cicero's *Aratea*, also excerpts from Hyginus with scholia and Isidore's *Etymology*. 76 folios, decorated with 38 illustrations.

ff. 1–76v: astronomical texts including *Aratea* of Cicero, *Scholia Strozziana*, Hyginus

ff. 2v–4v: cosmological diagrams

Incipit: *Duo igitur sunt extremi vertices mundi*

Miniatures: f.44 Draco, Ursae and Boötes; 44v Virgo and Corona; 45 Hercules; 45v Lyra; 46 Cygnus; 46v Cepheus; 47 Cassiopeia; 47v Andromeda; 48 Perseus; 48v Pleiades and Hyades; 49v Ophiuchus with Serpent; 50v Auriga and Sagitta; 51 Aquila; 51v Delphinus; 52 Pegasus; 52v Triangula; 53 Aries; 53v Taurus; 54 Gemini; 54v Cancer; 55 Leo; 55v Virgo; 56 Scorpio with Libra; 57 Sagittarius; 57v Capricorn; 58 Aquarius; 58v Pisces; 62 Orion; 62v Lepus; 63 Canis; 63v Anticanis; 64 Argo; 64v Eridanus; 65 Centaurus; 65v Cetus; 66 Pisces Magnus; 66v Ara; 67 Hydra, Corvus and Crater.

Literature: Saxl and Meier III, 345–46.

Göttweig, Stiftsbibliothek Codex Göttweig 7 Bl. 20 (15th C.)

Made at Koster Gottweig, modeled on Cicero's Harley 647. One of few manuscripts that retains Hyginus scholia written inside illustrations of constellation figures. Text written in prose form. Illustrations have no color added, but nicely modeled.

Literature: Blume, D., Haffner M., and Metzger W. (2016) *Sternbilder des Mittelalters: Der gemalte Himmel zwischen Wissenschaft und Phantasie*. Part Two: 1200–1500.

Manuscripts with Avienus Text

Manuscript tradition of extensive translation composed by Avienus is simplest of three Latin poets, as there are only a few surviving manuscripts. The *Aratea* of Avienus has no associated scholia and usually not illustrated. The oldest extant Avienus manuscript, Vienna, Osterreichische Nationalbibliotek, Codex Vindobonensis Pal.107, is dated to the ninth century, probably copied from an earlier insular source. (Reeve 1983: 19) The Vienna compendium contains the major part of poem, lines 1–1581. Another manuscript of Avienus' *Aratea* is found in Wolfenbüttel, Codex Gudianus 132, from tenth century, probably made in Corvey. The only other extant manuscript of Avienus' version is Milan, Ambrosiana D.52 inf., which was produced in the fifteenth century, sometime before 1477 by Milanese scholar Boninus Mombritius. This manuscript is not only a copy of Vienna 107, 'but a direct copy'. It is illustrated, also contains some text of Germanicus and Cicero. Milan manuscript adds lines 1750 to the end, also includes parts of his *Orbis terrae*. Edition published in Venice in 1488 which printed whole poem of Avienus. This volume also contains Latin versions of Cicero and Germanicus (Reeve 1983: 19).

Vienna, Osterreichische Nationalbibliotek Cod. Vind. Pal. 107 (9th C.)

Poem of Avienus on folios 1–8v, but not illustrated, microfilm at HMML.

Wolfenbüttel, Herzog Augustbibliothek Codex Gudianus 132 (10th C.)

Made at Corvey, text is Avienus' *Aratea*, not illustrated.

Milan, Biblioteca Ambrosiana Codex Ambrosiana D.52 inf. (15th C.)

Aratea of Avienus, also includes text of Germanicus and Cicero, includes illustrations. Listed twice, also with Cicero texts above.

Manuscripts with *Revised Aratus Latinus*

The *revised Aratus Latinus,* also called *Aratus Latinus recensio interpolata*, was explained in Chap. 4, a compressed version of Aratus' *Phaenomena* translated from Greek in the Middle Ages. There are three different types of constellation picture cycles that appear in various *revised Aratus Latinus* manuscripts. There is one variety of miniatures in Cologne, Dom.Cod. 83 II. The second variation appears in Paris BNF, MS. Nouv.acq. Lat.1614. A third different style is found in Dresden MS. Dc.183; Paris, BNF, MS. Lat. 12957; St Gall, MS 920 and St Gall MS 250. The last four codices are closely related to each other. McGurk has studied these manuscripts in relation to their illuminations. He states, "It is likely that the *revised Aratus Latinus*, which was a revision of the unillustrated *Aratus Latinus*, borrowed its cycles of illustrations from elsewhere. Such borrowings might explain both the variations in the cycles and the misunderstandings and inaccuracies in the illustrations themselves" (McGurk 1981: 320).

Cologne, Dombibliothek Cod. 83 II (ca. 805)

Miscellany compiled in Cologne, contains *Aratea* with works of Isidore and Bede's *De temporum ratione*, almost complete. Also has short treatises of various astronomical and computistical texts compiled from usual authors, giving impression that it was unified book. Displays uniform page layout, similar rubrication and quire signatures written by hands responsible for copying texts. Has one of first copies of Calendar of Lorsch, compiled in 789, also works by religious authors, such as St Jerome. Aratean text is exceptionally corrupt version of *revised Aratus Latinus* (McGurk 1981: 319).

Earliest known example of *Aratea* with Germanicus prose scholia or *revised Aratus Latinus*, and most beautiful of *revised Aratus Latinus* group. Not all constellations illustrated, images large, inserted between text sections with no frames. Displays uniform page layout, similar rubrication and quire signatures written by hands responsible for copying texts. Has one of first copies of Calendar of Lorsch, compiled in 789, also works by religious authors, such as St Jerome. Aratean text is exceptionally corrupt version of *revised Aratus Latinus*.[4]

Literature: Anton von Euw (1993) "Die künstlerische Gestaltung", *Science in Western and Eastern Civilization in Carolingian Times*, 251–272 and *Die Kolner Dombibliothek,* Munchen, (1998) Nr. 29, S. 136–156; L.W. Jones (1932) *The Script of Cologne from Hildebald to Hermann*, 23; Thiele (1898) *Antike Himmelsbilder*, 157; C. Heusgen (1947) "die komputistische Handschrift der Kölner Dombibliothek", *Colonia Sacra*, 1 11–18; Haffner M., and Metzger W. (2012) *Sternbilder des Mittelalters: Der gemalte Himmel zwischen Wissenschaft und Phantasie*. Part One: 800–1200.

Paris, Bibliothèque nationale de France Nouv. Acq. lat. 1614 (early 9th C.)

Miscellany compiled at Cologne or possibly Tours region. Part of dismembered Tours Public Library manuscript and has *revised Aratus Latinus*. Also extracts from '810 compilation', and report of 809 computistical colloquium. Contains cycle of constellation images, ff.146–171v, which are dissimilar to Cologne, MS 83, and page layout in constellation series different than most other *Aratea*. Folios divided almost in half from top to bottom, text written in one column. Illustrations run together, one below the other, no frame, no identifying label, or other manner of separating each image. Drawings crudely executed, bodily representations poorly done, proportions inaccurate. Includes writings of Isidore and Bede's *De temporum ratione*.

Dresden, Sächsische Landesbibliothek Dc.183 (early 9th C.)

Miscellany compiled in Cologne, contains *Aratea* with works of Isidore and Bede's *De temporum ratione*, almost complete. Also has short treatises of various astronomical and computistical texts compiled from usual authors, giving impression that it was unified book. Displays uniform page layout, similar rubrication and quire signatures written by hands responsible for copying texts. Has one of first copies of Calendar of Lorsch, compiled in 789, also works by religious authors, such as St Jerome. Aratean text is exceptionally corrupt version of *revised Aratus Latinus*.[5] Has *revised Aratus Latinus* and fragments of Cicero with Hyginus scholia. Based on Harley 647 or its archetype, completely destroyed during WWII. Was illustrated, compiled with other astronomical texts.

Literature: Bischoff, *Katalog der festlandischen Handschriften des Jahrhunderts* (mit Ausnahme der wisigotischen) I. Wiesbaden (1998) Nr. 1044, S. 225; Mütherich, "Book Illumination in the Court of Louis the Pious" (2004)107; A. Byvanck, *De platen in de Aratea van Hugo* de *Groot*, Afd. Letterkunde, Nieuwe Reeks, Deel 12, no. 2 (1949) 213–19.

Paris, Bibliothèque nationale de France MS latin 12957 (9th C.)

In delicate condition, compiled at St-Martin de Tours, 100 ff. Contains *revised Aratus Latinus* and fragments of Germanicus, ff. 57–74v. Text written in one column with 33 lines, brown ink, names of constellations in red. Forty-three

[4]McGurk (1981), p. 319.
[5]McGurk (1981), p. 319.

constellations pictured. First image is celestial diagram with two spheres, depicts band of Zodiac, top sphere begins with Libra, lower one begins with Aries. Next miniature is outstanding illustration of detailed celestial globe on pedestal within temple-like structure, seen in other Aratean manuscripts. Authentic globe even has 'handle' on top to turn it. Folio 71v shows miniature of five planets as encircled busts. Illustrations of Helios and Luna have same iconographical features found in many Aratean codices. On same folio, Sol stands in chariot with four horses, shares page with Luna and two oxen.

Literature: Byvanck, *De platen in de Aratea van Hugo de Groot.*1949; McGurk (1981) 319; E.K. Rand, *A Survey of the Manuscripts of Tours I*, Princeton NJ (1929) Nr. 172, S. 185 f; D. Kidd, 1997.

St Gall, Stiftsbibliothek Codex 902 (c.837)

Cicero's *Aratea* and *revised Aratus Latinus* part of astronomical anthology at St Gall; made at or near St Gall. Very large compilation with variety of writings; includes Bede's *De temporum ratione* and complete *De astronomia* of Hyginus, works of Hrabanus Maurus, computistical texts and excerpts of 'Seven-Book Computus'. Has two Plinian diagrams, but not excerpts from Pliny usually associated with diagrams. An example of popularity of invented diagrams; perhaps considered more interesting than texts they illustrated. Also has celestial map on folio 76. Map divided into two spheres, depicting northern and southern constellations; ecliptic shown as band with signs of zodiac drawn in little blocks. Number of constellations greatly abbreviated, and circumpolar constellations omitted.

Literature: A. Bruckner, "Scriptoria medii aevi Helvetica III", *Die Schreibschulen der Diözese Konstanz: Sankt Gallen 2*; Geneva, (1938) 122; Merton (1923) *Die Buchmalerei in St. Gallen vom 9. Bis zum 11. Jahrhundert*, Leipzig L.56057 49–51; McCluskey (1993) 167; Jones (1994) Bede, the Schools and the Computus, ed. Wesley M. Stevens, *Variorum* 78; Jones, "Bedae Opera Didascalica", 1 (DNR), p.181 no.95; Jones (1943) Cambridge: Medieval Academy of America, "Bedae Opera Didascalica", 2 (DTR), p. 252, no.191; Viré, (1992) *Hygini De Astronomia*, 172, no.64.

St Gall, Stiftsbibliothek Codex 250 (ca. 889)

Second St Gall manuscript is also large compilation, copy of St Gall 902; also made at or near St Gall. Has *revised Aratus Latinus* bound with fragments of Cicero, Hyginus and Bede's *De temporum ratione* and *De temporibus*. Like its model, also includes two Plinian diagrams without his text.

The following manuscripts are designated only with text of Aratus, mainly fragments, but they do not list to which Latin translation they belong, possibly the *revised Aratus Latinus* (or not).

Dijon, Bibliothèque Municipale MS 448 (11th C.)

Contains various astronomical and computistical treaties, including Isidore's *De natura rerum*, without illustrations. Many diagrams are independent of any text. Two different diagrams of heavens pictured. First diagram is circular band of each planet with Earth in middle; Sun, Moon and five planets are personified and drawn waist high aligned atop outer circle. Other celestial diagram shows zone of ecliptic filled in with illustrations of Signs of Zodiac; in center are Draco and Ursae. Illustrations accompany Ps-Bede text with 40 constellation images depicted, beginning with Draco between two Bears, ending with Hydra, Corvus, Crater, and Anti-canis. Each page has text and images interspersed in one or sometimes two columns. Latin text placed above llustrations in 'papyrus' format; text written in brown ink, drawings are reddish-orange. Stars included as dots strategically placed, but not accurately.

ff. 63v and 64: two diagrams of heavens

ff. 1–71: astronomical texts, annals of monastery of Saint-Benigne, Dijon. Pseudo-Bedan treatise, *De signis caeli*, accompanied by fragments of *Arati de astrologia*,

ff.71v–72: short treatise by Pseudo-Hyginus. Also Lorsch calendar, Bede's *De temporibus*; Helpericus Sancti Galli, *Ars calculatoria seu Liber de computo*. Concludes with several more Bede treatises, *De temporum ratione*, *Chronica maiora*, and *De natura rerum*.

Miniatures: f.64v Draco and Ursae; 65 Aries.

Literature: Zaluska (1991) *Manuscripts enluminés de Dijon*, Paris pp. 30–33; Kühnel (2003) *The End of Time in the Order of Things: Science and Eschatology in Early Medieval Art*, Regensburg: Schnell & Steiner, p. 108; Saxl and Meier (1966) *Verzeichnis astrologischer* I; Destombes, (1964)*Mappemondes A.D. 1200–1500*, Amsterdam: p. 47; Munk Olsen (1982) *L'étude des auteurs classiques*, I, Paris: p. 335.

Undetermined Text

Salzburg, Universitätsbibliothek M I 36 (15th C.)

Includes fragments of *Aratea*, has 178 ff. with both German and Latin text. Contains some diagrams, but no illustrations. On microfilm at HMML.

Harvard University, Houghton Library MS Typ 43 (ca. 1425)

Astronomical miscellany, 237 leaves, paper. Probably made in Vienna. Written in brown and red; miniatures and diagrams in pen and ink. Image placed after description. Only few have stars marked. Some illustrations squeezed together, juxtaposed on page. Miniatures have some colored washes applied. Bought by Harvard in 1946. All images are online at Hollis Catalog, Harvard Library.

ff.1–80: texts includes Alfonsine Tables with conversion factors for Vienna

ff. 153–58: illustrations of 46 constellations, several on page

ff. 166–72v: Michael Scot, *Liber de signis et imaginibus*, constellation illustrations, and signs of zodiac

ff. 176v: Aratus Solensis, *descriptio sphaerae et signorum zodiaci*

ff. 176v–178v: Bede, *De constellationibus* (probably Ps.-Bede)

ff. 179–237: Accursius of Parma with a few diagrams, *Astrolabium sphaericum*, Franco of Polonia, Dominicus de Clavasio, *Practica geometric* and others.

Literature: W. H. Bond and C .U. Faye, *Supplement to the Census of Medieval and Renaissance Manuscripts in the United States and Canada*, New York: The Bibliographical Society of America, (1962) 255; R. Wieck, *Late Medieval and Renaissance Illuminated Manuscripts*, 1350–1525- in the Houghton Library, Harvard College Library, Cambridge, MA (1983) 86–87.

Oxford, Bodl. MS Laud Misc. 644, (S.C. 1487) (ca. 1268)

Astronomical anthology with 226 ff. Includes fragments of *Aratea*, with illustrations on 8 folios; 5 to 8 constellations depicted on each, no frames, little color, red and light orange wash. Pictures integrated with text, similar to *Liber Floridus* constellation miniatures. First miniatures are Ursa Minor and then Ursa Major, not shown with Draco. Hercules, back view with club and lion skin. Small red dots for stars, the numbers agree with text in number, but not accurately placed.

ff. 1–7v: astronomical texts, *revised Aratus Latinus* mingled with Pseudo-Bede.

ff. 11–226: computus diagrams, Abu Ma'shar, Ptolemy's *Almagest*, and Grosseteste treatise.

Miniatures: f.8 Draco, Ursae, Hercules, Corona, Serpentarius and Boötes; 8v Virgo, Gemini, Cancer, Leo, Auriga, Taurus and Cepheus; 9 Cassiopeia, Andromeda, Pegasus, Aries, Triangula and Pisces; 9v Perseus, Lyra, Cygnus, Aquarius, Capricorn and Sagittarius; 10 Aquila, Delphinus, Orion, Canis, Lepus and Argo; 10v Cetus, Eridanus, Pisces Magnus, Ara, Centaurus, Hydra, Corvus, Crater and Anticanis.

Greek Text of Aratus With Scholia

Città del Vaticano, Biblioteca Apostolica Vaticana Codex Vat. Gr. 1087 (15th C.)

Byzantine manuscript of *Phaenomena* of Aratus with Greek text held in Vatican Library, obviously copied from ninth-century prototype (such as Cod. Vat. Gr. 1291, derived from Late Antique prototype).[6] (Panofsky and Saxl 1933: 233) Unlike most Greek Aratus manuscripts, is illustrated with constellations, text is missing, which must have been one of many commentaries to *Phaenomena*. Pen drawings look back to early model. Very similar to Rabanus Maurus illustrations. Contains zodiacal diagram with Sol and Luna personified, wearing classical garments. Also includes celestial map with curious type of projection. Amidst scholia on f. 42, diagram of circle encloses regular hexagon, with sun outside circumference. Northern and southern hemispheres not represented in two separate drawings as usual, they are divided by horizontal section through ecliptic. Whole globe is flattened out into one panorama, consisting of five concentric circles. As a result, constellations appear in grotesque distortion (Panofsky and Saxl 1933: 233). This

[6]Panofsky and Saxl 233

manuscript demonstrates that as late as fifteenth century, pictures such as these preserve fairly intact the feeling for ancient form.

Literature: K. Weitzmann, *Studies in Classical and Byzantine Manuscript Illumination*, University of Chicago Press (1971) 148; Weitzmann, *Roll and Codex* (1970) 96 and 144; Panofsky and Saxl, "Classical Mythology in Mediaeval Art", *Metropolitan Museum Studies* (1933) 228–280.

El Escorial, Scorialensis Σ III 3 (late 15th C.)

Byzantine manuscript, *Aratea* in Greek, two Aratean texts precede scholia and it is illustrated.

Madrid, Biblioteca Nacionale Codex Matritensis 4691 (1465) Greek, not illustrated.

Cambridge MA, Harvard University, Houghton Library MS Typ 18 (ca. 1490.)

Anthology of Greek poetry, 146 leaves. Written in Italy, Florence, by Joannes Rhosos, presented to Harvard Library in 1948. Aratus, *Phaenomena* (106–130v; vita and scholia 131–146v). Bound with Ps.-Orpheus, Argonautica and Hymns, Moschus, Musaeus, Hero, and Leander. All in Greek. Only one decorated page, f. 1, historiated initial with Orpheus playing a viol to animals, and at bottom a circular miniature of the Argo, these may be work of Benedetto de Silvestro. The only example I've come across that has Aratus bound with poetry.

Literature: Bond and Faye, 252–53; Wieck, (1983) 119; J. Martin, *Histoire du texte des Phénomènes d'Aratos*, Paris (1956) 258–59.

Appendix C
Description and Iconography of the Constellations

For it was Zeus himself who fixed the signs in the sky,
making them into distinct constellations, and organised the stars for the year
to give the most clearly defined signs of the seasonal round to men,
so that everything may grow without fail. *The Phaenomena* (Kidd 11–15)

Appendix C includes the listing, description and assorted illustrations of the 45 constellations in the Greek poem, the *Phaenomena*, following the order designated by Aratus as well as excerpts from his poetic interpretations of each constellation.[7] Since the *Phaenomena* comprises the oldest surviving account of the names and relative positions of the constellations, this celestial system embodied within its poetic verses played an enormous role in the early history of astronomy. Passed orally from generation to generation, the ancient star groups, as described by Aratus, are those constellations that are still used by modern astronomers, both amateur and professional.[8] The only constellation listed by Aratus that is not used by astronomers or celestial mapmakers today is Argo, or Navis; it has been subdivided into three smaller constellations, as it covered a vast and unwieldy area of the sky.

In addition to providing the individual names of the constellations in his poem, Aratus also includes information known to the Greeks on the structure and terminology of the celestial coordinates, the tropics, the equator and the ecliptic. Many ancient Greeks believed in Aristotle's assertion that the entire field of fixed stars

[7]The exact number of constellations was never standardized in the ancient or the medieval world. Ptolemy, whose *Almagest* is the 'bible' of astronomy, listed 48 constellations in his seventh and eighth books. Ptolemy includes another horse, Equus, another dog Canis Minor, the Southern Crown, and the Southern Fish. He also lists Ophiuchus and Serpens as two separate constellations; he does not list the Pleiades as a constellation, but includes them as part of Taurus. The 'star catalog' of Ptolemy is almost certainly the same one given by Hipparchus, edited by Ptolemy to correct the changes brought by precession. Hipparchus listed 49 constellations in his *Catalog* and 46 in his *Commentary on Eudoxus and Aratos*. Hyginus included only 42 constellations in his *Poetica Astronomica*.

[8]Astronomers now recognize 88 constellations, covering the entire sky with no overlaps and no gaps between them. This is a recent development in astronomy which systematized the celestial maps in modern times.

M. Dolan, *Astronomical Knowledge Transmission Through Illustrated Aratea Manuscripts*, Historical & Cultural Astronomy,
DOI 10.1007/978-3-319-56784-6

was imbedded into a crystalline sphere that rotated as a unit the entire 360° every 24 h. In addition each of the 'seven planets' was attached to its own sphere, these moved at variable speeds along a confined path, called the ecliptic. These celestial orbs encircled the Earth, a terrestrial sphere held stationary at the center of the entire universe. This body of fundamental astronomical knowledge from the classical scientific tradition was transmitted to the Middle Ages, not directly from Greek authors, but through Roman Latin intermediaries, such as Macrobius, Martianus Capella, and the *Aratea* manuscripts.

Origins of the Constellations

The exact origin of the figures represented in constellations is irretrievable from the greyness of antiquity, a date neither knowable nor of particular concern for this study. Of concern are the concepts that are represented in these forms and the role that artistic representation of heavenly forms plays in the transmission of astronomical knowledge. For the myths and figures firmly fixed in the constellation cycle are the elementary roots of astronomical science.

Considered by many as the most important study on the antiquity of astronomical knowledge is *Hamlet's Mill: an Essay Investigating the Origins of Human Knowledge and its Transmission through Myth* by G. DeSantillana and H. von Dechend, earlier historians of science at Harvard. Their lifelong research presents ample evidence to prove their startling theory that important myths in every culture have one common origin, drawn from their celestial cosmology. The Harvard scholars contend that all the tales and adventures of the gods, heroes and characters of worldwide myths are but symbols or ciphers for the movements of stars and planets. Myth becomes a simplified, easily retained language for the perpetuation of complex astronomical data. The ideas of the Harvard researchers, challenging the evolutionary biases of theoretical science, were too revolutionary to be accepted by most of the scientific community, and have yet to be fully absorbed. New studies in the history of science have demonstrated that the astronomical material found in the *Phaenomena* reaches much further into the past than the time of Eudoxus, and that the poem performs a vital link in the transmission of ancient knowledge through its myths, poetry and art.[9]

The formations of the individual constellations are not just separate creations inspired by a poet's imagination formulated gradually over time simply to provide entertainment. Surprisingly, the research of historians of astronomy have provided

[9]A recent investigation of Homer's *Iliad* applies an astronomical interpretation to the characters in the tale, *Homer's Secret Iliad* by Florence and Kenneth Wood.

important scientific evidence that the constellation figures were developed as a complete package and set into exact patterns in deep antiquity.[10] After a long and insightful investigation, astronomer Michael Ovenden states, "I think that there can be no doubt, in view of the results of these investigations, that far from being the idle fancies of a primitive people, the constellations represent an ordered mapping of the sky".[11] (Ovenden 1966: 13) He presents strong evidence that "the constellations were designed at a definite time and place, in a very logical way, for a definite purpose".[12] (Ovendon 1966: 2)

Ovenden and other astronomers as well had noticed discrepancies when studying the astronomy of Aratus/Eudoxus, which was perplexing since Eudoxus was a responsible and exact astronomer. These variations in the verses of the *Phaenomena* lead the astronomers to look for an earlier date when the positions stated in the poem would have been more accurate. Through mathematical calculations, Ovenden demonstrated how he was able to pinpoint the time period and the exact latitude at which the astronomy of Aratus would have been 'correct'. Ovenden credits Hipparchus for leading him to this conclusion; since Hipparchus himself noticed discrepancies in Aratus' descriptions when compared with his own observations of the positions of the constellations.[13] As discussed in Chap. 5, astronomers have established that the observations of the stars, as noted in the poem by Aratus, actually attest to a much earlier time, probably to observations recorded in Crete at about 2800 BCE.[14]

Furthermore beyond the early date and unexpected location of the original composition, Ovenden believes the organization of the heavens was purposefully created as a whole unit. He presents a convincing explanation of the intriguing theory that the figures, animals and objects represented by the constellations were not randomly designed simply as amusement, idle fantasies or even memory devices for primitive people. Instead, the careful layout of the pictorial program as an entire program provides important scientific knowledge. All areas of the sky have a random mixture of bright and dim stars with no perceptible relationship to

---

[10]Here I am summarizing information from a lecture given to the Royal Philosophical Society of Glasgow by M. W. Ovenden, "The Origin of the Constellations" and subsequently printed in their prestigious publication, *The Philosophical Journal*. His conclusions have been verified and expanded upon by later astronomers, Mary Blomberg and Göran Henriksson in "New Arguments for the Minoan Origin of the Stellar Positions in Aratos' *Phainomena*".

[11]Ovenden, M. W. (1966) "The Origin of the Constellations", *the Philosophical Journal*, Vol. 3: pp. 1–18.

[12]Astronomical details that prove his theories are rather technical and are better read in his journal article.

[13]Perhaps the astronomy of Aratus led Hipparchus to question and research celestial motions that resulted in the publication of his 'discovery' of the essential concept of the Precession of the Equinoxes.

[14]See also Mary Blomberg and Göran Henriksson, "Literary and Archaeoastronomical Evidence for the Origins of the Hellenic Calendar in the Aegean Bronze Age", in *Ad Astra per Aspera et per Ludum*, and Sergey Zhitomirsky, "Aratus" *Phainomena*: Dating and Analysing its Primary Sources", *Astronomical and Astrophysical Transactions* 17.

one another, but those, who designed the system of constellations which we have inherited, chose specific areas of the sky, and then oriented the figures in exact directions. Finally they applied well-known stories to signify particular areas, avoiding certain areas and mapping out others which fit into their design.

By determining the distribution of the constellations over the vault of the sky, certain conclusions can be drawn. It is no accident that the zodiacal constellations form a belt around the ecliptic, and that a ring of constellations centers on the north celestial pole. Ovenden shows how exceptional features of the constellations become apparent; in particular, significant points on the celestial sphere seem to be associated with serpentine or dragon-like forms. The constellation of Draco links the celestial pole[15] with the pole of the ecliptic, and the coils of the constellation Serpens turn through 90° at the autumn equinox. The sinuous figure of Hydra, the Water-snake, extends for 90° and marks, quite accurately, the celestial equator at about 3000 BCE. There are many more examples, but just these few furnish ample evidence that the constellations were not envisioned because of their resemblances to mythological figures, but rather as an elemental form of celestial co-ordinates. Creating these specific patterns in the sky manifests the earliest intermingling of 'science' and art. But why, and more importantly, how would early societies encapsulate this astronomical information, and then manage to pass it on orally, keeping it intact?

These current discoveries and a clearer understanding of the astronomy in Aratus' poem demonstrate the deep connections of science, poetry and art, inspiring even more questions concerning this long tradition. Was the *Phaenomena* the continuation or reincarnation of an even more ancient astronomical poetic tradition? Did an even earlier oral tradition have a significant role in the writings of Aratus, in addition to the scientific output of Eudoxus? These questions can probably not be answered, but has anyone really investigated this possibility?

Since most of the myths associated with the constellations refer to Greek and Roman mythological figures, how could this theory of a more ancient origin be adapted or explained? This problem could be answered more readily since the mythological stories are archetypal. Many early cultures have similar stories with the exploits of typical heroic and tragic figures, which are interchangeable with Greek myths. Although modern scholars have been re-interpreting this ancient knowledge, it was not part of average awareness during the Middle Ages. Medieval scholars, artists, or patrons maintained this body of astronomical information out of respect for antiquity, not because they recognized any deeper scientific implications.

Invigorated with this broader, more universal interpretation of Aratus' poem, it is helpful to examine the structure, contents, and arrangement of the poem. The 44 (or so) constellations of the poem will be discussed individually, both as described by Aratus and his translators, and as interpreted in artistic renderings. Additional

---

[15] That is, the celestial pole at the time when the constellations were first established.

descriptive information supplied by Hyginus' *De astronomica* is also included for comparison.

Structure of the *Phaenomena*

The verses of the *Phaenomena* explain the phenomena of the celestial sphere in a systematic order, first describing the positions of the constellations in relationship to each other, including some of their myths. Next Aratus recounts the motions of the constellations, as they rise and set, each in their designated turn and at their appointed times throughout the solar year. The poem credits the order and dependability of the cosmos to the beneficence of Zeus; this divine elemental force credited to Zeus corresponds with the Demiurge of Plato, the Prime Mover of Aristotle, and the Trinitarian God of Christianity. According to Hipparchus, Aratus designed this description of the heavens, not from his direct observation of the night sky, but instead followed the arrangement of Eudoxus in his writings and on his celestial globe. Aratus avows that the constellations were not randomly arranged in the sky simply as ornaments, but for the more practical purpose of forming recognizable patterns and to providing warnings of seasonal and weather changes for those on Earth.

Aratus arranges the constellations into two different groups, dividing those north of the ecliptic from those to the south with a short transitional passage between them. The northern area of the sky seems more populated, embracing all 12 signs of the zodiac. The constellations south of the ecliptic are much fewer in number; limited to those visible from the Mediterranean area. The poetic descriptions begin with the circumpolar stars, the two Bears and Dragon; then the focus radiates outward, following the system created by Eudoxus.[16] In manuscripts, they are usually combined as the first constellations illustrated. The following verses drawn from the *Phaenomena* are taken from the Greek original, as translated by Kidd; the arrangement of the constellations into separate groups is also after Kidd.

A. 1–18 Proem: hymn to Zeus as beneficent father-god who designed the constellations as a guide to the changing season. [1–14] invocation of Zeus and the Muses [15–18]
B. 19–461 The constellations and how to recognize them.
   19–26 The rotation sky, its axis and poles.
   26–62 The northern circumpolar constellations: Two Bears and Draco
   63–90 The Dragon's head group: the Kneeling Figure, Crown, Serpent-holder and
   Serpent and Claws.
   91–146 The Great Bear's tail group: Bootes, Maiden, nameless stars.

[16] Kidd's word by word analysis of the text can be read in his commentary 174–577.

147–178 The Great Bear's body group: the Twins, Crab, Lion, Charioteer, Bull.
179–267 The Cynosura group: Cepheus, Cassiepeia, Andromeda, Horse, Ram, Triangle, Fishes, Perseus, Pleiades.
268–318 The kneeling figure group: Lyre, Bird, Water-pourer and Capricorn, Archer,
Arrow, Eagle and Dolphin.
319–321 Transition to southern stars.
322–352 The Orion Group: Orion, Dog, Hare, Argo.
353–401 The Sea-monster group: Sea-monster and River, origin of the constellations, Southern Fish, the Water stars.
402–450 The Altar group: Altar, Centaur and Beast, Hydra, Bowl, Raven and Procyon.
451–453 Conclusion to the constellations.
454–461 The erratic movements of the five planets.

C. 462–757 The passage of time and how to estimate it by observing the constellations and the Sun and Moon.
D. 758–1141 Local weather signs observable in natural phenomena and the behavior of birds and animals.
758–777 Second proem on the weather at sea and the value of learning the signs.
778–908 Celestial signs, from the appearance of the Moon, the Sun, the Manager.
909–1043 Misc. signs of weather conditions: of wind, from birds and natural phenomena; of rain, of fair weather, from clouds, stars, lamps and birds; of storms, from stars, cloud, birds and domestic life.
1044–1103 Seasonal signs: from vegetation of winter from farm life, of drought from comets, of summer from birds.
1104–1141 Local signs of bad weather from animals.

Since the translations of Cicero and Germanicus survive in greater numbers than Avienus, the following is a comparison of the lines of verse in their respective poems. Aratus' description of the northern and southern constellations consists of lines 19–558; whereas Cicero's description of the same area of the heavens survives in lines 1–340. Cicero's poem is missing the northern circumpolar constellations; the Dragon's head group; the Great Bear's tail group; the Great Bear's body group; and the Cynosura group. Manuscripts containing Cicero's poem begin with the description of Aries (which compares to line 229 in Aratus) and then continues with the remaining northern and southern groups. Cicero mentions the planets (lines 223–36) and the circles (lines 237–340). The risings and settings of the constellations in Cicero's version are discussed in lines 341–480; the same section in Aratus is found on lines 559–757. Germanicus' descriptions of the constellations, planets and circles are covered in lines 17–572 and his rising and setting section is lines 573–725.

Appendix C has examined the appearance, descriptions and myths of the constellations, including the translations of appropriate sections of the original poem of Aratus. The iconography of each constellation is discussed. Examples of the various types of constellation miniatures that appear in the *Aratea* manuscripts are pictured along with other popular constellation illustrations, such as Hyginus and Michael Scot, for comparison.

★ The Circumpolar Group: The Bears, and the Dragon (19–62)

1. **Ursa Minor, Cynosura**

> The numerous stars, scattered in different directions
> sweep all alike across the sky every day continuously for ever.
> The axis, however, does not move even slightly from its place,
> but just stays forever fixed, holds the earth
> in the center evenly balanced, and
> rotates the sky itself. Two poles terminate it at the two ends;
> but one is not visible, while the opposite one in the north is high above the horizon.
> On either side of it two Bears wheel in unison,
> and so they are called the Wagons.
> They keep their heads for ever pointing to each other's loins,
> and for ever they move with shoulders leading,
> aligned towards the shoulders, but in opposite directions.
> If the tale is true, these Bears ascended to the sky from Crete
> by the will of great Zeus, because when he was a child
> then in fragrant Lyctus near Mount Ida,
> they deposited him in a cave and tended him for the year,
> while the Curetes of Dicte kept Cronus deceived.
> Now one of the Bears men call Cynosura by name, the other Helice (19–37).

Aratus launches his ode to the stars by introducing the two Bears, who are always visible and rotating in unison round the stationary pole star. With these primary stars, Aratus ushers in the first catasterism of the poem, but qualifies their story with a disclaimer, "if the tale is true". He recognizes the two faithful Bears who cared for the infant Zeus; in return, Zeus rewarded their year of nurturing by immortalizing them, and placing them where they will forever be navigational guides for those who sail the seas. Known more commonly today as the Big and Little Dippers, also called the Wains or Wagons, the Bears are probably the most familiar and easily found constellations when looking to the north.

After locating the Bears, stargazers are able to orient themselves, then by reciting the verses, follow the narrative thread as it guides the viewer through the sky naming one by one the other constellations. Of these two circumpolar star groups, Ursa Minor is much less significant, fainter and smaller, in contrast to its principal significance as provider of the current pole star, Polaris. Even though the two Bears possess quite similar patterns and consist of the same number of stars, Ursa Major dominates the sky on clear nights.

2. **Ursa Major, Helice**

> Helice is the one by which Greek men at sea
> judge the course to steer their ships,
> while Phoenicians cross the sea relying on the other.
> Now the one is clear and easy to identify, Helice,
> being visible in all its grandeur as soon as night begins;
> the other is slight, yet a better guide to sailors,
> for it revolves entirely in a smaller circle;
> so by it the Sidonians sail the straightest course (37–44).

The seven bright stars that form Ursa Major's ladle-like pattern make up the Bear's body and tail; adjacent stars fill in the head and shoulders to create the entire animal. Visible as soon as the sky darkens, the Big Dipper is found lying close to the Little Dipper and their 'cups' can be imagined as pouring into each other. While the two Bears are revolving round the fixed pole star, each holds its head toward the other's loins; they circle continuously, back to back, facing opposite directions, never setting. In illustrations, artists do not regularly follow this 'correct' form; the Bears are depicted in various positions. These first three constellations are almost always combined into a triple arrangement. In most cases, Draco appears more snake-like than dragon-like, wound round the Bears. The writings of Hyginus, *De astronomica*, present similar information, agreeing that the Bears were nurses of infant Zeus and that the Greeks applied the names Bear or Wagon. But he reports that the Romans referred to the Bears as 'oxen', and so provides an explanation for the name Boötes or plowman given to that nearby constellation.[17]

3. **Draco, the Dragon**

> Between the two Bears, in the likeness of a river, winds a great wonder,
> the Dragon, writhing around and about at enormous length;
> on either side of its coil the Bears move, keeping clear of the dark-blue ocean.
> It reaches over one of them with the tip of its tail,
> and intercepts the other with its coil.
> The tip of its tail end level with the head of the Bear Helice,

[17]The descriptions of the constellations and the myths connected with them are from source books by Theony Condos, Julius D. W. Staal, and Richard H. Allen.

and Cynosura keeps her head within its coil.
The coil winds past her very head, goes as far as her foot,
then turns back again and runs upwards.
In the Dragon's head there is not just a single star shining by itself,
but two on the temple and two on the eyes,

while the one below them occupies the jaw-point of the awesome monster.
Its head is slanted and looks altogether as if it is inclined towards the tip of Helice's tail:
the mouth and the right temple are in a very straight line with the tip of the tail.
The head of the Dragon passes through the point where the end of setting
and the start of risings blend with each other (45–62).

The sprawling constellation of the Dragon, or Draco, writhes between the two Bears, who lie on either side of the Dragon's tail, the tip of its tail ends near the head of Ursa Major. Legend records that originally the Dragon was placed in the land of the Hesperides by Hera, as the guardian of her Golden Apples. At the marriage of Hera and Zeus, the other gods had given them many gifts, Hera especially liked the Golden Apples brought by Gaea and ordered that the tree be planted in the garden of the gods, near Mount Atlas. A huge and fierce dragon guarded the special tree against the daughters of Atlas who were constantly stealing the valuable apples. The concept of a guardian serpent was common in Greek religion and folklore, where serpents guarded springs, houses, and sacred temples. The dragon was eventually killed as one of the tasks of Hercules, and then rewarded with a place in the sky by Hera. According to Aratus and Hyginus, Draco has a star on each side of its head, two on the eyes, one on the jaw; Hyginus adds that there are ten more scattered along the rest of the body, with 15 stars in all.[18] This precise placement of stars is rarely observed by medieval artists.

★ Dragon's Head Group: the Kneeler, the Crown, the Serpent-Holder and Serpent, the Scorpion, the Claws (63–90)

4. **Herakles or Engonasin**

Near it there circles a figure like a man toiling. No one is able to say definitely
what it is or on what task the man is intent, but they just call him
the man on his knees; again, laboring on its knees, the figure looks
like a man crouching. From both his shoulders arms are raised and

---

[18]Aratus gives no notice to the number of stars or their location within the constellation. The total number of stars in each constellation as recorded by Hyginus, Eratosthenes and Ptolemy usually differs slightly. Since Hyginus' version appears most often in illustrated manuscripts, his number for stars is listed for each constellation.

extend in different directions to a full stretch. He has the tip of his right foot above the mid-point of the tortuous Dragon's head (63–70).

The constellation described by Aratus next lies near to the Dragon's head, a figure on a bent knee working at an arduous task; the poem leaves him nameless, and simply refers to him as, the 'man on his knees'. This phantom male has become syncretized with the mythical image of Herakles, one of the most popular heroes from antiquity. The iconography of Herakles/Hercules was firmly established by the Classical period, as his image appeared frequently in Greek artwork, especially in relief carvings, painted vases and coins. Even before that time, he can be recognized in Greek art of the Geometric and early Archaic Periods by his principal identifying symbols, the tawny lion-skin (flayed from the Nemean lion), his club, and his bow and arrows. He is always associated with his Labors, a canonical 12 in number, but the particular tasks he performed vary through the different areas of the ancient world.

The iconography most often found in astronomical illustrations depicts Herakles as he threatens the snake-like dragon that guarded the Golden Apples of the Hesperides; quite appropriate since this Labor was set at the edge of the world. Miniatures of this constellation usually depict heroic Heracles with a lion's skin over his left arm, a club in his right hand, the left knee bent to the ground, and his right knee forward. The Leiden *Aratea* substitutes a crook for his club. In the Egerton manuscript, Hercules raises a spear.

This huge constellation spreads out over the night sky, appropriate for a man of great strength. Herakles was blessed, or cursed, with both mortal and immortal blood, as his father was Jupiter and his mother was earthborn Alcmene. The myth relates that the vengeful Juno, suspecting Hercules might be the consequence of her husband's infidelity, made life very difficult for him by bringing discord into his life.

As retribution, young Hercules was confronted by two female divinities, Pleasure and Virtue, who required him to make a difficult choice. Pleasure promised him a future life of happiness and enjoyment, and Virtue promised him a lifetime of hardship; but if he bore the hardships well, he would be richly rewarded and crowned with glory for eternity. He accepted Virtue's challenge, but was required to perform the Twelve Labors; some have associated the Labors with the 12 signs of the zodiac. According to Hyginus, Hercules has one star on his head, a bright star on each shoulder, and others outline the rest of his body with four stars on the left hand, which represent the lion's pelt. In mid-summer the constellation of Hercules toils at his trial upside down, high above the southern horizon; the mighty figure includes 19 stars over all.

5. Corona Borealis, the Crown of Ariadne

There too the famous Crown, which Dionysus established
to be an illustrious memorial to the departed Ariadne,
circles close to the back of the laboring figure (71–73).

Near the Kneeler's back sits a small, dim constellation, called Ariadne's Crown; even though lusterless, it is quite easy to recognize because of the distinctive curve of its closely set stars. It appears in the northern sky, opposite the head of Draco. Legends say that Hephaestus had forged the heavenly Crown, a circlet of stars formed of fiery gold and embellished with Indian gems. Ariadne was the daughter of King Minos, and her Crown was set among the stars by her lover Dionysus, who wished to honor himself before the gods. The magical light emitted from this Crown helped to guide Theseus while he sought to discover a path through the labyrinth. Hyginus says the Corona Borealis, or the Northern Crown, consists of eight stars lying in a circle, only three are very bright.

6. **Ophiuchus, Effaces or Serpentarius, the Serpent-Holder**

> The Crown is close to his back, but beside the top of his head
> observe the head of Ophiuchus, and from that you can identify
> the whole of Ophiuchus as clearly visible.
> So bright are the brilliant shoulders that appear lying below his head:
> even in the light of the full Moon these can be visible.
> But his hands are not equally bright;
> for faint is the light that runs along this side and that;
> nevertheless even these are visible, for they are not dim.
> Both hands struggle to hold the Serpent that writhes
> round Effaces' waist. He constantly, with a good firm stance,
> tramples with both his feet the great monster (74–84).

Serpentarius is part of a large, complex group of stars which is actually two constellations that are intertwined; a male figure and a snake. Ophiuchus stands huge and upright grasping the writhing body of the lengthy serpent with both hands, most often depicted as if wrapped around his waist. In the past Ophiuchus has sometimes been associated with Asclepius, the patron of medicine, since both figures are always accompanied by their iconographic symbol, a serpent. Of the four snake-like creatures that appear among the constellations, this one is considered The Serpent. Under the feet of Ophiuchus, the dangerous Scorpion or 'great monster' reclines; the great weight of this powerful snake-handler impresses upon the monster, as if to crush it. In manuscript illustrations, Ophiuchus is often pictured as standing atop Scorpio. According to Hyginus, there are a total of twenty-three stars in this sizeable double constellation, seventeen in the male figure and six in the wriggling serpent.

7. **Scorpio, the Scorpion**

> Scorpion, standing upright on its eye and its breast.
> But the Serpent writhes in his two hands, a short section in his right,
> while large and high it rises in the left (85–87).

Although receiving only scant attention by Aratus, Scorpio is a beautiful autumn constellation with a well-defined group of stars, actually appearing to form the distinctive shape of a scorpion ready for attack, with its claws extended and its tail curving upward tipped with a deadly stinger. Originally one large constellation, Scorpio has been divided into two; as its Claws now form the zodiacal sign, Libra. Scorpio contains numerous bright stars easy to pick out from its neighboring stars. Specialists regard it as one of the earliest named constellations. The first magnitude star, Antares, becomes the heart of Scorpio, befittingly reddish in color; it is the first zodiacal sign to be mentioned in Aratus' poem.

Legends say that Artemis enticed a scorpion from a hill on the island of Chios, and sent it to sting Orion, because he had used force against her on a hunt, and had also boasted of his hunting skills. Orion died from the creature's poisonous sting. As a reminder not to become arrogant, Zeus placed the Scorpion in a prominent position among the brightest stars to warn humanity of its strength and power. When visible, the star groups act out this story each night; when deadly Scorpio threatens by rising in the east, Orion perishes by setting in the west. But luckily, Orion was given an antidote for the poison in his heel, and was subsequently restored to life by Ophiuchus/Asclepius.

Ophiuchus can be seen standing atop Scorpion when the latter sets in the West, as he conquers the dangerous 'monster'. Recovered and renewed in the underworld however, Orion returns, rising again in the East. But Jupiter punished Ophiuchus for interfering in the fate of humanity, bestowed by the immortal gods, and sent Aquila after him with one of his thunderbolts. When Ophiuchus sets in the West, the Eagle can be seen in a threatening position over him; this implies that Ophiuchus is dying, mortally wounded by Jupiter's thunderbolt. So the battle between Orion/Osiris and Scorpio/Set can be followed day after day, year after year, symbolizing the eternal battle between light and dark, the ancient myth of Osiris, Isis and Set. Orion represents life and the Scorpion death.

Scorpio is one of the constellations that always remains comparatively low in the southern sky, creeping just above the horizon. When the Sun passes into the Sign of the Scorpion, who fittingly lives in dark places and is active at night, the winter season begins. Looking at a map of the heavens, the tail of Scorpio extends into a dark place in the Milky Way. The total number of stars in the sign of Scorpio is nineteen.

8. **The Claws or Libra, the Scales**

> Now the point of the Serpent's jar lies close to the Crown,
> but under its coil you can look for the great Claws,
> though these are lack-luster and not at all brilliant (88–90).

As mentioned, the constellation of Libra was originally part of Scorpio, but as a twelfth sign was needed when the Zodiac was created, the Claws were cut off from the body to form a separate star group. Therefore Libra was the last of the zodiacal signs to be recognized, not even named by Aratus. Libra is the sign which the Sun enters at the autumn equinox when the world is held in balance, an appropriate symbol for equal hours of daylight and night. The Scales weigh the length of the day and the night, finding them in perfect equilibrium. Its four stars are called 'lack-luster' by Aratus because of their deficiency in brightness, not an impressive constellation in size or brilliance. Sometimes the Scales of Libra are placed in the right hand of Virgo, who is personified as the Goddess of Justice.

★ Bear's Tail Group: Boötes, The Maiden and Justice (91–136)

9. **Boötes or Arctophylax, the Bear-guardian**

> Behind Helice there come, like a man driving, Arctophylax, whom men
> call Boötes, because he is seen to just touching the Wagon–Bear.
> All of his is very conspicuous; below his belt circles
> Arcturus itself, a star distinct from the rest (91–95).

Above is Kidd's translation of Aratus and below is Germanicus' version of the same passage.

> Next an old man follows Helice and threatens with his staff.
> Whether he is the Bear-Guardian or Icarus who was murdered
> on account of Bacchus' gift, he made up for a life.
> that was snatched away with a constellation.[19]

Boötes, sometimes called Arctophylax or the Bear Driver, is located next to Ursa Major and appears to follow the Bears. The 'old man' was associated with the Bears when they were known as the Wagons, which explains why he is 'like a man driving'. Hyginus associated the constellation with Arcas, son of Callisto and Jupiter, a myth concerned with human sacrifice. Boötes has four stars on his right hand which never set; he also has a bright star on his forehead and one on each shoulder. As the stars of Boötes are rather dim, attention is focused on the brilliant Arcturus which sits between his knees. The total number of stars in Boötes is fourteen.

10. **Virgo, the Maiden**

> Beneath the two feet of Boötes you can observe the Maiden,
> who carries in her hand the radiant Spica.
> Whether she is the daughter of Astraeus, who, they say,

[19]Possanza 202 and 207

was the original father of the stars, or of some other,
may her way be peaceful! (96–100)

Below Boötes, the winged goddess Virgo is pictured wearing full, flowing robes; she carries in her left hand a shaft of wheat, exemplified by the bright star Spica. The heliacal rising of Spica occurs at the time of harvest, which explains Virgo's iconographic symbol. The constellation lies on the ecliptic, not significant in its brightness, but in illuminated manuscripts the Maiden, the lone female zodiacal sign, often appears as the most attractive depiction.

The constellation of Virgo receives the most attention in the poem by Aratus and Germanicus, who both relate her long narrative. Several versions of the myth of the Ages of Man were associated with the constellation of Virgo. Aratus recognizes her as Astrea or Dike, the daughter of Astraeus and the Goddess of Justice; he continues his description of the Maiden with an extended passage on this theme. According to the myth, even though an immortal, Dike became associated with mortals of the Golden Race. She administered just laws, but the Golden Race did not know crime or war; peace reigned at that time. Eventually the Golden Race was replaced by the Silver Race, not quite as ideal as their predecessors; and so she had to reprimand them. Eventually the Silver Race died out as well, and the Bronze Race supplanted them. Changing for the worse, the Bronze Race was warlike and violent, no longer upholding justice; as a result, she detested them. So she departed the Earth forever and flew to the heavens becoming catasterized, which may explain why she acquired wings in artistic renderings.

Another chore assigned to Virgo, in her capacity as Dike, was to weigh the souls of those who died in order to determine whether they should go to the Elysian Fields of eternal happiness, or sent instead to places of punishment, a role assumed by St. Michael the Archangel in medieval art (A role assigned to Thoth in ancient Egypt.). In some illustrations, Virgo is conflated with Libra, as they are both associated with the scales of justice. She is the third sign of the zodiac to be described. The total number of stars in Virgo as reported by Hyginus is nineteen. In Aratus' poem, her story as Justice continues.

In 2013 Aratean scholar Gee issued a volume, *Aratus and the Astronomical Tradition*, in which she *discussed the Aratea of Avienus in light of popular religious thought of late antiquity, mainly Neoplatonism. She focuses her lengthy, detailed analysis of the Myth of Dike, goddess of Justice, by examining the writings of Hesiod, who transfigures Dike from a goddess to a heavenly constellation. Gee also shows how Plato's treatment of Dike as a prose philosophical myth was replaced by the poetic myth of Aratus who conflates her iconography with the constellation Virgo, claiming a cosmic truth, and also mediating astronomy and cosmology for the first time.*

Justice or Dike, an aspect of Virgo

There is, however, another tale current among men,
that once she actually lived on earth, and came face to face with men,
and did not ever spurn the tribes of ancient men and women,

but sat in their midst although she was immortal.
And they called her Justice: gathering together the elders,
either in the market-place or on the broad highway,
she urged them in prophetic tones to judgements for the good of the people.

At that time they still had no knowledge of painful strife
or quarrelsome conflict or noise of battle,
but lived just as they were; the dangerous sea was far from their thoughts,
and as yet no ships brought them livelihood from afar,
but oxen and ploughs and Justice herself, queen of the people
and giver of civilized life, provided all their countless needs (101–113).

....

Then Justice, conceiving a hatred for the generation of these men,
flew up to the sky and took up her abode in that place,
where she is still visible to men by night
as the Maiden near conspicuous Boötes (133–136).

★ The Bear's Body Group: The Twins, the Crab, the Lion, the Charioteer and the Bull (147–178)

11. Gemini, Castor and Pollux, the Twins or Dioscuri

Beneath the Bear's head are the Twins, (148)

Aratus' presentation of the fourth sign of the zodiac, the twins Castor and Pollux (Greek Polydeuces), is extremely brief, not even naming them or relating their mythological tale. This constellation is clearly visible from autumn to late spring, consisting of two elongated figures marked by a pair of bright stars, sparkling side by side. Golden in color, Pollux shines as the brighter of the pair. In drawings, these brilliant stars usually mark their foreheads, but sometimes the shoulders of the twin brothers. In ancient Greece the twin stars were variously identified with Apollo and Heracles, or Amphion and Zethus, but most often with Castor and Pollux. In astronomical picture cycles, Castor and Pollux, also called the Dioscuri, are commonly represented by two male warriors, most often shown naked, holding long spears or a club and a bow and arrow; but frequently they take on the iconography of a lesser-known mythological pair, Amphion and Zethus, who have been associated with Castor and Pollux in Greek literature. Amphion, a singer and musician, and Zethus, a hunter, were sons of Zeus and Antiope in Greek mythology. In their depiction in the Leiden *Aratea,* their identifying emblems are intermingled by substituting the iconography of Castor with the golden Lyre given to Amphion by Zeus, while Pollux still holds his long club. This alternative iconography of the Dioscuri, often adopted during the Middle Ages, appears quite often in constellation cycles.

During the Classical period, the Dioscuri were depicted as heroic nudes; no representation earlier than the second century CE shows them in military attire. In the Greco-Roman era, they were often depicted in sculptures associated with horses. The fact that, in the Leiden *Aratea,* the Gemini twins appear wearing small caps with Christian crosses atop is mentioned in Chap. 7, as it is the only insertion of Christianity into astronomical art. In Michael Scot astronomical manuscripts, the twins are usually depicted with large wings on their backs but sometimes small ones on their feet, similar to Mercury's iconography. In the al-Sufi manuscript tradition, Gemini are fully clothed and wear turbans, in many al-Sufi Latinus illustrations, Gemini return to their naked state but still sport a tight-fitting head covering.

Although some alternative stories exist, the myth most commonly associated with the Twins is the following. The mythological brothers Gemini were sons of Leda, but after she had already become pregnant with twins by her husband, was visited by Zeus in the guise of a swan. The result was a double twin birth, one of each pair was mortal, Castor and Clytemnestra and the other pair, Pollux and Helen, the offspring of Zeus, became immortal. Castor and Pollux, were known to have very different temperaments; but were the most loving of all brothers, never argued, were always in agreement and joined together for many heroic exploits. When the mortal twin Castor was killed in battle, Pollux unselfishly gave one-half of his immortality to his deceased brother, so they could share life and death alternately, possibly symbolizing the dual aspects of the sky, night and day; the constellation is visible only half the year. Zeus wished to memorialize the devotion of the loving brothers by placing them together in the sky, eternally side by side. Castor and Pollux were widely revered in many sanctuaries; the remains of a cult temple survive in Sicily. Beginning in the period of the Republic, the Dioscuri were honored by appearing regularly on the coinage of Rome wearing capes decorated with a star. The Dioscuri were a very popular subject for Imperial Roman art, particularly mosaic floorsThe constellation Gemini also became associated with St Elmo's fire and protecting sea voyagers.

12. **Cancer, the Crab**

> beneath her belly the Crab, and under her hind legs the Lion shines brightly (148).

Cancer, the fifth sign of the zodiac in the poem, is a rather insignificant and dim star group lying to the southwest of Ursa Major, obtaining barely a line from Aratus. The sign of the Crab marks the Tropic of Cancer, the most northerly reaches of the Sun. Its name may have come from Egypt, originating from the sacred scarab beetle. The constellation Cancer is usually identified with the infamous Crab that attacked the feet of Hercules while he was performing the second of his Labors, slaying the Lernaean nine-headed hydra. Enraged by the vicious attack, Hercules crushed the Crab with his foot; but Juno took pity on the crustacean and placed it among the stars. Like its appearance in the sky, the illustrations are rather plain and uninspired. There are 18 stars in the constellation of Cancer.

13. Leo, the Lion

This is where the Sun's track is hottest,
and the fields are seen bereft of their corn-ears when
the Sun first comes into conjunction with the Lion.

This is the time when the whistling etesian winds sweep strongly
across the broad sea, and it is no longer seasonable for ships to be under oars.
Then let broad-beamed ships be my pleasure,
and let helmsmen hold their steering-oars into the wind (149–155).

Leo, the sixth sign of the zodiac to be mentioned by Aratus, is easily recognized, as its outline actually resembles a lion, its head is in the form of a reversed question mark, facing west. It was said that the heat of summer doubles in Leo. Later mythographers name Leo as the lion that had been ravaging the valley of Nemea. The first labor of Hercules was to slay the Nemean Lion, and he succeeded, bravely killing it with his bare hands. His iconography includes the lion's skin either over his arm or tied as a cape, as a reminder of his famous exploit. Some associate the creation of this constellation with Jupiter, the king of the gods, as the lion is considered the king of the beasts. Seven stars near the tail of Leo form a triangle and are called the Lock of Berenice or Coma Berenices. Ptolemy included Coma Berenices in his star chart as the northern most of the nebulous mass between the edges of Leo and Ursa Major with a magnitude of five. The total number of stars in the constellation of Leo is 19 from Hyginus, but 33 in the *Almagest*.

14. Auriga, the Charioteer, including the Goat, Capella and the Kids

If you are minded to observe the Charioteer and the Charioteer's stars,
and rumor has reached you of the Goat herself and the Kids,
who have often looked down on men being tossed upon the heaving sea,
you will find the Charioteer lying large to the left of the Twin
while opposite Helice circles his head at that extremity.
Fastened to his left shoulder is the sacred Goat,
who is said to have tendered her breast to Zeus:
the interpreters of Zeus call her the Olenian Goat.
Now she is large and brilliant, but her Kids there
on the Charioteer's wrist shine faintly (156–166).

The Charioteer Auriga sits above the horns of Taurus, on the left side of the Twins, with the Goat on his left shoulder and the Kids glowing softly on his left wrist. Ancient Charioteers looked after the horses and chariots of their masters, and also watched over the livestock; the goat and kids may refer to this function. The shape of Auriga is defined by a pentagon that reaches partly into the Milky Way; its most brilliant star is Capella, the fifth brightest star in the sky. Associated with autumn, the appearance of Auriga forewarns sailors of the onset of rough sailing weather. The left hand of the western Gemini Twin points directly toward the Charioteer. In illustrative cycles, Auriga stands in his stellar chariot looking similar to and sometimes mistaken for Helios. But Auriga's garments are the clothes of a worker; he wears no crown, and does not hold the great torch that lights the Sun, but simply a whip. Germanicus suggests that the charioteer is either Erichthonis, son of Hephaestus and Gaea, the great innovator, who was said to be the first man to harness four horses to a chariot. Zeus saw him at work and marveled that he imitated Helios' great quadriga; thus he honored him by placing him among the stars. The constellation is probably of Babylonian origin, since similar images appear on Babylonian cylinder seals. The al-Sufi Auriga does not include the chariot, but he does hold a whip. There are eight stars in the Charioteer.

15. **Taurus, the Bull, including the Hyades**

> Near the feet of the Charioteer look for the horned Bull crouching.
> This constellation is very recognizable, so clearly defined is its head;
> one needs no other sign to identify the ox's head,
> so well do the stars themselves model both sides of it as they go round.
> Their name is also very popular: the Hyades are not just nameless.
> They are set out all along the Bull's face;
> the point of its left horn and the right foot of the adjacent Charioteer
> are occupied by a single star, and they are pinned together as they go.
> But the Bull is always ahead of the Charioteer in sinking to the horizon,
> though it rises simultaneously (167–178).

Next to the feet of the Charioteer and west of Orion lies the fierce Bull Taurus, but only the front half of the animal, the head, shoulders and forelegs are included. The constellation of Taurus includes two beautiful and fabled star clusters, the Hyades and the Pleiades. The Hyades form a recognizable v-shaped group of stars that 'gleam on his brow' tracing the shape of the Bull's head. In mythology the Hyades were the daughters of Atlas and Aethra, the half-sisters of the Pleiades. Ovid calls the Hyades, the Rain-Bringing Stars, because the sisters were so grief-stricken after their brother, Hyas, had drowned in a well that their tears caused heavy rainfall. Five bright stars outline each side of the face of Taurus; and the brilliant, first-magnitude star, Aldebaran, becomes the Bull's eye. Above the neck of the Bull are the Pleiades, the Seven Sisters. Aratus notes that Taurus and Auriga share the same star that forms the Bull's left horn and the Charioteer's right foot. Taurus has been identified with various mythological bulls; in particular, the tales of Pasiphae and Theseus, and with the white bull whose shape was assumed by Jupiter to attract and abduct Europa, the princess of Tyre.

According to myth, one day Europa was playing with friends close to her father's herd of bulls. She noticed an outstandingly beautiful white bull and went to stroke him, adorning him with flowers. Since the bull appeared agreeable, she climbed onto his back for a ride; and he rushed straight into the sea with his lovely rider, swimming all the way to Crete. In Crete, Europa gave birth to Minos, who became the King of Crete, and then later the father of Ariadne. This constellation was important in the ancient Babylonian era, when Taurus marked the astrological age, just before the arrival of the Age of Aries. A constellation in the form of a bull was discovered in both Egyptian and Babylonian records named as the 'First Sign' of the zodiac. It was called the 'First Sign' because Taurus coincided with the vernal equinox between the years 4000 and 1800 BCE approximately. The total number of stars in Taurus the Bull is eighteen.

Although the Hyades are mentioned by Aratus as part of the face of Taurus, I have seen them pictured in only one *Aratea* manuscript.

★ Cynosura Group: Cepheus, Cassiopeia, Andromeda, the Horse, the Ram, the Triangle, the Fishes, Perseus, The Pleiades (179–267)

16. **Cepheus, mythological king**

Nor will the suffering family of Cepheus, son of Iasius be just left unmentioned;
their name also has reached the sky, for they were akin to Zeus.
Behind the Bear Cynosura Cepheus himself is like a man stretching out both his arms.
The line that extends from the tip of her tail to each of his feet
equals the distance from foot to foot.
And you have only to look a little way past his belt
if you are searching for the first coil of the great Dragon (179–187).

The constellation of Cepheus is found in the northern part of the sky, partially circumpolar. From feet to chest he lies within the Arctic Circle, as he is upside down; the rest of his body lies between the Arctic Circle and the Tropic of Cancer. Cepheus is always depicted with long, outstretched arms; a straight row of faint stars serves as his left arm. The long-suffering royal family of Cepheus, king of Ethiopia, inhabits this same area of the sky; he faces his wife Cassiopeia, and standing below them is their daughter, Princess Andromeda.

According to the ancient tale, Cepheus' country was being punished, terrorized by the frightful sea monster Cetus. Since the king's efforts had been ineffective; he felt compelled to consult the great oracle of Ammon to determine a method for rendering the monster harmless. To Cepheus' horror, the oracle foretold that the only way to save his people and lands was to sacrifice his daughter Andromeda to the threatening creature. As a good and true king, he had no choice but to defend his kingdom; and so he was required to offer Andromeda as a sacrificial victim to appease the sea monster. The remorseful Cepheus extends his arms for eternity in supplication toward the gods to both save his people and spare the life of his daughter. Because Cepheus rules the Ethiopians, he is often depicted in oriental garb. Germanicus adds an interesting note in his translation that Cepheus' family began with Jupiter, and adds not too subtlety that 'the kingly nature of one's father is often an advantage' (His adopted father was Emperor Tiberius.). The stars in the constellation of Cepheus number nineteen in all, none are particularly bright.

17. **Cassiopeia, Wife of Cepheus**

In front of him revolves the tragic Cassiopeia, not very large,
but visible on the night of a full Moon;
only a few zigzagging stars adorn her, giving her all over a distinct outline.

Like to a key with which men attacking a double door
barred on the inside knock back the bolts,
such is the appearance of the individual stars that together comprise her.
She extends outstretched arms just from her small shoulders:
you would say she was grieving over her daughter (188–196).

Even though a haughty queen, Cassiopeia forms a modest, circumpolar constellation with only five semi-bright stars in the form of a wide-angled W. The royal couple Cepheus and Cassiopeia are unusual since they were not honored for bravery or duty to the gods, but as part of a favorite moralistic story. Like her stellar husband, Cassiopeia's arms are reaching outward, perhaps grieving the fate of her sacrificed daughter. She reigned as a beautiful queen, but her pride and arrogance caused their great misfortune. One day she foolishly boasted of being fairer than the Nereids, the water nymphs, who were celebrated for their exquisite beauty. The Nereids overheard her vain bragging and complained to their father Poseidon. Infuriated, Poseidon created a great flood and called up the monster Cetus from the depths of the ocean. Poseidon dispatched Cetus to the waters of Ethiopia to ravage the people and their herds. Only the sacrifice of Cassiopeia's daughter would appease her affront to the sea-nymphs.

In addition to losing her daughter, Cassiopeia was condemned by the gods to revolve around the North Celestial Pole forever, as a lesson in humility and as punishment for her offensive boasting. The constellation actually appears in the shape of a W when below the Pole, and as an M when above the Pole. When appearing below the Pole, Cassiopeia must hang on tightly, so that she does not tumble from her throne. Twelve hours later, the mourning queen is relieved to be returned to her full upright position. The tragic figure sits grieving in her chair or throne, which Aratus does not mention here; but he discusses later in the poem. Germanicus says, "Her face contorted in agony, she stretches out her hands as if bewailing abandoned Andromeda, unjustly atoning for the sin of her mother". The al-Sufi illustration of Cassiopeia shows the legs of the throne ending in lion's paws, which are often seen in the classical tradition, while the shaped knobs at the top are characteristic of Islamic work.

18. **Andromeda, the Woman Chained**

There too revolves that awesome figure of Andromeda, well defined
beneath her mother. I do not think you will have to look all round
the night sky in order to sight her very quickly,

> so clear are her head, the shoulders on either side,
> the feet at her extremity, and all her girdle.
> Even there, however, she is extended with outstretched arms,
> and bonds are laid on her even in the sky;
> those arms of hers are raised and out-spread there all the time (197–204).

Andromeda, the daughter of Cepheus and Cassiopeia, is place below her mother in the sky, a long line of fairly bright stars, her head coincides with Pegasus and her feet lie near Perseus. Stars mark her head, shoulders, feet and belt; her outstretched arms, chained forever, are much fainter. The ill-fated, innocent victim of her mother's arrogance suffered the punishment for her mother's vanity, her boast of beauty superior to the Nereids. Andromeda was carried to the edge of the sea, and there with her outstretched arms, wary from being chained to the rocks, she waited in anticipation, expecting to be devoured by the dreadful monster Cetus. That is how she has been immortalized, bound between hope and despair, waiting and wishing for rescue from her frightful fate. By chance the 'awesome figure' of Andromeda was noticed by the brave Perseus, who just happened to be passing by after beheading Medusa. He fell in love with her instantly, and managed to save her from the sea monster.[20] Upon being rescued by Perseus, Andromeda abandoned her parents, preferring to follow her rescuer to Argos. Andromeda is honored in the sky, not on her own behalf, but as a reminder of the foolishness of her mother and the courage and deeds of Perseus.

In Arab-influenced manuscripts, Andromeda can be pictured with two and sometimes three fish, covering her chest or at her feet, denoting the star system of the traditional astronomy of the Bedouin societies called the An 'war. The al-Sufi manuscript presents Andromeda as seen in the sky in three slightly different versions, one version depicts her in the usual depicted manner; the second version shows an outline of a fish at her feet and the third variation has two overlapping fish covering her chest. Both her hands are raised, but not chained; the bracelets worn on her wrists may be reminiscent of the chains.

In the Leiden *Aratea*, Andromeda is pictured partially clothed by the shore, arms chained to rocks. The group of six figures, Cepheus, Cassiopeia, Andromeda, Cetus, Pegasus and Perseus, represent all the principal figures of this moralistic tale, the only classical myth to be depicted so fully amidst the constellations. The Andromeda constellation includes 20 stars.

19. **Equus or Pegasus, the Winged Horse**

> Now the monster horse is actually pinned to her head by its lower belly;
> there shines a star that is common to its navel and the head at her extremity.
> The three other stars mark off lines of equal length upon the flanks

[20]See Perseus for more details of the rescue.

> and shoulders of the horse; they are beautiful and bright.
> Its head is not at all comparable, nor its neck,
> though it is long. But the end star on its shining mouth
> could well rival the former four, which outlive it so very conspicuously.
> But it is no quadruped; at its navel edge the sacred Horse is halved
> in the middle as it goes round. This was the Horse, they say,
> that from the height of Helicon produced the good water of fertilizing Hippocrene.
> The summit of Helicon was not then flowing with streams, but the Horse struck it
> and from that very spot a flood of water gushed out at the stroke of its forefoot;
> the shepherds were the first to call that draught the Horse's Spring.
> So the water wells out of a rock, and you can see it
> never far from the men of Thespiae; but the Horse revolves
> in the realm of Zeus and you may view it there (205–224).

The constellation of Pegasus represents the winged horse of heroic Perseus, born from the blood of Medusa's head as it combined with the foam of the sea. It is usually depicted upside down as a demi-horse, only the front half of his body appears, made up of bright stars near Andromeda. The Princess' head is attached to the navel of the Horse by a common star, this star plus three more brilliant stars form the Great Square in Pegasus. His head and the forelegs are formed by less bright stars, 18 stars in all. Aratus does not mention wings on the horse, but that mythical tradition dates back to Hesiod. After Perseus delivered the head of Medusa, Pegasus flew off to Mount Olympus where he became the bearer of thunder and lightning for Jupiter. In illustrations, the half-figured Pegasus is almost always drawn with colorful wings, large enough to lift the powerful horse in flight. In the al-Sufi illustration of Pegasus, his curled wings are no longer those of a bird, but of a fabulous Sassanian dragon.

As the first sign of the zodiac, Aries the Ram stands as one of the most important constellations for calendar and astrology; but its stars are faint and inconspicuous, except for one bright star in its head. The Ram is usually pictured with a full body, often running, its head to the right or the East, but turned and looking back. Frequently a band or circle is drawn around its mid-section, referring to the band of the zodiac or the equinoctial colure. This is important since Aries appears at the meeting place of the ecliptic and the celestial equator. The constellation of Aries rose in the background when the Sun reached the vernal equinox beginning about 2000 BCE. The Sun has precessed further along its extensive rhythmic cycle; presently the stars rising at the vernal equinox are the constellation Pisces. In approximately 21,000 years, the Sun will again introduce Aries as the indicator of the vernal equinox.

There are several mythic stories centered upon the celestial Ram. Briefly the most common myth concerns two children, Phrixus and Helle, who were raised by an evil stepmother. She convinces her husband, their father, to sacrifice his daughters to the gods in order to save the town and its crops from disease. Their real mother in heaven sends the Golden Ram to save the girls from their nasty step-mother. The children are told to climb on the Ram's back and hang on tightly, but not to look down from the great height. Of course one child named Helles did, and the spot where she fell is called the Hellespont in her honor. After the fall of Helles, the Ram brought the other child Phrixus to Colchis and the Ram flew off to become a constellation; its Golden Fleece was then nailed to a tree. The Golden Fleece was subsequently recovered by Jason. Hyginus adds two more versions of the Aries myth.

Neither Aratus nor Germanicus identifies the Ram or associates it with the Golden Fleece. Germanicus says that Aries moves through the longest circle in the sky. "His swiftness as hastens to touch the distant turning post with his horns" allows him to keep up with the other constellations that have a shorter distance to travel. Aries is the eighth sign of the zodiac to be discussed in the *Phaenomena*, and its total number of stars is seventeen.

21. **Triangula or Deltoton**

> There is yet another constellation formed near by beneath Andromeda:
> the Triangle is measured out on three sides, recognizably isosceles;
> the third side is shorter, but it is very easy to find,
> for it is well starred compared to the other two.
> Its stars are a little to the south of those of the Ram (233–238).

Deltoton is positioned below Andromeda and to the right of the Ram; it takes the form of the triangular shape of the Greek letter delta. The Triangle is not especially bright, but distinctive enough to be easily recognizable. There are at least four different stories surrounding the Triangle. Because the island of Sicily also takes the shape of a triangle, it is often associated with this small constellation. Ancient myths tell that Sicily received an honored place in the sky as a reward from Ceres for its devotion to her. The inhabitants had honored Ceres with many temples, thankful that the volcano, Mount Etna, continually created fertile soil for the island's farmers. Another account relates that perhaps the geometric form of the Triangle symbolizes the accomplishments of the Greeks and their advances in mathematics. Germanicus traces the origin of the Triangle to the Nile, perhaps referring to the river's delta where it meets the Mediterranean. According to Hyginus, Mercury placed this constellation, whose shape is that of the first letter of the word "Deus", in the sky to honor Jupiter. The constellation of Deltoton has only three major stars.

22. **Pisces, the Fishes**

> Still farther ahead and more in the approaches to the south are the Fishes;
> but one is always more prominent than the other and hears more
> the fresh onset of the north wind. From both of them stretch, as it were,
> chains from the tail-parts, coming together in an unbroken line on both sides.
> One beautiful bright star occupies this position, and they call it the celestial Knot.
> Let Andromeda's left shoulder be your guide
> to the more northerly Fish, for it is very close to it (239–247).

The two fish of the constellation Pisces are found south and west of Triangula; Andromeda's left shoulder is near the more northerly Fish. Pisces is a relatively large but not brilliant zodiacal constellation; both fish are equally faint, as are the two lines that join them. The celestially 'correct' depiction of Pisces is for each fish to face in opposite directions with a rope or line extending from their tails, but more often the line leads from their mouths. The cords extending from each of the fish converge and are tied together forming the 'Knot of the Heavens'. Hyginus (3.29) explains that the Knot lies at the crucial point where meridian meets the equator, and so can be considered as the Knot of the celestial sphere.

The Roman myth concerning Pisces relates that one day Venus and her son Cupid were startled and threatened by Typhon, the monster-dragon, who could survive in fire but would perish in water. Since Venus was born from the foam of the sea, she could easily escape from Typhon through the water. So both Venus and Cupid changed themselves into fishes and disappeared into the dark blue waters of the ocean. In order that they not lose each other in the dark water, they tied themselves together with a long rope. In this way they were able to escape the vicious Typhon; their escapade has been commemorated in the constellation Pisces, sometimes called 'Venus and Cupid'.

This constellation is significant for several reasons. In the year 7 B.C.E. a triple conjunction of Jupiter and Saturn and some of the stars in Pisces united, creating an extra bright star. This triple conjunction has often been regarded as the special star that shone at the time of Christ's birth. The Greek work for fish is *icthys*, the letters were thought to represent the phrase, "Jesus Christ son of God, the Redeemer". The Sun moved into the astrological Age of Pisces near to the birth of Christ; the fish therefore has many associations with Christianity. The 'more prominent' one is called the Northern Fish, which has 12 stars; the other is the Southern Fish which is comprised of 15 stars, the cord between them is formed by 12 stars.

23. **Perseus, with the head of Medusa**

> The two feet of Andromeda will be pointers to her suitor Perseus, as they move
> for ever above his shoulders. He runs taller than other figures in the north.
> His right hand is stretched out towards his bride's mother's chair-seat,
> and as if on some immediate pursuit he takes long strides,
> as he runs in the realm of his father Zeus (248–253).

Perseus, the airborne hero of classical mythology, was an obvious character to be catasterized in the sky. This constellation is relatively easy to find, having two second-magnitude and five third-magnitude stars. In his poem Aratus claims that Andromeda's two feet stand on the shoulders of Perseus, but actually only one foot comes close to his shoulder. The right hand of Perseus is extended and conveniently directs your eye toward the chair of Cassiopeia; he stands erect wearing winged shoes, his legs held widely apart, as if intent on taking 'long strides'. In his left hand, the bright star Algol represents the head of the evil Medusa, but this feature is not mentioned by Aratus. He does introduce the chair of 'his bride's mother', which he ignored in the previous description of the Queen. Aratus conjoins the key constellations of Perseus, Cepheus, Cassiopeia, and Andromeda and retells their tragic story explaining why the four figures are transfixed and captured for eternity in the sky.

Like many heroes in Greek mythology, Perseus was the son of a union between a god and a mortal, Jupiter and Danae. Danae's father kept her locked in a brass-lined tower to prevent her from marrying, because it had been prophesied that her son would one day murder him. Jupiter, enamored of the exceeding beauty of Danae, gained entry to the formidable tower by changing himself into a shower of gold coins. Her father was so frightened by the dire predictions that he had Danae placed in an open chest and set adrift in the sea, but she arrived safely at a nearby island where she raised Perseus.

The wise Minerva appeared to Perseus in a dream ordering him to slay the three sister Gorgons. Minerva gave Perseus a magical sword warning him not to look at the frightful face of Medusa which had the power to turn him to stone. Next Mercury furnished him winged sandals to carry him over land and sea, and later Atlas bestowed on him a magic helmet that made him invisible. Perseus was able to maneuver his shield so that he would see only the reflection of the three horrible sisters. In this way Perseus was able to sever Medusa's head, writhing with snakes, and restrain it in a leather sack.

After carrying out his appointed task, Perseus mounted the mighty winged-horse Pegasus, and they were passing over Ethiopia just as the unfortunate Andromeda was being chained to the rocks. Perseus stopped and offered to rescue the beautiful princess on the condition that King Cepheus would allow him to marry her. Cepheus agreed immediately, and Perseus took up his position by the rocks not far from Andromeda, warning her to keep her eyes firmly closed. As soon as the sea monster Cetus crawled out of the water and started toward the terrified Andromeda, Perseus jumped from his hiding place, yanked Medusa's severed head from the leather pouch and dangled it in front of Cetus. The sight of the revolting head halted the threatening sea monster; it was instantly transformed into a great rock which, it is said, can still be seen on the Levantine shore of the Mediterranean. Perseus quickly thrust the horrible head back in his bag and freed Andromeda from her chains. A great wedding followed combined with festivities to celebrate the town's release from the terror of the sea monster.

The Islamic manuscripts depict Perseus holding a head, but it is no longer Medusa's, instead it has become that of a male demon, perhaps the streams of blood from her neck or the snake-hair were mistaken for a beard. In the sky, Perseus towers protectively over Andromeda, standing with one foot on the Pleiades, who directed him to Medusa. The total number of stars comprising the constellation of Perseus is nineteen.

24. **The Pleiades or Seven Sisters**

> Near his left knee all in a cluster the Pleiades move.
> The space that holds them all is not great, and they are individually
> faint to observe. Seven in number they are in the lore of men,

> although there are only six apparent to the eye. No star at all
> has been lost from our ken in Zeus since our oral tradition began,
> but this is just what is said. Those seven are called by name
> Alcyone, Merope, Celaeno and Electra, Sterope, Taygete and honored Maia.
> All alike they are small and faint, but they are famous
> in their movements at morning and evening, and Zeus is the cause,
> in that he authorised them to mark the beginnings
> of summer and winter and the onset of ploughing time (254–267).

The Pleiades consist of a small, faint group of stars, but are easily found, bundled together in a tight cluster, occupying little more space in the heavens than the Moon. One of the oldest constellations, they are traditionally called the Seven Sisters 'in the lore of men'; but only six stars can be picked out in the night sky. They are positioned near Perseus, but are now considered as part of the constellation of Taurus. The Pleiades are acknowledged for their time-serving appearances in the morning and evening; their rising signals the onset of summer and their setting announces the coming of winter. Germanicus associates them with the Bull's back, and he lists the names of the Seven Sisters; but adds an identifier, calling them the daughters of Atlas, "born of a father who supports the sky", reflecting the poetic phrases of Hesiod, and Atlas' consort, Pleione, from whom they take their name. Germanicus names the sisters in a different order than Aratus; Electra, Alcyone, Celaeno, Merope, Asterope, Taygete and Maia. According to an ancient legend, six of the sisters consorted with immortal gods, while one married a mortal, explaining her invisibility.

In illustrated constellation cycles, the Seven Sisters are treated individually as seven female heads, the Leiden *Aratea* shows Six Sisters with heads uncovered, displaying sophisticated Late Antique hair styles, surrounding the central disgraced Sister whose head is covered with a veil.

25. **Lyra, the Lyre**

> The Tortoise too is small; when Hermes was actually still in his cradle,
> he hollowed out the shell and bade it be called a Lyre.
> He set it down in front of the unknown figure,
> when he had brought it to the sky. The figure, as he crouches,
> comes near it with his left knee, while the Bird's head
> at one extremity circles opposite it:
> the Lyre is set fast between the Bird's head and the knee (268–274).

Lyra lies on the western bank of the Milky Way, a small insignificant constellation having only one very bright star called Vega. The constellation is found near the head of the Bird Cygnus and the knee of the 'unknown figure'. According to the mythological tale, the Lyre was invented by Hermes/Mercury when he was still in his cradle. The Lyre had seven strings representing the seven planets or the seven daughters of Atlas. The Lyre was one of the earliest and most important musical instruments of the Greeks, used mostly to accompany songs. The Lyre became an attribute of Apollo, who adapted song to accompany the Lyre, and then it was passed on to Orpheus, the son of the Muse of epic poetry, Calliope, who taught him to play

Later Orpheus increased the number of strings to nine, to honor the number of the Muses. Playing the Lyre, Orpheus could charm the trees, rocks and wild animals, as well as humans, but Orpheus slighted the god Liber causing great anger toward him. In punishment, Liber dispatched the Bacchae upon him, tearing him to pieces. The sorrowful Muses gathered his dismembered body, buried him and then presented him with the greatest reward by placing his Lyre in the heavens to perpetuate his memory for eternity. Germanicus tells us the Lyre was "most welcome at the banquets of the gods". Total number of stars is only eight.

Cygnus, the Bird or Swan

> Yes, there is even a dappled Bird accompanying Zeus, hazy in some parts,
> while other parts on it bristle with stars, not very bright, though still not dim.
> Just like a bird in fair-weather flight, it glides on the breeze toward the
> stretching its right wing-tip in the direction of Cepheus' right hand,
> while close to its left wing lies the prancing Horse (275–281).

Cygnus, the Swan, is a mediocre summer constellation, "not very bright, though still not dim', its neck is long and its wings are extended similar to a bird drifting in flight. Cygnus, sometimes called the Northern Cross, takes the shape of a giant cross against the background of the silvery glow of the Milky Way. The gentle Swan is pursued through the heavens by the mighty Eagle, Aquila.

Cygnus was the best friend of Phaethon, the son of Helios, who one day attempted to steer the mighty Sun Chariot with fatal consequences. The inexperienced Phaethon was not capable of managing the four fiery horses, and so endangered the heavens and the Earth; his flight catapulted into a fiery disaster. He was ejected from the quadriga by Jupiter and fell to Earth like a meteor, where he landed in the river Eridanus. Because of this disastrous ride, the Milky Way is sometimes referred to as the scorched path, the remnant of the fateful ride of Phaethon. Germanicus identified Cygnus as either the bird of Apollo, or the swan whose form Jupiter assumed when he seduced Leda, wife of the King of Sparta, causing her to become the mother of twins, Pollux and Helen of Troy. The tail of the Swan is rich in stars, the most brilliant being first-magnitude Deneb, with a total of fourteen.

27. Aquarius, the Water-Bearer

> The two Fishes range about the Horse as it prances among them.
> Besides the Horse's head the right hand of the Water-pourer stretches out:
> He rises after Capricorn (282–284).

Aratus barely mentions Aquarius, his tenth sign of the zodiac; the Water-Bearer is found in an area of the sky which he calls the Water, which the Babylonians called the Sea. The constellation of Aquarius is represented by a standing man or sometimes a boy, who pours water from a jug or urn, often holding a towel in his left hand. Usually the liquid he pours is pictured as a silvery stream of stars running into the mouth of the Southern Fish, or sometimes merging with the river Eridanus. In Classical times, the Sun rose in the sign of Aquarius during the rainy season, a possible reason for its association with water.

Though neither Aratus nor Germanicus mention it, Aquarius is associated with the handsome youth Ganymede, who was abducted by Jupiter and taken to Olympus to be the Cup-bearer of the gods. For his faithful service to the gods, Ganymede was granted immortality. The liquid being poured from the jar is sometimes said to resemble nectar, the drink of the gods. The abundant stars of Aquarius are relatively faint; as a result, it is not a conspicuous star group. This constellation is generally associated with water, not only in Greek literature, but also in Ancient Egyptian and Babylonian records; again because of its association with the rainy season. Aquarius comprises a great number of dim stars, 17 in the male figure and 31 for the stream of liquid flowing from his jar.

28. **Capricorn, the Goat**

> Capricorn lies ahead and lower down, where the powerful Sun turns back.
> In that month I hope you will not be surged about by the sea
> through taking to open waters. Neither by day can you make much headway,
> for the days pass most swiftly then, nor in your terror by night
> will the dawn come soon, however much you cry out.
> It is then that the dread southerlies strike, when the Sun meets up with Capricorn;
> then the icy cold from Zeus is more cruel to the freezing sailor. But for that matter
> the sea surges under the stem all the year long; and we, like our ships,
> turning our eyes towards the beaches; but the surf there is still far off,
> and only a little timer keeps death away (285–299).

The constellation of Capricorn combines the head of a goat with the tail of a sea creature. The composite goat figure gained its curled fish's tail because it, like Aquarius, is connected with the rainy season; it too is located in the area of the sky that Aratus calls the Water. This constellation lies low in the sky, just above the southern horizon. Its chief importance resides in the fact that it marks the southernmost reach of the Sun at the winter solstice, providing its name to the Tropic of Capricorn.

One tale of how Capricorn reached the sky as a constellation was associated with the legend of the she-goat that suckled Zeus in Crete. Another myth explaining Capricorn's unusual form, states that the gods were taken by surprise by the demon Typhon. To save themselves from his attacks, they quickly shifted their shapes into all varieties of animals. The God Pan, in the form of a goat, was playing his pipes when he was startled by the sudden approach of Typhon, and in a panic (a term derived from his name) Pan instantly dove into the nearby river, changing himself into a fish in the process. But it happened so quickly, only his back half became a fish's tail. The front part of his pastoral disguise, a goat with horns and a beard, remained unchanged.

This is one of the few places in the poem where Aratus uses the first person; he considers himself in tune with the hardships of humanity, "I hope you will not be surged about by the sea." Capricorn is the eleventh sign of the zodiac named, and is occasionally pictured as emerging from a large seashell, another water reference. The number of stars in the constellation of Capricorn totals 24.

29. Sagittarius, the Archer

> After much suffering at sea even in the previous month,
> when the Sun inflames the bow and the Drawer of the bow,
> you should put ashore in the evening and not continue to trust the night.
> A sign of that season and that month will be the rising of the Scorpion
> at the end of night. The Archer actually draws his great bow near the sting;
> the rising Scorpion stands a little ahead of him and he then rises shortly after.
> At that time the head of Cynosura runs very high up at the end of night,
> and just before dawn Orion sets entirely, and Cepheus from hand to waist.
> (300–310)

Sagittarius, or the Archer, takes the form of a Centaur, front-half man and back-half horse, always armed for action with a bow and arrow. There are two Centaurs in the night sky, which occasionally leads to confusion. Sagittarius is sometimes called a Centaur, a more generic title; but the Centaur is never called Sagittarius. The two heavenly horse-men exhibit entirely different personality types. Sagittarius is a powerful warlike hunter; while the Centaur, also called Chiron, is peace-loving, kind and gentle, the personification of wisdom, strength and the art of medicine. The constellation of Sagittarius is outlined by a group of bright stars that suggest the shape of his bow, held aloft ready to shot an arrow toward the poisonous stinger in the raised tail of the Scorpion. Aratus warns his readers that when the Sun is in the constellation of the Archer, avoid seafaring at dark, "put ashore and do not trust the night".

In the poem, Sagittarius is the twelfth and last sign of the zodiac named. Neither Aratus nor Germanicus mention any legends associated with this constellation. Sagittarius is relatively difficult to find in the sky, since it lays low in the south, lost in the haze of the horizon. The best time for viewing Sagittarius is in June to August. Centaurs were sometimes confused with satyrs, who were human-shaped creatures with horse's tails, but not the body, and often goat's legs or horns. Satyrs were wild, lusty and fond of liquid celebrations and revelry. There are 15 stars included in Sagittarius himself and seven in the wreath at his feet.

It has been noted and accepted by some historians of astronomy that Sagittarius performs a very important and amazing function in the map of the cosmos. Its particular location and the distinctive iconography of this constellation were not randomly selected, but carefully chosen by those early astronomers who mapped the heavens and designed the constellations. Sagittarius and Scorpio are the only two constellations of the Zodiac that include features which point toward a specific direction. When viewing Sagittarius, his Bow and Arrow are pointed directly at the tail of Scorpio, both within the Milky Way. The tail of Scorpio is also represented as a pointer, usually ending in an arrow tip; the two constellations are obviously pointing to a particular place in the sky. That place to which they point is the Milky Way, and not to anywhere in the Milky Way, but at a rift in the clouds that if followed deeper into the cosmos would led exactly to the center of our Galaxy. It took modern astronomers with sophisticated instruments hundreds of years to determine what Sagittarius and Scorpio have been pointing to since ancient times.

30. **Sagitta, the Arrow**

> Ahead of the Archer is another Arrow lying all by itself without a bow (311).

Aratus groups together three small constellations, Sagitta, the Eagle and the Dolphin, to complete his sketch of the northern sky. Sagitta is a small constellation consisting of just four stars; two represent the feathered end of the Arrow, a faint star at the middle and one at the tip. It lies east of the Archer's bow, and since it stands alone in the sky, interpretations suggest that it was shot into its current position. The Arrow has become associated with the two most famous archers of Greek myth, Apollo and Hercules. Sagitta was said to be the arrow of Apollo, which he used to kill the Cyclopes, who had fashioned the thunderbolt of Zeus. This momentous arrow was placed in the heavens by Apollo to commemorate his struggle with the Cyclopes. Hyginus gives an alternative explanation for this constellation, writing instead that this was the important arrow used by Hercules to rescue the unfortunate Prometheus, killing the eagle that was attacking him.

31. **Aquila, the Eagle and Antinuous**

> The Bird is outstretched beside it, nearer to the north.
> Close to it another bird is wafted, inferior in size,
> but, stormy when it rises from the sea at the departure of night;
> and men call it the Eagle (312–315).

Aquila is an undistinguished constellation, described as outstretched, both lengthwise and cross-wise; it spreads its wings, and flies toward the east, across the eastern bank of the Milky Way. Its brightest star is Altair, classified as first magnitude. An imaginary line drawn from Deneb in the Swan, to Altair in the Eagle, to Vega in the Lyre forms a distinct geometrical shape called the Summer Triangle. Aquila disappears below the western horizon near the end of November. The Eagle is honored as the only bird which boldly flies in a direct path toward the brilliant Sun, and thus holds first place among all the birds. Sometimes the Eagle is pictured with the Arrow held in its powerful talons. Aquila, the king of birds, was appropriately the royal bird of the king of the gods, Zeus/Jupiter, and performed many difficult tasks at his demand. The giant bird was ordered to swoop down upon Ophiuchus and kill him with one of Jupiter's arrows. On another day, the gods needed someone to serve at their table, thus Aquila was ordered to fly down to Earth and select the most handsome youth he could find to serve as the cup-bearer for the gods. Aquila appears in the sky flying just above Aquarius. Ultimately Jupiter placed the Eagle in the sky as a reward for his long and devoted service.

Aratus does not mention the Eagle as the bird of Zeus at this point of the poem, but does speak of it later. Germanicus combines the Eagle and the Arrow in his translation.[21] Aquila is usually pictured with either both wings closed or with just one wing out-stretched, symbolizing the fact that he is falling mortally wounded. Hyginus attributes only four stars to the constellation of the Eagle; Ptolemy gives it nine.

32. **Delphinus, the Dolphin**

> The very slight Dolphin runs just above Capricorn, dark in the center,
> but four jewels outline it, two lying parallel to two.
> Now these are the stars that are broadcast between the north
> and the Sun's wandering path. But many others rise below this,
> between the south and the track of the Sun (316–321).

Delphinus is another tiny star group, north of Capricorn. The Dolphin can be easily recognized by its compact geometric shape with no other nearby stars to blur the pattern. Aratus relates no myth connected with Delphinus, but Germanicus writes a brief account of the events behind the creation of this constellation. He states that Neptune wished to make Amphitrite his wife; but in shyness she had run away and taken refuge either with Atlas or near Mount Atlas. The compassionate Delphinus, "taking pity on a lover," found her, convinced her of Neptune's sincerity, and carried her back to wed Neptune. In gratitude for this service, the god transported the little Dolphin to the heavens. When represented in artworks, Neptune is often pictured with his favorite tiny Dolphin either in his hand or under his foot. In the head of Delphinus are four stars, equal in brightness, arranged in a parallelogram or a diamond, with a total number of stars reaching nine. At this point in the *Phenomena*, Aratus begins the transition to the stars of the Southern Constellations.

---

[21]During the reign of Hadrian, an attempt was made to form a constellation out of six or seven stars in the lower section of Aquila. The newly formed constellation was called Antinous, in honor of the Emperor's lover who had drowned as they were sailing on the Nile. The constellation of Antinuous persisted until the late eighteenth century, but these stars have now been reassigned back to Aquila.

★ The Orion Group: Orion, the Dog, the Hare and Argo (322–352)

33. Orion, the Hunter

> Aslant from the cut-off figure of the Bull lies Orion himself.
> Anyone whose glance misses him when he is positioned
> high up on a clear night may be sure he can never sight anything better
> to identify when he gazes up at the sky (322–325).

Orion, the great Hunter, is a conspicuous group of bright stars that can without much difficulty be likened to an actual human form with stars at each shoulder, at his feet, with a distinctive three-starred belt. Orion, one of the most-celebrated constellations, appears in the morning hours in September and is due south in December; it is among the easiest star groups to locate in the winter sky. The Egyptians recognized this same star group and associated the constellation of Orion with the god Osiris. The primary Egyptian goddess, Osiris' sister-wife Isis, is associated with Orion's companion, the most brilliant star in the sky, Sirius.

Orion is one of the earliest noted Greek constellations, appearing on the Shield of Achilles along with the Pleiades, the Hyades and the Bear.

Orion was the son of Neptune and the nymph Euryale, of gigantic size and strength, and a most handsome hunter with no fear of any animal. It is said that he was so tall that he could walk through the deepest water without getting his head wet. According to one Greek legend, Orion attempted to seduce Pleione, the consort of Atlas, and her daughters, the Pleiades. The women fled, and the gods immortalized their flight by turning Orion and the women alike into constellations, arranged so that Orion pursues his prey for eternity, never reaching them. Some say the gods blinded him, which relates to the very faint stars in his head, making him appear to be sightless. The blinding of Orion was said to occur on Chios, residence of Homer who, according to tradition, was also blind.

In the sky, Orion is provided with all his needs; Canis Major, his favorite hunting dog at his feet, a Hare to hunt, a sword in one hand and a club in the other. Myths say that Orion died not from hazards of the hunt, but from the sting of the great Scorpion, for which he was given an antidote by Ophiuchus/Asclepius. As mentioned with Scorpio, when Ophiuchus sets in the West, Orion rises in the East fully recovered. Germanicus expresses, "No other constellation more accurately represents the figure of a man than the stars scattered throughout his body".

In *De astronomia*, Hyginus relates two versions of the Orion myth, in one he is mortal and in the other immortal. The bright stars of Orion can be utilized as guides, a line drawn downward through the belt of Orion leads directly to Sirius, a line drawn upward points to Aldebaran, the eye of Taurus. The shoulder of Orion is marked by the brilliant reddish star, Betelgeuse, derived from the Arabic *yad-al-jawz*, which means 'arm of the giant'; it is the twelfth brightest star in the heavens. His foot is represented by a blue-white star Rigel, *rijl* means 'foot' in Arabic, the seventh brightest star.

In al-Sufi manuscripts, Orion is called *al-jabbâr* which means 'the giant', and his description says the outlying line of stars are on 'the sleeve of the figure' resulting in a rather bizarre sleeve that is greatly lengthened hanging limply beyond his hand. The familiar belt of Orion is translated to *mintaqat al-jawza*, the 'belt of the giant'; the star of the right side of the belt is call Mintaku The total number of stars assigned to Orion is seventeen.

34. **Canis Major, the Dog including Sirius**

> Such is also his guardian Dog, seen standing on its two legs
> below the soaring back of Orion, variegated, not bright over all,
> but dark in the region of the belly as it moves round;
> but the tip of its jaw is inset with a formidable star,
> that blazes most intensely: and so men call it the Scorcher.
> When Sirius rises with the Sun, trees can no longer outwit it
> by feebly putting forth leaves. For with its keen shafts it easily pierces
> their ranks, and strengthens some, but destroys all the growth of others.
> We also hear of it at its setting. The other stars
> lying round about Sirius define the legs more faintly (326–337).

Canis Major, Orion's faithful dog, sits to the southeast at his feet; this constellation's most brilliant star Sirius, in fact the brightest star in the sky, marks the dog's jaw, fainter stars form the rest of the body. His loyal dog follows the mighty Orion as he passes across the sky, just as he had followed him on the hunt. In some instances, Sirius denotes destruction since it brings the scorching heat of mid-summer that devastates crops. The Dog-star Sirius was particularly important to the Egyptians, as its heliacal rising announced the time of the annual flooding of the Nile, as well as the hottest time of the year, the dog-days of summer. The earliest Greek reference to this constellation as Orion's Dog is found in Homer, although he may be referring to just Sirius not the entire constellation. Canis Major holds 19 stars.

35. **Lepus, the Hare**

> Under the two feet of Orion, the Hare is hunted constantly all the time:
> Sirius moves forever behind it as if in pursuit,
> rises after it and watches it as it sets (338–341).

Lepus, the Hare, is located below the feet of Orion and is constantly watched and pursued by Sirius, says Aratus. This constellation was probably created to complete the hunting theme associated with Orion. Lepus is a tiny star group, designed to face west as if running ahead of the stars of the Dog constellation. Later mythographers tell that Mercury himself placed the little Hare among the constellations because of its swiftness. The number of stars comprising the small constellation of Lepus is only seven.

36. **Argo or Navis**

> Close to the great Dog's tail is Argo towed stern first.
> Its course is not that of a ship proceeding on its normal business,
> but its movement is backward-turned, like that of real ships
> when the sailors have already turned the stern about on entering harbor;
> all the crew quickly back water, and the ship surging astern makes fast to the land.
>
> So this Argo of Jason is towed stern first. Dark and starless
> from the prow as far as the actual mast she goes, but the rest is all bright.
> The steering-oar is detached and set fast under
> the Dog's hind legs as it runs ahead (342–352).

Argo or Navis, the image of ship in the sky, is found below Canis Major and according to Aratus, proceeds backward like a ship 'entering harbor', its stern turned by the sailors. The constellation of Argo was not entirely visible from Greece, as it lies entirely in the Southern Hemisphere, covering an enormous portion of the sky. Navis appears to have no bow, as that area of the sky is dark, only the stern is represented by its stars; and so like Equus, Pegasus and Taurus comprises only half a figure.

Hyginus states that the sight of half a ship would encourage seafarers not to fear shipwrecks. This ship of mythology is associated with the vessel built by Argos which carried his expedition in search of the Golden Fleece manned by Jason and the Argonauts, the fifty famous Greek heroes. Some say it was called Argo because of its speed, or that its builder was Argus who used special wood for its construction. Through time the Argo acquired a supernatural quality, and it became endowed with speech through a speaking plank that was added by Minerva. In classical literature, this magical ship had in its prow a special piece of wood from the sacred oak groves of the oracle of Zeus at Dodona. This extraordinary plank could not only speak, but could even see into the future. Argo was considered the first ship to cross the untraversed sea. Germanicus tells that the Argo was damaged as it passed between the treacherous Clashing Rocks. After the sea voyage had accomplished its mission, Athena placed the great ship in the sky to immortalize it.

Navis is no longer recognized as a single constellation today; it has been broken up into three smaller constellations. The total number of stars in the ship was 27, three stars in a line constituted the upright mast; its brightest star is Canopus.

★ Sea-Monster Group: the Sea-Monster, the River, the Southern Fish (353–401)

37. **Cetus, the Whale or Sea Monster**

> Although she lies no small distance away, Andromeda is threatened
> by the approach of the great Sea Monster. For in her course she lies
> exposed to the blast of the north wind from Thrace, while the southerly
> brings against her the hostile Monster, below the Ram and the two Fishes,
> and positioned a little above the starry River (353–358).

The sinister sea monster Cetus crawls low in the sky over the southern horizon in September and October, threatening Andromeda. Cetus demarcates the Celestial Equator, its sinister head protrudes above the Equator; and its elongated, bulky body stretches out south of the Equator. Cetus is amphibious and consists of a group of stars that make up its head, a body group and a tail, with the Fishes' Knot close to its back fin. Though sometimes called the Whale, it was more likely based on a fictitious sea monster of myths and legends. Cetus was sent by Poseidon to provoke King Cepheus, when his boastful wife Cassiopeia claimed to rival the Nereids in beauty. Living deep in the dark waters of the sea, Cetus would have devoured Andromeda, if she had not been rescued by her hero Perseus.

The frightful monster came to symbolize all the evils of the world; sin, immorality, hatred and envy. The monstrous beast was placed in the sky not to reward him, but to commend the courage of Perseus in destroying him. Cetus lives in the realm of the sky where all its adjoining constellations have some association with water. Eridanus the Great River; Pisces the Fishes; Aquarius the Water-Bearer; Capricorn the Sea-goat; and Piscis Austrinus, the Southern Fish, all inhabit this area of the celestial sphere. In artistic renderings, Cetus sometimes resembles a sea-dragon, or alternatively is pictured with the head of a dog and the body of a fish. The number of stars included in this constellation is thirteen.

38. **Eridanus, the River**

> For under the gods' feet, that too moves as a separate group,
> a remnant of Eridanus, river of much weeping. It extends
> below the left foot of Orion. The tail-chains, by which the extremities of the Fishes
> are held, both come together as they descend from the tail-parts,
> and behind the Monster's back-fin move jointly
> as they converge, and terminate in a single star
> that lies close to the top of the Monster's spine (359–366).

The River called Eridanus is formed by a long course of dim stars, winding west from Orion's foot; it curves round to flow east, then west again and south. Its origin seems to be Greek since neither Babylonian nor Egyptian myths speak of a river in this section of the sky. Hyginus associates this river of stars with the Nile since, viewing it from Greece, appears in the direction of Egypt, in addition, it flows from the south to the north like the Nile. Several variations of its name appear in myths, sometimes it is referred to as Flumen or Flumena, Germanicus calls it Amnis.

Eridanus is featured in the tale of Phaethon, the son of Helios, who was thrown from the Sun Chariot and fell into the River, where he drowned. Eridanus became a 'river of much weeping' as the tears of Phaethon's sisters, the Heliades, flowed into the Eridanus and were transformed into drops of amber. The constellation of Eridanus is always drawn as a personification, a reclining river god holding a vessel that flows with water, rather than as an actual flowing stream of water. This manner of depicting a river is a classical convention taken over from antique models, and continues through the Middle Ages. The River is composed of 13 stars, three at the first bend, three at the second and seven spread out from the third bend to the last one.

At this point of the poem, Aratus renews his discussion of the constellation Pisces. Chains from the tails of Pisces 'converge' and 'terminate' at line 365 and 1/4, the exact length of a year. They 'converge' in a 'single star', **the Knot**, the first indication of the Sun's precessional movement from the Age of Aries into the Age of Pisces, this occurred in approximately 100 BCE.[22]

39. The Southern Fish, Piscis Australis

> Below Capricorn and exposed to the winds of the
> south hovers the Fish, turned to face the Monster,
> distinct from the former two: they call it the Southern Fish ((386–388).

The Southern Fish is a single, unattached fish in the southern sky, distinct from the two joined fish of Pisces located in the north. Aratus supplies just its position, orientation and name. This great Fish is often illustrated with its mouth open, taking in the water pouring from the jar of Aquarius. In constellation cycles, it is drawn in a similar manner to Pisces, but usually a bit larger. Germanicus says, "There is a fish that swims alone, apart from the twin fishes". Hyginus identified this single Fish as the one which rendered a special kindness to the Egyptian goddess Isis. In return, Isis secured a place in the heavens for it, and the twin fishes of Pisces are thought to be its offspring. This constellation consists of 12 stars in all.

The Water

> Other stars lying scattered below the Water-pourer hang in the sky
> between the celestial Monster and the Fish, but they are faint and nameless.

---

[22]For more information on this topic, see G. De Callatay, "The Knot of the Heavens", *Journal of the Warburg and Courtauld Institutes*, Vol. 59, (1996) 1–13.

> Close to the, like a light spray of water being sprinkled this way and that
> from the right hand of the illustrious Water-pourer, some pale and
> feeble stars go round. Among them go two rather brighter stars,
> not so very far apart nor yet very close, one beautiful and bright star
> beneath the feet of the Water-pourer, the other below the dark Monster's tail.
> Men call them collectively the Water. But a few others below the Archer,
> under his forefeet, are curved in a ring as they go circling round (389–401).

This area of faint stars does not give rise to any recognizable figure, therefore it is not considered a constellation, and is not illustrated. Aratus simply calls this area the Water. Cicero and Avienus follow Aratus in describing the group as Aqua, but Germanicus omits the name entirely.

★ The Altar Group: The Altar, the Centaur, the Beast, the Hydra, the Bowl, the Raven, Procyon (402–451)

40. The Altar or Ara

> Now below the blazing sting of the great monster Scorpion,
> Close to the south hovers the Altar. This constellation you will observe
> only for a short time above the horizon, for its period of visibility is
> the opposite of Arcturus'. In fact the paths of Arcturus come fully overhead,
> whereas the Altar goes quickly under the western sea.
> Yet even round that Altar ancient Night, sad for the suffering of men,
> has set an important sign of storm at sea; for ships in distress
> are not to her liking, and she displays different signs in different ways
> in her pit for storm-tossed men. So pray, I beg you, that at sea
> this constellation be not visible in mid-sky overarched by clouds,
> everywhere else, cloudless itself and brilliant, but higher up
> oppressed by billowing clouds, as they are often packed
> when an autumn northerly piles them up (402–417).

The Altar is found below 'the great monster Scorpio', far south; not fully visible above 23 degrees of latitude and shines for only a brief period above the horizon, as mentioned by Aratus. Ara lies only a short distance from the southern horizon, therefore it is difficult to pick out because of haze, clouds and the unsteadiness of the atmosphere. Ara, a winter constellation, is a sign of storms when seen 'over-arched by clouds', put there by Night to guide sailors and 'storm-tossed men'. Aratus gives the Altar a large portion of space in his poem, more lines than would be expected for its minimal importance as a constellation, but the gods of Olympus needed an Altar to swear allegiance to Jupiter. The Leiden *Aratea* pictures the Altar as a tripod censer with incense burning. Other illustrations of Ara usually show fire rising from the altar, as it was filled with burning incense, the smoke usually drifts to the north, represented by the soft, hazy whiteness of the Milky Way. Germanicus describes the burning altar as one of the signs that Nature has given man for protection. Ara is a small constellation with only four stars, two at the top and two at the bottom.

41. Centaurus, the Centaur

> If the Centaur's shoulder should be as far from the western horizon
> as from the eastern, and a slight haze shroud the constellation,
> while behind it Night is fashioning recognizable signs on the radiant altar,
> you must certainly be on the lookout not for a southerly
> but for an easterly wind. You will find this constellation lying below two others:
> part of it, resembling a man, lies beneath the Scorpion, and the Claws have
>
> the horse's hind-part under them. He looks as if he is always stretching
> his right hand towards the round Altar, and fast in his hand is tightly gripped
> another constellation, the Beast; for so our predecessors named it (431–442).

The Centaur is a large and shapeless constellation in the southern section of the sky, occasionally identified with Chiron, the best-known Centaur. Chiron was immortal, while other centaurs were said to be mortal. Chiron has a reputation for wisdom and justice, loving music and art; he was said to have tutored Asclepius, Achilles and many other important figures, not only their teacher, but friend and advisor. Attitudes toward centaurs and their reception in the Middle Ages is the subject of an article in which the author discusses hybrid creatures. The author reports that 'in Greek and Latin literature Chiron is considered an outsider in the family of centaurs'. 'Other centaurs are portrayed as barbaric brutes, Chiron is a wise tutor.'[23]

Sometimes the Centaur tends to be confused with Sagittarius, but he holds 'the Beast' in his hand rather than the bow and arrow indicative of Sagittarius. The human or front-half of the Centaur lies below Scorpio and the horse-like hind quarters are found below Libra, 'the Claws', his right hand points toward the Altar. The wild 'Beast' gripped tightly in his left hand is depicted more like a rabbit, Lupus. The animal that he is sacrificing is a manifestation of his piety; its shape is actually square. Some say this is a wine skin from which he is pouring a libation on the Altar. The Centaur died when talking with Hercules; because one of Hercules' arrows slipped from his quiver, striking Chiron on the foot. The al-Sufi manuscript illustrates the Centaur in the Western manner, but the 'thyrsus' or spear wreathed with vine leaves usually held by this figure has become simply a bunch of leaves. The Centaur has a large number of stars, totaling 24, the sacrificial offering or the 'wild beast' that he holds has ten.

---

[23]Ludo Jongen, "Do Centaurs Have Souls?" in *Animals and the Symbolic in Mediaeval Art and Literature*, L.A.J.R. Houwen, ed. Egbert Forsten (1997) pp. 139–154.

42. **Hydra, the Snake**

> But yet another constellation sweeps across the horizon:
> they call it the Hydra. Like a living thing it winds at great length,
> its head comes below the middle of the Crab, its coil under the Lion's body,
> and its tail hangs over the Centaur himself (443–447).

Hydra is a lengthy series of stars, stretching roughly parallel to the ecliptic, with one bright star of second-magnitude, six of third-magnitude and nine that are only fourth-magnitude. Hydra slithers low on the horizon below Libra, Virgo, Leo and Cancer. The last four constellations are closely related and the three, Hydra, Crater, and Corvus are always combined into one illustration. All three constellations are linked together in a well-known myth of Apollo. The god sent a raven, said to be one of his servants, to fetch a cup of pure water for a sacrifice. The bird became distracted by a fig tree and waited for the figs to ripen in order to eat them. When he finally returned to Apollo, the raven brought along a snake and claimed that the snake had prevented him from getting the water from the spring. The legend states that Apollo punished the tardy raven by making it unable to drink in springtime. Aratus describes Hydra accurately for it is quite long, a winding snake 'like a living thing' lying with its head below the Crab, its coil below the Lion, and its tail over the Centaur

Hydra is also associated with the myth of Hercules, as the snake he had to combat, one of his Twelve Labors. The snake had nine heads, every time Hercules chopped off a head, two new ones grew. Eventually, Hercules burned the stump of each decapitated head to prevent the new ones from growing. Juno made Hercules task more difficult by sending a Crab, Cancer, after him, which can be seen in the sky above the snake's head. Hercules crushed the Crab under his feet. When the Sun rises in Cancer, the head of Hydra also rises, meaning that Hercules has managed to cut off the first head, because the stars of Hydra begin to fade in the morning light. This goes on until the end of October. This lengthy transit is symbolic of the long battle of Hercules, as he overpowered the snake bit by bit. Every time Hercules cuts off another head, a little more of Hydra disappears in the morning Sunlight. Hydra includes 27 stars visible to the naked eye, a straggling collection of faint stars not worthy of a constellation figure, but important as it designates the celestial equator.

43. **Crater, the Cup or the Bowl**

> On its middle coil lies the Bowl, and on the last one (448).

Crater is actually shaped like a cup and has ten stars. It is always illustrated balanced atop the coils of Hydra, the snake, the third serpent found among the constellations.

44. Corvus, the Raven

> The figure of a Raven that looks like one pecking the coil (449).

Crater and Corvus are small constellations, but distinct, and depicted in illustrations as actually standing on the Hydra. The combination was probably invented by the Greeks, since the Raven is an attribute of Apollo. Once Corvus had white feathers and a beautiful singing voice, but lost these as punishment for not following the orders of Apollo. The god changed his shiny white feathers into dirty black ones and altered his voice to a dull screech. As they were immortalized, Hydra was to make sure that Corvus never reached the Beaker to quench his thirst. The Crater always remains well in front of Corvus. Corvus has a total of seven stars.

45. Procyon or Canis Minor

> Yes, and there too Procyon shines brightly beneath the Twins.
> These you can see as the years pass returning in succession;
> For these figures of the passing night are all well fixed
> in the sky just as they are (450–453).

Procyon, the second dog in the sky, is the name of both the constellation and a single bright star within it. Procyon is located just below Castor and Pollux, a short distance to the west of Hydra's head, behind the shoulder of Orion. The Lesser Dog was important in ancient calendars as the forerunner of the Greater Dog, Canis Major and its primary star Sirius; both are associated with the hottest time of the year. Aratus and Germanicus mention only a single star of this constellation, Procyon, the "Forerunner of the Dog", its name derives from the fact that it rises one hour before Sirius. Canis Minor has one first-magnitude star; the total number is only three. This completes Aratus' description of the northern and southern constellations.

Ptolemy adds Corona Australis and Lupus.

The Iconography of the Sun, Moon and Planets

The planets are mentioned by Aratus, but not listed by name. They are not relevant to his purposes as they provide no practical use for measuring time or predicting the change of seasons. Since the Sun and Moon are relevant, they are discussed in the final section of the poem, especially 778–891. In his poem, Germanicus does name the five planets, and also discusses the Four Seasons.

In early Renaissance astrological manuscripts, a new concept concerning the planets developed in illustrations which was named 'children of the planets' including the Sun and Moon. This pictorial convention combined the astrological aspects assigned to each of the 'seven planets' with the images of their labors, an erstwhile linking of the Labors of the Month, which were idealized views of seasonal duties. The traditional activities of laborers were pictured along with personifications of the traditional human traits such as love, war and death. The first appearance of these representations occurs in an illustrated manuscript of Christine de Pisan's *Epitre d'Othea a Hector* (1399–1400). Christine's father was court physician and astrologer to France's King Charles V; she was undoubted adept at astrological theories. Her writings defined and illustrated the positive and negative influences of each planet visualizing what a good knight should work towards or avoid (Page 2002: 49). This knowledge was disseminated through calendar illuminations in Books of Hours and the Secreta Secretorum. The practices and influences of astrology were spread among the population most effectively through astrological medicine. An especially beautiful astrological manuscript produced in Italy c. 1470 is *De Sphaera* attributed to Cristoforo de Predis and bearing the coat of arms of the Visconti and Sforza families contains illuminations of the planets and their children (Four facsimiles of *De Sphaera* have been produced.).

Helios, the Sun God

The brilliant Sun God, Helios, was the supreme cosmic power, the symbol of light and truth, identified with Apollo. The golden young man, portrayed beardless with the ideal form of the physical male body in the prime of life, brought light to the darkest corners of the world. The Sun God ripens the fruits of the fields, but his beneficence can scorch and burn as well. His iconography often shows him encircled as if enclosed in the fiery Sun disc. Crowned with brilliant rays of light and gripping a glowing torch, the powerful god, his billowing cape lifted by the wind, steers the Sun Chariot pulled by four mighty horses bringing illumination and invigorating life. 'Children' of the Sun were valued leaders, with strong wills, endless energy, strength and wisdom.

Luna or Selene, the Moon Goddess

The Moon had many mythological aspects, usually as the female consort to the Sun, just as powerful but in a passive role, as her silvery light is but a reflection of the Sun's golden light. Luna's influence operated principally at night, and the waxing and waning of her monthly phases evoked growth in plants and movement with animals. The Moon symbolized dreams, magic, madness and love, with a close connection with women through their menstrual cycle. Her ebb and flow was powerful enough to move the seas and all of nature with her powerful rhythms. Luna is personified in classical and medieval illustrations as the counterpart of Helios. She controls the path of the Moon through the heavens by directing from her Lunar Chariot pulled by two powerful oxen. Her image is crowned by a crescent Moon or the horns of a cow, and she holds a pale torch. The three days when the Moon is dark is the archetypal period of the descent of the hero to the underworld, from which like the Moon, he arises reborn. The 'children' of the Moon were often represented as 'lunatics', unstable, moody; the positive traits were kindness and sympathy.

The Planets

But there are five other stars among them that circulate
all the way through the twelve figures of the zodiac.
You cannot in this case identify where these lie by looking at other stars,

for they all change their positions. The years of their orbits are long,
and at long intervals are their configurations when they come
from afar into conjunction. I am not at all confident in dealing with them:
I hope I may be adequate in expounding the circles of the fixed stars
and their guide-constellations in the sky (454–462).

Mercury

The visual appearance and motions of each planet indicate just how its characteristics and iconography were generated in the antiquity. Mercury/Hermes shines merely a tiny point of light but moves in rapid, tight loops always nearest to the Sun. Therefore the qualities attributed to Mercury consisted of a nimble mind, sharp cunning wit, with exceptional skill in communication and speech. As the messenger of the gods, Mercury kept them informed of events in the world. His chameleon-like actions symbolize the mercurial aspects of human nature, a restless youth listening and advising where needed. In artworks his appearance carries evidential emblems of his various functions; the caduceus represents all the skills for holding information, from magic to mathematics; wearing a brimmed hat and winged sandals, he flits round the Sun in 88 days. His 'children' are writers, scribes, and students known for their eloquence and communication skills.

Venus

The brightest planet, Venus/Aphrodite brings the light (Lucifer) in medieval manuscripts) as the morning or evening star (Vesperis) never straying far from the Sun. The archetypal image of Venus/Aphrodite represents a powerful, attractive and alluring young woman with attributions of pleasure and sensuality. Abhorring deprivation and ugliness, Venus prefers luxury, comfort and beautiful things. Most often Venus is personified as naked revealing her curvaceous body and long flowing hair, sometimes admiring her image in a mirror. She orbits the Sun in just 224 days. Her 'children' in Renaissance manuscripts are shown as lovers, musicians, women in gardens with artworks and good food.

Mars

The planet Mars/Aries glows fiery red in the night sky, and as a result is usually cast in the role of the god of war. Carrying a spear Mars is depicted as energetic, disciplined and powerful, armed and ready for battle. His attributes include courage and stamina with a concern for judgement, discipline, justice and passionate activities. Renaissance manuscripts depict his 'children' as armies, individual soldiers, anger and hostility.

Jupiter

The largest and most majestic of the planets, Jupiter/Zeus was considered the esteemed king of the gods and the benefactor of humanity. Like all other gods, Jupiter represented both positive and negative aspects of human life. Bestowing favors or hurling thunderbolts with equal zeal, Jupiter was magnanimous, aiding humanity but not able to control his wife. Jupiter's orbital period takes 11.9 years. His children in astrological manuscripts represented trade, governing, mathematics, and royalty.

Saturn

The slowest and most distant planet known in antiquity, Saturn/Chronos was the god of time, the destroyer. His deliberate and lengthy period of orbit gave him the image of an old man, characterizing age, rigidity and gravity. His emblem, the scythe, not only cut away time, but also cleared away the old and useless parts of life. Since he was the least erratic of the planets, Saturn was recognized for constancy and dependability, known for endurance and the rewards of hard work. The wisdom of old age made him the distinguished teacher of the young, demanding but just. Originally the king of the solar system, he was banished by Jupiter to the outer realms of the universe and sometimes personifies melancholia and depression.[24] According to ancient myths, Saturn ate each of his children after they were born. His 'children' were wise teachers honored elders, but thought to have limitations and boundaries on their goals. This concludes the mythological descriptions of the constellations included in the *Phaenomena* of Aratus.

[24]See Panofsky and Saxl (1923) *Dürer, Melancolia I: eine quellen- und typengeschichtliche Untersuchung*, Studien der Bibliothek Warburg, II. Leipzig.

Index

M. Dolan, *Astronomical Knowledge Transmission Through Illustrated Aratea Manuscripts*, Historical & Cultural Astronomy,
DOI 10.1007/978-3-319-56784-6

MIX
Papier aus verantwortungsvollen Quellen
Paper from responsible sources
FSC® C105338

FSC
www.fsc.org

If you have any concerns about our products,
you can contact us on
ProductSafety@springernature.com

In case Publisher is established outside the EU,
the EU authorized representative is:
Springer Nature Customer Service Center GmbH
Europaplatz 3, 69115 Heidelberg, Germany

Printed by Libri Plureos GmbH
in Hamburg, Germany